Gas
Production
Operations

Gas Production Operations

H. Dale Beggs

OGCI Publications
Oil & Gas Consultants International Inc.
Tulsa

4554 South Harvard Avenue
Tulsa, Oklahoma 74135

Printed in the United States of America

Library of Congress Catalog Card Number: 83-063489
International Standard Book Number: 0-930972-06-6

First Printing — July, 1984
Second Printing — November, 1985

Contents

Nomenclature

		Dimensions
A	area	L^2
B_g	gas formation volume factor	
B_{gb}	gas formation volume factor at bubble-point conditions	
B_o	oil formation volume factor	
B_{ob}	oil formation volume factor at bubble-point conditions	
B_t	total (two-phase) formation volume factor	
B_w	water formation volume factor	
c_f	formation (rock) compressibility	Lt^2/m
c_g	gas compressibility	Lt^2/m
c_o	oil compressibility	Lt^2/m
c_{pr}	pseudoreduced compressibility	
c_w	water compressibility	Lt^2/m
C	coefficient of gas-well back-pressure curve	$L^{3-2n}t^{4n}/m^{2n}$
C	concentration	various
C_L	condensate or natural gas liquids content	various
d	diameter	L
D	depth	L
E	efficiency	
E_A	areal efficiency	
E_D	displacement efficiency	
E_I	invasion (vertical) efficiency	
E_P	pattern sweep efficiency	
E_R	reservoir recovery efficiency, overall	
E_V	volumetric efficiency	
f	fraction	
f	friction factor	
f	fugacity	m/Lt^2
F	force	mL/t^2

		Dimensions
F_{wo}	instantaneous producing water-oil ratio	
F_{wop}	cumulative water-oil ratio	
g	acceleration of gravity	L/t^2
g_c	conversion factor in Newton's Second Law of Motion	mL/Ft^2
G	total initial gas in place in reservoir	L^3
G_i	cumulative gas injection	L^3
G_L	initial condensate liquids in place in reservoir	L^3
G_{Lp}	cumulative condensate liquid produced	L^3
G_p	cumulative gas produced	L^3
G_{wgp}	cumulative wet gas produced	L^3
ΔG_p	gas produced during an interval	L^3
h	thickness (general and individual bed)	L
H	enthalpy (always with phase or system subscripts)	mL^2/t^2
i	injection rate	L^3/t
J	productivity index	L^4t/m
J_s	specific productivity index	L^3t/m
k	absolute permeability (fluid flow)	L^2
k_g	effective permeability to gas	L^2
k_o	effective permeability to oil	L^2
k_{rg}	relative permeability to gas	
k_{ro}	relative permeability to oil	
k_{rw}	relative permeability to water	
k_w	effective permeability to water	L^2
K	equilibrium ratio (y/x)	
ln	natural logarithm, base e	
log	common logarithm, base 10	
L	length	L
L	moles of liquid phase	
m	mass	m
m	ratio of initial reservoir free-gas volume to initial reservoir oil volume	
m	slope	various
M	mobility ratio	
M	molecular weight	m
n	exponent of back-pressure curve, gas well	
n	total moles	
n_j	moles of component j	
N	initial oil in place in reservoir	L^3
N_p	cumulative oil produced	L^3
N_{Re}	Reynolds number (dimensionless number)	
ΔN_p	oil produced during an interval	L^3
p	pressure	m/Lt^2
p_a	atmospheric pressure	m/Lt^2
p_b	bubble-point (saturation) pressure	m/Lt^2
p_c	critical pressure	m/Lt^2

		Dimensions
p_{cf}	casing pressure, flowing	m/Lt^2
p_{cs}	casing pressure, static	m/Lt^2
p_d	dew-point pressure	m/Lt^2
p_D	dimensionless pressure	
p_e	external boundary pressure	m/Lt^2
p_i	initial pressure	m/Lt^2
p_{pc}	pseudocritical pressure	m/Lt^2
p_{pr}	pseudoreduced pressure	
p_r	reduced pressure	
p_{sc}	pressure, standard conditions	m/Lt^2
p_{sp}	separator pressure	m/Lt^2
p_{tD}	dimensionless pressure function at dimensionless time t_D	
p_{tf}	tubing pressure, flowing	m/Lt^2
p_{ts}	tubing pressure, static	m/Lt^2
p_w	bottom-hole pressure, general	m/Lt^2
p_{wf}	bottom-hole pressure, flowing	m/Lt^2
p_{ws}	bottom-hole pressure, static	m/Lt^2
$\bar{p}$	average pressure	m/Lt^2
$\bar{p}_R$	pressure, reservoir average	m/Lt^2
P_c	capillary pressure	m/Lt^2
q	production rate or flow rate	L^3/t
q_D	dimensionless production rate	
q_{sc}	gas production rate	L^3/t
q_{gD}	production rate, gas, dimensionless	
q_o	oil production rate	L^3/t
q_{oD}	production rate, oil, dimensionless	
q_w	water production rate	L^3/t
q_{wD}	production rate, water, dimensionless	
Q_{tD}	dimensionless fluid influx function at dimensionless time t_D	
r	radial distance	L
r_d	radius of drainage	L
r_D	dimensionless radial distance	
r_e	external boundary radius	L
r_s	radius of well damage or stimulation (skin)	L
r_w	well radius	L
r_{wa}	apparent or effective wellbore radius (includes effects of well damage or stimulation)	L
R	producing gas-oil ratio	
R	universal gas constant (per mole)	mL^2/t^2T
R_p	cumulative gas-oil ratio	
R_s	solution gas-oil ratio (gas solubility in oil)	
R_{sb}	solution gas-oil ratio at bubble-point conditions	
R_{si}	initial solution gas-oil ratio	
R_{sw}	gas solubility in water	

		Dimensions
s	skin effect	
S	saturation	
S_g	gas saturation	
S_{gc}	critical gas saturation	
S_{gr}	residual gas saturation	
S_L	total (combined) liquid saturation	
S_o	oil saturation	
S_{og}	interstitial-oil saturation in gas cap	
S_{or}	residual oil saturation	
S_w	water saturation	
S_{wc}	critical water saturation	
S_{wg}	interstitial-water saturation in gas cap	
S_{wr}	residual water saturation	
t	time	t
t_D	dimensionless time	
t_s	time for stabilization of a well	t
T	temperature	T
T_c	critical temperature	T
T_f	formation temperature	T
T_{pc}	pseudocritical temperature	T
T_{pr}	pseudoreduced temperature	
T_r	reduced temperature	
T_R	reservoir temperature	T
T_{sc}	temperature, standard conditions	T
v	specific volume	L^3/m
v	velocity	L/t
V	moles of vapor phase	
V	volume	L^3
V_b	bulk volume	L^3
V_M	volume per mole	L^3
V_p	pore volume	L^3
w	mass flow rate	m/t
W	initial water in place in reservoir	L^3
W	water (always with identifying subscripts)	various
W	work	mL^2/t^2
W_e	cumulative water influx (encroachment)	L^3
W_i	cumulative water injected	L^3
W_p	cumulative water produced	L^3
ΔW_e	water influx (encroachment) during an interval	L^3
ΔW_i	water injected during an interval	L^3
ΔW_p	water produced during an interval	L^3
x	mole fraction of a component in liquid phase	
y	mole fraction of a component in vapor phase	
z	gas deviation factor (compressibility factor, $z = pV/nRT$)	
z	mole fraction of a component in mixture	

			Dimensions
Z	elevation referred to datum		L
α	alpha	angle	
γ	gamma	specific gravity	
γ_g	gamma	gas specific gravity	
γ_o	gamma	oil specific gravity	
γ_w	gamma	water specific gravity	
Δ	delta	difference ($\Delta x = x_2 - x_1$ or $x_1 - x_2$)	$\{x\}$
η	eta	hydraulic diffusivity ($k/\phi c\mu$ or $\lambda/\phi c$)	L^2/t
θ	theta	angle	
λ_g	lambda	gas mobility	$L^3\mathrm{t}/m$
λ_o	lambda	oil mobility	$L^3\mathrm{t}/m$
λ_w	lambda	water mobility	$L^3\mathrm{t}/m$
μ	mu	viscosity	m/Lt
μ_a	mu	air viscosity	m/Lt
μ_g	mu	gas viscosity	m/Lt
μ_o	mu	oil viscosity	m/Lt
μ_w	mu	water viscosity	m/Lt
ν	nu	kinematic viscosity	L^2/t
ρ	rho	density	m/L^3
ρ_f	rho	fluid density	m/L^3
ρ_g	rho	gas density	m/L^3
ρ_{ma}	rho	matrix (solids, grain) density	m/L^3
ρ_o	rho	oil density	m/L^3
ρ_w	rho	water density	m/L^3
σ	sigma	surface tension (interfacial tension)	m/t^2
τ	tau	tortuosity	
ϕ	phi	porosity	

1 Introduction

ALTHOUGH natural gas has been utilized as a fuel for more than 150 years, the large demand for it has developed fairly recently. The principal reason for this is the greater difficulty of storing and transporting gas as compared to liquid fuels. Initially, natural gas was used only in the areas in which it was produced, with excess production being vented to the air or flared. This was especially true of gas produced along with the oil in oil fields.

The development of large diameter, high pressure pipelines and compressors, along with the technology of gas storage in reservoirs, has spurred both the demand for natural gas and the development of the technology required to produce and transport it. This is illustrated in the United States by the fact that natural gas supplied more than 30% of the total energy demand in 1980 as compared to 18% in 1950 and less than 4% in 1920. Natural gas supplied about 20% of the energy worldwide in 1980.

The increased demand has also greatly increased the price obtained for the gas. In 1950 the average price for natural gas in the United States was about \$0.07 per thousand standard cubic foot (Mscf), and as late as 1970 the price averaged only about \$0.17/Mscf. In 1980 the average price had increased to more than \$0.90/Mscf with gas in some areas selling for as much as \$9.00/Mscf. The large difference in the average and maximum selling price is due to the huge quantities being sold at very low prices under long term contracts made many years ago.

Natural gas is used primarily as a fuel for space heating and for generating steam for electric power plants, although its use as a feedstock for petrochemical plants is increasing rapidly. Because of the necessity of using very high pressure to store significant quantities of natural gas in small spaces, its use as a fuel for motor vehicles is very limited. However, as the supply of liquid fuels such as gasoline diminishes, it is likely that technology will be developed to overcome this problem.

GEOGRAPHICAL OCCURRENCE OF NATURAL GAS

Geological conditions necessary for commercial accumulations of natural gas exist in various locations worldwide. The tables and graphs presented in this section locate and quantify the known reserves and the production rates of various countries. These data are also presented for various states in the United States.

Worldwide Occurrence of Natural Gas

The worldwide reserves of natural gas has steadily increased during recent years, while the United States reserves declined from 1967 to 1983.

Table 1-1 shows that the United States' share of the total world reserves decreased from 27.8% to 6.7% during the 1967–1983 period. As of 1983, the proved world gas reserve was 3,033 trillion cubic feet. The changes in reserves by area from 1967 to 1983 are shown in Table 1-1. These statistics are also reported by country for the twenty leading gas producing countries in Table 1-2.

Although the United States has less than 7% of the total world reserves, it currently produces more than 34% of the gas produced worldwide. Production statistics by area for the 1975–1981 period are presented in Table 1-3. The production statistics for the twenty leading countries are presented in Table 1-4 for 1980, 1981 and 1984. At the 1981 worldwide production rate of 160 billion cu ft/

TABLE 1-1
Estimated Proved World Reserves of Natural Gas Annually As of January 1 (Billions of Cubic Feet)

	WESTERN HEMISPHERE												
Year	United States	Canada	Latin America	Western Hemisphere Total	Middle East	Africa	Asia-Pacific	Western Europe	Total Free World	Communist Nations	Total World	U.S. as a % of Total World	Year
1967	289,333	43,450	64,550	397,333	215,070	158,155	32,450	88,582	891,590	150,000	1,041,590	27.8%	1967
1968	292,908	45,682	67,101	405,691	220,670	167,223	40,050	133,965	967,599	215,500	1,183,099	24.8	1968
1969	287,350	47,666	62,900	397,916	223,775	168,345	52,724	141,176	983,936	343,000	1,326,936	21.7	1969
1970	275,109	51,951	163,150	490,210	235,275	197,143	67,500	150,800	1,140,928	350,000	1,490,928	18.5	1970
1971	290,746[1]	53,376	73,100	417,222	354,262	191,516	56,330	147,731	1,167,061	440,000	1,607,061	18.1	1971
1972	278,806[1]	55,462	72,700	406,968	343,930	193,018	69,800	163,250	1,176,966	558,000	1,734,966	16.1	1972
1973	266,085[1]	52,936	79,218	398,239	344,150	189,015	101,236	178,400	1,211,040	664,400	1,875,440	14.2	1973
1974	249,950[1]	52,457	91,321	393,728	413,325	187,720	114,200	193,797	1,302,770	735,400	2,038,170	12.3	1974
1975	237,132[1]	56,708	100,214	394,054	672,670	314,974	115,880	202,826	1,700,404	846,000	2,546,404	9.3	1975
1976	228,200[1]	56,975	90,487	375,662	538,648	207,152	111,560	180,875	1,413,897	835,000	2,248,897	10.1	1976
1977	216,026[1]	58,282	90,325	364,633	536,460	209,077	120,010	141,905	1,372,085	953,000	2,325,085	9.2	1977
1978	208,878[1]	59,472	108,480	376,830	719,660	207,504	122,725	138,190	1,564,909	955,000	2,519,909	8.3	1978
1979	200,302[1]	59,000	112,950	372,252	730,660	186,290	119,850	143,260	1,552,312	945,000	2,497,312	8.0	1979
1980	194,917[1]	85,500	144,500	424,917	740,330	210,350	128,815	135,376	1,639,158	935,000	2,574,158	7.6	1980
1981	199,021[1]	87,300	159,811	446,132	752,415	208,470	126,290	159,315	1,692,622	953,900	2,646,522	7.5	1981
1982[r]	201,730[1]	89,900	176,323	467,953	762,490	211,667	127,616	150,650	1,720,376	1,194,700	2,915,076	6.9	1982[r]
1983[p]	204,000[1]	97,000	186,591	487,591	769,730	189,423	146,247	156,736	1,749,727	1,283,800	3,033,527	6.7	1983[p]

[r] Revised
[p] Preliminary
[1] Figures include 26 trillion cubic feet in Prudhoe Bay, Alaska (discovered in 1968) for which transportation facilities are not yet available

Source: 1967–1980: United States—American Gas Association, Committee on Natural Gas Reserves
1981–1982: United States—Department of Energy
1983: Oil and Gas Journal
Rest of the World—The Oil & Gas Journal "Worldwide Report" Issues.

Courtesy the American Petroleum Institute

day, known reserves exist to sustain this rate for another forty-five years.

Occurrence of Natural Gas in the United States

The United States is not in the same condition as the total world with respect to supply of natural gas. As shown in Table 1-5, almost eleven years would be required to deplete the 1981 known reserves at 1981 production rates. This does not mean that the gas will be depleted in ten years, however, as new supplies are being found continuously. The changing reserve situation in the United States is illustrated graphically in Figure 1-1. Table 1-6 shows estimates of the total remaining resources of natural gas in the United States. Depending on the source of the estimate, the resources remaining in 1979 range from 502 to 1202 trillion cubic feet.

Table 1-7 presents salient statistics related to the gas industry in the United States for the period 1970–1981. Approximately 5% of the gas consumed is imported, primarily from Canada. Some liquefied natural gas is imported from Algeria.

Of the gas produced in the United States, almost 90% comes from only five states. These are Kansas, Louisiana, New Mexico, Oklahoma, and Texas. The state of Louisiana is by far the largest producer, accounting for about 34% of the total gas produced in the United States in 1981. Much of this gas is produced offshore. Table 1-8 lists the gas production by states for the period 1972–1981.

Consumption of natural gas is of course more evenly distributed among the states, with Texas being the largest consumer. Table 1-9 shows where the United States-produced gas was used in the period 1977–1981.

The large growth of the gas industry in the United States is attributable to the comprehensive transmission system that has developed since the late 1940's. In 1977 the network of gas pipelines consisted of more than 250,000 miles and served almost 45 million gas customers. The pipeline grid is shown in Figure 1-2, and the distribution of pipelines and customers among the states is listed in Table 1-10.

GEOLOGICAL OCCURRENCE OF NATURAL GAS

Certain requirements must be met for a commercial deposit of petroleum to exist. These are:

TABLE 1-2
Estimated Natural Gas Reserves— Leading Nations (As of January 1)(Billion Cubic Feet)

1982[r]			1983[p]			1984		
Nation	Reserves	% of Total World Reserves	Nation	Reserves	% of Total World Reserves	Nation	Reserves	% of Total World Reserves
1. U.S.S.R	1,160,000	39.8	1. U.S.S.R	1,240,000	40.9	1. U.S.S.R.	1,450,000	42.62
2. Iran	484,000	16.6	2. Iran	482,600	15.9	2. Iran	478,600	14.07
3. United States	201,730	6.9	3. U.S.A.	204,000	6.7	3. United States	198,000	5.02
4. Algeria	130,900	4.5	4. Saudi Arabia	117,000	3.9	4. Qatar	150,000	4.41
5. Saudi Arabia	114,000	3.9	5. Algeria	111,250	3.7	5. Saudi Arabia	123,270	3.62
6. Canada	89,900	3.1	6. Canada	97,000	3.2	6. Algeria	109,100	3.21
7. Mexico	75,350	2.6	7. Mexico	75,850	2.5	7. Canada	92,300	2.71
8. Qatar	60,000	2.1	8. Qatar	62,000	2.0	8. Norway	89,000	2.62
9. Netherlands	55,700	1.9	9. Norway	58,000	1.9	9. Mexico	77,000	2.26
10. Norway	49,370	1.7	10. Venezuela	54,079	1.8	10. Netherlands	68,480	2.01
11. Venezuela	47,000	1.6	11. Netherlands	51,920	1.7	11. Venezuela	55,370	1.63
12. Nigeria	40,500	1.4	12. Malaysia	34,000	1.1	12. Malaysia	50,000	1.47
13. Kuwait	30,500	1.0	13. Nigeria	32,400	1.1	13. Indonesia	40,000	1.18
14. Indonesia	27,400	0.9	14. Kuwait	29,900	1.0	14. Nigeria	35,600	1.05
15. Iraq	27,300	0.9	15. Indonesia	29,600	1.0	15. Kuwait	32,500	0.96
16. United Kingdom	26,000	0.9	16. Iraq	28,800	0.9	16. China	30,900	0.91
17. People's Republic of China	24,400	0.8	17. United Kingdom	25,400	0.8	17. Iraq	28,800	0.85
18. Argentina	23,400	0.8	18. Argentina	25,200	0.8	18. United Kingdom	27,800	0.82
19. Libya	23,200	0.8	19. Libya	21,500	0.7	19. Argentina	24,630	0.72
20. Abu Dhabi	19,500	0.7	20. Abu Dhabi	19,260	0.6	20. Libya	21,200	0.62
						21. Abu Dhabi	20,750	0.61
						22. Australia	17,850	0.52
						23. Pakistan	15,760	0.46
						24. India	15,000	0.44
						Subtotal	3,251,910	95.59
						Others	150,120	4.41
TOTAL WORLD	2,915,076	100.0	TOTAL WORLD	3,033,527	100.0	TOTAL WORLD	3,402,030	100.0

[r] Revised
[p] Preliminary

Source: 1982: United States — Department of Energy, Rest of World — Oil and Gas Journal, "Worldwide Report" issue.
1983: Oil and Gas Journal, "Worldwide Report" issue.
1984: Oil and Gas Journal, "Worldwide Report" issue.

Courtesy the American Petroleum Institute

TABLE 1-3
World Marketed Production of Natural Gas by Area (Millions of Cubic Feet)

	WESTERN HEMISPHERE													
Year	United States	Canada	Latin America	Western Hemisphere Total	Middle East	Africa	Asia-Pacific	Western Europe	Total Free World	Communist Nations	Total World	U.S. as a % of Free World Total	U.S. as a % of World Total	Year
1975	20,108,661	3,075,693	1,735,060	24,919,414	1,436,210	670,402	938,365	5,966,659	33,931,050	13,276,275	47,207,325	59.3	42.6	1975
1976	19,952,438	3,067,353	1,722,106	24,741,897	1,492,772	905,792	1,173,878	6,343,593	34,657,932	14,801,281	49,459,213	57.6	40.3	1976
1977	20,025,463	3,230,672	1,865,734	25,121,869	1,506,[illegible]	[illegible]19,464	1,329,016	6,502,502	35,379,833	14,720,257	50,100,090	56.6	40.0	1977
1978	19,974,033	3,128,056	2,012,912	25,115,001	1,579,[illegible]	[illegible]17,416	1,469,154	6,649,855	35,930,904	15,818,398	51,749,302	55.6	38.6	1978
1979	20,471,260	3,646,500	2,610,000	26,727,760	1,624,700	1,027,900	1,827,900	6,866,800	38,075,060	19,591,400	57,666,460	53.8	35.5	1979[1]
1980[r]	20,378,787	2,668,300	3,297,500	26,344,587	1,221,300	884,700	2,558,400	6,668,700	37,677,687	21,069,900	58,747,587	54.1	34.7	1980[r]
1981[r]	20,177,701	2,623,000	3,227,000	26,027,701	1,542,000	1,947,000	2,765,000	6,906,000	39,187,701	19,210,000	58,397,701	51.5	34.6	1981[r]

[r] Revised

Source: 1975–78: U.S. Energy Information Administration, World Natural Gas, Annuals.
1979–1981: U.S. Energy Information Administration, United States only; Rest of World, Oil and Gas Journal.

Courtesy the American Petroleum Institute

TABLE 1-4
World Natural Gas Production—Twenty Leading Nations (Billion Cubic Feet)

1980[1]

Nation	Marketed Production[1,2] Total	% of Total World Production
1. United States	20,378.8	34.7
2. U.S.S.R.	15,355.5	26.3
3. People's Republic of China	3,469.0	5.9
4. Netherlands	2,799.6	4.8
5. Canada	2,668.3	4.6
6. United Kingdom	1,500.0	2.6
7. Mexico	1,190.5	2.0
8. Romania	1,176.4	2.0
9. Indonesia	1,028.4	1.8
10. West Germany	738.6	1.3
11. Norway	705.2	1.2
12. Pakistan	600.0	1.0
13. Italy	524.6	0.9
14. Venezuela	518.0	0.9
15. Algeria	517.0	0.9
16. Saudia Arabia	310.2	0.5
17. Argentina	296.9	0.5
18. Iran	292.0	0.5
19. Kuwalt	291.3	0.5
20. Brunei	282.0	0.5
TOTAL WORLD	58,747.7	100.0

1981[r]

Nation	Marketed Production[1,2] Total	% of Total World Production
1. United States	20,177.7	34.6
2. U.S.S.R.	16,390.0	28.1
3. Netherlands	3,054.0	5.2
4. Canada	2,623.0	4.5
5. Mexico	1,486.0	2.5
6. Romania	1,440.0	2.5
7. United Kingdom	1,427.0	2.4
8. Algeria	1,149.0	2.0
9. Indonesia	1,075.0	1.8
10. Norway	923.0	1.6
11. Libya	674.0	1.2
12. West Germany	636.0	1.1
13. Venezuela	602.0	1.0
14. Pakistan	600.0	1.0
15. Italy	500.0	0.9
16. People's Republic of China	459.0	0.8
17. Saudi Arabia	435.0	0.7
18. Australia	377.0	0.7
19. Kuwait	340.0	0.6
20. Argentina	330.0	0.6
TOTAL WORLD	58,397.7	100.0

1984

Nation	Marketed Production[1,2] Total	% of Total World Production
1. U.S.S.R.	20,721.1	34.57
2. United States	18,068.0	30.15
3. Netherlands	2,677.5	4.47
4. Canada	2,651.8	4.42
5. United Kingdom	1,423.3	2.37
6. Mexico	1,373.3	2.29
7. Algeria	1,260.0	2.10
8. Romania	1,240.0	2.07
9. Norway	922.6	1.54
10. Indonesia	732.0	1.22
11. Argentina	653.2	1.09
12. West Germany	648.2	1.08
13. Italy	638.9	1.07
14. Venezuela	609.6	1.02
15. Australia	444.4	0.74
16. China	420.0	0.70
17. Iran	360.0	0.60
18. Pakistan	355.2	0.59
19. Brunei	312.0	0.52
20. France	312.0	0.52
Top 20	55,884.5	93.25
Others	4,047.8	6.75
TOTAL WORLD	59,932.3	100.0

[r] Revised
[1] Comprises all gas collected and utilized as a fuel or as a chemical industry raw material, including gas used in oil and/or gas fields as a fuel by producers, even though it is not actually sold.
[2] United States is reported marketed production; all others may include some gross production.

Source: 1982: United States—Department of Energy, Rest of World—Oil and Gas Journal, "Worldwide Report" issue.
1983: Oil and Gas Journal, "Worldwide Report" issue.
1984: Oil and Gas Journal, "Worldwide Report" issue.

Courtesy the American Petroleum Institute

TABLE 1-5
The Reserves of United States Natural Gas (Millions of Cubic Feet—14.72 psia at 60°F.)

Year	Proved Reserves at Start of Year		Reserve Revisions, Extensions and Discoveries During Year	Net Change In Underground Storage[7]	Production During Year[1]	Proved Reserves at Year-end		Net Change, In Reserves, During Year	Indicated, Years Supply, of Year-end, Proved Reserves	Year
1975	211,132,497[3]	237,132,497[2]	10,483,688	302,561	19,718,570	202,200,176[3]	228,200,176[4]	−8,932,321	10.2[3] 11.6[4]	1975
1976	202,200,176[3]	228,200,176[3]	7,555,468	−187,550	19,542,020	190,026,074[3]	216,026,047[4]	−12,174,102	9.7[3] 11.1[4]	1976
1977	190,026,074[3]	216,026,074[4]	11,851,924	446,930	19,447,050	182,877,878[3]	208,877,878[4]	−7,148,196	9.4[3] 10.7[4]	1977
1978	182,877,878[3]	208,877,878[4]	10,586,144	148,733	19,311,048	174,301,707[3]	200,301,707[4]	−8,576,171	9.0[3] 10.4[4]	1978
1979	174,301,707[3]	200,301,707[4]	14,285,947	239,323	19,910,353	168,916,624[3]	194,916,624[4]	−5,385,083	8.5[3] 9.8[4]	1979
1980	200,997,000[4]		16,723,000	NA	18,699,000	199,021,000[4]		−1,976,000	10.6[4]	1980
1981	199,021,000[4]		21,446,000		18,737,000	201,730,000[4]		+2,709,000	10.8[4]	1981

[1] Estimated
[3] Figures exclude 26 trillion cubic feet in Prudhoe Bay, Alaska (discovered in 1968) which are not yet available for market due to the lack of transportation facilities.
[4] Figures include 26 trillion cubic feet in Prudhoe Bay, Alaska.
[7] The net difference between gas stored in and gas withdrawn from underground storage reservoirs, inclusive of adjustments and native gas transferred from other reserve categories. (Adjustments include change of reporting basis starting in 1973 to report only gas reserves considered recoverable, in effect, reducing gas reserves by 1,024,140 MMCF that would have been reported since 1972 using former basis.)

Source: 1975–1979—American Gas Association, Committee on Natural Gas Reserves.
1980–1981—U.S. Department of Energy, Energy Information Administration.

Courtesy the American Petroleum Institute

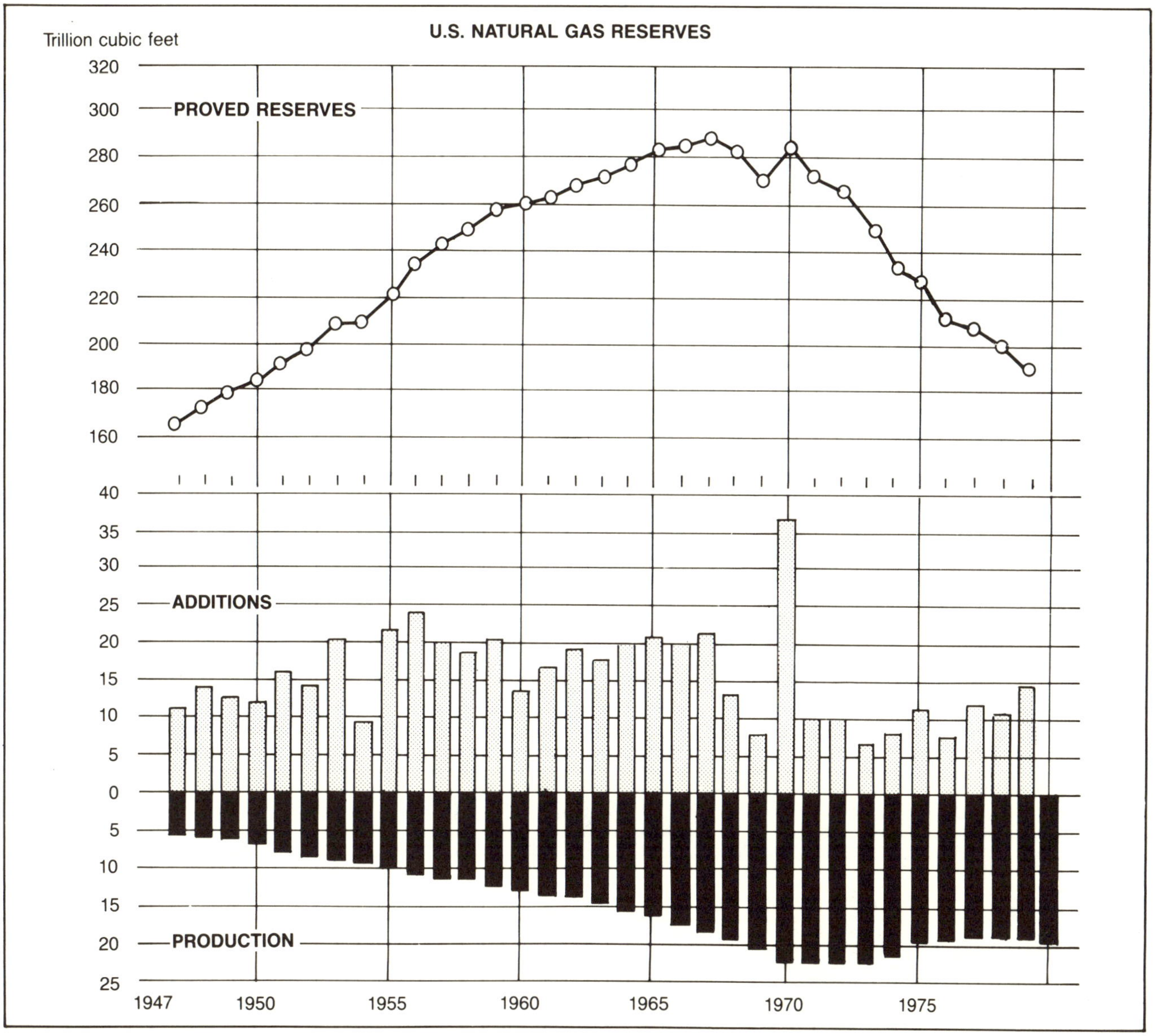

Fig. 1-1. U.S. natural-gas reserves. Used by permission of the copyright holder, American Gas Association.

TABLE 1-6
Estimated Gas Resources and Reserves (Trillion Cubic Feet) Including Alaska

Source of Estimate	Year of Estimate	Potential Resources[1] New Fields	Old Fields	Total	1979 Proved Reserves	Total Remaining Resources[2]
U.S. Geological Survey	1980	594	162	756	195	922
National Academy of Sciences	1974	530	118	648	195	717
Exxon Base	1974	342–942	56–321	423–1143	195	502–1202
Potential Gas Committee	1980	720	193	913	195	1089

[1]Does not include possible resources from unconventional sources such as coal-bed degasification. Devonian shale, Rocky Mountain tight-gas formation, geopressured resources, and biomass and coal gasification.
[2]As of December 31, 1981. Estimates are corrected for gas consumed since the date of resource estimate.
Used by permission of the copyright holder, American Gas Association.

TABLE 1-7
Salient Statistics of Natural Gas in the U.S. (Million Cubic Feet)

Year	Marketed Production	Exports	Consumption	Imports	VALUE AT WELLHEAD Total Thousand Dollars	Average Cents Per Mcf	Exports as a % of Production	Imports as a % of Consumption	Withdrawn from Storage	Stored	Lost in Trans-mission[1]	Vented and Flared	Year
1970	21,920,642	69,813	22,045,799	820,780	$ 3,745,680	17.1¢	0.3%	3.7%	1,458,607	1,856,767	227,650	489,460	1970
1971	22,493,012	80,212	22,676,581	934,548	4,085,482	18.2	0.4	4.1	1,507,630	1,839,398	338,999	284,561	1971
1972	22,531,698	78,013	23,009,445	1,019,496	4,180,462	18.6	0.3	4.4	1,757,218	1,892,952	328,002	248,119	1972
1973	22,647,549	77,169	22,965,914	1,032,901	4,894,072	21.6	0.3	4.5	1,532,820	1,974,324	195,863	248,292	1973
1974	21,600,522	76,789	22,110,623	959,284	6,573,402	30.4	0.4	4.3	1,700,546	1,784,209	288,731	169,381	1974
1975	20,108,661	72,675	20,409,875	953,008	8,945,062	44.5	0.4	4.7	1,759,565	2,103,619	235,065	133,913	1975
1976	19,952,438	64,711	20,800,582	963,768	11,571,776	58.0	0.3	4.6	1,921,017	1,755,690	216,240	131,930	1976
1977	20,025,463	55,626	19,520,581	1,011,001	15,833,719	79.0	0.3	5.2	1,749,884	2,306,515	41,063	136,807	1977
1978	19,974,033	52,532	19,627,478	965,545	18,084,914	90.5	0.3	4.9	2,157,765	2,278,002	287,201	153,350	1978
1979	20,471,260	55,673	20,240,761	1,253,383	24,113,634	117.8	0.3	6.1	2,047,000	2,295,034	372,330	167,019	1979
1980	20,378,787	48,731	19,877,293	984,767	32,052,182	159.0	0.2	4.8	1,909,902	1,896,284	644,072	125,451	1980[r]
1981	20,177,701	59,372	19,403,858	903,949	39,561,874	198.0	0.3	4.5	1,886,940	2,179,683	505,026	98,017	1981

[1] Also includes changes in above ground storage and gas unaccounted for.
[r] Revised

Source: U.S. Energy Information Administration, Natural Gas Annuals

Courtesy the American Petroleum Institute

TABLE 1-8
U.S. Marketed Production[1] of Natural Gas by State (Million Cubic Feet)

State	1972	1973	1974	1975	1976	1977	1978	1979	1980	1981
Alabama	3,644	11,271	27,865	37,814	41,427	57,227	85,399	85,815	105,527	119,334
Alaska	125,596	131,007	128,935	160,270	166,072	187,889	203,088	220,754	230,588	242,564
Arizona	442	125	224	208	262	240	286	247	214	187
Arkansas	166,522	157,529	123,975	116,237	109,533	104,096	106,792	109,452	111,808	92,986
California	487,278	449,369	365,354	318,308	354,334	311,462	311,084	248,206	309,783	380,730
Colorado	116,949	137,725	144,629	171,629	183,972	188,792	183,693	191,239	190,814	199,146
Florida	15,521	33,857	38,137	44,383	43,165	48,171	51,595	50,190	45,744	35,938
Illinois	1,194	1,638	1,436	1,440	1,556	1,003	1,159	1,585	1,574	1,295
Indiana	355	276	176	346	192	183	163	350	463	330
Kansas	889,268	893,118	886,782	843,625	829,170	781,289	854,484	797,762	735,035	640,114
Kentucky	63,648	62,396	71,876	60,511	66,137	60,902	70,044	59,520	57,180	61,312
Louisiana	7,972,678	8,242,423	7,753,631	7,090,645	7,006,596	7,215,006	7,476,497	7,266,217	6,939,924	6,780,184
Maryland	244	298	133	93	75	82	88	28	68	56
Michigan	34,221	44,579	69,133	102,113	119,262	129,954	148,047	159,731	158,302	152,593
Mississippi	103,989	99,706	78,787	74,345	70,762	82,995	106,579	144,077	185,469	211,371
Missouri	9	33	33	30	29	20	—	—	—	—
Montana	33,474	56,175	54,873	40,734	42,563	46,819	46,522	53,888	51,867	56,565
Nebraska	3,478	3,836	2,538	2,565	2,511	2,789	2,882	3,208	2,550	2,519
New Mexico	1,216,061	1,218,749	1,244,779	1,217,430	1,230,976	1,202,973	1,174,198	1,181,363	1,149,781	1,134,113
New York	3,679	4,539	4,990	7,628	9,235	10,682	13,900	15,500	15,643	16,074
North Dakota	32,472	27,703	31,206	24,786	31,470	29,173	30,499	18,468	42,346	42,573
Ohio	89,995	93,610	92,055	84,960	88,891	99,327	114,098	123,181	138,856	141,134
Oklahoma	1,806,887	1,770,980	1,638,942	1,605,410	1,726,513	1,769,519	1,773,582	1,835,366	1,891,824	2,019,199
Oregon	—	—	—	—	—	—	—	2	5	5
Pennsylvania	73,958	78,514	82,637	84,676	89,386	91,717	97,763	96,313	97,439	122,454
South Dakota	—	—	—	—	—	—	—	914	1,193	1,155
Tennessee	25	20	17	27	47	263	468	941	1,241	1,719
Texas	8,657,840	8,513,850	8,170,798	7,485,764	7,191,859	7,051,027	6,548,184	7,174,623	7,251,879	7,050,207
Utah	39,474	42,715	50,522	55,354	57,416	60,696	58,416	58,605	87,766	91,191
Virginia	2,787	5,101	7,096	6,723	6,937	8,220	8,492	8,544	7,812	8,903
West Virginia	214,951	208,676	202,306	154,484	153,322	152,767	148,564	150,505	156,551	161,251
Wyoming	375,059	357,731	326,657	316,123	328,768	330,180	357,267	414,416	409,541	410,449
Total	22,531,698	22,647,549	21,600,522	20,108,661	19,952,438	20,025,463	19,974,033	20,471,260	20,378,787	20,177,701

[1] Marketed production of natural gas represents gross withdrawals less gas used for repressuring and quantities vented and flared.

Source: U.S. Energy Information Administration, Natural Gas Annuals. January 1983.

Courtesy the American Petroleum Institute

TABLE 1-9
U.S. Consumption of Natural Gas by State (Million Cubic Feet)

State	1977	1978	1979	1980	1981[4]
Alabama	241,237	237,258	283,435	269,240	271,280
Alaska	116,278	145,025	157,236	153,345	121,727
Arizona	167,092	175,041	172,738	165,850	182,533
Arkansas	229,558	220,699	250,747	273,720	264,787
California	1,772,041	1,563,172	1,810,381	1,807,731	1,858,178
Colorado	282,215	267,838	292,272	256,363	211,780
Connecticut	64,068	65,191	67,957	72,583	76,829
Delaware	15,787	20,626	24,981	29,797	30,643
Dist. of Col.	(2)	(2)	(2)	27,868	29,023
Florida	302,344	318,344	344,177	316,588	337,557
Georgia	264,665	278,272	312,299	315,201	316,613
Idaho	45,537	44,210	54,237	49,047	44,960
Illinois	1,167,099	1,174,934	1,142,732	1,089,720	1,061,957
Indiana	398,268	440,955	504,185	489,254	496,349
Iowa	280,246	238,229	291,521	269,581	253,231
Kansas	506,910	519,346	584,236	488,338	428,130
Kentucky	219,521	212,913	218,689	202,284	199,445
Louisiana	2,190,848	2,249,172	1,978,382	1,794,007	1,781,716
Maine	(1)	(1)	(1)	2,268	2,326
Maryland	158,838	161,695	201,584	160,244	175,289
Massachusetts	160,343	160,503	156,459	182,587	184,575
Michigan	741,295	789,594	875,726	865,137	800,873
Minnesota	293,381	313,229	334,202	285,846	266,282
Mississippi	197,689	203,719	254,366	263,530	242,987
Missouri	366,970	358,905	347,257	318,451	283,514
Montana	70,956	72,649	69,805	60,724	52,452
Nebraska	188,804	163,051	170,013	163,079	138,322
Nevada	71,052	64,506	84,433	58,454	73,182
New Hampshire	13,609	14,039	14,881	9,474	10,220
New Jersey	247,120	228,843	260,579	339,825	390,121
New Mexico	229,813	213,698	211,182	221,793	196,212
New York	561,698	569,702	623,891	737,443	760,463
North Carolina	72,527	81,989	130,884	153,316	152,496
North Dakota	37,650	39,067	29,236	22,848	34,445
Ohio	847,497	929,593	898,029	896,983	870,341
Oklahoma	766,986	770,249	824,980	722,318	670,971
Oregon	72,521	86,140	93,707	78,703	75,562
Pennsylvania	668,337	673,770	740,818	775,715	784,691
Rhode Island	25,631	23,042	27,219	27,630	29,202
South Carolina	138,585	117,573	119,319	142,228	141,924
South Dakota	36,097	35,423	25,724	24,078	22,056
Tennessee	202,135	184,048	226,477	229,661	223,529
Texas	4,143,023	4,211,432	4,001,355	4,091,099	3,926,687
Utah	106,315	118,513	126,047	115,092	102,239
Vermont	(1)	(1)	(1)	3,991	4,383
Virginia	118,020	134,144	134,088	158,479	151,737
Washington	142,657	127,280	158,515	128,770	124,942
West Virginia	144,535	151,839	148,538	143,047	148,527
Wisconsin	349,160	370,726	367,517	351,857	324,611
Wyoming	83,623	87,292	93,797	68,975	69,056
U.S. Total	19,520,581	19,627,478	20,240,761	19,877,293[3]	19,403,858[3]

(1) Included with New Hampshire
(2) Included with Maryland
(3) Total includes 3,131 MMcf from Hawaii in 1980 and 2,899 MMcf in 1981.
(4) Totals may not add due to independent rounding.

Source: U.S. Energy Information Administration, Natural Gas Annuals. January 1983.

Courtesy the American Petroleum Institute

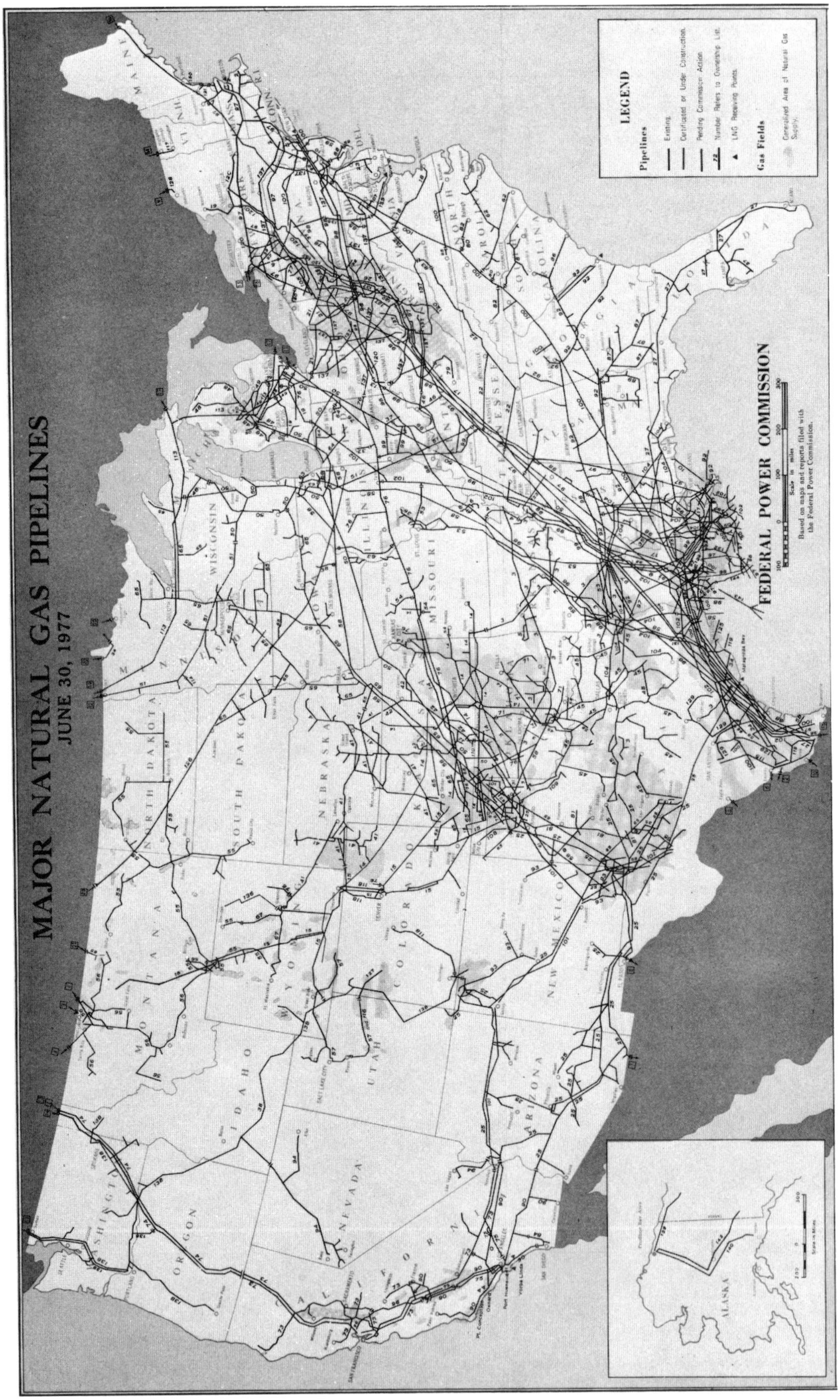

Fig. 1-2. Major natural-gas pipelines in the United States in 1977.

TABLE 1-10
Transmission Pipelines and Natural-Gas Customers by State

State	Transmission Pipelines (Miles)	Gas Customers (Thousands)
Alabama	5,301	644.4
Alaska	115	30.4
Arizona	5,041	561.0
California	7,813	6,632.2
Colorado	6,219	726.6
Connecticut	538	390.2
Delaware	227	82.4
District of Columbia	23	153.2
Florida	3,101	409.9
Georgia	4,982	927.3
Hawaii	0	34.5
Idaho	1,312	109.0
Illinois	9,955	3,158.3
Indiana	5,655	1,196.3
Iowa	6,142	702.6
Kansas	15,547	679.0
Kentucky	6,950	627.4
Louisiana	21,697	943.3
Maine	538	17.6
Maryland	730	773.8
Massachusetts	789	1,052.4
Michigan	6,891	2,244.4
Minnesota	3,983	727.3
Mississippi	9,355	383.3
Missouri	4,153	1,177.5
Montana	3,041	176.1
Nebraska	7,448	411.2
Nevada	1,236	118.7
New Hampshire	129	46.2
New Jersey	1,310	1,823.9
New Mexico	6,023	288.7
New York	4,232	3,996.3
North Carolina	2,291	342.2
North Dakota	1,111	74.9
Ohio	10,218	2,756.3
Oklahoma	11,136	753.9
Oregon	1,167	250.8
Pennsylvania	11,983	2,327.1
Rhode Island	53	160.8
South Carolina	2,315	285.9
South Dakota	942	94.7
Tennessee	5,116	494.0
Texas	38,295	3,029.1
Utah	909	310.0
Vermont	60	16.6
Virginia	2,352	526.1
Washington	1,707	348.8
West Virginia	5,830	396.1
Wisconsin	3,166	959.0
Wyoming	2,885	103.5
Total U.S.	252,012	44,941.5

Source: Interstate Natural Gas Association of America, Washington, D.C.

1. A source; that is, material from which the petroleum is formed.
2. Porous and permeable beds in which the petroleum may migrate and accumulate after being formed.
3. A trap or subsurface condition restricting further movement so that it may accumulate in commercial quantities.

Natural gas and crude oil are generated from organic matter under the influence of increasing temperature and time. Both the type of organic matter and the temperature it experiences have a role in controlling whether oil or gas is formed. Organic matter can be divided into two broad categories depending on whether it was derived from organisms growing on the land surface or growing in water, that is, whether it is terrestrial or aquatic. It is generally thought that the terrestrial matter produces natural gas and some waxy crudes, whereas aquatic material produces normal crudes.

This distinction is important in estimating the maximum depth to which crude oil or natural gas may occur. Rivers have played a critical role in transporting terrestrial material to the depositional environment. Therefore,

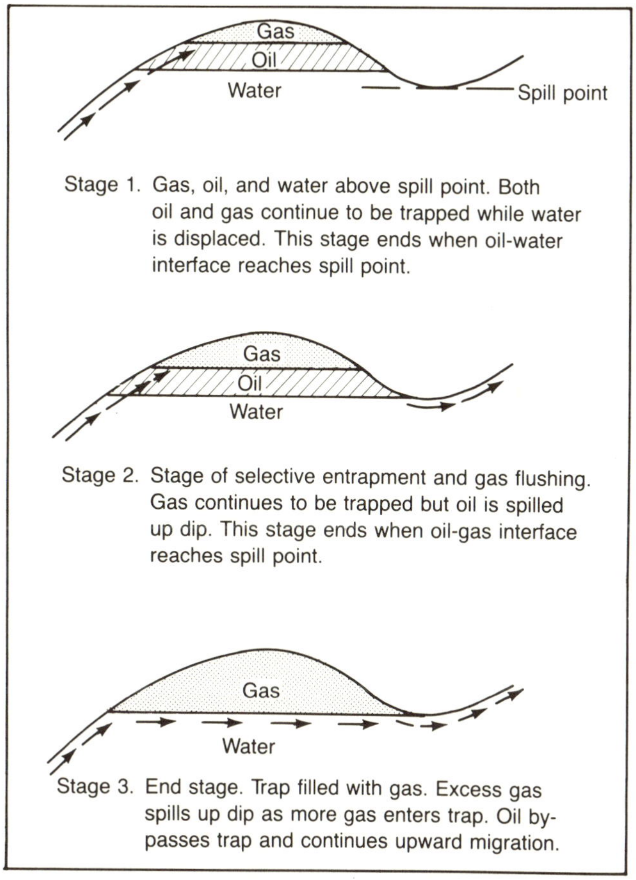

Fig. 1-3. Illustration of differential entrapment principle, showing various stages of hydrocarbon accumulation in an anticline. Solid and dashed arrows denote oil and gas movement respectively. (After Gussow) Courtesy American Association of Petroleum Geologists.

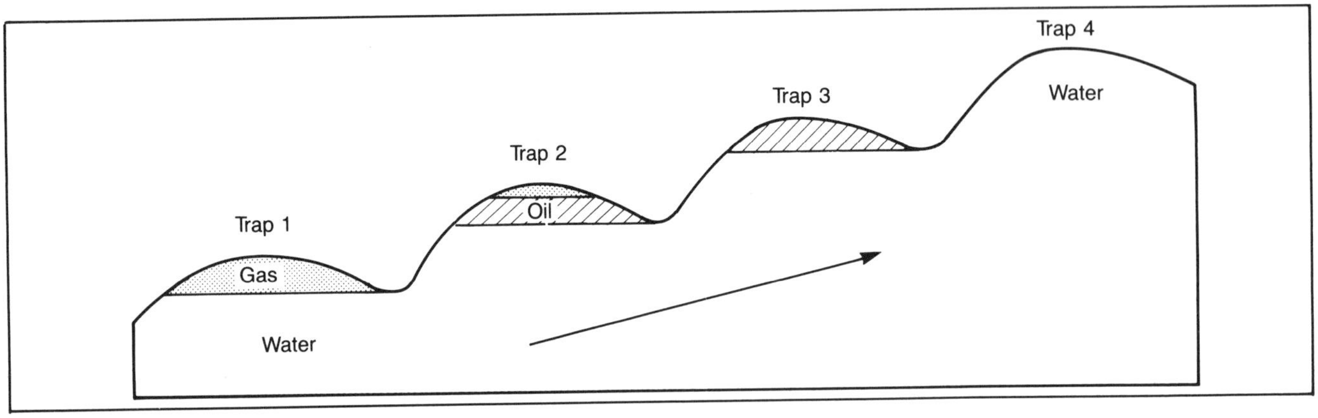

Fig. 1-4. Final condition of differential entrapment in a series of interconnected traps. (After Gussow) Courtesy American Association of Petroleum Geologists.

deltas are very gas-prone depositional environments. The oldest and deepest sediments were deposited in the continental rift and are rich in terrestrial organic matter. They are overlain by increasingly marine sediments containing greater amounts of aquatic matter so that a vertical sequence develops with the gas-generating organic matter at the bottom and the oil-generating material at the top.

Modification by Migration and Burial

Oil and gas may be remobilized after it has formed in reservoirs. This secondary migration is most frequently caused by regional tipping and may lead to a marked separation of oil and gas. Gussow[1] used the term "differential entrapment" for the case in which a full trap with a gas cap spills oil from the bottom into the next higher trap. This can lead to adjacent traps with gas, oil, or varying mixtures of the two. The gas-filled reservoirs can be downdip from the oil-filled reservoirs. The process is illustrated in Figures 1-3 and 1-4.

As can be seen from Figure 1-4, natural gas can occur either associated or not associated with oil. Trap 1 in Figure 1-4 is a non-associated gas reservoir while Trap 2 is an associated gas reservoir; that is, the gas occurs as a gas cap in contact with crude oil. In some associated gas reservoirs the oil exists as a thin rim at the bottom of the trap.

The temperature increase that accompanies increasing depth of burial has a major role in determining the composition of the trapped petroleum. Crude oils are not equilibrium mixtures, and as temperature increases the hydrocarbons readjust toward equilibrium at an increasing rate. This readjustment causes a redistribution of hydrogen, ultimately giving methane and a solid carbon-rich residue. The oil changes first to condensate, then to wet gas, and finally to dry gas. Gas developed in this manner is frequently associated with hydrogen sulfide and carbon dioxide.

TABLE 1-11
Total Estimated Potential Supply of Natural Gas by Depth Increments As of December 31, 1982 (Trillion Cubic Feet)

	Probable	Possible	Speculative
Onshore (Drilling Depth)			
Lower 48 States			
less than 15,000 feet	103	133	83
15,000 to 30,000 feet	48	130	96
	151	263	179
Alaska			
less than 15,000 feet	6	16	28
15,000 to 30,000 feet	—	—	—
	6	16	28
Total Onshore	157	279	207
Offshore (Water Depth)			
Lower 48 States			
less than 200 meters	30	41	30
200 to 1,000 meters	3	22	23
	33	63	53
Alaska			
less than 200 meters	2	13	59
200 to 1,000 meters	—	—	10
	2	13	69
Total Offshore	35	76	122
Total Lower 48 States	184	326	232
Total Alaska	8	29	97
Total United States	192	355	329

Courtesy Potential Gas Agency, Colorado School of Mines

Although the maximum temperature for oil to exist is fairly well established, the temperature and thus depth limitations on the occurrence of natural gas have yet to be determined. Barker and Kemp[2] have published the results of a computer study to determine this depth limitation and have found that under certain conditions methane can remain stable at depths beyond 40,000 ft. The amount of methane surviving is strongly influenced by reservoir lithology with fairly cool, clean sandstones being the most favorable reservoirs. It was found that

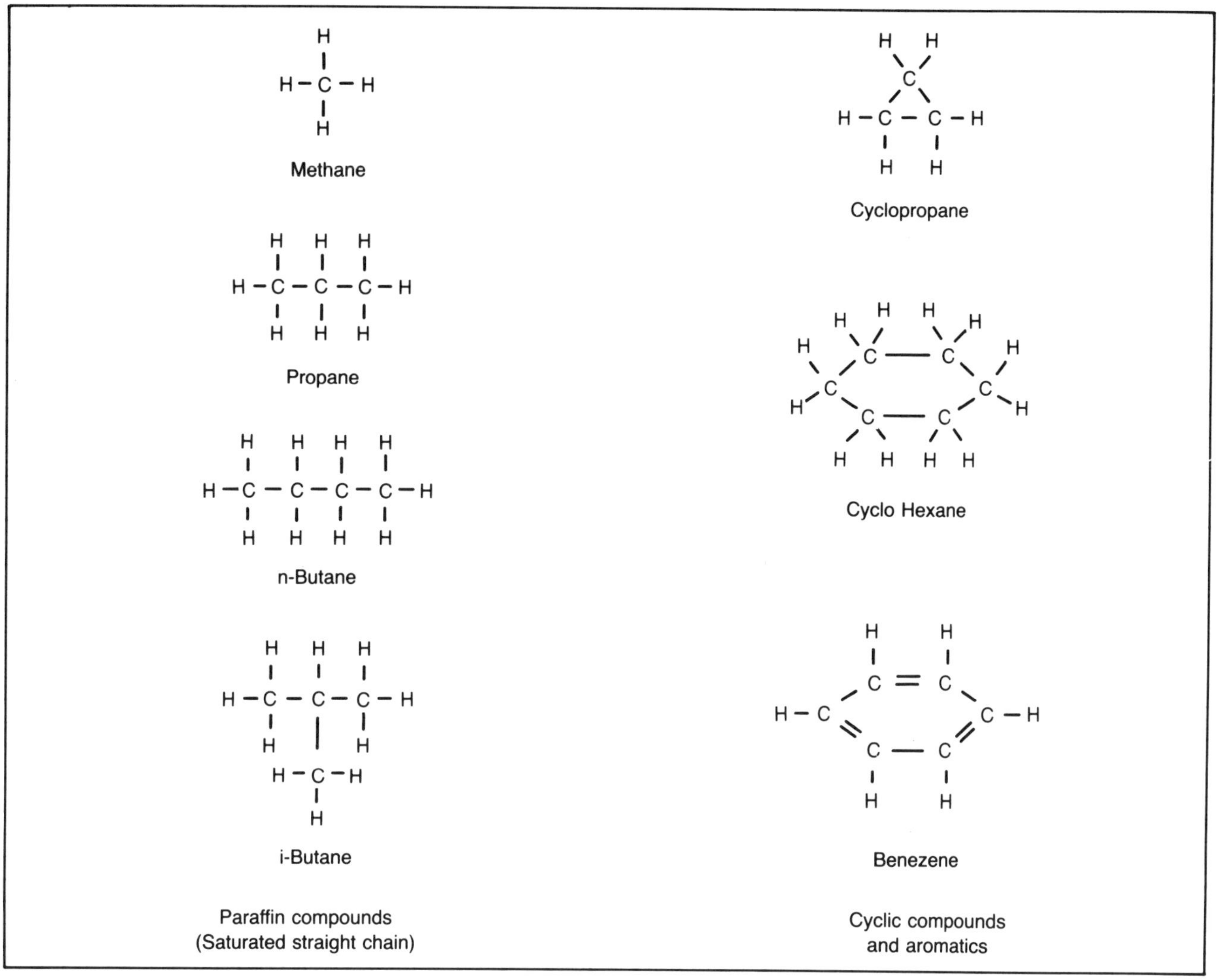

Fig. 1-5. Hydrocarbon gas molecule structures.

methane has less stability in deep carbonates.

The Barker and Kemp study implies that natural gas may exist at depths greater than have been presently explored if porous and permeable reservoirs exist at these depths. Table 1-11 shows that considerable reserves may exist at depths between 15,000 and 30,000 ft.

CHARACTERISTICS OF NATURAL GAS

Natural gas is a mixture of hydrocarbon gases with some impurities, mainly nitrogen (N_2), hydrogen sulfide (H_2S), and carbon dioxide (CO_2). Gases containing significant amounts of H_2S or CO_2 or both are called sour or acid gases. These impurities must be removed before the gas is used as a fuel. The hydrocarbon gases are methane, ethane, propane, butanes, pentanes, and small amounts of hexanes, heptanes, and some heavier fractions. In gas used for fuel, methane is the largest component, usually 95 to 98%.

Natural gas is normally considered to be a mixture of straight chain or paraffin hydrocarbon compounds. However, occasionally cyclic and aromatic compounds occur in a natural gas. The molecular structures of some of these are shown in Figure 1-5. The general formula for the paraffin hydrocarbons is C_nH_{2n+2}, where n is the number of carbon atoms.

Gas Composition

The actual composition of a natural gas can vary over wide ranges. Even two gas wells producing from the same reservoir may have different compositions. Table 1-12 shows typical hydrocarbon compositions of the gas produced from different types of reservoirs. The composition of the gas produced from a given reservoir may

TABLE 1-12
Typical Natural Gas Compositions

Component	Symbol	Mol Percent: Associated Gas	Wet Gas	Dry Gas
Methane	C_1	27.52	59.52	97.17
Ethane	C_2	16.34	5.36	1.89
Propane	C_3	29.18	4.71	0.29
i-Butane	i-C_4	5.37	2.03	0.13
n-Butane	n-C_4	17.18	2.39	0.12
i-Pentane	i-C_5	2.18	1.80	0.07
n-Pentane	n-C_5	1.72	1.61	0.05
Hexane	C_6	0.47	2.60	0.04
Heptanes Plus	C_{7+}	0.04	19.98	0.24
		100.00	100.00	100.00

change with time if liquids condense in the reservoir as pressure declines. This occurs in a so-called retrograde condensate reservoir.

Although natural gases contain small fractions of hydrocarbon components much heavier than heptane, most analyses group all of the heavier components into a category called Heptanes plus or C_{7+}. Table 1-13, from Katz[3], lists some of the components of petroleum and the commercial products that contain these components. Only the constituents in the range of condensate well effluents and lighter will be considered in this text.

OTHER SOURCES OF GASEOUS FUEL

Alternatives to conventional sources of natural gas are importation of liquefied natural gas, substitute natural gas, coal gasification, and gas from shale reservoirs. The technology required for handling and transporting these gases once they are produced is essentially the same as for natural gases.

Liquefied Natural Gas. Liquefied natural gas (LNG) is essentially methane that is liquefied at atmospheric pressure by cooling it to −260°F. The phase change reduces the volume by 623:1, and approximately 15% of the energy is consumed in the phase change. Most of the LNG imported into the United States comes from Algeria and is transported in large tankers that hold approximately 750,000 barrels of LNG. This is equivalent to about 2.5 Bcf of gas.

Coal Gasification. Synthetic gas from coal has been investigated in several pilot plants in the United States, but the commercial viability has not been demonstrated. The gas derived from coal is usually low in heating value as compared to natural gas.

Substitute Natural Gas. Substitute natural gas (SNG) can be made from liquid petroleum feedstock such as naphtha, crude oil, propane, and butane. Facilities exist

TABLE 1-13
Constituents of Petroleum

Name	Formula	Product of Commerce Containing Constituent
Methane	CH_4	Natural gas
Ethane	C_2H_6	Natural gas
Propane	C_3H_8	Natural gas, propane
Isobutane	*i*-C_4H_{10}	Natural gasoline, butane
n-Butane	*n*-C_4H_{10}	Natural gasoline, motor fuel, butane
Pentanes	C_5H_{12}	Natural gasoline, motor fuel
Hexanes	C_6H_{14}	Natural gasoline, motor fuel
Heptanes	C_7H_{16}	Natural gasoline, motor fuel
Octanes	C_8H_{18}	Natural gasoline, motor fuel
Decanes	$C_{10}H_{22}$	Motor fuel
Tetradecane	$C_{14}H_{20}$	Kerosene, light furnace oil
Hexadecane	$C_{16}H_{34}$	Mineral seal oil, furnace oil
Triacontane	$C_{30}H_{62}$	Light lubricating oil, heavy fuel oil
Tetracontane	$C_{40}H_{82}$	Lubricating oil, heavy fuel oil
Asphaltene	$C_{80}H_{162}+$	Asphalt, road oil, bunker fuel oil

General Range of Constituents in Field Stream

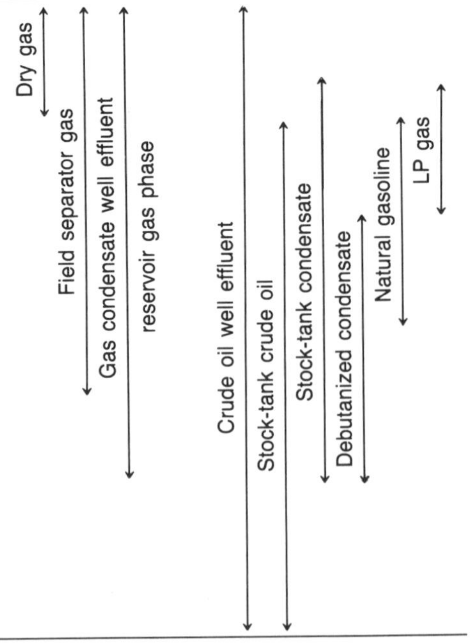

for this process, but the cost is extremely high in comparison to natural gas at its present price. SNG may well become economical in the near future as demand for all forms of energy increases.

Gas from Devonian Shale. Gas from Devonian shale formations is a potential source of gaseous fuel that could be extracted from an area of approximately 250,000 square miles of formations underlying the middle and eastern portions of the United States. Massive fracturing, advanced recovery techniques, and a higher price for the gas will have to occur before this source contributes a substantial amount of energy to the United States.

Tight Formation Gas. Gas from very low permeability formations is being produced from deposits in the Rocky Mountain states of the United States. In order to recover this gas in commercial quantities, some method is required for increasing the flow capacity of the wells. One method is to fracture the formation with a nuclear blast, but so far this technique has not proved feasible. The other method is massive hydraulic fracturing, which is the subject of much research of the present time. If either of these stimulation techniques can be perfected, recovery of the gas in place could reach 40 to 50%.

Gas from Geopressured Aquifers. High pressure brine in geopressured aquifers may contain 30 to 40 scf of natural gas per barrel of water. In the United States, these aquifers are located in a band that extends from Florida to Texas along the Gulf of Mexico. Estimates of the gas in place range from 1000 to 3000 Tcf, but in 1981 no commercial method of recovering this gas had been developed.

GAS PRODUCTION OPERATIONS

The engineer involved in gas production operations has one principal objective: to move the gas from some location in an underground reservoir to a pipeline that may be used to transport it to its final destination. Figure 1-6 shows that this involves moving the gas through a porous medium or rock formation, to the surface through casing or tubing, to separation facilities through a surface piping system, through a compressor if one is necessary to maintain sufficient flow rates, and finally through a surface line to the point of utilization.

The engineer in charge of this operation must understand both reservoir engineering and production engineering concepts, as both are included in the total producing system. He must be able to determine the gas

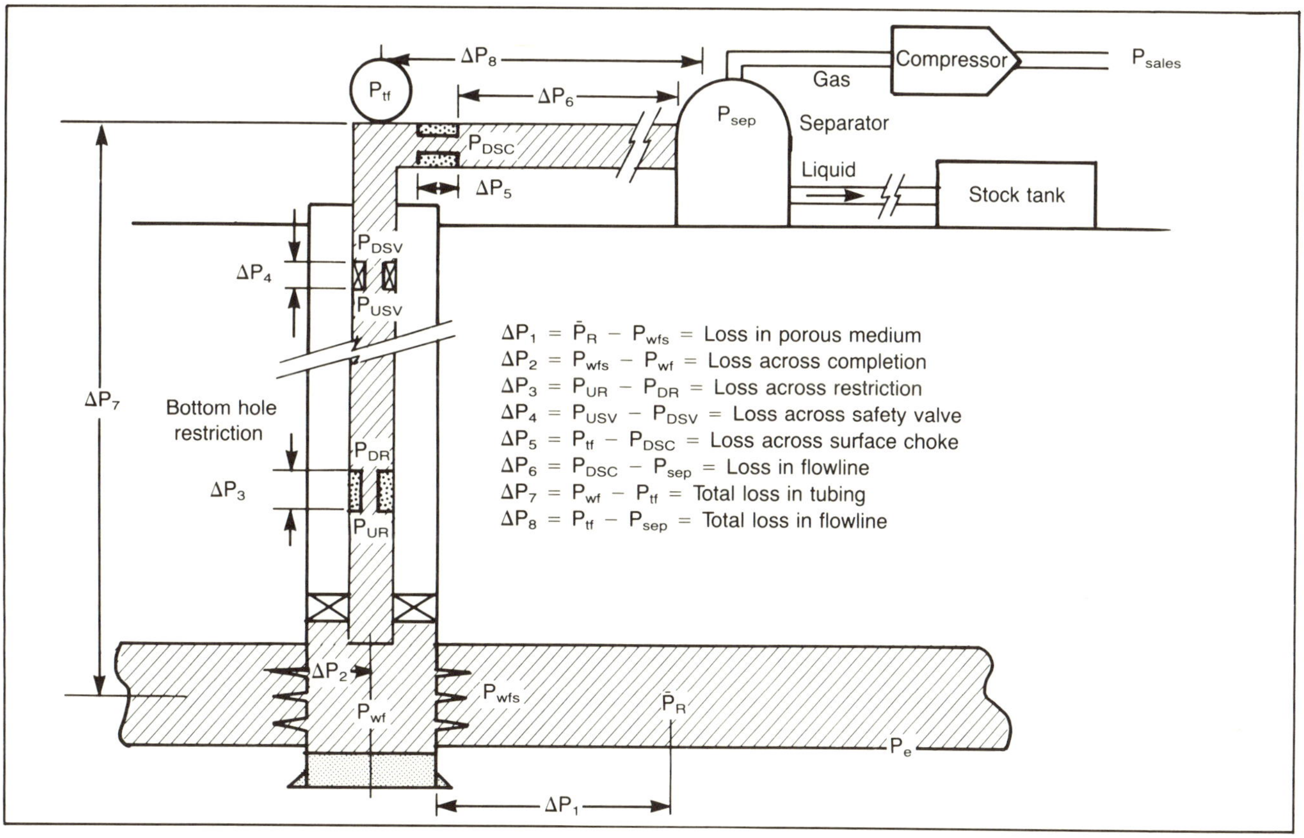

Fig. 1-6. Possible pressure losses in complete system.

recoverable in a reservoir and the time that will be required to recover the gas. This requires analysis of individual well performance and how this performance changes as gas is depleted from the reservoir and pressure declines. He must be capable of determining the relationship between flow rate and pressure drop in all parts of the system, the rock as well as all segments of the piping system. As pressure declines he must be able to determine the size and power of compression required to maintain production rates, and to handle liquids that form in the system, both condensate and water. All of the above design considerations require accurate values for flow rate through the system, which means that the engineer must be familiar with accurate methods to measure flow rates.

Since demand often exceeds producing or transportation capacity during the winter months, the engineer may be required to determine the most feasible method for storing gas until it is needed. This requires the calculation of the efficiency with which gas displaces water in aquifers and the performance of injection wells.

The following chapters contain the technology to handle all of these engineering problems. Each component of the system is treated individually and then combined for a total system analysis. Numerous example calculations are made in order to clarify the application of the theory presented. Although some operations involved in the gas industry are not covered, such as unsteady state flow in pipelines and highly fractured reservoirs, the technology for handling day-to-day gas production operations is included in a complete and practical form.

REFERENCES

1. Gussow, W. C.: "Differential Entrapment of Oil and Gas: A Fundamental Principle," *Bull.* AAPG (1954) 816–853.
2. Barker, C. and Kemp, M. K.: "Generation of Natural Gas and its Survival in the Deep Subsurface," Presented at the Natural Gas Res. Development In Mid-Continent Basins: Production and Exploration Techniques, The Univ. of Tulsa, Tulsa, Oklahoma, March 11–12, 1980.
3. Katz, D. L. *et al: Handbook of Natural Gas Engineering,* McGraw-Hill Book Co., New York (1959).

2 Gas Properties

THE ability to calculate the performance of a gas producing system, including the reservoir and the piping system, requires knowledge of many gas properties at various pressures and temperatures. If the natural gas is in contact with liquids, such as condensate or water, the effect of the liquids on gas properties must be evaluated.

This chapter presents the best and most widely used methods to perform the necessary calculations. Some of the information presented in this chapter will be used only in reservoir calculations and some will be used only in the piping system design chapter; therefore, this chapter will be referred to frequently in the subsequent chapters.

Numerous example problems are worked and graphs are presented for empirical correlations. Application of some of the methods requires a computer, and FORTRAN subroutines are included in the appendix if available.

IDEAL GASES

The understanding of the behavior of gases with respect to pressure and temperature changes is made clearer by first considering the behavior of gases at conditions near standard conditions of pressure and temperature; that is:

$$p = 14.7 \text{ psia} = 101.325 \text{ kPa (SPE uses 100 kPa)}$$

$$T = 60°\text{F} = 520°\text{R} = 288.72°\text{K (SPE uses 288°K)}.$$

At these conditions the gas is said to behave ideally, and most of the early work with gases was conducted at conditions approaching these conditions. An ideal gas is defined as one in which: (1) the volume occupied by the molecules is small compared to the total gas volume; (2) all molecular collisions are elastic; and (3) there are no attractive or repulsive forces among the molecules.

The basis for describing ideal gas behavior comes from the combination of some of the so-called gas laws proposed by early experimenters.

Early Gas Laws

Boyle's Law. Boyle observed experimentally that the volume of an ideal gas is inversely proportional to the pressure for a given weight or mass of gas when temperature is constant. This may be expressed as

$$V \alpha \frac{1}{p} \quad \text{or} \quad pV = \text{constant}.$$

Charles' Law. While working with gases at low pressures, Charles observed that the volume occupied by a fixed mass of gas is directly proportional to its absolute temperature, or

$$V \alpha T \quad \text{or} \quad \frac{V}{T} = \text{constant}.$$

Avogadro's Law. Avogadro's Law states that under the same conditions of temperature and pressure, equal volumes of all ideal gases contain the same number of molecules. This is equivalent to the statement that at a given temperature and pressure one molecular weight of any ideal gas occupies the same volume as one molecular weight of another ideal gas. It has been shown that there are 2.73×10^{26} molecules/lb-mole of ideal gas and that one molecular weight in pounds of any ideal gas at 60°F and 14.7 psia occupies a volume of 379.4 cu ft. One mole of a material is the quantity of that material whose mass, in the system of units selected, is numerically equal to the molecular weight. This means that one mole of any ideal gas, that is, 2.73×10^{26} molecules of any gas, will occupy the same volume at a given pressure and temperature.

The Ideal Gas Law

The three gas laws described previously can be combined to express a relationship among pressure, volume, and temperature, called the ideal gas law.

In order to combine Charles' Law and Boyle's Law to describe the behavior of an ideal gas when both temperature and pressure are changed, assume a given mass of gas whose volume is V_1 at pressure p_1 and temperature T_1, and imagine the following process through which the gas reaches volume V_2 at pressure p_2 and temperature T_2:

$$(V_1 \text{ at } p_1, T_1) \xrightarrow[T_1 = \text{constant}]{\text{Step 1}} (V \text{ at } p_2, T_1)$$

$$(V \text{ at } p_2, T_1) \xrightarrow[p_2 = \text{constant}]{\text{Step 2}} (V_2 \text{ at } p_2, T_2).$$

In the first step the pressure is changed from a value of p_1 to a value of p_2 while temperature is held constant. This causes the volume to change from V_1 to V. In Step 2, the pressure is maintained constant at a value of p_2, and the temperature is changed from a value of T_1 to a value of T_2.

The change in volume of the gas during the first step may be described through the use of Boyle's Law since the quantity of gas and the temperature are held constant. Thus

$$p_1V_1 = p_2V \quad \text{or} \quad V = \frac{p_1V_1}{p_2}, \tag{2-1}$$

where V represents the volume at pressure p_2 and temperature T_1. Charles' Law applies to the change in the volume of gas during the second step since the pressure and the quantity of gas are maintained constant; therefore

$$\frac{V}{T_1} = \frac{V_2}{T_2} \quad \text{or} \quad V = \frac{V_2T_1}{T_2}. \tag{2-2}$$

Elimination of volume, V, between Equations 2-1 and 2-2 gives

$$\frac{p_1V_1}{p_2} = \frac{V_2T_1}{T_2}$$

or

$$\frac{p_1V_1}{T_1} = \frac{p_2V_2}{T_2}. \tag{2-3}$$

Thus for a given quantity of gas, pV/T = a constant. The constant is designated with the symbol R when the quantity of gas is equal to one molecular weight. That is,

$$\frac{pV_M}{T} = R$$

where V_M is the volume of one molecular weight of the gas at p and T.

In order to show that R is the same for any gas, Avogadro's Law is invoked. In symbolic form, this law states

$$V_{MA} = V_{MB},$$

where V_{MA} represents the volume of one molecular weight of gas A and V_{MB} represents the volume of one molecular weight of gas B, both at pressure, p, and temperature, T. This implies that

$$\frac{pV_{MA}}{T} = R_A \quad \text{or} \quad V_{MA} = \frac{R_AT}{p},$$

and

$$\frac{pV_{MB}}{T} = R_B \quad \text{or} \quad V_{MB} = \frac{R_BT}{p},$$

where R_A represents the gas constant for gas A and R_B represents the gas constant for gas B. The combination of the above equations reveals that

$$\frac{R_AT}{p} = \frac{R_BT}{p} \quad \text{or} \quad R_A = R_B.$$

Thus, the constant R is the same for all ideal gases and is referred to as the universal gas constant. Therefore, the equation of state for one molecular weight of any ideal gas is

$$pV_M = RT. \tag{2-4}$$

For n moles of ideal gas this equation becomes

$$pV = nRT, \tag{2-5}$$

where V is the total volume of n moles of gas at temperature, T, and pressure, p. Since n is the mass of gas divided by the molecular weight, the equation can be written as

$$pV = \frac{m}{M}RT$$

or, since m/V is the gas density,

$$\rho = \frac{pM}{RT}. \tag{2-6}$$

This expression is known by various names such as the ideal gas law, the general gas law, or the perfect gas law. This equation has limited practical value since no known gas behaves as an ideal gas; however, the equation does describe the behavior of most real gases at low pressure and gives a basis for developing equations of state which more adequately describe the behavior of real gases at elevated pressures.

The numerical value of the constant R depends on the units used to express temperature, pressure, and volume. As an example, suppose that pressure is expressed in psia, volume in cubic feet, temperature in degrees Rankin, and moles in pound moles. Avogadro's Law states that 1 lb-mole of any ideal gas occupies 379.4 cu ft at 60°F and 14.7 psia. Therefore,

$$R = \frac{pV}{nT} = \frac{(14.7 \text{ psia})(379.4 \text{ cu ft})}{(1 \text{ lb-mole})(520°\text{R})}$$

$$= 10.73 \text{ psia cu ft/lb-mole °R}.$$

Table 2-1 gives numerical values of R for various systems of units.

Example 2-1:

Calculate the mass of methane gas contained at 1,000 psia and 68°F in a cylinder with volume of 3.20 cu ft. Assume that methane is an ideal gas.

Solution:

$$m = \frac{pMV}{RT}$$

$$m = \frac{(1000 \text{ psia})\left(16.0\dfrac{\text{lbm}}{\text{lb-mole}}\right)(3.20 \text{ cu ft})}{\left(10.73\dfrac{\text{psia cu ft}}{\text{lb-mole°R}}\right)(528°\text{R})}$$

$$m = 9.0 \text{ lbm}$$

Example 2-2:

Calculate the density of methane at standard conditions.

$$\rho_g = \frac{pM}{RT} = \frac{(101.325 \text{ kPa})\left(16\dfrac{\text{kg}}{\text{kg-mole}}\right)}{8.314\left(\dfrac{\text{kPa-}m^3}{\text{kg-mole-°K}}\right)(288.72°\text{K})}$$

$$\rho_g = 0.675\frac{\text{kg}}{m^3} = 0.0421\frac{\text{lbm}}{\text{ft}^3}$$

TABLE 2-1
Values of Gas Constant R in Various Units

Units	R
atm, cc/g-mole, °K	82.06
BTU/lb-mole, °R	1.987
psia, cu ft/lb-mole, °R	10.73
lb/sq ft abs, cu ft/lb-mole, °R	1544
atm, cu ft/lb-mole, °R	0.730
mm Hg, liters/g-mole, °K	62.37
in. Hg, cu ft/lb-mole, °R	21.85
cal/g-mole, °K	1.987
kPa, m^3/kg-mole, °K	8.314
J/kg-mole, °K	8314

Ideal Gas Mixtures

The previous treatment of the behavior of gases applies only to single component gases. As the gas engineer rarely works with pure gases, the behavior of a multicomponent mixture of gases must be treated. This requires the introduction of two additional ideal gas laws.

Dalton's Law. Dalton's Law states that each gas in a mixture of gases exerts a pressure equal to that which it would exert if it occupied the same volume as the total mixture. This pressure is called the partial pressure. The total pressure is the sum of the partial pressures. This law is valid only when the mixture and each component of the mixture obey the ideal gas law. It is sometimes called the Law of Additive Pressures.

The partial pressure exerted by each component of the gas mixture can be calculated using the ideal gas law. Consider a mixture containing n_A moles of component A, n_B moles of component B and n_C moles of component C. The partial pressure exerted by each component of the gas mixture may be determined with the ideal gas equation:

$$p_A = n_A\frac{RT}{V}, \qquad p_B = n_B\frac{RT}{V}, \qquad p_C = n_C\frac{RT}{V}.$$

According to Dalton's Law, the total pressure is the sum of the partial pressures

$$p = p_A + p_B + p_C$$

$$p = n_A\frac{RT}{V} + n_B\frac{RT}{V} + n_C\frac{RT}{V}$$

$$p = \frac{RT}{V}\Sigma n_j = \frac{RT}{V}n.$$

It follows that the ratio of the partial pressure of component j, p_j, to the total pressure of the mixture p is

$$\frac{p_j}{p} = \frac{n_j}{\Sigma n_j} = \frac{n_j}{n} = y_j, \tag{2-7}$$

where y_j is defined as the mole fraction of the jth component in the gas mixture. Therefore, the partial pressure of a component of a gas mixture is the product of its mole fraction times the total pressure.

Amagat's Law. Amagat's Law states that the total volume of a gaseous mixture is the sum of the volumes that each component would occupy at the given pressure and temperature. The volumes occupied by the individual components are known as partial volumes. This law is correct only if the mixture and each of the components obey the ideal gas law.

The partial volume occupied by each component of a gas mixture consisting of n_A moles of component A, n_B moles of component B, and so on, can be calculated using the ideal gas law.

$$V_A = n_A \frac{RT}{p}, \quad V_B = n_B \frac{RT}{p}, \quad V_C = n_C \frac{RT}{p} \ldots$$

Thus, according to Amagat, the total volume is

$$V = V_A + V_B + V_C + \ldots$$

$$V = n_A \frac{RT}{p} + n_B \frac{RT}{p} + n_c \frac{RT}{p} + \ldots$$

$$V = \frac{RT}{p} \Sigma n_j = \frac{RT}{p} n.$$

It follows that the ratio of the partial volume of component j to the total volume of the mixture is

$$\frac{V_j}{V} = \frac{n_j \frac{RT}{p}}{n \frac{RT}{p}} = \frac{n_j}{n} = y_j. \quad (2\text{-}8)$$

This implies that for an ideal gas the volume fraction is equal to the mole fraction.

Apparent Molecular Weight

Since a gas mixture is composed of molecules of various sizes, it is not strictly correct to say that a gas mixture has a molecular weight. However, a gas mixture behaves as if it were a pure gas with a definite molecular weight. This molecular weight is known as an apparent molecular weight and is defined as

$$M_a = \Sigma y_j M_j. \quad (2\text{-}9)$$

Example 2-3:

Dry air is a gas mixture consisting essentially of nitrogen, oxygen, and small amounts of other gases. Compute the apparent molecular weight of air given its approximate composition. The molecular weight of each component may be found in Table 2-2.

Component	*Mole fraction, y_j*
Nitrogen	0.78
Oxygen	0.21
Argon	0.01
	1.00

Solution:

$$M_a = y_{N2}M_{N2} + y_{O2}M_{O2} + y_A M_A$$

$$M = (0.78)(28.01) + (0.21)(32.00) + (0.01)(39.94)$$

$$= 28.97$$

A value of 29.0 is usually considered sufficiently accurate for engineering calculations.

The specific gravity of a gas is defined as the ratio of the density of the gas to the density of dry air taken at standard conditions of temperature and pressure. Symbolically,

$$\gamma_g = \frac{\rho_g}{\rho_{air}}. \quad (2\text{-}10)$$

Assuming that the behavior of both the gas and air may be represented by the ideal gas law, specific gravity may be given as

$$\gamma_g = \frac{\frac{M_g p}{RT}}{\frac{M_{air} p}{RT}} = \frac{M_g}{M_{air}} = \frac{M_g}{29},$$

where M_{air} is the apparent molecular weight of air. If the gas is a mixture, this equation becomes

$$\gamma_g = \frac{M_a}{M_{air}} = \frac{M_a}{29}, \quad (2\text{-}11)$$

where M_a is the apparent molecular weight of the gas mixture.

Example 2-4:

Calculate the gravity of a natural gas of the following composition.

Component	*Mole fraction y_j*
Methane	0.85
Ethane	0.09
Propane	0.04
n-butane	0.02
	1.00

TABLE 2-2
PHYSICAL CONSTANTS OF HYDROCARBONS

No.	Compound	Formula	Molecular weight	Boiling point °F., 14.696 psia	Vapor pressure, 100°F., psia	Freezing point, °F., 14.696 psia	Critical constants: Pressure, psia	Critical constants: Temperature, °F.	Critical constants: Volume, cu ft/lb
1	Methane	CH_4	16.043	−258.69	(5000)	−296.46	667.8	−116.63	0.0991
2	Ethane	C_2H_6	30.070	−127.48	(800)	−297.89	707.8	90.09	0.0788
3	Propane	C_3H_8	44.097	−43.67	190.	−305.84	616.3	206.01	0.0737
4	n–Butane	C_4H_{10}	58.124	31.10	51.6	−217.05	550.7	305.65	0.0702
5	Isobutane	C_4H_{10}	58.124	10.90	72.2	−255.29	529.1	274.98	0.0724
6	n–Pentane	C_5H_{12}	72.151	96.92	15.570	−201.51	488.6	385.7	0.0675
7	Isopentane	C_5H_{12}	72.151	82.12	20.44	−255.83	490.4	369.10	0.0679
8	Neopentane	C_5H_{12}	72.151	49.10	35.9	2.17	464.0	321.13	0.0674
9	n–Hexane	C_6H_{14}	86.178	155.72	4.956	−139.58	436.9	453.7	0.0688
10	2–Methylpentane	C_6H_{14}	86.178	140.47	6.767	−244.63	436.6	435.83	0.0681
11	3–Methylpentane	C_6H_{14}	86.178	145.89	6.098	—	453.1	448.3	0.0681
12	Neohexane	C_6H_{14}	86.178	121.52	9.856	−147.72	446.8	420.13	0.0667
13	2,3–Dimethylbutane	C_6H_{14}	86.178	136.36	7.404	−199.38	453.5	440.29	0.0665
14	n–Heptane	C_7H_{16}	100.205	209.17	1.620	−131.05	396.8	512.8	0.0691
15	2–Methylhexane	C_7H_{16}	100.205	194.09	2.271	−180.89	396.5	495.00	0.0673
16	3–Methylhexane	C_7H_{16}	100.205	197.32	2.130	—	408.1	503.78	0.0646
17	3–Ethylpentane	C_7H_{16}	100.205	200.25	2.012	−181.48	419.3	513.48	0.0665
18	2,2–Dimethylpentane	C_7H_{16}	100.205	174.54	3.492	−190.86	402.2	477.23	0.0665
19	2,4–Dimethylpentane	C_7H_{16}	100.205	176.89	3.292	−182.63	396.9	475.95	0.0668
20	3,3–Dimethylpentane	C_7H_{16}	100.205	186.91	2.773	−210.01	427.2	505.85	0.0662
21	Triptane	C_7H_{16}	100.205	177.58	3.374	−12.82	428.4	496.44	0.0636
22	n–Octane	C_8H_{18}	114.232	258.22	0.537	−70.18	360.6	564.22	0.0690
23	Diisobutyl	C_8H_{18}	114.232	228.39	1.101	−132.07	360.6	530.44	0.0676
24	Isooctane	C_8H_{18}	114.232	210.63	1.708	−161.27	372.4	519.46	0.0656
25	n–Nonane	C_9H_{20}	128.259	303.47	0.179	−64.28	332.	610.68	0.0684
26	n–Decane	$C_{10}H_{22}$	142.286	345.48	0.0597	−21.36	304.	652.1	0.0679
27	Cyclopentane	C_5H_{10}	70.135	120.65	9.914	−136.91	653.8	461.5	0.059
28	Methylcyclopentane	C_6H_{12}	84.162	161.25	4.503	−224.44	548.9	499.35	0.0607
29	Cyclohexane	C_6H_{12}	84.162	177.29	3.264	43.77	591.	536.7	0.0586
30	Methylcyclohexane	C_7H_{14}	98.189	213.68	1.609	−195.87	503.5	570.27	0.0600
31	Ethylene	C_2H_4	28.054	−154.62	—	−272.45	729.8	48.58	0.0737
32	Propene	C_3H_6	42.081	−53.90	226.4	−301.45	669.	196.9	0.0689
33	1–Butene	C_4H_8	56.108	20.75	63.05	−301.63	583.	295.6	0.0685
34	Cis–2–Butene	C_4H_8	56.108	38.69	45.54	−218.06	610.	324.37	0.0668
35	Trans–2–Butene	C_4H_8	56.108	33.58	49.80	−157.96	595.	311.86	0.0680
36	Isobutene	C_4H_8	56.108	19.59	63.40	−220.61	580.	292.55	0.0682
37	1–Pentene	C_5H_{10}	70.135	85.93	19.115	−265.39	590.	376.93	0.0697
38	1,2–Butadiene	C_4H_6	54.092	51.53	(20.)	−213.16	(653.)	(339.)	(0.0649)
39	1,3–Butadiene	C_4H_6	54.092	24.06	(60.)	−164.02	628.	306.	0.0654
40	Isoprene	C_5H_8	68.119	93.30	16.672	−230.74	(558.4)	(412.)	(0.0650)
41	Acetylene	C_2H_2	26.038	−119	—	−114.	890.4	95.31	0.0695
42	Benzene	C_6H_6	78.114	176.17	3.224	41.96	710.4	552.22	0.0531
43	Toluene	C_7H_8	92.141	231.13	1.032	−138.94	595.9	605.55	0.0549
44	Ethylbenzene	C_8H_{10}	106.168	277.16	0.371	−138.91	523.5	651.24	0.0564
45	o–Xylene	C_8H_{10}	106.168	291.97	0.264	−13.30	541.4	675.0	0.0557
46	m–Xylene	C_8H_{10}	106.168	282.41	0.326	−54.12	513.6	651.02	0.0567
47	p–Xylene	C_8H_{10}	106.168	281.05	0.342	55.86	509.2	649.6	0.0572
48	Styrene	C_8H_8	104.152	293.29	(0.24)	−23.10	580.	706.0	0.0541
49	Isopropylbenzene	C_9H_{12}	120.195	306.34	0.188	−140.82	465.4	676.4	0.0570
50	Methyl Alcohol	CH_4O	32.042	148.1(2)	4.63(22)	−143.82(22)	1174.2(21)	462.97(21)	0.0589(21)
51	Ethyl Alcohol	C_2H_6O	46.069	172.92(22)	2.3(7)	−173.4(22)	925.3(21)	469.58(21)	0.0580(21)
52	Carbon Monoxide	CO	28.010	−313.6(2)	—	−340.6(2)	507.(17)	−220.(17)	0.0532(17)
53	Carbon Dioxide	CO_2	44.010	−109.3(2)	—	—	1071.(17)	87.9(23)	0.0342(23)
54	Hydrogen Sulfide	H_2S	34.076	−76.6(24)	394.0(6)	−117.2(7)	1306.(17)	212.7(17)	0.0459(24)
55	Sulfur Dioxide	SO_2	64.059	14.0(7)	88.(7)	−103.9(7)	1145.(24)	315.5(17)	0.0306(24)
56	Ammonia	NH_3	17.031	−28.2(24)	212.(7)	−107.9(2)	1636.(17)	270.3(24)	0.0681(17)
57	Air	N_2O_2	28.964	−317.6(2)	—	—	547.(2)	−221.3(2)	0.0517(3)
58	Hydrogen	H_2	2.016	−423.0(24)	—	−434.8(24)	188.1(17)	−399.8(17)	0.5167(24)
59	Oxygen	O_2	31.999	−297.4(2)	—	−361.8(24)	736.9(24)	−181.1(17)	0.0382(24)
60	Nitrogen	N_2	28.013	−320.4(2)	—	−346.0(24)	493.0(24)	−232.4(24)	0.0514(17)
61	Chlorine	Cl_2	70.906	−29.3(24)	158.(7)	−149.8(24)	1118.4(24)	291.(17)	0.0281(17)
62	Water	H_2O	18.015	212.0	0.9492(12)	32.0	3208.(17)	705.6(17)	0.0500(17)
63	Helium	He	4.003	—	—	—	—	—	—
64	Hydrogen Chloride	HCl	36.461	−121(16)	925.(7)	−173.6(16)	1198.(17)	124.5(17)	0.0208(17)

( ) estimated values

Specific gravity 60°F./60°F.a,b	Density of liquid; 60°F., 14.696 psia: lb/gal*a (Wt in vacuum)	lb/gal*a,c (Wt in air)	Gal/lb Mole*	Temperature Coefficient of density*a	Pitzer acentric factor	Compressibility factor of real gas, Z 14.696 psia, 60°F.	Gas density, 60°F. 14.696 psia Ideal gas*: Specific gravity Air = 1*	cu ft gas/lb*	cu ft gas/gal liquid*	Specific heat 60°F., 14.696 psia Cp, BTU/lb/°F.: Ideal gas	Liquid	No.
0.3	2.5	2.5	6.4	—	0.0104	0.9981	0.5539	23.65	59.	0.5266	—	1
0.3564	2.971	2.962	10.12	—	0.0986	0.9916	1.0382	12.62	37.5	0.4097	0.9256	2
0.5077	4.233	4.223	10.42	0.00152	0.1524	0.9820	1.5225	8.606	36.43	0.3881	0.5920	3
0.5844	4.872	4.865	11.93	0.00117	0.2010	0.9667	2.0068	6.529	31.81	0.3867	0.5636	4
0.5631	4.695	4.686	12.38	0.00119	0.1848	0.9696	2.0068	6.529	30.65	0.3872	0.5695	5
0.6310	5.261	5.251	13.71	0.00087	0.2539	0.9549	2.4911	5.260	27.67	0.3883	0.5441	6
0.6247	5.208	5.199	13.85	0.00090	0.2223	0.9544	2.4911	5.260	27.39	0.3827	0.5353	7
0.5967	4.975	4.965	14.50	0.00104	0.1969	0.9510	2.4911	5.260	26.17	(0.3866)	0.554	8
0.6640	5.536	5.526	15.57	0.00075	0.3007	—	2.9753	4.404	24.38	0.3864	0.5332	9
0.6579	5.485	5.475	15.71	0.00078	0.2825	—	2.9753	4.404	24.15	0.3872	0.5264	10
0.6689	5.577	5.568	15.45	0.00075	0.2741	—	2.9753	4.404	24.56	0.3815	0.507	11
0.6540	5.453	5.443	15.81	0.00078	0.2369	—	2.9753	4.404	24.01	0.3809	0.5165	12
0.6664	5.556	5.546	15.51	0.00075	0.2495	—	2.9753	4.404	24.47	0.378	0.5127	13
0.6882	5.738	5.728	17.46	0.00069	0.3498	—	3.4596	3.787	21.73	0.3875	0.5283	14
0.6830	5.694	5.685	17.60	0.00068	0.3336	—	3.4596	3.787	21.57	(0.390)	0.5223	15
0.6917	5.767	5.757	17.38	0.00069	0.3257	—	3.4596	3.787	21.84	(0.390)	0.511	16
0.7028	5.859	5.850	17.10	0.00070	0.3095	—	3.4596	3.787	22.19	(0.390)	0.5145	17
0.6782	5.654	5.645	17.72	0.00072	0.2998	—	3.4596	3.787	21.41	(0.395)	0.5171	18
0.6773	5.647	5.637	17.75	0.00072	0.3048	—	3.4596	3.787	21.39	0.3906	0.5247	19
0.6976	5.816	5.807	17.23	0.00065	0.2840	—	3.4596	3.787	22.03	(0.395)	0.502	20
0.6946	5.791	5.782	17.30	0.00069	0.2568	—	3.4596	3.787	21.93	0.3812	0.4995	21
0.7068	5.893	5.883	19.39	0.00062	0.4018	—	3.9439	3.322	19.58	(0 3876)	0.5239	22
0.6979	5.819	5.810	19.63	0.00065	0.3596	—	3.9439	3.322	19.33	(0.373)	0.5114	23
0.6962	5.804	5.795	19.68	0.00065	0.3041	—	3.9439	3.322	19.28	0.3758	0.4892	24
0.7217	6.017	6.008	21.32	0.00063	0.4455	—	4.4282	2.959	17.80	0.3840	0.5228	25
0.7342	6.121	6.112	23.24	0.00055	0.4885	—	4.9125	2.667	16.33	0.3835	0.5208	26
0.7504	6.256	6.247	11.21	0.00070	0.1955	0.9657	2.4215	5.411	33.85	0.2712	0.4216	27
0.7536	6.283	6.274	13.40	0.00071	0.2306	—	2.9057	4.509	28.33	0.3010	0.4407	28
0.7834	6.531	6.522	12.89	0.00068	0.2133	—	2.9057	4.509	29.45	0.2900	0.4332	29
0.7740	6.453	6.444	15.22	0.00063	0.2567	—	3.3900	3.865	24.94	0.3170	0.4397	30
—	—	—	—	—	0.0868	0.9938	0.9686	13.53	—	0.3622	—	31
0.5220	4.352	4.343	9.67	0.00189	0.1405	0.9844	1.4529	9.018	39.25	0.3541	0.585	32
0.6013	5.013	5.004	11.19	0.00116	0.1906	0.9704	1.9372	6.764	33.91	0.3548	0.535	33
0.6271	5.228	5.219	10.73	0.00098	0.1953	0.9661	1.9372	6.764	35.36	0.3269	0.5271	34
0.6100	5.086	5.076	11.03	0.00107	0.2220	0.9662	1.9372	6.764	34.40	0.3654	0.5351	35
0.6004	5.006	4.996	11.21	0.00120	0.1951	0.9689	1.9372	6.764	33.86	0.3701	0.549	36
0.6457	5.383	5.374	13.03	0.00089	0.2925	0.9550	2.4215	5.411	29.13	0.3635	0.5196	37
0.658[h]	5.486	5.470	9.86	0.00098	0.2485	(0.969)	1.8676	7.016	38.49	0.3458	0.5408	38
0.6272	5.229	5.220	10.34	0.00113	0.1955	(0.965)	1.8676	7.016	36.69	0.3412	0.5079	39
0.6861	5.720	5.711	11.91	0.00086	0.2323	(0.962)	2.3519	5.571	31.87	0.357	0.5192	40
0.615	—	—	—	—	0.1803	0.9925	0.8990	14.57	—	0.3966	—	41
0.8844	7.373	7.365	10.59	0.00066	0.2125	0.929(15)	2.6969	4.858	35.82	0.2429	0.4098	42
0.8718	7.268	7.260	12.68	0.00060	0.2596	0.903(21)	3.1812	4.119	29.94	0.2598	0.4012	43
0.8718	7.268	7.259	14.61	0.00054	0.3169	—	3.6655	3.574	25.98	0.2795	0.4114	44
0.8848	7.377	7.367	14.39	0.00055	0.3023	—	3.6655	3.574	26.37	0.2914	0.4418	45
0.8687	7.243	7.234	14.66	0.00054	0.3278	—	3.6655	3.574	25.89	0.2782	0.4045	46
0.8657	7.218	7.209	14.71	0.00054	0.3138	—	3.6655	3.574	25.80	0.2769	0.4083	47
0.9110	7.595	7.586	13.71	0.00057	—	—	3.5959	3.644	27.67	0.2711	0.4122	48
0.8663	7.223	7.214	16.64	0.00054	0.2862	—	4.1498	3.157	22.80	0.2917	(0.414)	49
0.796(3)	6.64	6.63	4.83	—	—	—	1.1063	11.84	78.6	0.3231 (24)	0.594(7)	50
0.794(3)	6.62	6.61	6.96	—	—	—	1.5906	8.237	54.5	0.3323 (24)	0.562(7)	51
0.801 (8)	6.68	6.67	4.19	—	0.041	0.9995(15)	0.9671	13.55	—	0.2484(13)	—	52
0.827 (6)	6.89	6.88	6.38	—	0.225	0.9943(15)	1.5195	8.623	59.5	0.1991(13)	—	53
0.79 (6)	6.59	6.58	5.17	—	0.100	0.9903(15)	1.1765	11.14	73.3	0.238(4)	—	54
1.397 (14)	11.65	11.64	5.50	—	0.246	—	2.2117	5.924	69.0	0.145(7)	0.325 (7)	55
0.6173(11)	5.15	5.14	3.31	—	0.255	—	0.5880	22.28	114.7	0.5002(10)	1.114 (7)	56
0.856 (8)	7.14	7.13	4.06	—	—	0.9996(15)	1.0000	13.10	—	0.2400(9)	—	57
0.07 (3)	—	—	—	—	0.000	1.0006(15)	0.0696	188.2	—	3.408(13)	—	58
1.14 (3)	9.50	9.49	3.37	—	0.0213	—	1.1048	11.86	—	0.2188(13)	—	59
0.808 (3)	6.74	6.73	4.16	—	0.040	0.9997(15)	0.9672	13.55	—	0.2482(13)	—	60
1.414(14)	11.79	11.78	6.01	—	—	—	2.4481	5.352	63.1	0.119(7)	—	61
1.000	8.337	8.328	2.16	—	0.348	—	0.6220	21.06	175.6	0.4446(13)	1.0009(7)	62
—	—	—	—	—	—	—	—	—	—	—	—	63
0.8558(14)	7.135	7.126	5.11	0.00335	—	—	1.2588	10.41	74.3	0.190(7)	—	64

No.	Compound	Calorific value, 60° F.* — Net: BTU/cu ft, Ideal gas, 14.696 psia (20)*	Gross: BTU/cu ft, Ideal gas, 14.696 psia	Gross: BTU/lb liquid (wt in vacuum)	Gross: BTU/gal liquid	Heat of vaporization, 14.696 psia at boiling point, BTU/lb	Refractive indexa, nD 68° F.	Air required for combustion ideal gas cu ft/cu ft	Flammability limits, vol % in air mixture — Lower	Higher	ASTM octane number — Motor method D-357	Research method D-908
1	Methane	909.1	1009.7	—	—	219.22	—	9.54	5.0	15.0	—	—
2	Ethane	1617.8	1768.8	—	—	210.41	—	16.70	2.9	13.0	+.05	+1.6
3	Propane	2316.1	2517.4	21513	91065	183.05	—	23.86	2.1	9.5	97.1	+1.8
4	n-Butane	3010.4	3262.1	21139	102989	165.65	1.3326	31.02	1.8	8.4	89.6	93.8
5	Isobutane	3001.1	3252.7	21091	99022	157.53	—	31.02	1.8	8.4	97.6	+.10
6	n-Pentane	3707.5	4009.5	20928	110102	153.59	1.35748	38.18	1.4	8.3	62.6	61.7
7	Isopentane	3698.3	4000.3	20889	108790	147.13	1.35373	38.18	1.4	(8.3)	90.3	92.3
8	Neopentane	3682.6	3984.6	20824	103599	135.58	1.342	38.18	1.4	(8.3)	80.2	85.5
9	n-Hexane	4403.7	4756.1	20784	115060	143.95	1.37486	45.34	1.2	7.7	26.0	24.8
10	2-Methylpentane	4395.8	4748.1	20757	113852	138.67	1.37145	45.34	1.2	(7.7)	73.5	73.4
11	3-Methylpentane	4398.7	4751.0	20768	115823	140.09	1.37652	45.34	(1.2)	(7.7)	74.3	74.5
12	Neohexane	4382.6	4735.0	20710	112932	131.24	1.36876	45.34	1.2	(7.7)	93.4	91.8
13	2,3-Dimethylbutane	4391.7	4744.0	20742	115243	136.08	1.37495	45.34	(1.2)	(7.7)	94.3	+0.3
14	n-Heptane	5100.2	5502.9	20681	118668	136.01	1.38764	52.50	1.0	7.0	0.0	0.0
15	2-Methylhexane	5092.1	5494.8	20658	117627	131.59	1.38485	52.50	(1.0)	(7.0)	46.4	42.4
16	3-Methylhexane	5095.2	5497.8	20668	119192	132.11	1.38864	52.50	(1.0)	(7.0)	55.8	52.0
17	3-Ethylpentane	5098.2	5500.9	20679	121158	132.83	1.39339	52.50	(1.0)	(7.0)	69.3	65.0
18	2,2-Dimethylpentane	5079.4	5482.1	20620	116585	125.13	1.38215	52.50	(1.0)	(7.0)	95.6	92.8
19	2,4-Dimethylpentane	5084.3	5487.0	20636	116531	126.58	1.38145	52.50	(1.0)	(7.0)	83.8	83.1
20	3,3-Dimethylpentane	5085.0	5487.6	20638	120031	127.21	1.39092	52.50	(1.0)	(7.0)	86.6	80.8
21	Triptane	5081.0	5483.6	20627	119451	124.21	1.38944	52.50	(1.0)	(7.0)	+0.1	+1.8
22	n-Octane	5796.7	6249.7	20604	121419	129.53	1.39743	59.65	0.96	—	—	—
23	Diisobutyl	5781.3	6234.3	20564	119662	122.8	1.39246	59.65	(0.98)	—	55.7	55.2
24	Isooctane	5779.8	6232.8	20570	119388	116.71	1.39145	59.65	1.0	—	100.	100.
25	n-Nonane	6493.3	6996.6	20544	123613	123.76	1.40542	66.81	0.87	2.9	—	—
26	n-Decane	7188.6	7742.3	20494	125444	118.68	1.41189	73.97	0.78	2.6	—	—
27	Cyclopentane	3512.0	3763.7	20188	126296	167.34	1.40645	35.79	(1.4)	—	84.9	+0.1
28	Methylcyclopentane	4198.4	4500.4	20130	126477	147.83	1.40970	42.95	(1.2)	8.35	80.0	91.3
29	Cyclohexane	4178.8	4480.8	20035	130849	153.0	1.42623	42.95	1.3	7.8	77.2	83.0
30	Methylcyclohexane	4862.8	5215.2	20001	129066	136.3	1.42312	50.11	1.2	—	71.1	74.8
31	Ethylene	1499.0	1599.7	—	—	207.57	—	14.32	2.7	34.0	75.6	+.03
32	Propene	2182.7	2333.7	—	—	188.18	—	21.48	2.0	10.0	84.9	+0.2f
33	1-Butene	2879.4	3080.7	20678	103659	167.94	—	28.63	1.6	9.3	80.8	97.4
34	Cis-2-Butene	2871.7	3073.1	20611	107754	178.91	—	28.63	(1.6)	—	83.5	100.
35	Trans-2-Butene	2866.8	3068.2	20584	104690	174.39	—	28.63	(1.6)	—	—	—
36	Isobutene	2860.4	3061.8	20548	102863	169.48	—	28.63	(1.6)	—	—	—
37	1-Pentene	3575.2	3826.9	20548	110610	154.46	1.37148	35.79	1.4	8.7	77.1	90.9
38	1,2-Butadiene	2789.0	2940.0	20447	112172	(181.)	—	26.25	(2.0)	(12.)	—	—
39	1,3-Butadiene	2730.0	2881.0	20047	104826	(174.)	—	26.25	2.0	11.5	—	—
40	Isoprene	3410.8	3612.1	19964	114194	(153.)	1.42194	33.41	(1.5)	—	81.0	99.1
41	Acetylene	1422.4	1472.8	—	—	—	—	11.93	2.5	80.	—	—
42	Benzene	3590.7	3741.7	17992	132655	169.31	1.50112	35.79	1.3	7.9	+2.8	—
43	Toluene	4273.3	4474.7	18252	132656	154.84	1.49693	42.95	1.2	7.1	+0.3	+5.8
44	Ethylbenzene	4970.0	5221.7	18494	134414	144.0	1.49588	50.11	0.99	6.7	97.9	+0.8
45	o-Xylene	4958.3	5210.0	18445	136069	149.1	1.50545	50.11	1.1	6.4	100.	—
46	m-Xylene	4956.8	5208.5	18441	133568	147.2	1.49722	50.11	1.1	6.4	+2.8	+4.0
47	p-Xylene	4956.9	5208.5	18445	133136	144.52	1.49582	50.11	1.1	6.6	+1.2	+3.4
48	Styrene	4828.7	5030.0	18150	137849	(151.)	1.54682	47.72	1.1	6.1	+0.2	>+3.
49	Isopropylbenzene	5661.4	5963.4	18665	134817	134.3	1.49145	57.27	0.88	6.5	99.3	+2.1
50	Methyl Alcohol	—	—	9760	64771	473.(2)	1.3288(8)	7.16	6.72(5)	36.50	—	—
51	Ethyl Alcohol	—	—	12780	84600	367.(2)	1.3614(8)	14.32	3.28(5)	18.95	—	—
52	Carbon Monoxide	—	321.(13)	—	—	92.7(14)	—	2.39	12.50(5)	74.20	—	—
53	Carbon Dioxide	—	—	—	—	238.2 (14)	—	—	—	—	—	—
54	Hydrogen Sulfide	588.(16)	637.(16)	—	—	235.6(7)	—	7.16	4.30(5)	45.50	—	—
55	Sulfur Dioxide	—	—	—	—	166.7(14)	—	—	—	—	—	—
56	Ammonia	359.(16)	434.(16)	—	—	587.2(14)	—	3.58	15.50(5)	27.00	—	—
57	Air	—	—	—	—	92.(3)	—	—	—	—	—	—
58	Hydrogen	274.(13)	324.(13)	—	—	193.9(14)	—	2.39	4.00(5)	74.20	—	—
59	Oxygen	—	—	—	—	91.6(14)	—	—	—	—	—	—
60	Nitrogen	—	—	—	—	87.8(14)	—	—	—	—	—	—
61	Chlorine	—	—	—	—	123.8(14)	—	—	—	—	—	—
62	Water	—	—	—	—	970.3(12)	1.3330(8)	—	—	—	—	—
63	Helium	—	—	—	—	—	—	—	—	—	—	—
64	Hydrogen Chloride	—	—	—	—	185.5(14)	—	—	—	—	—	—

Courtesy Gas Processors Suppliers Association

Solution:

Component	Mole fraction, y_j	Molecular weight, M_j	y_jM_j
C_1	0.85	16.0	13.60
C_2	0.09	30.1	2.71
C_3	0.04	44.1	1.76
n-C_4	0.02	58.1	1.16
	1.00		19.23 = M_a

$$\gamma_g = \frac{M_a}{29} = \frac{19.23}{29} = 0.66$$

REAL GASES

Several assumptions were made in formulating the equation of state for ideal gases. Since these assumptions are not correct for gases at pressures and temperatures that deviate from ideal or standard conditions, corrections must be made to account for the deviation from ideal behavior. The most widely used correction method in the petroleum industry is the gas compressibility factor, more commonly called the *Z*-factor. It is defined as the ratio of the actual volume occupied by a mass of gas at some pressure and temperature to the volume the gas would occupy if it behaved ideally. That is,

$$Z = \frac{V_{\text{actual}}}{V_{\text{ideal}}} \quad \text{or} \quad V_{\text{actual}} = Z\,V_{\text{ideal}}.$$

The equation of state is

$$p\,V_{\text{ideal}} = nRT \quad \text{or} \quad p\,\frac{V_{\text{actual}}}{Z} = nRT.$$

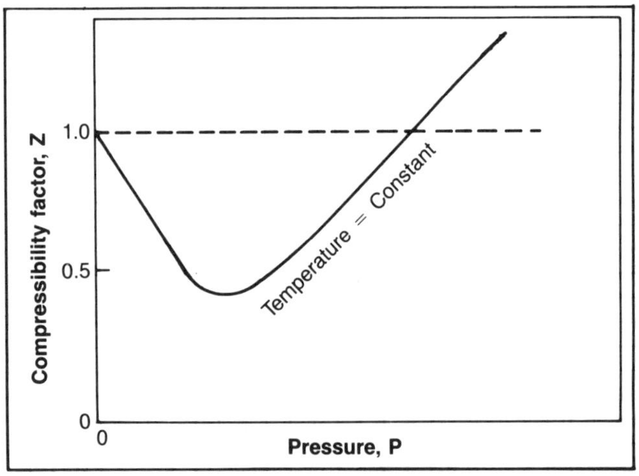

Fig. 2-1. Typical plot of the compressibility factor as a function of pressure at constant temperature. Courtesy Gas Processors Suppliers Association.

Therefore, the equation of state for any gas becomes

$$pV = ZnRT \tag{2-12}$$

where, for an ideal gas, $Z = 1$.

The compressibility factor varies with changes in gas composition, temperature, and pressure. It must be determined experimentally. The results of experimental determinations of compressibility factors are normally given graphically and usually take the form shown in Figure 2-1. The shape of the curve is consistent with present knowledge of the behavior of gases; at very low pressure the molecules are relatively far apart and the conditions of ideal gas behavior are more likely to be met. At low pressure the compressibility factor approaches a value of 1.0, which would indicate that ideal gas behavior does in fact occur. Compressibility factors for several hydrocarbon gases are given in Figures 2-2, 2-3, and 2-4.

Example 2-5:

Calculate the mass of methane gas contained at 1,000 psia and 68°F in a cylinder with volume of 3.20 cu ft.

Solution:

$$m = \frac{pMV}{ZRT}$$

$Z = 0.89$ (from Fig. 2-2)

$$m = \frac{(1{,}000 \text{ psia})\left(16.0\,\dfrac{\text{lb}}{\text{lb-mole}}\right)(3.2 \text{ cu ft})}{(0.89)\left(10.73\,\dfrac{\text{psia cu ft}}{\text{lb-mole°R}}\right)(528°\text{R})}$$

$m = 10.2$ lb

If ideal behavior had been assumed, the mass calculated would have been $m = 10.2\,(.89) = 9.08$ lbm.

Real Gas Mixtures

Compressibility factor charts are available for most of the single component light hydrocarbon gases, but in practice a single component gas is rarely encountered. In order to get *Z*-factors for natural gas mixtures, the law of corresponding states is used. This law states that the ratio of the value of any intensive property to the value of that property at the critical state is related to the ratios of the prevailing absolute temperature and pressure to the critical temperature and pressure by the same function for all similar substances. This means that all pure gases have the same *Z*-factor at the same values of reduced pressure and temperature, where the reduced values are defined as

$$T_r = \frac{T}{T_c}, \qquad p_r = \frac{p}{p_c}$$

where T_c and p_c are the critical temperature and pressure for the gas, respectively. The values must be in absolute units.

Example 2-6:

Calculate the density of ethane at 900 psia and 110°F.

The critical properties, from Table 2-2, are T_c = 550°R, p_c = 708 psia, M = 30.1 lbm/lb-mole.

Solution:

$$T_r = \frac{T}{T_c} = \frac{(110 + 460)}{550} = 1.04$$

$$p_r = \frac{p}{p_c} = \frac{900}{708} = 1.27$$

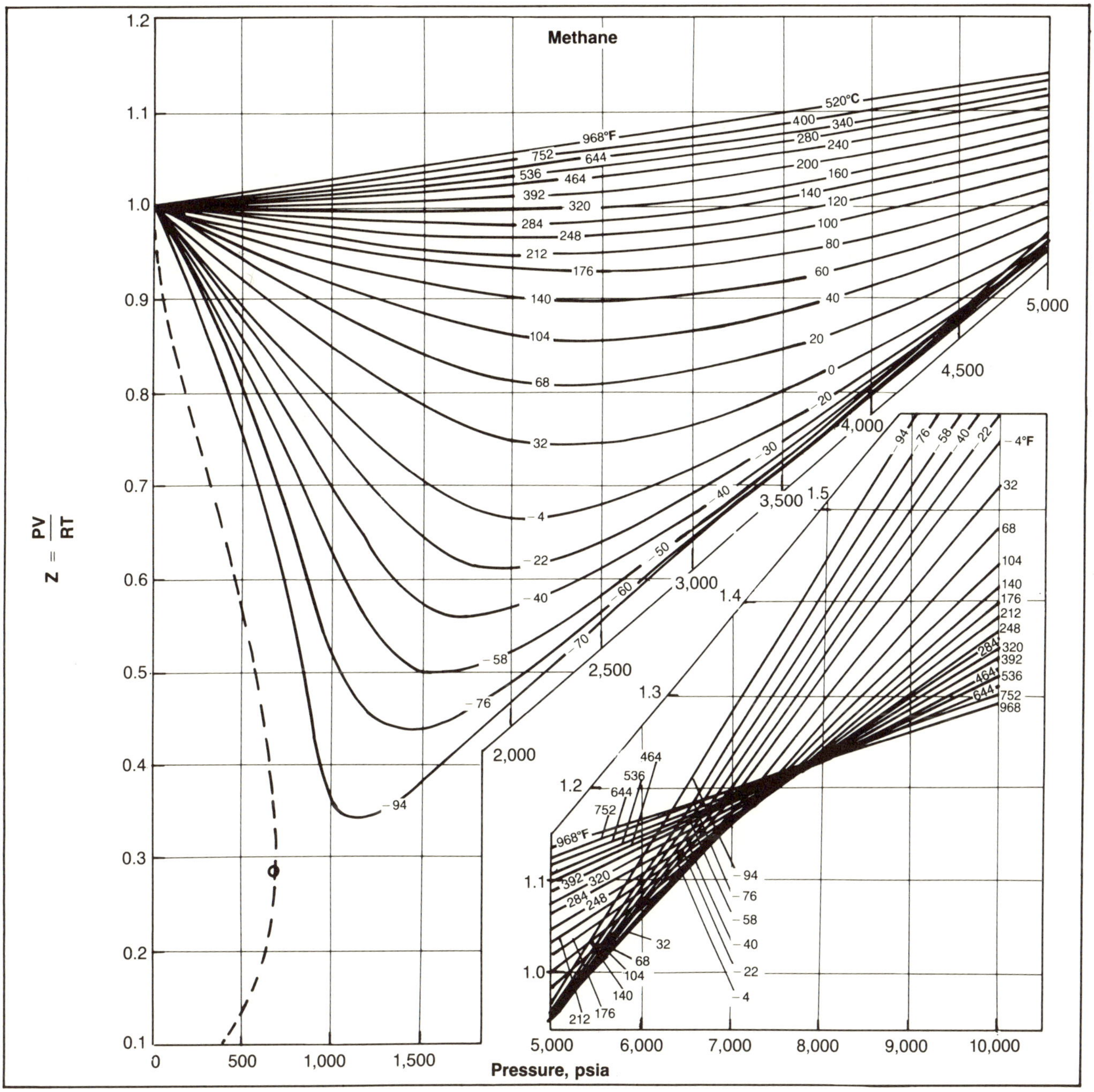

Fig. 2-2. Compressibility factors for methane. Courtesy Gas Processors Suppliers Association.

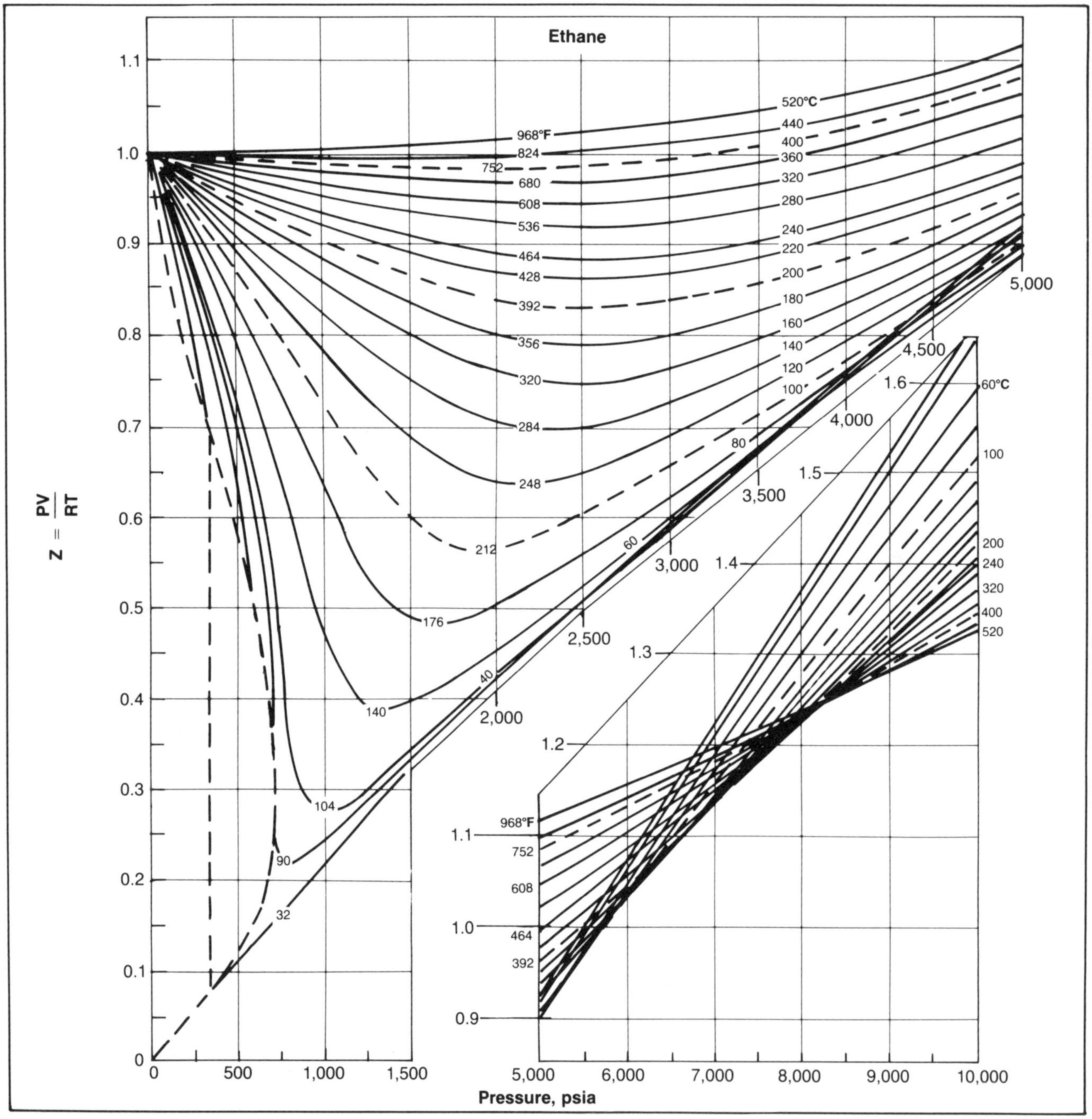

Fig. 2-3. Compressibility factors for ethane. Courtesy Gas Processors Suppliers Association.

(From Figure 2-5, $Z \approx 0.34$.)

$$\rho = \frac{pM}{ZRT} = \frac{(900 \text{ psia})\left(30.1 \dfrac{\text{lbm}}{\text{lb-mole}}\right)}{(0.34)\ 10.73 \left(\dfrac{\text{psia} - \text{ft}^3}{\text{lb-mole}°\text{R}}\right)(570°\text{R})}$$

$$= 13.03 \frac{\text{lbm}}{\text{ft}^3}$$

Assuming ideal behavior, the value calculated would be $\rho = 13.03\ (.34) = 4.43\ \text{lbm/ft}^3$.

It has been shown that the Law of Corresponding States works better for gases of similar molecular characteristics. This is fortunate since most of the gases that the petroleum engineer deals with are composed of molecules of the same class of organic compounds known as paraffin hydrocarbons. The law of corresponding states

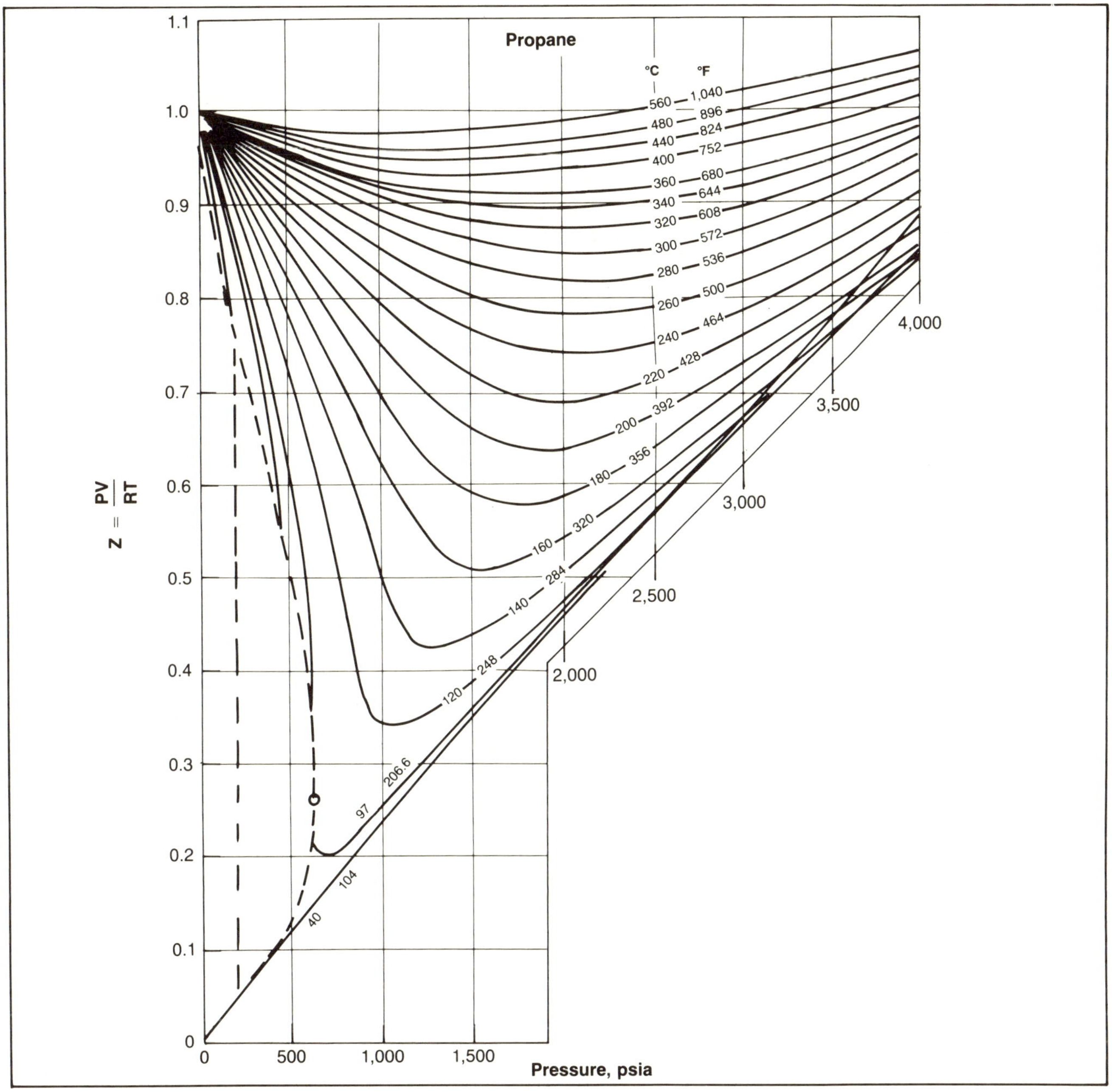

Fig. 2-4. Compressibility factors for propane. Courtesy Gas Processors Suppliers Association.

has been extended to cover mixtures of gases that are closely related chemically. Since it is somewhat difficult to obtain the critical point for multicomponent mixtures, the quantities of pseudocritical temperature and pseudocritical pressure have been conceived. These quantities are defined as

$$T_{pc} = \Sigma y_j T_{cj} \quad \text{and} \quad p_{pc} = \Sigma y_j p_{cj}. \tag{2-13}$$

These pseudocritical quantities are used for mixtures of gases in exactly the same manner as the actual critical temperatures and critical pressures are used for pure gases. It must be understood, however, that these pseudocritical properties were devised simply for use in correlating compressibility factors and in no way relate to the actual critical properties of the gas mixture.

Example 2-7:

Calculate the pseudocritical temperature and pseudocritical pressure of the following natural gas mixture. Use the critical constants given in Table 2-2.

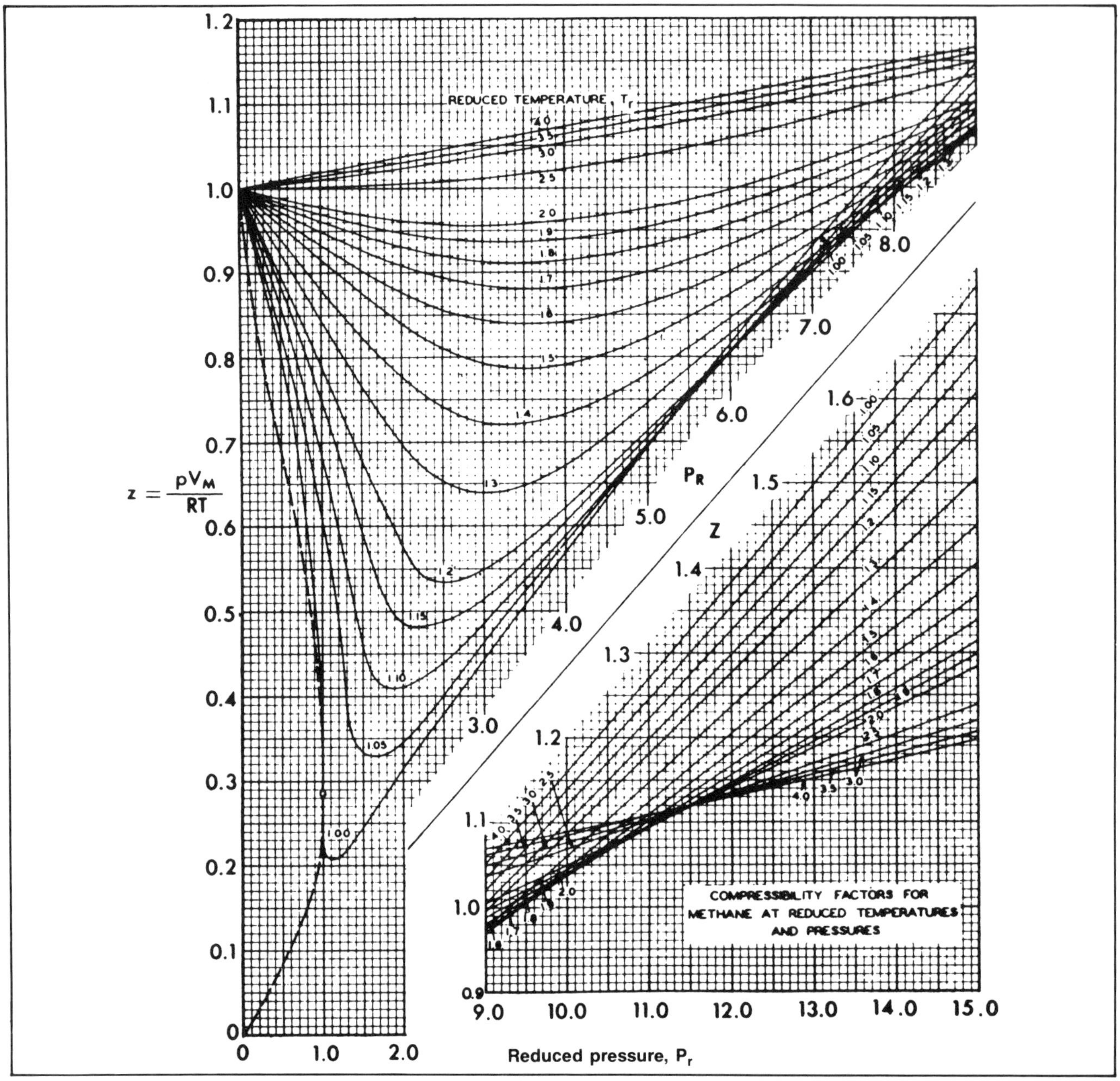

Fig. 2-5. Compressibility factors for pure hydrocarbon gases as a function of reduced pressure and temperature. Courtesy Gas Processors Suppliers Association.

Solution:

Component	Mole fraction y_i	Critical temperature, °R. T_{cj}	y_jT_{cj}	Critical pressure, psia p_{cj}	y_jp_{cj}
C_1	0.85	343.1	291.6	667.8	567.6
C_2	0.09	549.8	49.5	707.8	63.7
C_3	0.04	665.7	26.6	616.3	24.7
n-C_4	0.02	765.3	15.3	550.7	11.0
	1.00		T_{pc} = 383.0°R		p_{pc} = 667.0 psia

$T_{pc} = 383°R \qquad p_{pc} = 667$ psia

The compressibility factors for natural gases have been correlated using pseudocritical properties and are presented in Figures 2-6, 2-7, and 2-8. Compressibility factors are a function of composition as well as temperature and pressure. It has been pointed out that the components of most natural gases are hydrocarbons of the same family, and therefore a correlation of this type is possible.

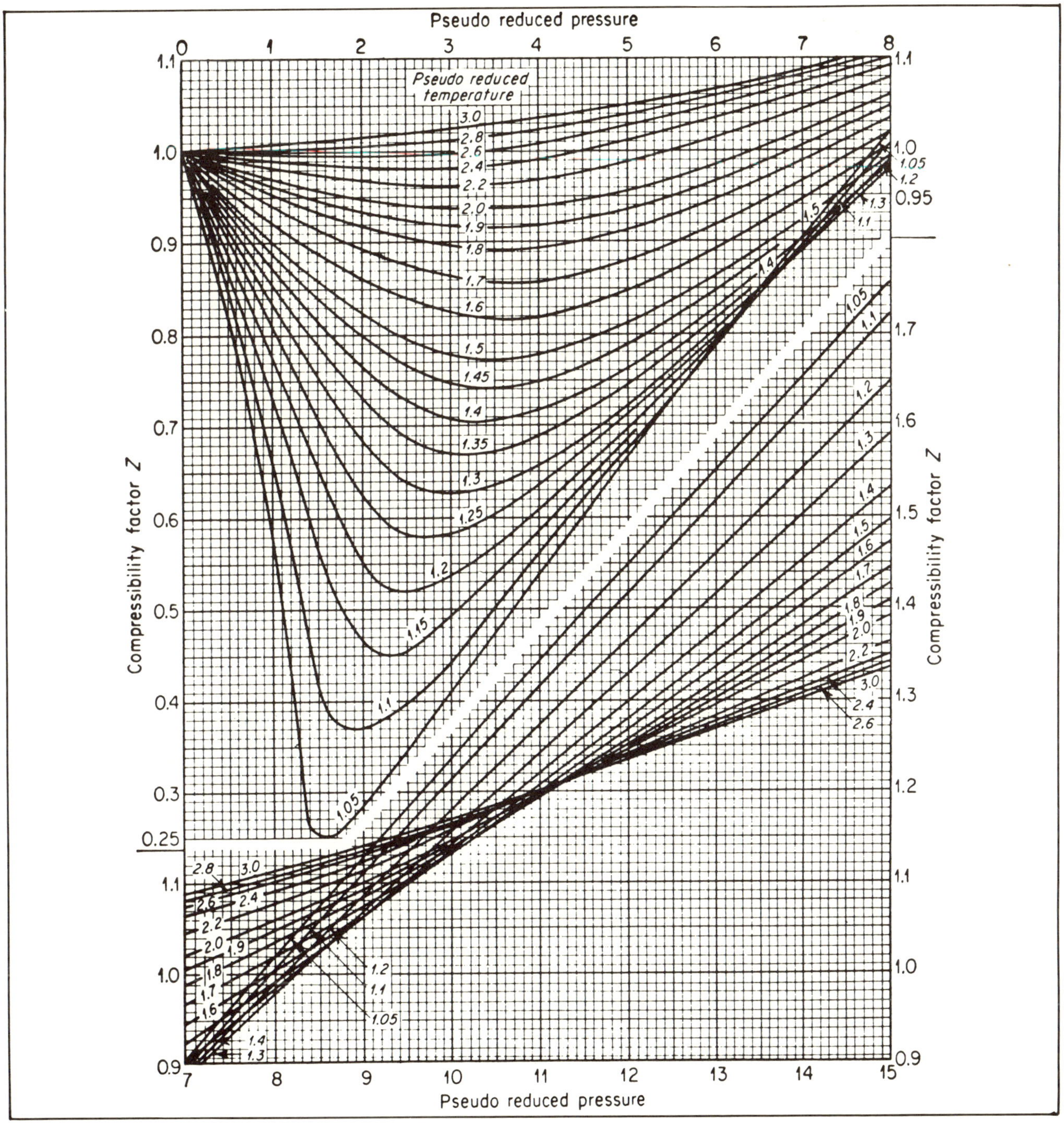

Fig. 2-6. Compressibility factors for natural gases. Courtesy Gas Processors Suppliers Association.

Example 2-8:

Calculate the volume occupied by 1 lb-mole of the natural gas given in Example 2-7 at 120°F and 1,500 psia.

Solution:

$$T_{pr} = \frac{T}{T_{pc}} = \frac{580°\text{R}}{383°\text{R}} = 1.51$$

$$p_{pr} = \frac{p}{p_{pc}} = \frac{1{,}500 \text{ psia}}{667 \text{ psia}} = 2.25$$

$$Z = 0.813 \text{ (from Fig. 2-6)}$$

$$V = \frac{ZnRT}{p}$$

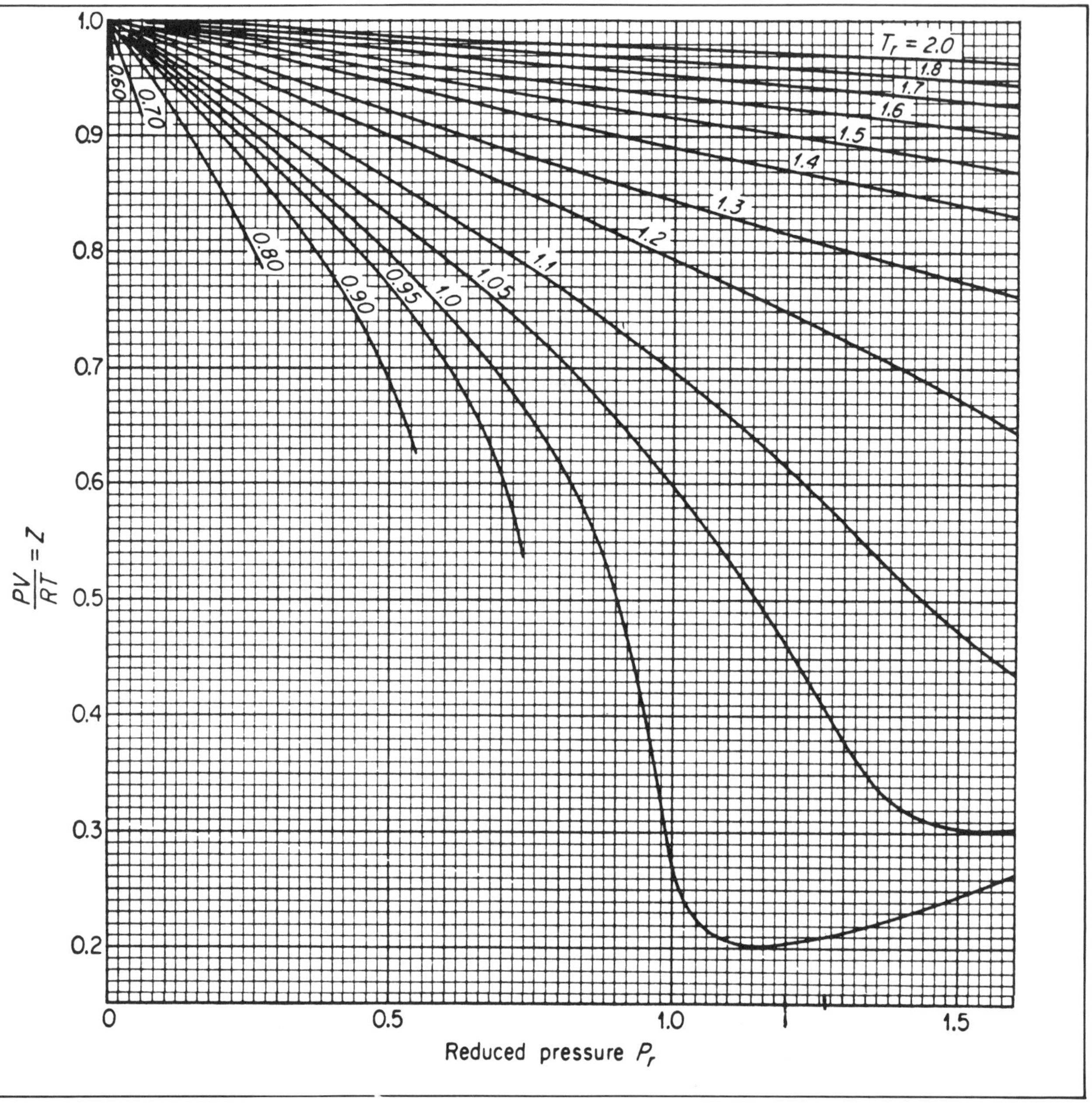

Fig. 2-7. Generalized plot of compressibility factors at low reduced pressures. Courtesy Gas Processors Suppliers Association.

$$V = \frac{(0.813)(1 \text{ lb-mole})\left(10.73\dfrac{\text{psia cu ft}}{\text{lb-mole°R}}\right)(580°\text{R.})}{1{,}500 \text{ psia}}$$

$$V = 3.37 \text{ cu ft}$$

A computer subroutine for calculating Z-factor is given in the Appendix. This subroutine reproduces Figure 2-6.

In some cases the composition of a gas will be given in weight or mass percent rather than mole percent. In this event, the composition must be first converted to mole fraction or percent before the mixture properties can be calculated. The following example illustrates this conversion.

Example 2-9:

A gas mixture consists of 50% C_1, 30% C_2, and 20% C_3 by weight. Calculate the apparent molecular weight and specific gravity of this mixture.

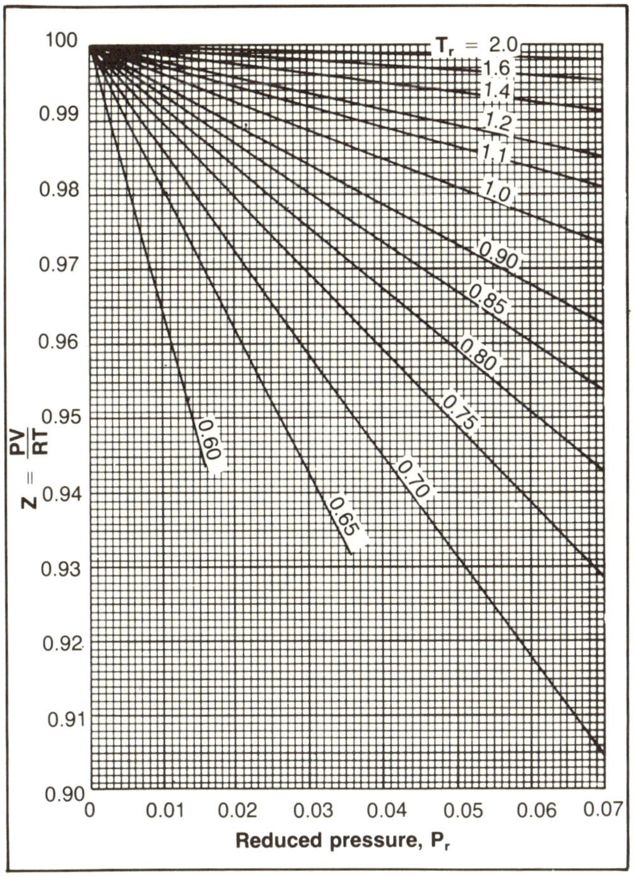

Fig. 2-8. Compressibility factors for gases near atmospheric pressure. Courtesy Gas Processors Suppliers Association.

Solution:

Assume 100 lbm of gas as a basis.

Component	Mass, m_j lbm	M_j lbm/ lb-mole	n_j, lb-mole	y_j	y_jM_j
C_1	50	16	3.13	0.68	10.9
C_2	30	30	1.00	0.22	6.6
C_3	20	44	0.45	0.10	4.4
	100		n = 4.58	1.00	21.9

$$n_j = \frac{m_j}{M_j}, \quad \text{for } C_1,\ n_1 = \frac{50 \text{ lbm}}{16 \dfrac{\text{lbm}}{\text{lb-mole}}} = 3.13 \text{ lb-mole}$$

$$M_a = \Sigma y_j M_j = 21.9 \frac{\text{lbm}}{\text{lb-mole}}$$

$$\gamma_g = \frac{M_a}{29} = 0.76$$

If the volume fraction is given at conditions other than standard, the volume fraction must be converted to a mole fraction basis, taking into account the deviation from ideal behavior. The following example illustrates this procedure.

Example 2-10:

A gas has the following composition measured at 2,500 psia and 300°F. Calculate the composition in mole fraction. The volume percents at these conditions are: C_1 = 67%, C_2 = 30%, C_3 = 1%, CO_2 = 2%.

Solution:

To convert from volume to moles, assume 100 ft^3 as a basis. Then use

$$n = \frac{pV}{ZRT} = \frac{2500V}{10.73\,(760)\,Z} = 0.3062\ V/Z$$

Component	V_j, ft^3	T_r	p_r	Z	V/Z	n_j	$y_j = n_j/n$
C_1	67	2.20	3.72	0.963	69.6	21.30	.595
C_2	30	1.40	3.53	0.708	42.5	13.00	.363
C_3	1	1.14	4.06	0.585	1.8	0.55	.015
CO_2	2	1.30	2.32	0.658	3.2	0.48	.027
	100					n = 38.83	1.000

In most cases the composition of a natural gas will be known and the apparent molecular weight and critical properties can be calculated as previously described. Occasionally, however, only the gas gravity will be known. Also, it is very easy to measure the gas gravity in the field. If the composition is unknown, or if accuracy requirements do not justify the longer calculations, Figure 2-9 can be used to estimate the pseudocritical properties. The properties can also be calculated using the following equations.

$$T_{pc} = 170.5 + 307.3\ \gamma_g \tag{2-14}$$

$$p_{pc} = 709.6 - 58.7\ \gamma_g \tag{2-15}$$

For condensate fluids:

$$T_{pc} = 187 + 330\ \gamma_g - 71.5\ \gamma_g^2$$

$$p_{pc} = 706 - 51.7\ \gamma_g - 11.1\ \gamma_g^2$$

Example 2-11:

Using the empirical correlation (Fig. 2-9) for critical properties, calculate T_{pc} and p_{pc} for the gas in Example 2-4. Compare these values with those obtained using the composition (Example 2-7).

Solution:

$$T_{pc} = 170.5 + 307.3\ \gamma_g = 170.5 + 307.3\ (0.66)$$

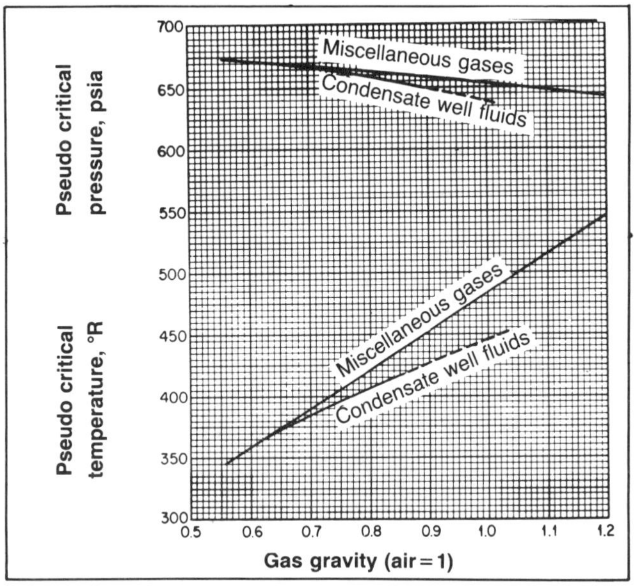

Fig. 2-9. Pseudocritical properties of natural gases. Courtesy Gas Processors Suppliers Association.

$$= 373°R$$

$$p_{pc} = 709.6 - 58.7\ \gamma_g = 709.6 - 58.7\ (0.66)$$

$$= 671 \text{ psia}$$

The values from Example 2-7 were 383°R and 667 psia.

GAS FORMATION VOLUME FACTOR

In most operations involving gas production the flow rates and quantities produced are measured at standard conditions such as scf/day or scf. Reservoir engineering and pipeline flow calculations require the volumes at in situ conditions of pressure and temperature, and therefore a convenient conversion factor from standard conditions to in situ conditions is needed. This conversion factor is called the gas formation volume factor and is defined as the actual volume occupied by the gas at some pressure and temperature divided by the volume that the gas would occupy at standard conditions. That is,

$$B_g = \frac{V_{p,T}}{V_{sc}} = \frac{\dfrac{Z\,n\,RT}{p}}{\dfrac{Z_{sc}\,n\,R\,T_{sc}}{p_{sc}}} = \frac{Z\,T\,p_{sc}}{Z_{sc}\,T_{sc}\,p}\ \frac{\text{vol}}{\text{std vol}}.$$

Using T_{sc} = 520°R, p_{sc} = 14.7 psia and Z_{sc} = 1, gives

$$B_g = \frac{ZT(14.7)}{(1)\,(520)p} = 0.0283\,\frac{ZT}{p}\ \frac{\text{vol}}{\text{std vol}},\ \text{e.g.}\ \frac{\text{ft}^3}{\text{scf}} \qquad (2\text{-}16)$$

It is sometimes convenient to express B_g in barrels of volume, or

$$B_g = 0.00504\,\frac{ZT}{p}\ \frac{\text{bbls}}{\text{scf}} \qquad (2\text{-}17)$$

For the constants given above, pressure is in psia and temperature is in °R. For the *SI* system (p = kPa, T = °K)

$$B_g = \frac{Z\,T\,(101.325)}{(1)\,(288.72)P} = 0.351\,\frac{ZT}{p} \qquad (2\text{-}18)$$

Example 2-12:

Calculate the formation volume factor for a natural gas having a gravity of 0.7 at a temperature of 93°C and an absolute pressure of 10343 kPa.

Solution:

$$T_{pc} = 170.5 + 307.3(.7) = 385.6°R = 214 \text{ K}$$

$$p_{pc} = 709.6 - 58.7(.7) = 668.5 \text{ psia} = 4609 \text{ kPa}$$

$$p_r = \frac{p}{p_{pc}} = \frac{10343}{4609} = 2.24 \qquad T_r = \frac{(93 + 273)}{214} = 1.71$$

From Figure 2-6, Z = .88

$$B_g = \frac{0.351(.88)(366)}{10343} = 0.0109\,\frac{\text{vol}}{\text{std vol}}$$

CORRECTION FOR NONHYDROCARBON IMPURITIES

Natural gases frequently contain materials other than hydrocarbons, such as nitrogen (N_2), carbon dioxide (CO_2), and hydrogen sulfide (H_2S). The presence of these impurities affects the value obtained from the *Z*-factor chart.

A procedure for adjusting the critical properties of the gas was proposed by Wichert and Aziz[2] in 1970. The adjusted critical properties are then used in calculating the reduced properties, and the *Z*-factor is then obtained from Figure 2-6.

The procedure for obtaining the *Z*-factor for sour gases is:

1. Determine p_{pc} and T_{pc} for the gas using the gas composition or Figure 2-9 and
2. Calculate the adjusted critical properties.

$$T'_{pc} = T_{pc} - \epsilon,$$

$$p'_{pc} = \frac{p_{pc}\,T'_{pc}}{T_{pc} + \epsilon\,(B - B^2)}$$

where

$$\epsilon = 120\,(A^{0.9} - A^{1.6}) + 15\,(B^{0.5} - B^4)$$

B = mole fraction H_2S,

A = mole fraction $CO_2 + B$

ϵ = correction factor, °F

3. Calculate the reduced properties using the corrected critical properties.

$$T_r = \frac{T}{T'_{pc}}, \qquad p_r = \frac{p}{p'_{pc}}$$

4. Find Z from Figure 2-6 or using the correlation given in the Appendix.

Example 2-13:

A gas containing 2.87% CO_2 and 23.27% H_2S has a critical pressure of 822 psia and a critical temperature of 465°R. Find the gas compressibility factor, Z, for p = 1000 psia, T = 100°F.

Solution:

$B = 0.2327 \qquad A = 0.0287 + 0.2327 = 0.2614$

$$\epsilon = 120\,[(0.2614)^{0.9} - (0.2614)^{1.6}] + 15\,[(0.2327)^{0.5} - (0.2327)^4]$$

$\epsilon = 29.0$°F

$T'_{pc} = T_{pc} - \epsilon = 465 - 29 = 436$°R

$$p'_{pc} = \frac{p_{pc}\,T'_{pc}}{T_{pc} + \epsilon(B - B^2)} = \frac{(822)(436)}{465 + 29\,(0.2327 - 0.0541)} = 762 \text{ psia}$$

$$T_r = \frac{T}{T'_{pc}} = \frac{560}{436} = 1.28 \qquad p_r = \frac{1000}{762} = 1.31$$

From Figure 2-6, Z = 0.78.

OTHER EQUATIONS OF STATE

One of the limitations in the use of the compressibility equation to describe the behavior of gases is that the compressibility factor is not constant, and therefore mathematical manipulations cannot be made directly but must be accomplished through graphical or numerical techniques. Most of the other commonly used equations of state were devised so that the correction factors, which correct the ideal gas law for nonideality, may be assumed constant, thus permitting the equations to be used in mathematical calculations involving differentiation or integration.

Many equations of state have been proposed for describing gas behavior and many modifications and improvements have been made. Only two of the most commonly used equations will be described. These will be used later in the calculation of phase behavior.

Benedict-Webb-Rubin Equation (BWR)

An earlier equation presented by Beattie and Bridgeman[3] was modified and resulted in an equation with eight empirical constants.

$$p = \frac{RT}{V_M} + \frac{(B_oRT - A_o - C_o/T^2)}{V_M{}^2} + \frac{(bRT - a)}{V_M{}^3} + \frac{a\alpha}{V_M{}^6} + \frac{c}{T^2V_M{}^3}\left(1 + \frac{\gamma}{V_M{}^2}\right)\text{EXP}\left(\frac{\gamma}{V_M{}^2}\right) \qquad (2\text{-}19)$$

The parameters B_o, A_o, C_o, a, b, c, α, and γ are constants for pure compounds and are functions of composition for mixtures. The constants for pure compounds are given in Table 2-3. These constants may be combined for use with mixtures of gases according to the following mixture rules.

$$A_o = (\Sigma y_j A_{oj}{}^{1/2})^2$$

$$B_o = \Sigma y_j B_{oj}$$

$$C_o = (\Sigma y_j C_{oj}{}^{1/2})^2$$

$$a = (\Sigma y_j a_j{}^{1/3})^3$$

TABLE 2-3
Benedict-Webb-Rubin Constants

Substance	A_0	B_0	$C_0 \times 10^{-5}$	a	b	$c \times 10^{-6}$	$a \times 10^3$	$\gamma \times 10^2$
Methane	6,995.25	0.682401	275.763	2,984.12	0.867325	498.106	511.172	153.961
Ethane	15,670.7	1.00554	2,194.27	20,850.2	2.85393	6,413.14	1,000.44	302.790
Propane	25,915.4	1.55884	6,209.93	57,248.0	5.77355	25,247.8	2,495.77	564.524
Isobutane	38,587.4	2.20329	10,384.7	117,047	10.8890	55,977.7	4,414.96	872.447
n-Butane	38,029.6	1.99211	12,130.5	113,705	10.2636	61,925.6	4,526.93	872.447
Isopentane	4,825.36	2.56386	21,336.7	226,902	17.1441	136,025	6,987.77	1,188.07
n-Pentane	45,928.8	2.51096	25,917.2	246,148	17.1441	161,306	7,439.92	1,218.86
n-Hexane	5,443.4	2.84835	40,556.2	429,901	28.0032	296,077	11,553.9	1,711.15
n-Heptane	66,070.6	3.18782	57,984.0	626,106	38.9917	483,427	17,905.6	2,309.42
Ethylene	12,593.6	0.891980	1,602.28	15,645.5	2.20678	4,133.60	731.661	236.844
Propylene	23,049.2	1.36263	5,365.97	46,758.6	4.79997	20,083.0	1,873.12	469.325

$$b = (\Sigma y_j b_j^{1/3})^3$$

$$c = (\Sigma y_j c_j^{1/3})^3$$

$$\alpha = (\Sigma y_j \alpha_j^{1/3})^3$$

$$\gamma = (\Sigma y_j \gamma_j^{1/2})^2$$

V_M is the molar volume in cu ft/lb-mole.

Redlich-Kwong Equation (RK)

The Redlich-Kwong Equation[4] involves only two empirical constants as opposed to the eight required in the BWR equation. The original RK equation is

$$p = \frac{RT}{V_M - b} - \frac{a}{V_M(V_M + b)T^{0.5}} \tag{2-20}$$

where

$$a = \frac{C_a R^2 T_c^{2.5}}{p_c}$$

$$b = \frac{C_b R T_c}{p_c}$$

To simplify the calculations with the RK equation, especially for application to mixtures, other constants have been defined as

$$A = \left(\frac{a}{R^2 T^{2.5}}\right)^{0.5} = \left(\frac{C_a T_c^{2.5}}{p_c T^{2.5}}\right)^{0.5}$$

$$B = \frac{b}{RT} = \frac{C_b T_c}{p_c T}$$

where, for temperature in °R and pressure in psia, $C_a = 0.42748$, $C_b = 0.08664$.

The RK equation was further modified by Soave[5] to improve its accuracy when applied to mixtures as follows:

$$p = \frac{RT}{V_M - b} - \frac{a\,\alpha}{V_M(V_M + b)} \tag{2-21}$$

where α is a function of temperature. Using the following definitions, a gas compressibility factor can be calculated

$$A = \frac{C_a \alpha p T_c^2}{p_c T^2}$$

$$B = \frac{C_b P T_c}{p_c T}$$

$$Z^3 - Z^2 + Z(A - B - B^2) - A B = 0 \tag{2-22}$$

The constants A and B are calculated for mixtures by

$$A_m = C_a \frac{p}{T^2}\left(\Sigma(y_i T_{ci} \propto_i^{0.5}/p_{ci}^{0.5})\right)^2 \tag{2-23}$$

$$B_m = C_b \frac{p}{T} \Sigma\, y_i T_{ci}/p_{ci} \tag{2-24}$$

where y_i is the mole fraction of the ith component of the mixture. Evaluation of α involves calculation of the acentric factor, ω

$$\alpha = \left(1 + S\,(1 - T_R^{0.5})\right)^2, \tag{2-25}$$

$$S = 0.480 + 1.574\,\omega - 0.176\,\omega^2, \tag{2-26}$$

$$\omega = C_a T_B \left[\frac{\log p_c - 1.167}{T_c - T_B}\right] - 1.0 \tag{2-27}$$

where T_B is the boiling point of the component at 14.7 psia in °R. Values of ω can also be obtained from Table 2-2.

Example 2-14:

Using the Soave modification to the RK equation, calculate the molar volume of the following gas mixture at p = 250 psia, T = 100°F.

Component	y_i
C_1	0.75
C_2	0.15
C_3	0.10
	1.00

Solution:

Component	y_j	p_{cj}	T_{cj}	ω_j	S_j	α_j
C_1	0.75	667.8	343.4	0.0104	0.496	0.744
C_2	0.15	707.8	550.1	0.0986	0.633	0.989
C_3	0.10	616.3	666.1	0.1524	0.716	1.123

Component	$y_j T_{cj} \propto_{\alpha j}^{0.5}/p_{cj}^{0.5}$	$y_j T_{cj}/p_{cj}$
C_1	8.597	0.38567
C_2	3.084	0.11658
C_3	2.843	0.10808
	14.524	0.61033

$$A_m = \frac{0.42748\ (250)}{(560)^2}(14.524)^2 = 0.07189$$

$$B_m = \frac{0.08664\ (250)}{560}(0.61033) = 0.02361$$

$$Z^3 - Z^2 + Z\,(0.07189 - 0.02361 - (0.02361)^2) - 0.07189\ (0.02361) = 0$$

$$Z^3 - Z^2 + 0.04772Z - 0.00170 = 0$$

$$Z = 0.952$$

$$V_M = ZRT/p = 0.952(10.73)(560)/250$$

$$= 22.88 \frac{\text{cu ft}}{\text{lb-mole}}$$

Calculations for C_1:

From Table 2-2, $\omega = 0.0104$.

$$S = 0.480 + 1.574(0.0104) - 0.176(0.0104)^2 = 0.496$$

$$T_r = 560/343.3 = 1.63 \qquad T_r^{0.5} = 1.277$$

$$\alpha = [1 + 0.496(1 - 1.277)]^2 = 0.744$$

GAS COMPRESSIBILITY

The isothermal compressibility of a gas is the measure of the change in volume per unit volume with pressure change at constant temperature. In equation form

$$C = -\frac{1}{V}\left(\frac{\partial V}{\partial p}\right)_T \tag{2-28}$$

This should not be confused with the gas compressibility factor, Z. The compressibility is required in many reservoir gas flow equations and may be evaluated in the following manner.

Ideal Gas Compressibility

For an ideal gas,

$$V = \frac{nRT}{p}, \left(\frac{\partial V}{\partial p}\right)_T = -\frac{nRT}{p^2}$$

and therefore,

$$C = -\frac{1}{V}\frac{\partial V}{\partial p} = -\frac{p}{nRT}\left(-\frac{nRT}{p^2}\right) = \frac{1}{p}$$

Real Gas Compressibility

For a real gas,

$$V = \frac{ZnRT}{p}$$

and since $Z = f(p)$, it must be included in the derivative

$$\left(\frac{\partial V}{\partial p}\right)_T = nRT\frac{p\frac{\partial Z}{\partial p} - Z}{p^2}.$$

Substituting these expressions into the equation defining compressibility gives

$$C = \frac{-p}{ZnRT}\left[\frac{nRT}{p^2}\left(p\frac{\partial Z}{\partial p} - Z\right)\right] = \frac{1}{p} - \frac{1}{Z}\frac{\partial Z}{\partial p} \tag{2-29}$$

Evaluation of C for real gases requires determining how the Z-factor varies with pressure at the pressure and temperature of interest. Because most of the charts and equations predict Z as a function of reduced pressure and temperature, a reduced compressibility has been defined as $C_r = C\,p_c$. This can be expressed as a function of p_r at a fixed value of T_r by

$$C_r = \frac{1}{p_r} - \frac{1}{Z}\left(\frac{\partial Z}{\partial p_r}\right)_{T_r}.$$

Values of $(\partial Z/\partial p_r)_{T_r}$ can be obtained from the slope of a constant T_r curve from Figure 2-6 at the Z-factor of interest. Values of $C_r T_r$ as a function of p_r and T_r have been presented graphically by Mattar, et al.[6] in Figures 2-10 and 2-11. The change of Z with p can also be calculated using an analytical expression by calculating the Z-factor at pressures slightly above and below the pressure of interest. That is,

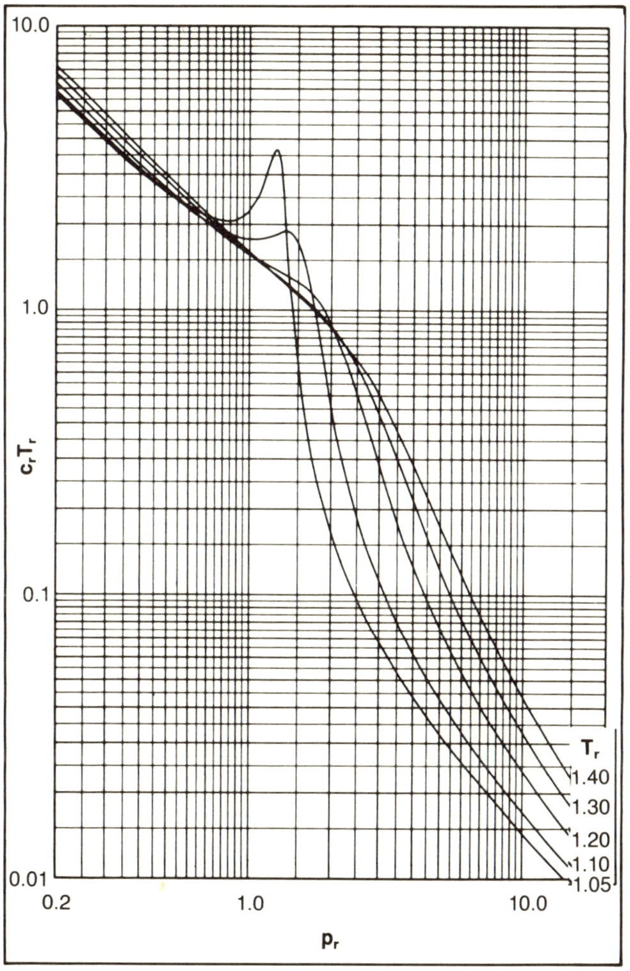

Fig. 2-10. Variation of c_rT_r with reduced temperature and pressure. ($1.05 \le T_r \le 1.4$; $0.2 \le p_r \le 15.0$). Courtesy The Journal of Canadian Petroleum Technology, Canadian Institute of Mining and Metallurgy.

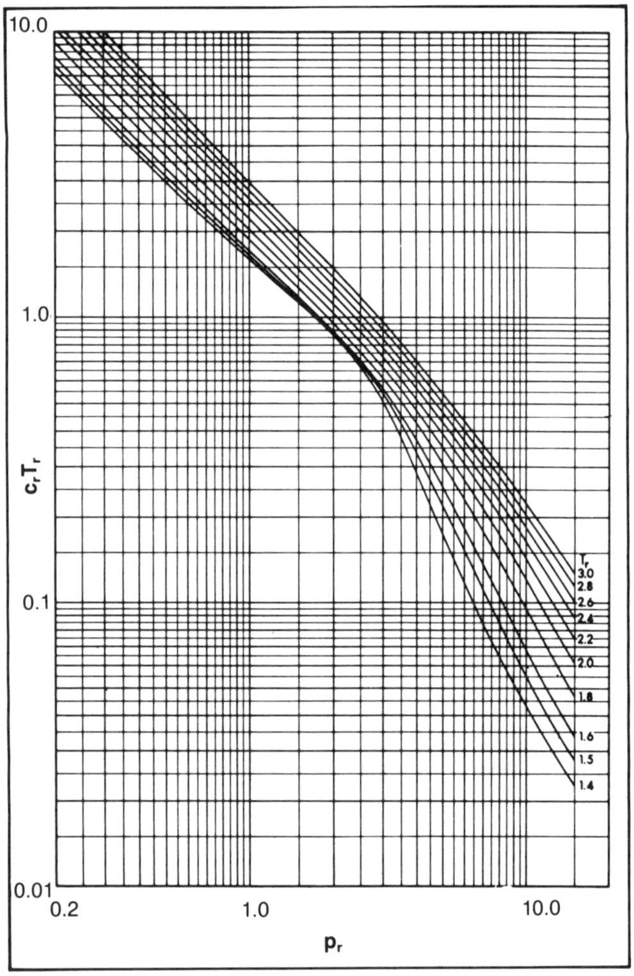

Fig. 2-11. Variation of c_rT_r with reduced temperature and pressure. ($1.4 \leq T_r \leq 3.0$; $0.2 \leq p_r \leq 15.0$). Courtesy The Journal of Canadian Petroleum Technology, Canadian Institute of Mining and Metallurgy.

$$\left(\frac{\partial Z}{\partial p_r}\right)_{T_r} \approx \left(\frac{Z_1 - Z_2}{p_{r1} - p_{r2}}\right)_{T_r}$$

Example 2-15:

Calculate the compressibility of a gas at a temperature of 40°F and a pressure of 665 psia if T_c = 357°R, p_c = 674 psia.

Solution:

$$T_r = \frac{T}{T_c} = \frac{460 + 40}{357} = 1.40$$

$$p_r = \frac{p}{p_c} = \frac{665}{674} = 0.987$$

From Figure 2-11, $C_r\ T_r = 1.62$

$$C_r = \frac{1.62}{T_r} = \frac{1.62}{1.4} = 1.16$$

$$C = \frac{C_r}{p_c} = \frac{1.16}{674} = 0.00172 \text{ psi}^{-1}$$

The problem can also be solved using the approximation for $\partial Z/\partial p_r$:

p, psia	p_r	Z
615	0.912	0.880
665	0.986	0.870
715	1.060	0.861

$$C_r = \frac{1}{p_r} - \frac{1}{Z}\left(\frac{Z_1 - Z_2}{p_{r1} - p_{r2}}\right) = \frac{1}{0.986} - \frac{1}{0.870}\left(\frac{0.880 - 0.861}{0.912 - 1.060}\right)$$

$$C_r = 1.168$$

$$C = \frac{C_r}{p_c} = \frac{1.168}{674} = 0.00173 \text{ psi}^{-1}$$

GAS VISCOSITY

The viscosity of a fluid is a measure of the fluid's ability to flow, or the ratio of the shearing force to the shearing rate. The viscosity is usually expressed in centipoises or poises, but can be converted to other units for unit compatibility.

$$1 \text{ poise} = 100 \text{ centipoises} = 6.72 \times 10^{-2} \text{ lbm/ft-sec}$$

$$= 2.09 \times 10^{-3} \text{ lbf-sec/ft}^2 = 0.1 \text{ kg/m-sec}$$

Gas viscosity is difficult to measure experimentally and for engineering purposes can be determined accurately enough from empirical correlations. The most widely accepted correlation used in the past has been that of Carr, et al.[7]. The Carr method is presented in Figures 2-12 and 2-13. The viscosity at one atmosphere is obtained first from Figure 2-12 and corrected for nonhydrocarbon impurities if necessary. This viscosity is a function of molecular weight or gravity and temperature only. The correction for pressure is obtained from Figure 2-13 as a function of reduced pressure and temperature.

An analytical expression for viscosity of hydrocarbon gas was presented by Lee, et al.[8] in 1966. The equation is

$$\mu_g = K.\ 10^{-4} \exp\left(X \rho_g^y\right) \tag{2-30}$$

where

$$K = \frac{(9.4 + 0.02M)\,T^{1.5}}{209 + 19M + T}$$

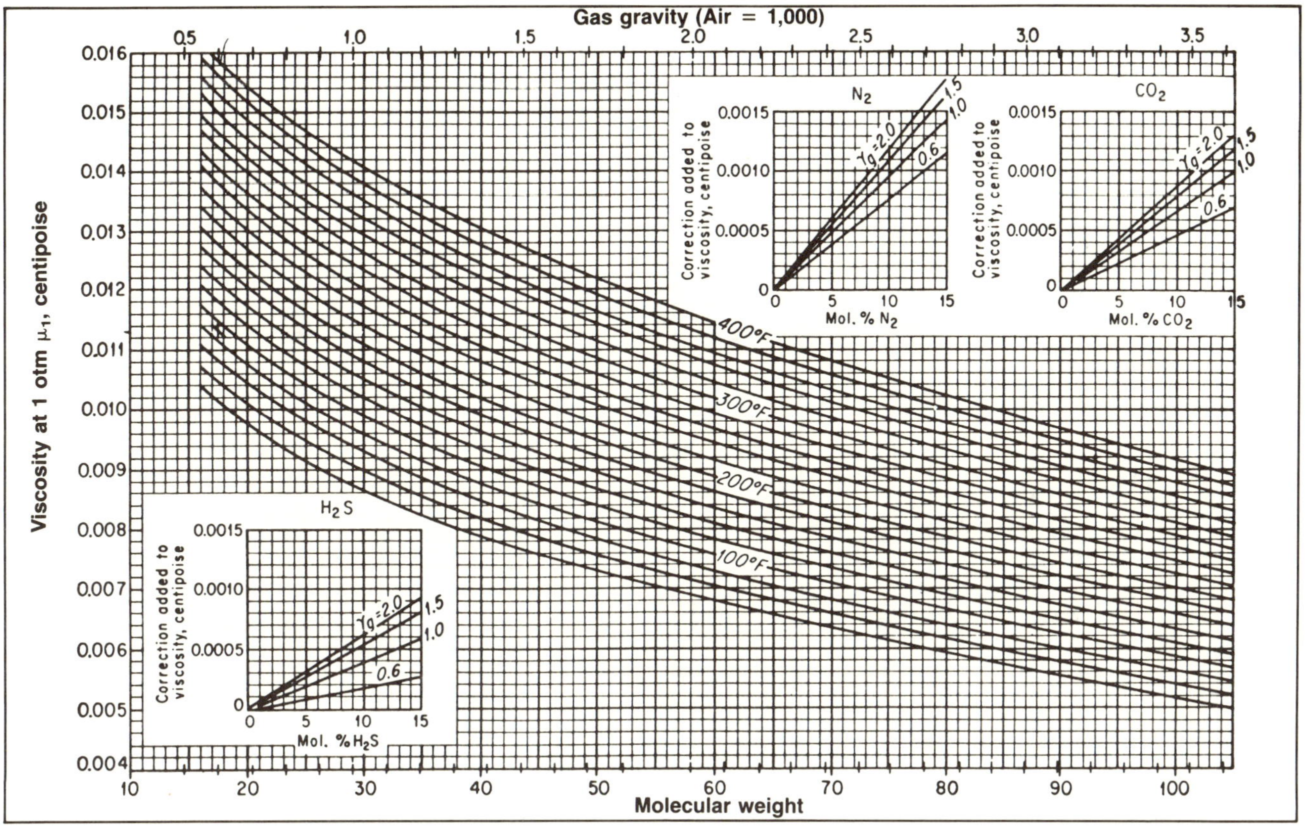

Fig. 2-12. Viscosity of paraffin hydrocarbon gases at atmosphere. Permission to publish by the Society of Petroleum Engineers of AIME. Copyright 1954 SPE-AIME.

$$X = 3.5 + \frac{986}{T} + 0.01\,M$$

$$y = 2.4 - 0.2X$$

In these equations, $T = °R$, $\mu_g = cp$, M = molecular weight and $\rho_g = gm/cm^3$.

Example 2-16:

Using both the Carr and Lee methods, calculate the viscosity of a 0.8 gravity gas at a pressure of 2000 psia and a temperature of 150°F. The gas contains 10% H_2S and 10% CO_2.

Solution:

Carr Method

From figure 2-12, $\mu_1 = 0.0111$ cp
Correction for 10% H_2S = +0.0003
Correction for 10% CO_2 = +0.0006

$$\mu_1 = 0.0111 + 0.0003 + 0.0006 = 0.0120\ cp$$

$$T_c = 170.5 + 307.3\ (.8) = 416°R$$

$$p_c = 709.6 - 58.7\ (.8) = 663 \text{ psia}$$

$$T_r = (460 + 150)/416 = 1.47$$

$$p_r = 2000/663 = 3.02 \qquad Z = 0.791$$

From Figure 2-13, $\dfrac{\mu}{\mu_1} = 1.61$

$$\mu = \mu_1\left(\frac{\mu}{\mu_1}\right) = (0.012)(1.61) = 0.0193 \text{ cp}$$

Lee Method

The Lee method does not include a method for correcting for nonhydrocarbon impurities; therefore, the value obtained will be for a pure hydrocarbon gas. However, if the Z-factor used in calculating the gas density has been corrected, the Lee method is valid for sour gases.

$$M = \gamma_{gf}(M_{air}) = (0.8)(29) = 23.2$$

$$K = \frac{(9.4 + 0.02M)\ T^{1.5}}{209.0 + 19.0\ M + T}$$

$$= \frac{[9.4 + (0.02)(23.2)](610)^{1.5}}{209.0 + (19.0)(23.2) + 610}$$

$$= 117.96$$

$$X = 3.5 + \frac{986}{T} + 0.01\ M$$

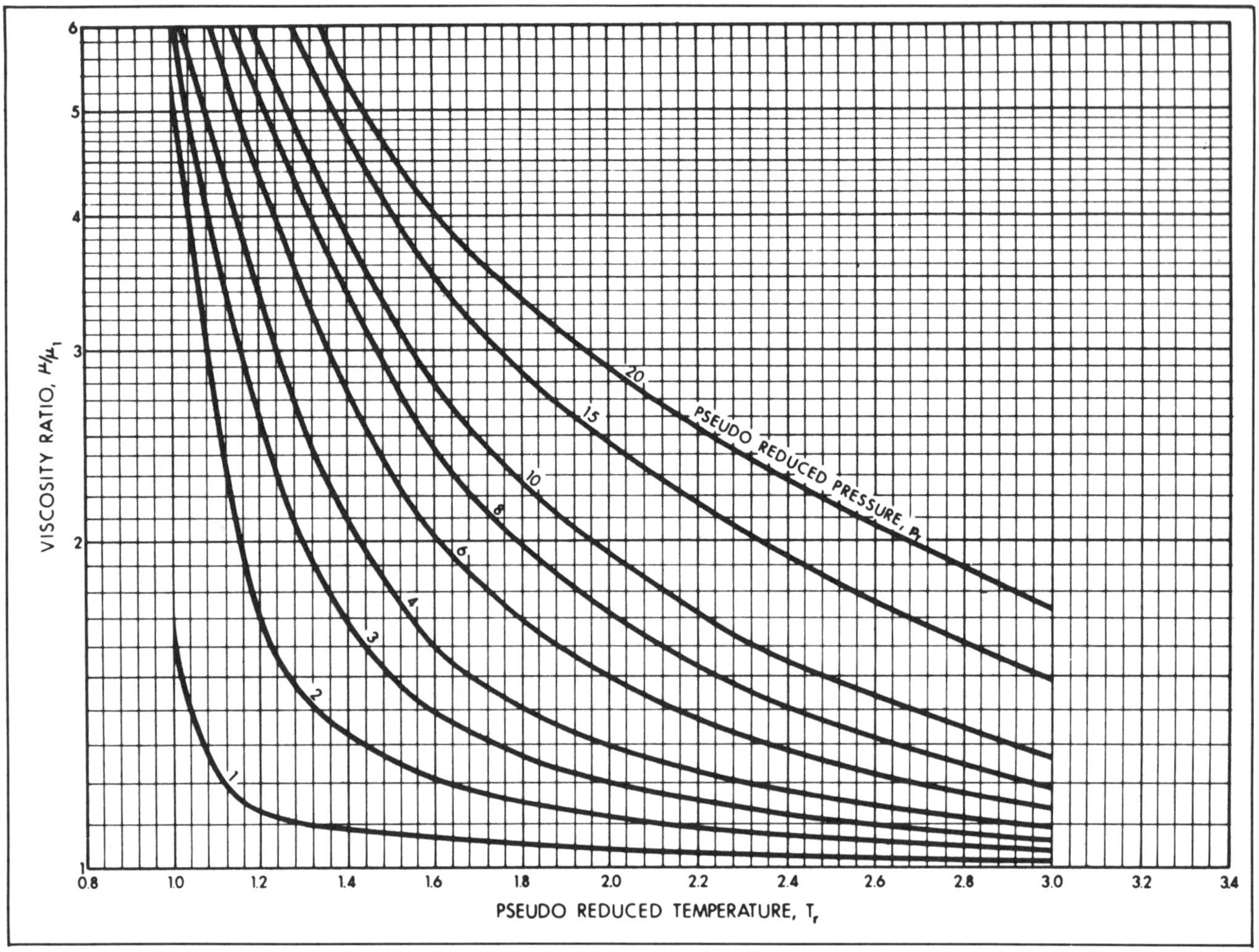

Fig. 2-13. Viscosity ratio verses pseudoreduced temperature. Permission to publish by the Society of Petroleum Engineers of AIME. Copyright 1954 SPE-AIME.

$$= 3.5 + \frac{986}{610} + (0.01)(23.2)$$

$$= 5.35$$

$$y = 2.4 - 0.2\,X$$

$$= 2.4 - (0.2)(5.35)$$

$$= 1.33$$

$$\rho_g = \gamma_{gf}\,(0.0764)\,\frac{(520)}{(14.7)}\,\frac{p}{ZT}\cdot\frac{1\text{ gm/cm}^3}{62.4\text{ lbm/ft}^3}$$

$$= \frac{(0.8)(0.0764)(520)(2000)}{(14.7)(0.791)(610)(62.4)} = 0.1436\text{ gm/cm}^3$$

$$\mu_g = K.\,10^{-4}\exp\,(X\rho_g^y)$$

$$= (117.96)(10^{-4})\exp\,[(5.35)(0.1436)^{1.33}]$$

$$= 0.0177\text{ cp}$$

It is sometimes convenient to express the viscosity of a fluid as a kinematic viscosity, ν. The relationship between viscosity and kinematic viscosity is

$$\nu = \frac{\mu}{\rho} \qquad (2\text{-}31)$$

where ρ is the fluid density.

GAS-WATER SYSTEMS

In many cases the engineering design of gas production operations will involve natural gas in contact with water. This water may be connate reservoir water or water produced from some other zone. In any event, it is necessary to determine the water contained in the gas in the vapor state, the gas dissolved in the water, and under what conditions of pressure and temperature gas hydrates will be formed.

Solubility of Natural Gas in Water

The solubility of natural gas in water is very low at most pressures and temperatures of interest in gas production engineering. The primary factors affecting the amount of gas that will be evolved from water saturated with the gas depends on pressure, temperature, and solids content of the water. The relationship is illustrated in Figure 2-14. The correction factor for salinity may be calculated from

$$\frac{R_{sw}}{R_{swp}} = 1 - X\,Y \times 10^{-4} \tag{2-32}$$

where

Y = salinity of water, ppm

$X = 3.471/T^{0.837}$

T = temperature, °F

R_{sw} = gas solubility in brine

R_{swp} = gas solubility in pure water

Example 2-17:

Calculate the scf of gas dissolved in a brine containing 50,000 ppm at a pressure of 5000 psia and a temperature of 200°F.

Solution:

From Figure 2-14, R_{swp} = 20 scf/STB

$$X = \frac{3.471}{(200)^{0.837}} = 0.041$$

$$R_{sw} = R_{swp}\,(1 - X\,Y \times 10^{-4})$$

$$= 20\,(1 - (0.041)(50000) \times 10^{-4})$$

$$R_{sw} = 20(0.795) = 15.9 \text{ scf/STB}$$

Solubility of Water in Natural Gas

The amount of water vapor contained in natural gas at various temperatures and pressures must be known in order to relate surface water production to equivalent volumes in the reservoir. The water content is also important in calculations involving hydrate formation. The water content depends on pressure, temperature, and water salinity. Figure 2-15 shows the mass or weight of fresh water contained in gas as a function of pressure and temperature. The following equation can be used to correct for water salinity.

$$\frac{W_s}{W_{sp}} = 1 - 2.87 \times 10^{-8}\,Y^{1.266} \tag{2-33}$$

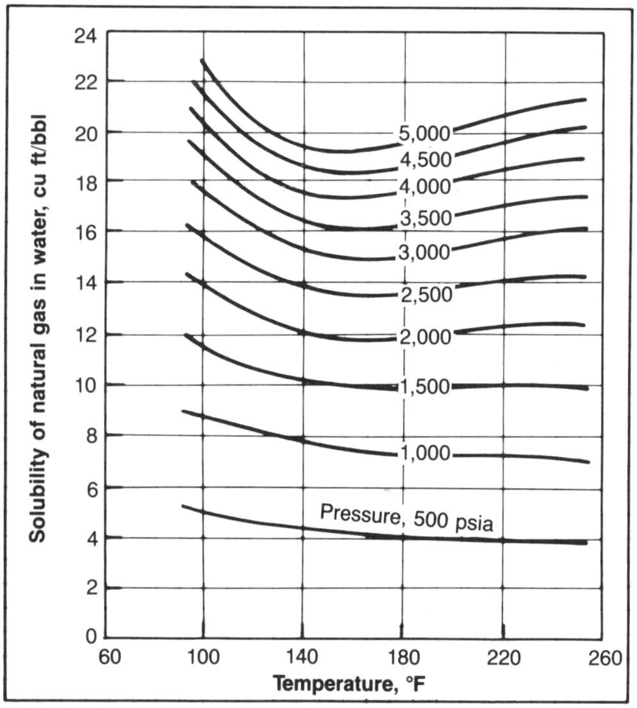

Fig. 2-14. Gas solubility in pure water. Courtesy Gas Processors Suppliers Association.

where

W_s = brine content, lbm/MMscf,

W_{sp} = pure water content from Figure 2-15,

Y = salinity of water, ppm.

Example 2-18:

Determine the water content in a natural gas in contact with a 100,000 ppm brine at 3000 psia and 200°F.

Solution:

From Figure 2-15, W_{sp} = 280 lbm/MMscf

$$W_s = W_{sp}\,(1 - 2.87 \times 10^{-8}(100{,}000)^{1.266}) = 280(0.939)$$

$$= 263\frac{\text{lbm}}{\text{MMscf}}$$

Gas Hydrates

Gas hydrates are crystalline compounds, formed by the chemical combination of natural gas and water under pressure at temperatures considerably above the freezing point of water. In the presence of free water, hydrates will form when the temperature of the gas is below a certain temperature, called the "hydrate temperature." Hydrate formation is often confused with condensation,

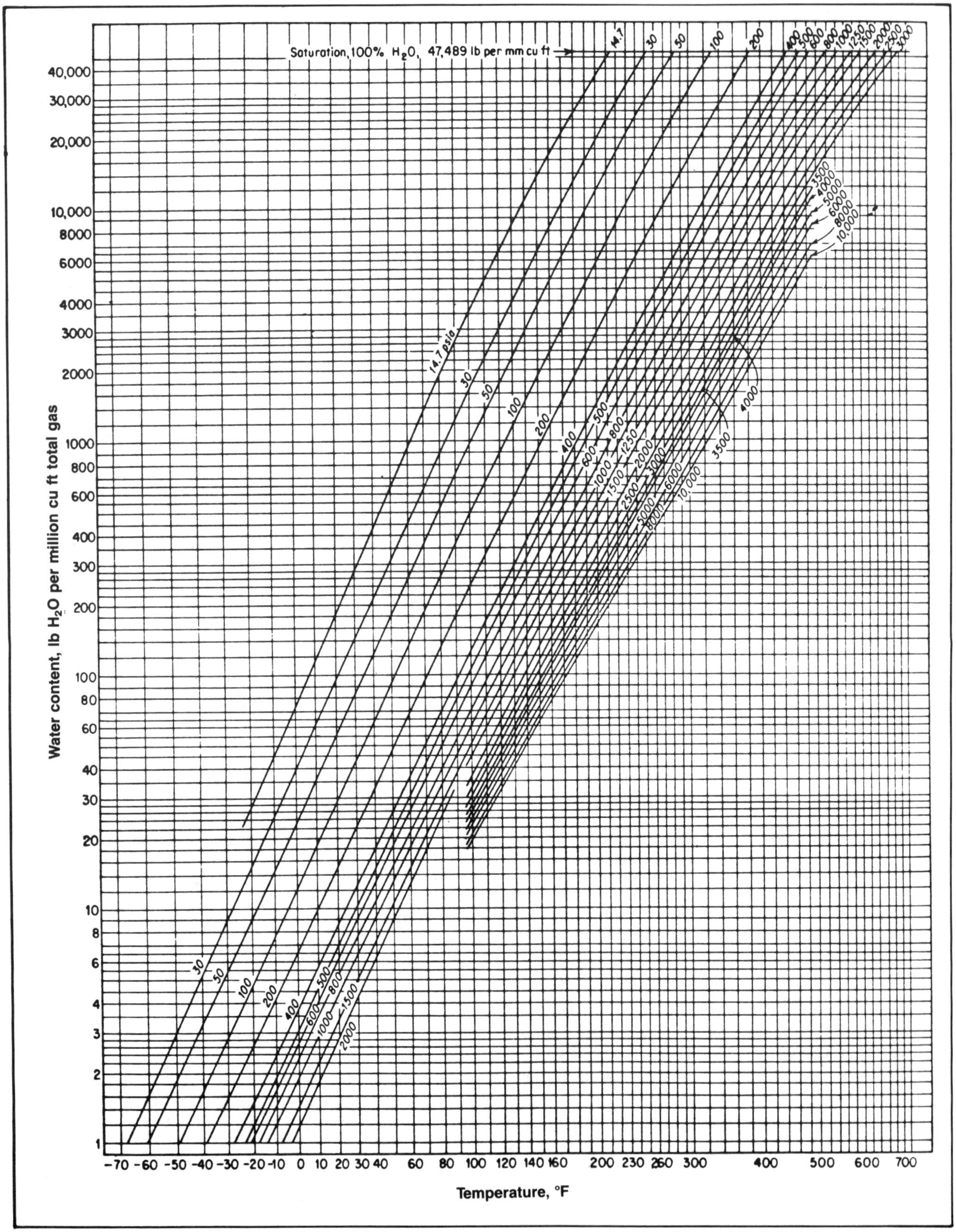

Fig. 2-15. Water content of natural gas in equilibrium with liquid water. Courtesy Gas Processors Suppliers Association.

and the distinction between the two must be clearly understood. Condensation of water from a natural gas under pressure occurs when the temperature is at or below the dew point at that pressure. Free water obtained under such conditions is essential to the formation of hydrates that will occur at or below the hydrate temperature at the same pressure. Hence, the hydrate temperature would be below, and perhaps the same as, but never above the dew point temperature.

In solving gas production problems, it becomes necessary to define, and thereby avoid, conditions that promote the formation of hydrates. Hydrates may choke the flow string, surface lines, and well testing equipment. Hydrate formation in the flow string would result in a lower value for measured wellhead pressures. In a flow rate measuring device, hydrate formation would result in lower flow rates. Excessive hydrate formation may also completely block flow lines and surface equipment.

Conditions promoting hydrate formation are:

1. Gas at or below its water dew point with "free" water present,
2. low temperature, and
3. high pressure.

A further discussion of hydrate formation and prevention is given in Chapter 9.

GAS-CONDENSATE SYSTEMS

As the exploration for natural gas is extended to deeper horizons, more reservoirs containing gas condensates are discovered. The gas may be in the gaseous phase at initial reservoir conditions but may condense to form some liquid at some point in the path to the separator.

Engineering design of these production systems requires some understanding of phase behavior. This section deals with the phenomenon of phase change and with methods to calculate these phase changes.

Phase Behavior

Phase behavior is simple for single component systems but becomes more complicated as more components are added to the system. A discussion of the simplest system will lead to an understanding of the more complex systems.

Single Component Fluid. The phase behavior of a fluid can be described by determining its response to pressure and temperature changes. In a liquid the molecules are very close together, but in a gas the molecules are widely separated. Certain forces exist that tend to either confine or disperse the molecules. Confining forces are primarily pressure and molecular attraction. Dispersing forces are kinetic energy and molecular repulsion. The relative magnitudes of the confining and dispersing forces dictate whether the fluid is a liquid or a gas.

An increase in temperature increases the kinetic energy of the molecules and thus the dispersing forces, while an increase in pressure increases the confining forces.

The behavior of a single component fluid can be illustrated graphically on a plot of pressure versus temperature, called a phase diagram. A line joining all of the points where the confining and repelling forces are equal can be plotted. This is a vapor pressure line and is illustrated in Figure 2-16. At any point above the vapor pressure line the fluid is a liquid, while below the line the fluid is a gas. At the critical point, Point *C*, there is no distinction between liquid and gas. A phase change at constant pressure occurs when temperature changes, as illustrated by the path 1-2. A phase change will also occur at constant temperature as pressure is changed (path 3-4).

Multicomponent Fluids. When more than one fluid is present, the difference in molecule size and energy has an influence on the phase change. Figure 2-17 illustrates the phase diagram for a two-component mixture and Figure 2-18 shows the phase diagram for a typical hydrocarbon multicomponent mixture. There is no sharp transition from liquid to vapor or from vapor to liquid, but the molecules are able to escape from the liquid or gas at different pressures and temperatures because of molecular attraction. The locus of all points where the first bubble of gas appears in a liquid as pressure and temperature conditions are changed is called the bubble point line. The locus of all points where the first droplet of

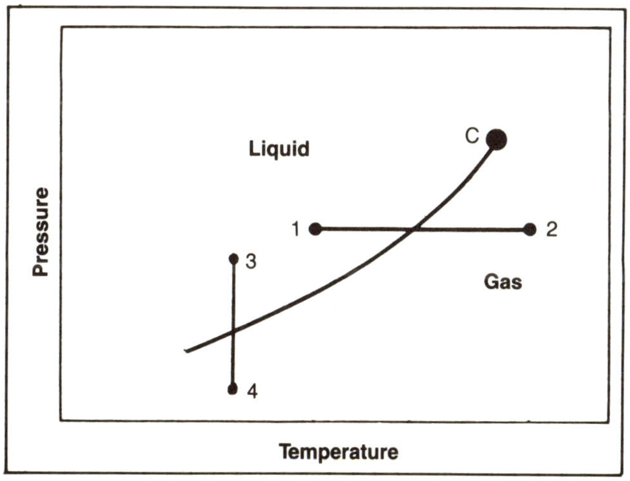

Fig. 2-16. Phase diagram for a pure substance showing lines of isobaric temperature change. Reprinted from McCain, Properties of Petroleum Fluids, copyright 1974, PennWell Books.

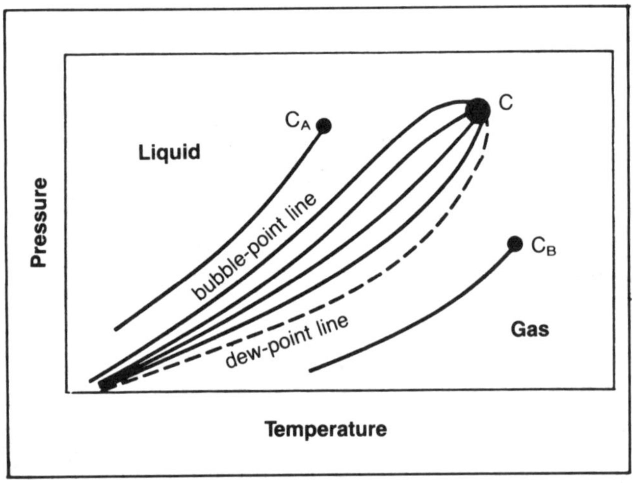

Fig. 2-17. Vapor pressure curves for two pure components and phase diagram for a 50:50 mixture of the same components. Reprinted from McCain, Properties of Petroleum Fluids, copyright 1974, PennWell Books.

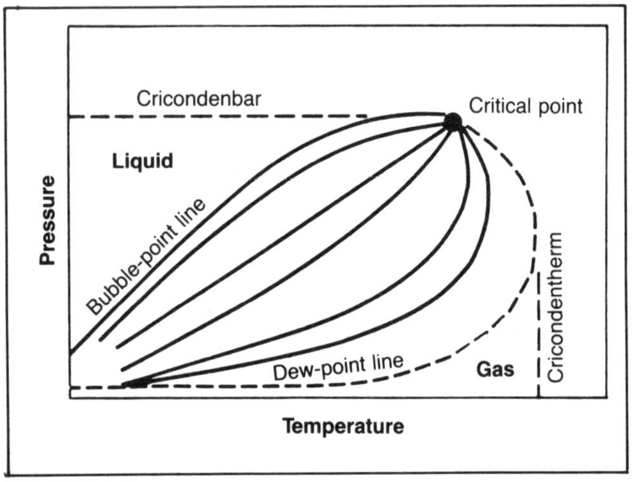

Fig. 2-18. Phase diagram showing cricondentherm and cricondenbar. Reprinted from McCain, Properties of Petroleum Fluids, copyright 1974, PennWell Books.

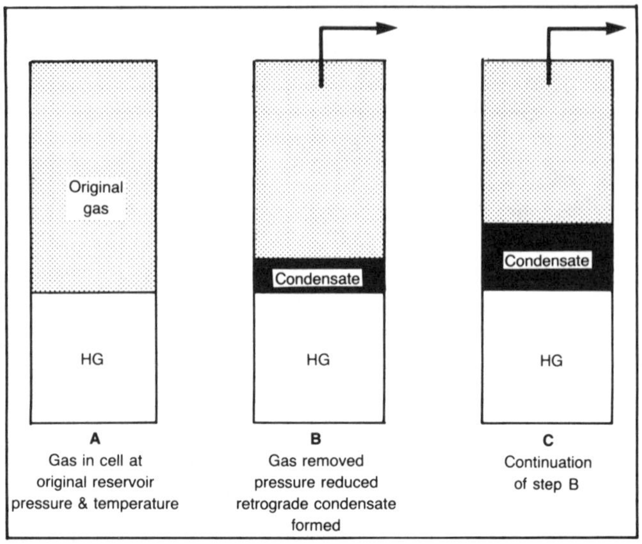

Fig. 2-19. Differential retrograde condensation of liquid from gas. Reprinted from Norman J. Clark, World Oil, April 1953. Courtesy Gulf Publishing Company.

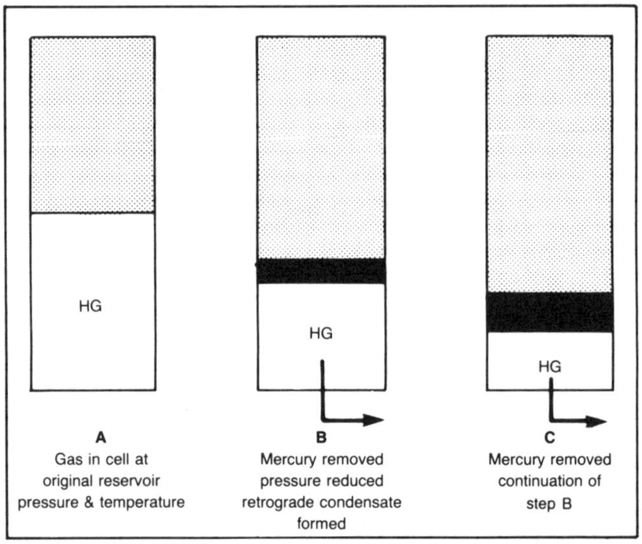

Fig. 2-20. Equilibrium retrograde condensation of liquid from gas. Reprinted from Norman J. Clark, World Oil, April 1953. Courtesy Gulf Publishing Company.

liquid appears as the conditions for a gas are changed is the dew point line. The highest pressure at which a gas can exist is called the cricondenbar while the highest temperature at which a liquid can exist is the cricondentherm.

Separation Processes. The two types of separation processes that can occur as pressure or temperature on a fluid changes are called differential and flash or equilibrium. These processes are illustrated in Figures 2-19 and 2-20. In the differential process the gas is removed from contact with the liquid as it is separated, which means that the composition of the remaining liquid changes constantly. In the flash process the gas remains in contact with the liquid, which means that the process is a constant composition process. Both types of separation, or a mixture of the two, can occur as gas is produced from a reservoir to the separator.

Types of Gas Reservoirs

Reservoirs that yield natural gas can be classified into essentially four categories. These are:

1. dry gas—the fluid exists as a gas both in the reservoir and the piping system. The only liquid associated with the gas from a dry gas reservoir is water. A phase diagram of a dry gas reservoir is illustrated in Figure 2-21;

2. wet gas—the fluid initially exists as a gas in the reservoir and remains in the gaseous phase as pressure declines at reservoir temperature. However, in being produced to the surface, the temperature also drops, causing condensation in the piping system and separator. This behavior is illustrated in Figure 2-22;
3. retrograde condensate gas—the fluid exists as a gas at initial reservoir conditions. As reservoir pressure declines at reservoir temperature, the dew point line is crossed and liquid forms in the reservoir. Liquid also forms in the piping system and separator. Figure 2-23 depicts the behavior of a retrograde gas system;
4. associated gas—many oil reservoirs exist at the bubble point pressure of the fluid system at initial conditions. Free gas can be produced from the gas cap of such a system. Gas which is initially dissolved in the oil can also be produced as free gas at the surface. The phase diagram of such a system will depend on the properties of the oil associated with the gas. An example phase diagram is shown in Figure 2-24.

The fact that hydrocarbon liquids are frequently in contact with natural gas makes it imperative that methods be available to calculate the volumes or masses of each phase and also the composition of each phase existing at various conditions of pressure and temperature. The procedure for accomplishing these calculations is described in the following section.

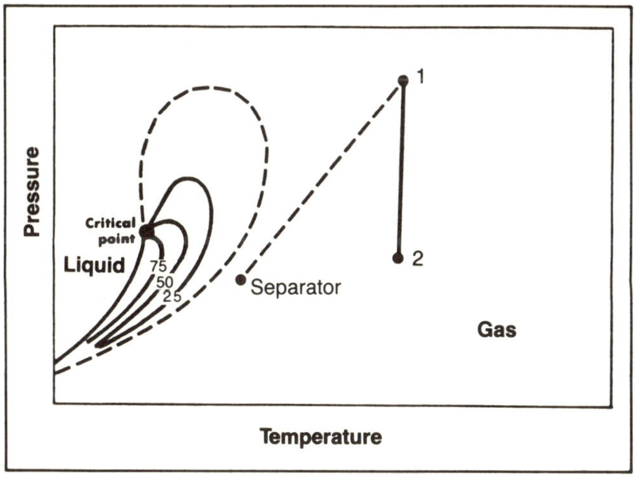

Fig. 2-21. Phase diagram of a dry gas. Reprinted from McCain, Properties of Petroleum Fluids, copyright 1974, PennWell Books.

Flash or Equilibrium Separation Calculations

The phase behavior of an ideal solution can be described by combining the laws of Raoult and Dalton to obtain the quantity of gas and liquid existing at given conditions. Raoult's Law states that the partial pressure of a component in the gas is equal to the product of the mole fraction of that component in the liquid multiplied by the vapor pressure of the pure component. Raoult's Law is valid only if both the gas and liquid mixtures are ideal solutions. The mathematical statement of Raoult's Law is

$$p_j = x_j p_{vj},$$

where p_j represents the partial pressure of component j

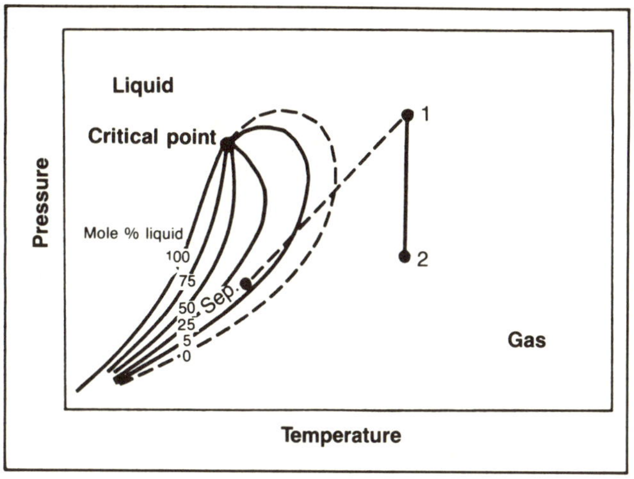

Fig. 2-22. Phase diagram of a wet gas. Reprinted from McCain, Properties of Petroleum Fluids, copyright 1974, PennWell Books.

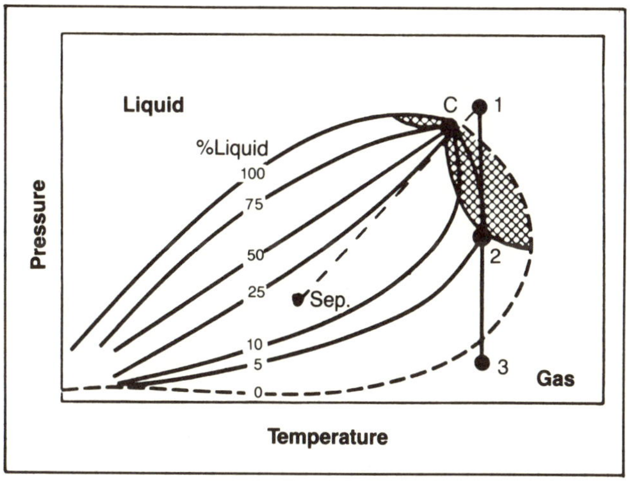

Fig. 2-23. Phase diagram showing regions of retrograde condensation. Reprinted from McCain, Properties of Petroleum Fluids, copyright 1974, PennWell Books.

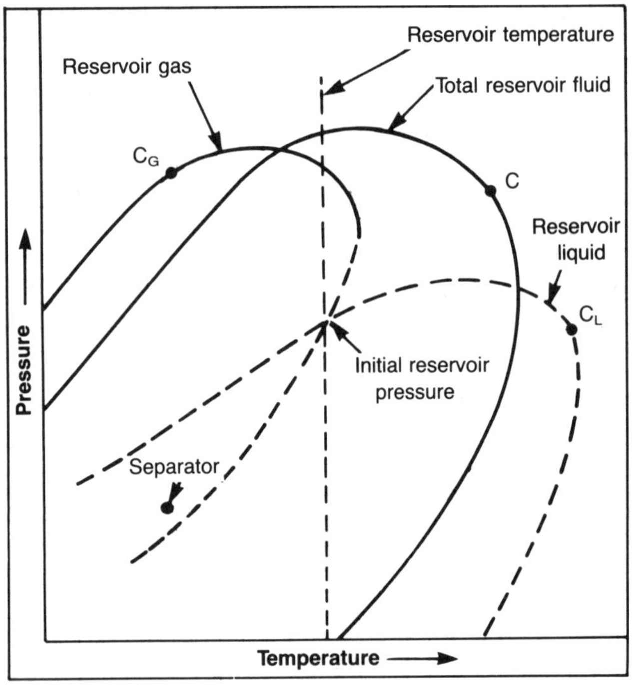

Fig. 2-24. Equilibrium relationship of reservoir containing an oil accumulation with a gas cap. Courtesy Gas Processors Suppliers Association.

in the gas in equilibrium with a liquid of composition x_j. The quantity p_{vj} represents the vapor pressure that pure component j would exert at the temperature of interest.

Dalton's Law can also be used to calculate the partial pressure exerted by a component of an ideal gas mixture:

$$p_j = y_j p.$$

Raoult's and Dalton's Laws both represent the partial pressure of a component in a gas. In the case of Raoult's Law, the gas must be in equilibrium with a liquid. These laws may be combined by eliminating partial pressure to give an equation that relates the compositions of the gas and liquid phases in equilibrium to the pressure and temperature at which the gas-liquid equilibrium exists, such as

$$y_j p = x_j p_{vj}$$

or (2-34)

$$\frac{y_j}{x_j} = \frac{p_{vj}}{p}.$$

Equation 2-34 can be used to calculate the ratio of the mole fraction of a component in the gas to its mole fraction in the liquid. To determine values of y_j and x_j, this equation must be combined with another equation relating these two quantities. Such an equation can be developed through consideration of a material balance on the jth component of the mixture.

Let n represent the total number of moles in the mixture,
n_L represent the total number of moles in the liquid,
n_g represent the total number of moles in the gas,
z_j represent the mole fraction of the jth component in the total mixture including both liquid and gas phases,
x_j represent the mole fraction of the jth component in the liquid,
y_j represent the mole fraction of the jth component in the gas,
$z_j n$ represent the moles of the jth component in the total mixture,
$x_j n_L$ represent the moles of the jth component in the liquid, and
$y_j n_g$ represent the moles of the jth component in the gas.

A material balance on the jth component results in

$$z_j n = x_j n_L + y_j n_g. \qquad (2\text{-}35)$$

Combination of Equations 2-34 and 2-35 to eliminate y_j results in

$$x_j = \frac{z_j n}{n_L + \dfrac{p_{vj}}{p} n_g}.$$

Since by definition $\Sigma x_j = 1$,

$$\Sigma x_j = \Sigma \frac{z_j n}{n_L + \dfrac{p_{vj}}{p} n_g} = 1. \qquad (2\text{-}36)$$

Equations 2-34 and 2-35 can also be combined to eliminate x_j with the result,

$$\Sigma y_j = \Sigma \frac{z_j n}{n_g + \dfrac{p}{p_{vj}} n_L} = 1. \qquad (2\text{-}37)$$

Either equation 2-36 or 2-37 can be used to calculate the compositions of the gas and liquid phases of a mixture at equilibrium. In either case a trial-and-error solution is required. The calculation is simplified if one mole of total mixture is taken as a basis so that $\bar{n}_L = n_L/n$, $\bar{n}_g = n_g/n$ and $\bar{n}_L + \bar{n}_g = 1$. This results in the reduction of Equations 2-36 and 2-37 to

$$\Sigma x_j = \Sigma \frac{z_j}{1 + \bar{n}_g \left(\dfrac{p_{vj}}{p} - 1\right)} = 1 \qquad (2\text{-}38)$$

and

$$\Sigma y_j = \sum \frac{z_j}{1 + \bar{n}_L \left(\dfrac{p}{p_{vj}} - 1\right)} = 1. \qquad (2\text{-}39)$$

The use of Equation 2-38 or 2-39 to predict gas-liquid equilibria behavior is severely restricted since Dalton's Law is based on the assumption that the gas behaves as an ideal solution of ideal gases. For practical purposes the ideal-gas assumption limits the use of the equations to pressures below about 50 psia.

Raoult's Law is also based on the assumption that the liquid behaves as an ideal solution. Ideal-solution behavior is approached only if the components of the liquid mixture are very similar chemically and physically.

Another restriction is that a pure compound does not have a vapor pressure at temperatures above its critical temperature. Thus, the use of these equations is limited to temperatures less than the critical temperature of the most volatile compound in the mixture. For example, if methane is a component of the mixture, this method cannot be used at temperatures above −116°F. These restrictions have led to the use of correlations based on experimental observations. These correlations usually involve the use of the equilibrium ratio K, which is defined as

$$K_j = \frac{y_j}{x_j}.$$

From equation 2-34, it is seen that the K-value is defined by the ratio of the vapor pressure of the component to the system pressure, or

$$\frac{p_{vj}}{p} = \frac{y_j}{x_j} = K_j.$$

Substituting this relationship into Equations 2-38 and 2-39 results in

$$\Sigma x_j = \Sigma \frac{z_j}{1 + \bar{n}_g(K_j - 1)} = 1 \qquad (2\text{-}40)$$

$$\Sigma y_j = \Sigma \frac{z_j}{1 + \bar{n}_L\left(\dfrac{1}{K_j} - 1\right)} = 1. \qquad (2\text{-}41)$$

The choice between Equations 2-40 and 2-41 is arbitrary; either can be used with equal efficiency. Either equation requires a trial-and-error solution. Normally, successive trial values of $\bar{n}_g$ or $\bar{n}_L$ are selected until the summation equals one. These equations can be used to calculate the quantity and composition of a mixture by the following procedure. The required data are the feed composition, the pressure and temperature at which the separation is to take place, and K-factors for each component at this pressure and temperature.

1. Determine the appropriate K-value for each component at p and T of separation.
2. Assume a value of $\bar{n}_g$ and calculate an assumed value of $\bar{n}_L$ from

   $$\bar{n}_L + \bar{n}_g = 1$$

   for one mole of feed.
3. Calculate the sum of the mole fractions of all the components using the value of $\bar{n}_g$ assumed in Step 2.

   $$\Sigma x_j = \Sigma \frac{z_j}{1 + \bar{n}_g (K_j - 1)}$$

4. If the summation calculated in Step 3 is equal to one, the assumption for $\bar{n}_g$ was correct. If not, assume a new value for $\bar{n}_g$ and repeat Step 3 until the summation is equal to one.

The x_j values calculated in Step 3 represent the composition of the $\bar{n}_L$ moles of liquid resulting from the separation. The composition of the $\bar{n}_g$ moles of gas can be calculated from

$$y_j = K_j x_j.$$

Step 3 can be modified to calculate the sum of the mole fractions in the gas phase by assuming a value of $\bar{n}_L$ and using Equation 2-41.

Example 2-19:

Calculate the quantity and composition of the gas and liquid phase resulting from a flash separation of the fluid at a pressure of 200 psia and a temperature of 150°F. Assume that the charts for a convergence pressure of 5000 psia are applicable.

Solution:

Component	z_j	K_j	$\bar{n}_g = 0.5$ x_j	$\bar{n}_g = 0.4$ x_j	$\bar{n}_g = 0.42$ x_j	y_j
C_3	0.61	1.520	0.484	0.505	0.501	0.761
n-C_4	0.28	0.595	0.351	0.334	0.337	0.201
n-C_5	0.11	0.236	0.178	0.158	0.162	0.038
	1.00		1.013	0.997	1.000	1.000

Since $\Sigma x_j = 1$ for $\bar{n}_g = 0.42$, this means that 42% of the total feed is in the gas phase and 58% in the liquid phase at 200 psia and 150°F. Also, for example, 50.1% of the liquid phase is C_3 and 76.1% of the gas phase is C_3.

The previous equations can also be used to determine bubble point and dew point pressures of a mixture at a given temperature. At the bubble point, the amount of gas is negligible and

$$z_j \approx x_j,$$

and

$$y_j = x_j k_j = z_j K_j \tag{2-42}$$

$$\Sigma y_j = \Sigma z_j K_j = 1.$$

At the dew point pressure,

$$z_j \approx y_j,$$

and

$$\Sigma x_j = \Sigma \frac{z_j}{K_j} = 1. \tag{2-43}$$

The bubble point or dew point pressures can be calculated by trial-and-error by assuming various pressures and determining the pressure that will produce K-values satisfying Equations 2-42 or 2-43.

The following relationships will aid in determining the bubble point or dew point.

a. If $\Sigma\ z_j K_j$ and $\Sigma \frac{z_j}{K_j}$ are both > 1, p is between p_b and p_d.
b. If $\Sigma\ z_j K_j < 1$, $p > p_b$.
c. If $\Sigma \frac{z_j}{K_j} < 1$, $p < p_d$

Example 2-20:
Determine the bubble point pressure of the following mixture at a temperature of 150°F.

Solution:
From the previous example we know that the bubble point pressure is greater than 200 psia. As a first guess, use p_b = 250 psia.

Trial 1 (p_b = 250 psia)

Component	z_j	K_j	z_jK_j
C_3	0.61	1.270	0.775
n-C_4	0.28	0.500	0.140
n-C_5	0.11	0.203	0.022
	1.00		0.937

The assumed pressure is greater than the bubble point pressure since $\Sigma z_j K_j < 1$.

Trial 2 (p_b = 220 psia)

Component	z_j	K_j	z_jK_j
C_3	0.61	1.490	0.909
n-C_4	0.28	0.580	0.162
n-C_5	0.11	0.230	0.025
			1.096

The fact that $\Sigma z_j K_j > 1$ implies that 220 psia is below the bubble point pressure. By linear interpolation, select the pressure for the third trial as 230 psia.

Trial 3 (p_b = 230 psia)

Component	z_j	K_j	z_jK_j
C_3	0.61	1.350	0.824
n-C_4	0.28	0.534	0.150
n-C_5	0.11	0.223	0.025
			0.999

The bubble point pressure is therefore 230 psia at a temperature of 150°F.

Determination of Equilibrium Ratios

In order to apply the methods described in the preceding section, it is necessary to have access to K-values for each component as a function of pressure and temperature. The most widely used method of determining K-values is to use the experimentally determined charts published by the NGPSA[9]. The problem with using these charts is that the convergence pressure, p_k, must be known in order to select the appropriate chart. At relatively low pressures the K-values are almost independent of convergence pressure, but for separation pressures approaching convergence pressure, they are very sensitive to convergence pressure. The determination of the proper convergence pressure to use for a given system is iterative. A suggested procedure is:

1. Estimate a value for convergence pressure, p_k'.
2. Read the K-values for each component from the charts prepared at p_k'.
3. Perform a flash calculation using the K-values from Step 2. This gives the liquid composition.
4. Identify the lightest hydrocarbon component in the liquid phase which comprises at least 0.001 mole fraction; that is, $x_j \geq 0.001$. This will usually be C_1 or C_2 for gas condensate systems.
5. Calculate the *weight* average critical temperature for the remaining heavier components in the liquid phase. These remaining heavier components represent a pseudo-heavy component in the pseudo binary system.
6. Enter Figure 2-25 at the critical temperature calculated in Step 5 and determine which actual hydrocarbon component has a critical temperature closest to the pseudo-heavy component. If the value is between two real components, interpolation is required.
7. Trace the critical locus between the light and pseudo-heavy component to the separation temperature and read the convergence pressure p_k at this point. If p_k

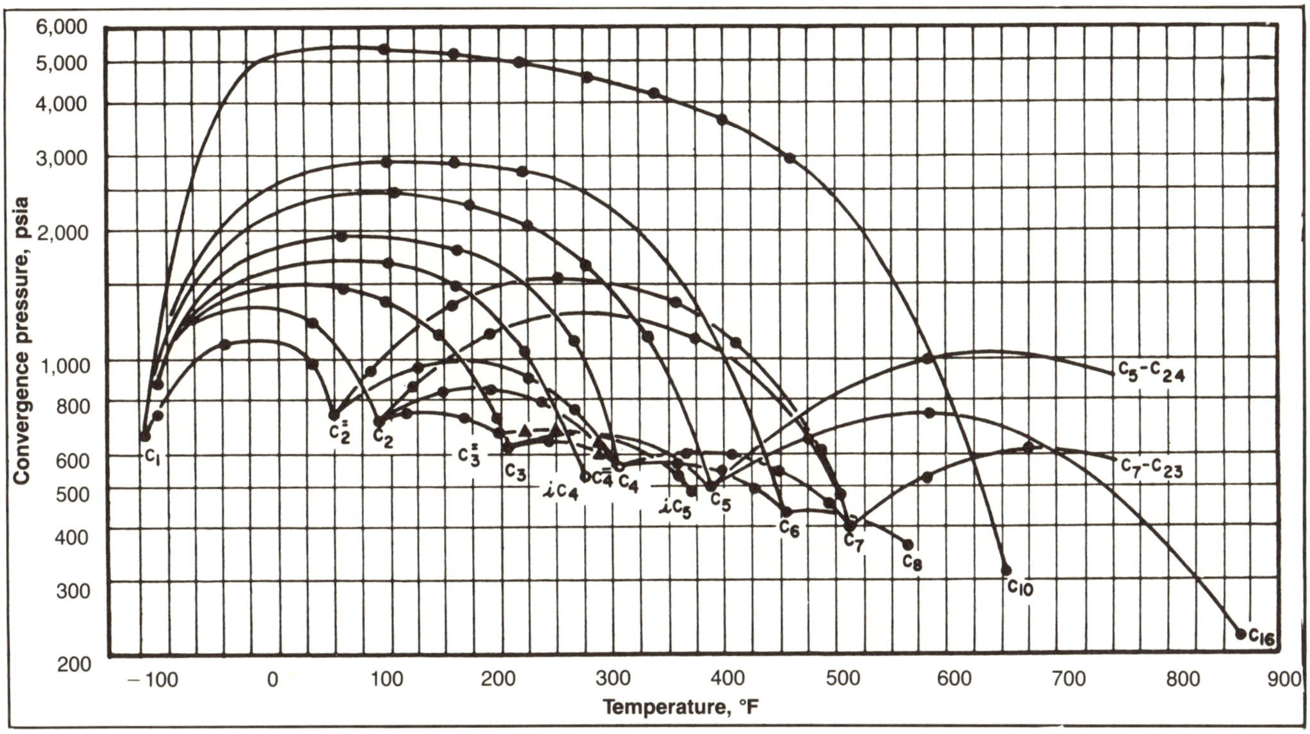

Fig. 2-25. Convergence pressure of binary hydrocarbon systems. Courtesy Gas Processors Suppliers Association.

$\neq p_k'$ from Step 1, set $p_k' = p_k$ and go to Step 2. Repeat until $p_k' \approx p_k$.

K-Values from Equations of State

Equilibrium ratios may be defined in terms of the fugacity of the component in a mixture. The fugacity, f, is a thermodynamic quantity defined in terms of the change in free energy in passing from one state to another. It has been shown that the fugacity of a component in a phase of a mixture is equal to the fugacity of that component in the same phase in the pure state multiplied by the mole fraction of the component in the mixture. That is, for the gas and liquid phases,

$$f_{gj}\,(\text{mix}) = y_j f_{gj}\,(\text{pure})$$

$$f_{Lj}\,(\text{mix}) = x_j f_{Lj}\,(\text{pure})$$

If the phases are in equilibrium, the fugacities of a component in the liquid and gas phases are equal, or

$$y_j f_{gj} = x_j f_{Lj}$$

or

$$\frac{y_j}{x_j} = K_j = \frac{f_{Lj}}{f_{gj}}. \qquad (2\text{-}44)$$

Therefore, the K-values can be approximated if the fugacity of each component can be calculated at a given pressure and temperature. This can be accomplished by the use of an equation of state, such as the Soave-Redlich-Kwong equation described earlier. The fugacity coefficient of a component in a mixture is given by

$$\ln \frac{f_j}{p x_j} = \frac{b_j}{b}(Z - 1) - \ln(Z - B) - \frac{A}{B}\left(2\frac{a_i^{0.5}}{a^{0.5}} - \frac{b_i}{b}\right) \ln\left(1 + \frac{B}{Z}\right) \qquad (2\text{-}45)$$

where

$$\frac{a_j^{0.5}}{a^{0.5}} = \frac{\alpha_j^{0.5}\, T_{cj}/p_{cj}^{0.5}}{\Sigma x_j\, \alpha_j^{0.5}\, p_{cj}^{0.5}}$$

$$\frac{b_j}{b} = \frac{T_{cj}/p_{cj}}{\Sigma x_j\, T_{cj}/p_{cj}}$$

The values for Z, A, B, and α are calculated by Equations 2-22, 2-23, 2-24, and 2-25, respectively. The x_j represents either the liquid or gas mole fraction.

The use of an equation of state to calculate phase behavior involves numerous calculations and is usually applied only by utilization of a computer. Some simplified programs have been published for use with hand calculators by Weber[10].

Adjustment of Properties for Condensate Mixtures

In many cases the properties of the total well or reservoir fluid will not be measured, but the gas and condensate properties after separation will be known. Many reservoir and piping system calculations require properties such as specific gravity for the reservoir or well fluid. Methods for making the calculations are given below.

Specific Gravity of Mixtures. By forming a molal balance on the separator gas and liquid effluents, the following equation can be derived for calculating the gravity of a mixture:

$$\gamma_{gm} = \frac{\gamma_g + 4584\,\gamma_o/R}{1 + V_o/R} \qquad (2\text{-}46)$$

where

γ_{gm} = specific gravity of the mixture (air = 1),

γ_g = specific gravity of the separator gas,

γ_o = specific gravity of the condensate,

V_o = condensate vaporizing volume, and

R = producing gas-condensate ratio.

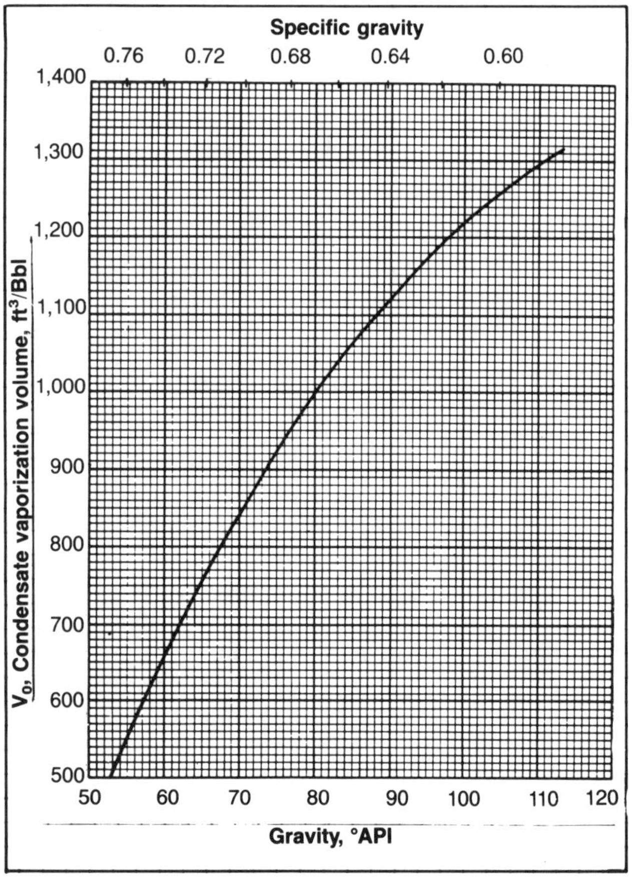

Fig. 2-26. Relationship between condensate vaporization volume ratio and API gravity. Courtesy Energy Resources Conservation Board, Calgary.

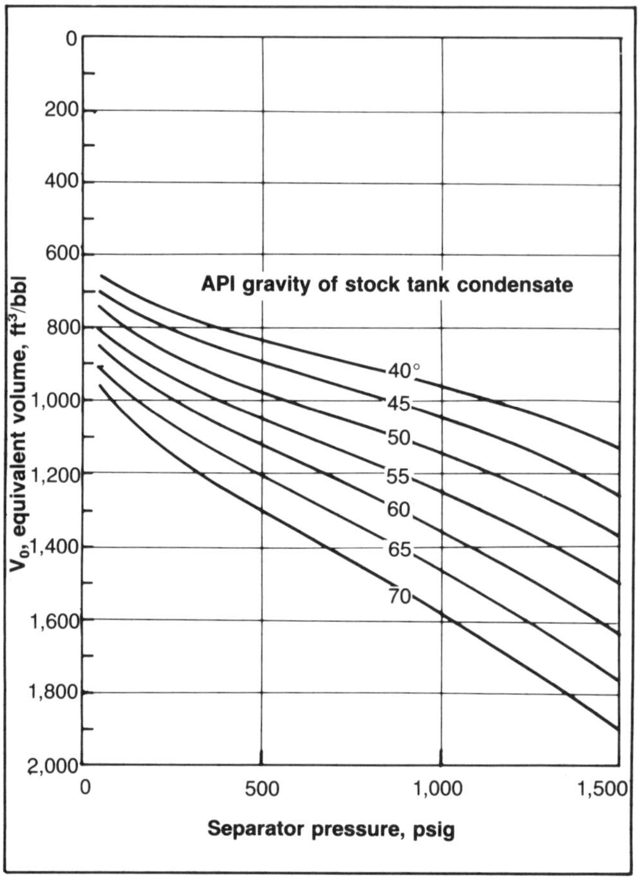

Fig. 2-27. Equivalent gas volume of stock tank condensate. Reprinted from Leshikan, World Oil, January 1961. Courtesy Gulf Publishing Company.

The condensate specific gravity is related to the API gravity by

$$\gamma_o = \frac{141.5}{131.5 + API}$$

V_o is obtained from Figure 2-26 for cases in which γ_o is measured at a high pressure separator. If γ_o is measured at stock tank conditions, the value of V_o depends on separator pressure and is obtained from Figure 2-27.

Example 2-21:

Calculate the specific gravity of a reservoir gas given the following:

γ_g = 0.65 γ_o from S.T. = 0.78 (50° API)

R = 30,000 scf/STB

Separator pressure = 900 psig

Solution:

From Figure 2-27, V_o = 1110 scf/STB

$$\gamma_{gm} = \frac{0.65 + 4584\,(.78)/30000}{1 + 1110/30000} = 0.74$$

Equation 2-46 can be used to calculate the specific gravity of a gas containing water if the dry gas gravity and water content are known. The liquid vaporizing volume for water is

$$V_w = \frac{m\,R\,T_{sc}}{M\,p_{sc}} = \frac{350\,\dfrac{\text{lbm}}{\text{STB}}\,(10.73)(520)}{18\,\dfrac{\text{lb}}{\text{lb-mole}}\,(14.7)}$$

$$V_w = 7380 \text{ scf/STB}.$$

To calculate R_w in scf/STB from the water content in lbm/MMscf, use

$$R_w = \frac{350 \times 10^6}{W_s}.$$

REFERENCES

1. Brown, G. G., Katz, D. L., Oberfell, G. G., and Alden, R. C.: *Natural Gasoline and Volatile Hydrocarbons*, NGAA, (1948).
2. Wichert, E., and Aziz, K.: "Calculate Z's for Sour Gases," *Hydrocarbon Processing*, 51 (1972).
3. Beattie and Bridgeman: *J. Am. Chem. Soc.*, 49 (1927) and 50 (1928).
4. Redlich-Kwong: *Chem. Rev.*, (1949) 44.
5. Soave, G.: "Equilibrium Constants from a Modified Redlich-Kwong Equation of State," *Chem. Eng. Sci.* **27** (1972).
6. Mattar, L., Brar, G. S., and Aziz, K.: "Compressibility of Natural Gases," *J. Can. Pet. Tech.* (Oct.–Dec., 1975).
7. Carr, N. L., Kobayashi, L. R., and Burrows, D. B., "Viscosity of Hydrocarbon Gases," *Trans.* AIME, (1954) **201**.
8. Lee, A. L., Gonzalez, M. H., and Eakin, B. E.: "The Viscosity of Natural Gases," *J. Pet. Tech.* (Aug., 1966).
9. *NGPSA Engineering Data Book*, Gas Processors Suppliers Assoc., Tulsa (1972).
10. Weber, J. H.,: "Predicting Properties of Gas Mixtures," *Chemical Eng.* (May 19, 1980).

3 Gas Reservoir Performance

IN traveling from its original location in the reservoir to the final point of consumption, the gas must first travel through the reservoir rock or porous medium. A certain amount of energy is required to overcome the resistance to flow through the rock, which is manifested in a pressure decrease in the direction of flow, toward the well. This pressure drop or decrease depends on the gas flow rate, properties of the reservoir fluids, and properties of the rock. The fluid properties were discussed in Chapter 2, and a brief discussion of the rock properties is given in this chapter.

The engineer involved in gas production operations must be able to predict not only the rate at which a well or field will produce, but also how much gas is originally in the reservoir and how much of it can be recovered economically. This requires the ability to relate volumes of gas existing in the reservoir to reservoir pressure. Because the flow capacity of a well depends on the reservoir pressure, both reservoir gas flow and reserve estimates are discussed in this chapter.

RESERVOIR GAS FLOW

Determination of the inflow performance or reservoir flow capacity for a gas well requires a relationship between flow rate coming into a well and the sand-face pressure or flowing bottom-hole pressure. This relationship may be established by the proper solution of Darcy's Law, which is the accepted expression relating pressure drop and fluid velocity in a porous medium, provided that the flow is laminar. Solution of Darcy's Law depends on the conditions of flow existing in the reservoir or the flow regime. The flow type or regime may be independent of time or steady-state, or if conditions at a particular location change with time, the flow regime is transient or unsteady-state. Under certain conditions of transient flow, conditions change at a constant rate at all locations in the reservoir. This condition is called pseudo-steady-state and may be analyzed more simply than the transient condition.

The flow regimes will be discussed qualitatively first, the equations for each regime will be presented, and then the application of the equations for determining inflow performance or well flow capacity will be presented.

Flow Regime Characteristics

When a well is opened to production from a shut-in condition, the pressure disturbance created at the well travels outward through the rock at a velocity governed by the rock and fluid properties. The various flow regimes are discussed with respect to the behavior of this pressure disturbance.

Steady-State Flow. Figure 3-1 illustrates the pressure and flow rate distribution occurring during radial, steady-state flow into a well. This pressure distribution will remain constant as long as the radius being drained by the well remains constant.

For such a situation to be strictly true it is necessary that the flow across the external drainage radius r_e, be equal to the flow across the well radius at r_w. This is never strictly met in a reservoir other than a strong water drive, whereby the water influx rate equals the producing rate. Pressure maintainance by water injection downdip or by gas injection updip would also approximate steady-state conditions as would most pattern waterfloods after the initial stages of injection have passed.

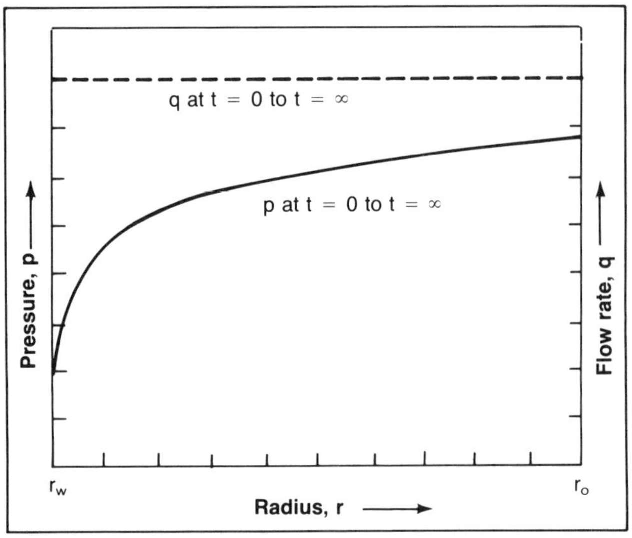

Fig. 3-1. Radial steady-state flow. From Slider, Practical Reservoir Engineering Methods, copyright 1976, PennWell Books.

Steady-state equations are also useful in analyzing the conditions near the wellbore because even in an unsteady-state system the flow rate near the wellbore is almost constant so that the conditions around the wellbore are almost constant. Thus, steady-state flow equations can be applied to this portion of the reservoir without any significant error.

Unsteady-State Flow. Figure 3-2 shows the pressure and rate distributions for a radial system at various times for a closed reservoir (no flow across r_e). In this case all of the production is due to the expansion of the fluid in the reservoir. This causes the rate at r_e to be zero and the rate increases to a maximum at the well radius, r_w. In the steady-state case the flow across the outer boundary, r_e, was equal to the flow across r_w, the well radius. With flow across r_e zero, the only energy causing the flow of fluid is the expansion of the fluids themselves. Initially the pressure is uniform throughout the reservoir at p_i. This represents the zero producing time.

The production rate is controlled so that the pressure at the well is constant. A pressure distribution shown as p at t_1 is obtained after a short period of time of producing the well at such a rate that the well pressure remains constant. At this time only a small portion of the reservoir has been affected or has had a significant pressure drop.

For a closed reservoir, flow occurs due to expansion of the fluid. Consequently, if no pressure drop exists in the reservoir at a particular point, or outside of that point, no flow could be taking place at that particular radius. The fluid could not expand without a drop in pressure. Thus, as shown in the plot of q at t_1 the rate at r_e is zero and increases with a reduction in radius until the maximum rate in the reservoir is obtained at r_w. The pressure and rate distributions at time t_1 represent an instant in time, and the pressure and rate distributions move on through these positions immediately as the production continues to affect more and more of the reservoir. That is, more and more of the reservoir continues to experience a significant pressure drop and is subjected to flow until the entire reservoir is affected as shown by the pressure at t_2. The rate, q, at t_2 indicates that the flow rate at this time extends throughout the reservoir since all of the reservoir has been affected and has had a significant pressure drop.

Notice that the rate at the well has declined somewhat from time t_1 to t_2 since the same pressure drop $(p_i - p_w)$, is effective over a much larger volume of the reservoir. Once the pressure in the entire reservoir has been affected the pressure will drop throughout the reservoir as production continues so that the pressure distribution might be as shown for p at t_3. The rate will have declined somewhat during time t_1 to t_2 due to the increase in the radius over which flow is taking place, and it will continue to decline from t_2 to t_3 due to the decline of the total pressure drop from r_e to r_w, $(p_e - p_w)$.

Note that from time $t = 0$ to time t_2, when a pressure drop is finally affected throughout the entire reservoir, the pressure and rate distributions would not be affected by the size of the reservoir or the position of the external drainage radius r_e. During this time the reservoir is said to be infinite-acting because during this period the outer drainage radius, r_e, could mathematically be infinite. Even in reservoir systems that are dominated by steady-state

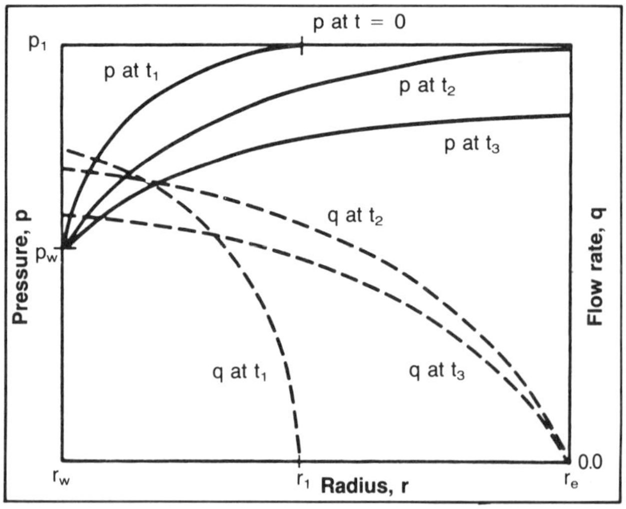

Fig. 3-2. Unsteady-state radial flow with constant well pressure. From Slider, Practical Reservoir Engineering Methods, copyright 1976, PennWell Books.

flow, the effect of changes in well rates or well pressures at the well will be governed by unsteady-state flow equations until the changes have been in effect for a sufficient length of time to affect the entire reservoir and have the reservoir again reach a steady-state condition.

Pseudo-Steady-State Flow. Figure 3-3 illustrates the pressure and rate distribution for the same unsteady-state system except that in this case the rate at the well, q_w, is held constant. This might be comparable to a prorated well or one that is pumping at a constant rate. Again at time $t = 0$ the pressure throughout the reservoir is uniform at p_i. Then after some short producing time t_1 at a constant rate, only a small portion of the reservoir will have experienced a significant pressure drop, and consequently the reservoir will be flowing only out to a radius r_1. As production continues at the constant rate, the entire reservoir will eventually experience a significant pressure drop as shown at t_2. Shortly after the entire reservoir pressure has been affected, the change in the pressure with time at all radii in the reservoir becomes uniform so that the pressure distributions at subsequent times are parallel as illustrated by the pressure distributions at times t_3, t_4, and t_5. This situation will continue with constant changes in pressure with time at all radii and with subsequent parallel pressure distributions until the reservoir is no longer able to sustain a constant flow rate at the wellbore. This will occur when the pressure at the well, p_w, has reached its physical lower limit.

Pseudo-steady-state flow occurs in the reservoir after it has been produced at a constant rate for a long enough period of time to cause a constant change in pressure at all radii, resulting in parallel pressure distributions and corresponding constant rate distributions. Pseudo-steady-state flow is a specialized case of unsteady-state flow and is sometimes referred to as stabilized flow. Most of the life of a reservoir will exist in pseudo-steady-state flow.

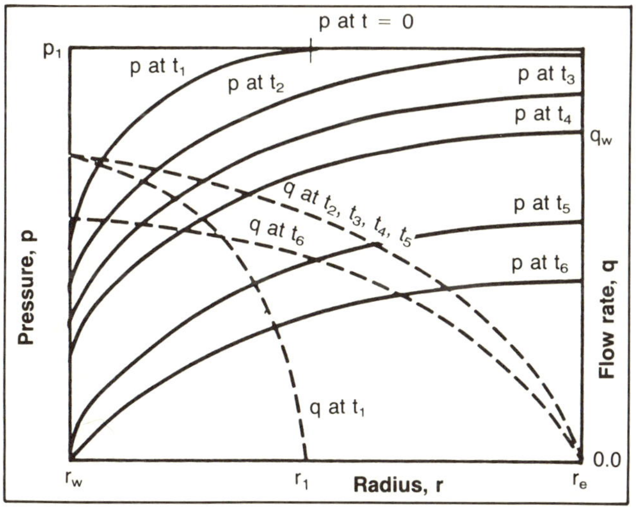

Fig. 3-3. Unsteady-state radial flow with constant producing rate pseudo-steady-state t_2 to t_5. From Slider, Practical Reservoir Engineering Methods, copyright 1976, PennWell Books.

Flow Equations

From the previous description of the various flow regimes it is obvious that a particular well will be operating in each of these regimes at some time in the life of the well. The applicable equations for each flow regime will be derived or presented in this section.

Steady-State Flow

Darcy's Law for flow in a porous medium is

$$v = -\frac{k}{\mu}\frac{dp}{dx} \quad \text{or,}$$

$$q = vA = -\frac{kA}{\mu}\frac{dp}{dx} \tag{3-1}$$

where

v = fluid velocity,

q = volumetric flow rate,

k = effective permeability,

μ = fluid viscosity, and

$\frac{dp}{dx}$ = pressure gradient in the direction of flow.

For radial flow in which the distance is defined as positive moving away from the well, the equation becomes

$$q = \frac{k(2\pi rh)}{\mu}\frac{dp}{dr} \tag{3-2}$$

where

r = radial distance, and

h = reservoir thickness.

Darcy's Law describes the pressure loss due to viscous shear occurring in the flowing fluid. If the formation is not horizontal, the hydrostatic or potential energy term must be included. This is usually negligible for gas flow in reservoirs. Equation 3-2 is a differential equation and must be integrated for application. Before integration the flow equation must be combined with an equation of state and the continuity equation. The continuity equation is

$$\rho_1 q_1 = \rho_2 q_2 = \text{constant} \tag{3-3}$$

From Chapter 2, the equation of state for a real gas is

$$\rho = \frac{pM}{ZRT}. \quad (3\text{-}4)$$

The flow rate for a gas is usually desired at some standard conditions of pressure and temperature, p_{sc} and T_{sc}. Using these conditions in Equation 3-3 and combining Equations 3-3 and 3-4:

$$\rho q = \rho_{sc}\, q_{sc},$$

or

$$q\frac{pM}{ZRT} = q_{sc}\frac{p_{sc}M}{Z_{sc}RT_{sc}}.$$

Solving for q_{sc} and expressing q with Equation 3-2 gives

$$q_{sc} = \frac{p\,T_{sc}}{p_{sc}\,Z\,T}\,\frac{2\pi rhk}{\mu}\,\frac{dp}{dr}$$

The variables in this equation are p and r. Separating the variables and integrating:

$$\int_{p_w}^{p_e} p\,dp = \frac{q_{sc}\,p_{sc}\,T\,\bar{\mu}\,\bar{Z}}{T_{sc}\,2\pi kh}\int_{r_w}^{r_e}\frac{dr}{r}$$

$$\frac{p_e^2 - p_w^2}{2} = \frac{q_{sc}\,p_{sc}\,T\,\bar{\mu}\,\bar{Z}}{T_{sc}\,2\pi kh}\ln(r_e/r_w)$$

or

$$q_{sc} = \frac{\pi kh\,T_{sc}\,(p_e^2 - p_w^2)}{p_{sc}\,T\,\bar{\mu}\,\bar{Z}\ln(r_e/r_w)}. \quad (3\text{-}5)$$

In this derivation it was assumed that μ and Z were independent of pressure. They may be evaluated at reservoir temperature and average pressure in the drainage area,

$$\bar{p} = \frac{p_e + p_w}{2}.$$

Equation 3-5 is applicable for any consistent set of units. In so-called conventional oil field units the equation becomes

$$q_{sc} = \frac{703 \times 10^{-6}kh\,(p_e^2 - p_w^2)}{T\,\bar{\mu}\,\bar{Z}\ln(r_e/r_w)} \quad (3\text{-}6)$$

where

q_{sc} = Mscf/day,
k = permeability in millidarcies,
h = formation thickness in feet,
p_e = pressure at r_e, psia,
p_w = wellbore pressure at r_w, psia, and
$\bar{\mu}$ = gas viscosity, *cp*.

This equation incorporates the following values for standard pressure and temperature:

$$p_{sc} = 14.7 \text{ psia},$$

$$T_{sc} = 60°\text{F} = 520°\text{R}.$$

These units will be used in all equations in the text unless otherwise stated.

Example 3-1:

Given the following data, determine the wellbore pressure required for an inflow rate of 3900 Mscfd assuming steady-state flow.

k = 1.5 md	h = 30 ft
p_e = 4625 psia	$\bar{\mu}$ = 0.027 cp
T = 712°R = 252°F	γ_g = 0.76
r_e = 550 ft	r_w = 0.333 ft

Solution:

The solution is iterative since

$$\bar{Z} = f(\bar{p}), \text{ where } \bar{p} = \frac{p_e + p_w}{2}, \text{ and}$$

p_w is the unknown. As a first estimate, assume $\bar{Z}$ = 1.0.

First Trial

$$p_w^2 = p_e^2 - \frac{\bar{\mu}\,T\ln(r_e/r_w)\,q_{sc}\,\bar{Z}}{703 \times 10^{-6}\,kh}$$

$$p_w^2 = 2.139 \times 10^7 - \frac{(0.027)(712)(7.41)(3900)\,\bar{Z}}{703 \times 10^{-6}\,(1.5)(30)}$$

$$p_w^2 = 2.139 \times 10^7 - 1.756 \times 10^7\,\bar{Z}$$

$$= 2.139 \times 10^7 - 1.756 \times 10^7\,(1)$$

$$p_w^2 = 3.83 \times 10^6 \qquad p_w = 1957 \text{ psia}$$

$\bar{p} = \frac{4625 + 1957}{2} = 3291$ psia. Evaluation of $\bar{Z}$ at 3291 psia and 712°R gives $\bar{Z}$ = 0.88

Second Trial

$$p_w^2 = 2.139 \times 10^7 - 1.756 \times 10^7\,(0.88)$$

$$= 5.937 \times 10^6$$

p_w = 2436 psia $\quad \bar{p}$ = 3530 psia

At 3530 psia and 712°R, $\bar{Z}$ = 0.89

Third Trial

$$p_w^2 = 2.139 \times 10^7 - 1.756 \times 10^7\,(0.89)$$

$$= 5.762 \times 10^6$$

p_w = 2400 psia $\quad \bar{p}$ = 3512 psia $\quad \bar{Z}$ = 0.89

Since the value for $\bar{Z}$ is the same as for Trial 2, the solution has converged and the required well pressure is

2400 psia. The solution would have been more complicated if a constant value for μ had not been assumed.

The above treatment of steady-state flow assumes no turbulent flow in the formation and no formation or skin damage around the wellbore. The effects of turbulence and skin will be examined in a following section.

Although steady-state flow in a gas reservoir is seldom reached, the conditions around the wellbore can approach steady-state. The steady-state equation including turbulence is

$$p_e^2 - p_w^2 = \frac{1422\, T\, \bar{\mu}\, \bar{Z}\, q_{sc} \ln(r_e/r_w)}{kh} \cdots$$

$$\cdots + \frac{3.161 \times 10^{-12}\, \beta\, \gamma_g\, \bar{Z}\, q_{sc}^2\, T}{h^2}\left(\frac{1}{r_w} - \frac{1}{r_e}\right). \quad (3\text{-}7)$$

The first term on the right hand side is the pressure drop from laminar or Darcy flow, while the second term gives the additional pressure drop due to turbulence. If the fluid properties are known and the permeability is known from some source such as a drawdown test, the turbulent effects can be calculated using the results of a test. This will be used later to distinguish between actual formation damage and turbulence. Values of the velocity coefficient β for various permeabilities and porosities can be obtained from Figure 3-4[1] or calculated from Equation 3-8.

$$\beta = \frac{2.33 \times 10^{10}}{k^{1.2}} \quad (3\text{-}8)$$

where k is in millidarcies.

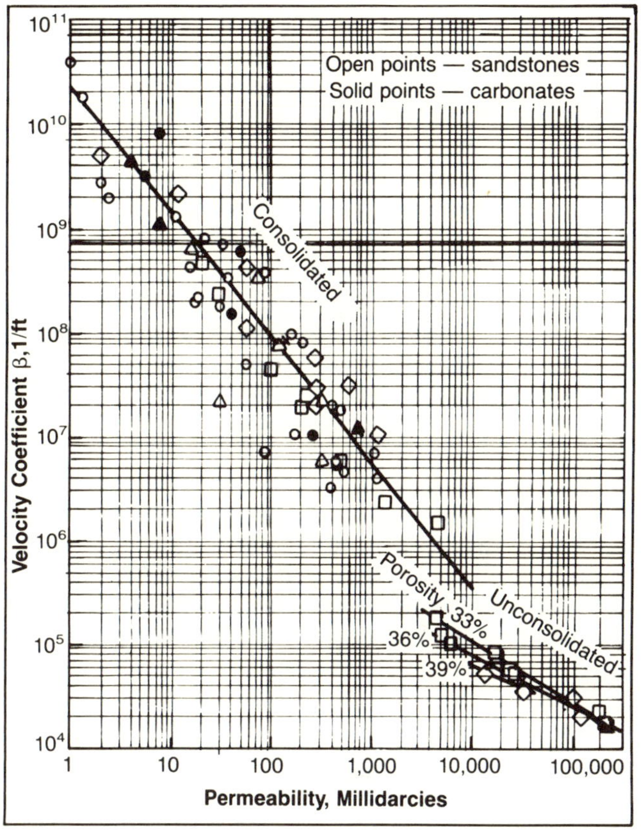

Fig. 3-4. Gas velocity coefficient. Permission to publish by the Society of Petroleum Engineers of AIME. Copyright 1979 SPE-AIME.

Pseudo-Steady-State Flow

An equation for pseudo-steady-state flow can be derived that will show that

$$q_{sc} = \frac{703 \times 10^{-6}\, kh(\bar{p}_R^2 - p_w^2)}{T\, \bar{\mu}\, \bar{Z} \ln(0.472 r_e/r_w)}. \quad (3\text{-}9)$$

Although time does not appear explicitly in Equation 3-9, it should be remembered that both $\bar{p}_R$ and p_w will be declining at the same rate for a constant q once the pressure disturbance has reached the reservoir boundary.

The effects of skin damage and turbulence are sometimes included in Equation 3-9 as follows:

$$q_{sc} = \frac{703 \times 10^{-6}\, kh\,(\bar{p}_R^2 - p_w^2)}{T\, \bar{\mu}\, \bar{Z}\,[\ln(.472 r_e/r_w) + S + D\, q_{sc}]} \quad (3\text{-}10)$$

where

S = dimensionless skin factor, and

D = turbulence coefficient.

It is frequently necessary to solve Equation 3-10 for pressure or pressure drop for a known q_{sc}.

$$\bar{p}_R^2 - p_w^2 = \frac{1422\, T\, \bar{\mu}\, \bar{Z}\, q_{sc}}{kh} \cdot [\ln(.472\, r_e/r_w) + S + D\, q_{sc}] \quad (3\text{-}11)$$

Unsteady-State Flow

It was stated earlier that any well flows in the unsteady-state or transient regime until the pressure disturbance reaches a reservoir boundary or until interference from other wells takes effect. Although the flow capacity of a well is desired for pseudo-steady-state or stabilized conditions, much useful information can be obtained from transient tests. This information includes permeability, skin factor, turbulence coefficient, and average reservoir pressure. The procedures are developed in the section on transient testing. The relationships among flow rate, pressure, and time will be presented in this section for various conditions of well performance and reservoir types. It will be seen that the steady-state and pseudo-steady-state equations can be obtained from solution of the diffusivity equation as special cases.

The diffusivity equation can be derived by combining

TABLE 3-1
Dimensionless Variables

Dimensionless Variable	Symbol	Definition
Time	t_D	$\dfrac{2.64 \times 10^{-4}\, kt}{\phi\, \bar{\mu}\, \bar{C}\, r^2}$ (3-13)
Radius	r_D	$\dfrac{r}{r_w}$ (3-14)
Flow rate	q_D	$\dfrac{1422\, T\, q_{sc}\, \bar{Z}\, \bar{\mu}}{kh\, p_i^2}$ (3-15)
Pressure	p_D	$\dfrac{p^2}{p_i^2\, q_D}$ (3-16)
Pressure drop	Δp_D	$\dfrac{p_i^2 - p^2}{p_i^2\, q_D}$ (3-17)

an unsteady-state continuity equation with Darcy's Law and the gas equation of state. The equation is

$$\frac{\partial^2 p^2}{\partial r^2} + \frac{1}{r}\frac{\partial p^2}{\partial r} = \frac{\phi\, \mu\, C}{k}\frac{\partial p^2}{\partial t} \tag{3-12}$$

This equation can be solved for pressure as a function of flow rate and time, but the solutions and application of the solutions are simplified if the diffusivity equation is written in dimensionless form. This is accomplished by defining the following dimensionless variables:

The following units are to be used in calculating values for the dimensionless numbers in Table 3-1:

k = millidarcies C = psi^{-1}
t = hours r = ft
μ = cp T = °R
q_{sc} = Mscfd h = ft
p = psia

The diffusivity equation in dimensionless variables becomes

$$\frac{\partial^2 \Delta p_D}{\partial r_D^2} + \frac{1}{r_D}\frac{\partial \Delta p_D}{\partial r_D} = \frac{\partial \Delta p_D}{\partial t_D}. \tag{3-18}$$

Solutions to Equation 3-18 depend on the reservoir type and boundary conditions. The following solutions will be presented:

1. Constant rate at well, infinite-acting reservoir (transient)
2. Constant rate at well, finite-acting (closed) reservoir (pseudo-steady-state)
3. Constant rate at well, constant pressure at outer boundary (steady-state)
4. Constant well pressure

Case 1

The most useful solution for transient flow is the so-called line source solution. The solution is

$$\Delta p_D = -0.5\, E_i\left(-\frac{1}{4t_D}\right). \tag{3-19}$$

Values for the E_i or exponential integral term as a function of t_D can be found in various mathematics handbooks, but for all practical purposes the function may be represented by a logarithmic approximation. That is,

$$\Delta p_D = 0.5\,(\ln t_D + 0.809) \tag{3-20}$$

Once a value of the dimensionless pressure drop Δp_D is obtained, the actual pressures may be calculated by using the definition of Δp_D from Table 3-1.

Example 3-2:

Using the following data and assuming the well is still in the transient regime, calculate the pressure at the well after a flowing time of 1.5 days.

h = 36 ft T = 580°R
ϕ = 0.15 q_{sc} = 7000 Mscfd
k = 20 md $\bar{Z}$ = 0.85
p_i = 2000 psia $\bar{\mu}$ = 0.0152 cp
r_w = 0.4 ft $\bar{C}$ = 0.00061 psi^{-1}
r_e = 2000 ft

Solution:

Calculate t_D for $r = r_w$:

$$t_{Dw} = \frac{2.64 \times 10^{-4}\, kt}{\phi\, \bar{\mu}\, \bar{C}\, r_w^2}$$

$$= \frac{2.64 \times 10^{-4}\,(20)(36)}{0.15\,(0.0152)(0.00061)(0.4)^2}$$

$$t_{Dw} = 8.54 \times 10^5$$

$$\Delta p_D = 0.5\,(\ln (8.54 \times 10^5) + 0.809) = 7.23$$

$$q_D = \frac{1422\, T\, q_{sc}\, \bar{Z}\, \bar{\mu}}{kh\, p_i^2}$$

$$= \frac{1422(580)(7000)(0.85)(0.0152)}{20\,(36)(2000)^2}$$

$$q_D = 0.0259$$

$$p_w^2 = p_i^2 - p_i^2\, q_D\, \Delta p_D$$

$$= (2000)^2 - (2000)^2\,(0.0259)(7.23)$$

$$p_w^2 = 3250972 \qquad p_w = 1803 \text{ psia}$$

Equation 3-20 applies for values of dimensionless time based on the well's drainage radius, t_{De} less than 0.25. That is, the well will still be infinite-acting if

$$t_{De} = \frac{2.64 \times 10^{-4}\, kt}{\phi\, \bar{\mu}\, \bar{C}\, r_e^2} < 0.25 \qquad (3\text{-}21)$$

Another restriction on the validity of Equation 3-20 is that t_D should be greater than 100. If t_D is less than 100, the E_i solution (Equation 3-19) must be used. For most practical cases t_D will be greater than 100.

Equation 3-20 may also be used to calculate pressure at a location other than the well. That is, r need not always be r_w. For the solution to be valid, the dimensionless time based on the radius of interest must be greater than 100. That is,

$$t_D = \frac{2.64 \times 10^{-4}\, kt}{\phi\, \bar{\mu}\, \bar{C}\, r^2} > 100. \qquad (3\text{-}22)$$

Case 2

The solution for wells that have reached pseudo-steady-state was presented by Van Everdingen and Hurst[2] in 1949. The solution can be applied to calculate the pressure at any radius where the flow rate is known, which effectively limits its application to calculating well pressures. The solution is presented both in graphical form and equation form. Values of Δp_D versus t_{Dw} are presented in Figure 3-5 for various reservoir sizes, that is for various values of r_{De}. The equation form of the solution is

$$\Delta p_{Dw} = \frac{2t_{Dw}}{r_{De}^2} + \ln(0.472\, r_{De}) \qquad (3\text{-}23)$$

Example 3-3:

Using the data given in Example 3-2, find the pressure at the well after the well has been flowing for 1800 hours.

Solution:

Check for finite or pseudo-steady-state validity:

$$t_{De} = \frac{2.64 \times 10^{-4}\, kt}{\phi\, \bar{\mu}\, \bar{C}\, r_e^2}$$

$$= \frac{2.64 \times 10^{-4}\,(20)(1800)}{0.15(0.0152)(0.00061)(2000)^2} = 1.71$$

Since $t_{De} > 0.25$, the well is finite acting.

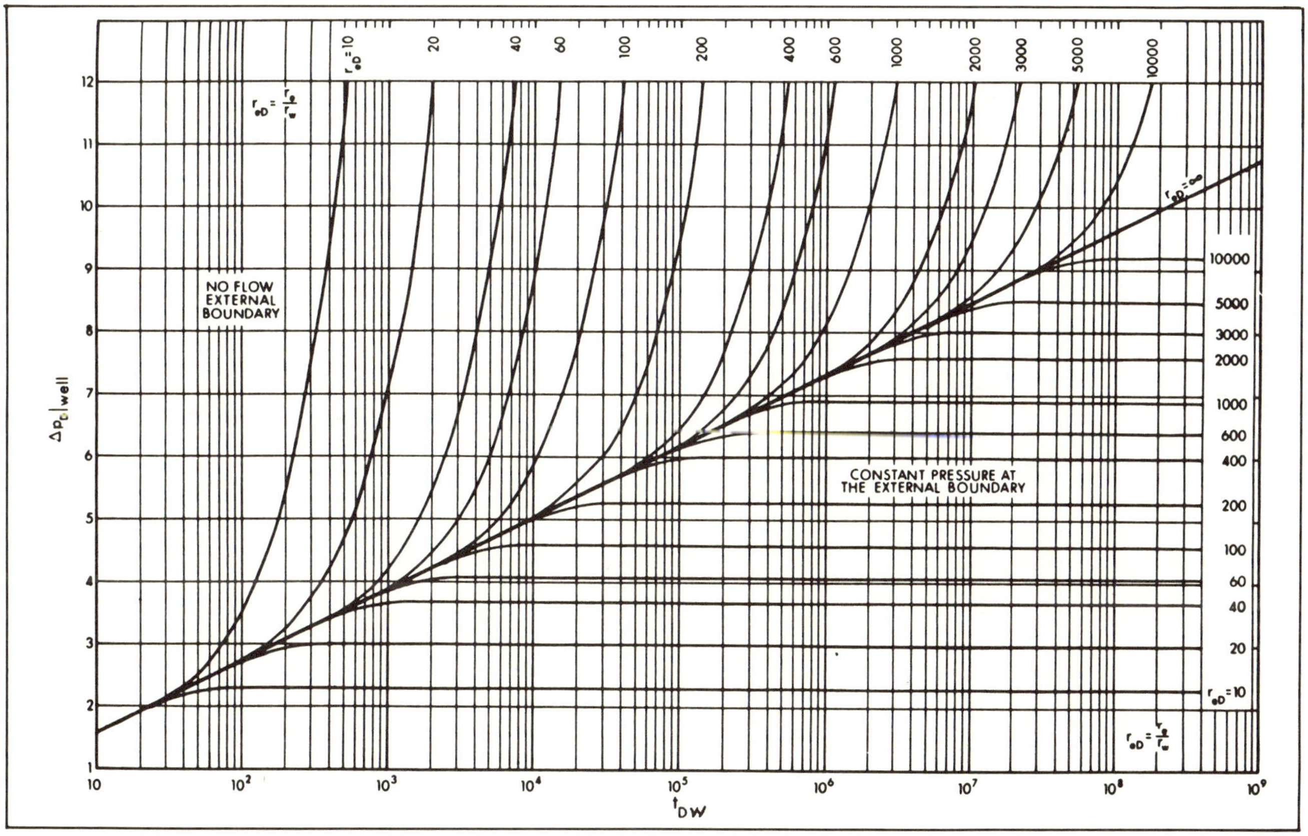

Fig. 3-5. Values of Δp_D/well for infinite reservoirs, for finite circular reservoirs with no flow at the external boundary, and for finite circular reservoirs with constant pressure at the external boundary. After Aziz and Flock (1963). Courtesy The Journal of Canadian Petroleum Technology, Canadian Institute of Mining and Metallurgy.

$$t_{Dw} = \frac{t_{De}\, r_e^2}{r_w^2} = 4.271 \times 10^7 \ (>100)$$

$$r_{De} = \frac{r_e}{r_w} = \frac{2000}{0.4} = 5000$$

$$\Delta p_{Dw} = \frac{2t_{Dw}}{r_{De}^2} + \ln(0.472\, r_{De})$$

$$= \frac{2\,(4.271 \times 10^7)}{(5000)^2} + \ln(0.472\,(5000))$$

$$= 3.417 + 7.766$$

$$\Delta p_{Dw} = 11.183$$

From Example 3-2, $\quad q_D = 0.0259$

$$p_w^2 = p_i^2\,(1 - q_D\,\Delta\, p_{Dw})$$

$$= (2000)^2\,(1 - 0.0259\,(11.183))$$

$$p_w^2 = 2.841 \times 10^6 \qquad p_w = 1686 \text{ psia}$$

Case 3

If a well is producing from a reservoir with constant pressure at the outer boundary, that is, in the steady-state condition, the solution to the diffusivity equation is

$$\Delta p_D = \ln r_{De}. \tag{3-24}$$

In this case the fixed pressure used in the definition of Δp_D is p_e rather than p_i, where p_e is the constant pressure at the outer boundary. Substituting the definitions of the dimensionless variables will result in Equation 3-6 presented earlier for steady-state flow.

Case 4

The constant pressure solution to the diffusivity equation can be expressed as a function of a dimensionless cumulative production, Q_{tD}. Q_{tD} is defined as

$$Q_{tD} = \frac{G_p\, \bar{Z}\, T}{0.111\, \phi\, h\, r_w^2\, C\,(p_i^2 - p_w^2)}, \tag{3-25}$$

where G_P is the cumulative gas produced in Mscf. Values of Q_{tD} as a function of dimensionless time and radius have been presented both graphically and in table form. Figures 3-6 and 3-7 show values of Q_{tD} versus t_D.

Noncircular Reservoirs

All of the previous equations were based on a single well draining a circular reservoir, which is rarely ever

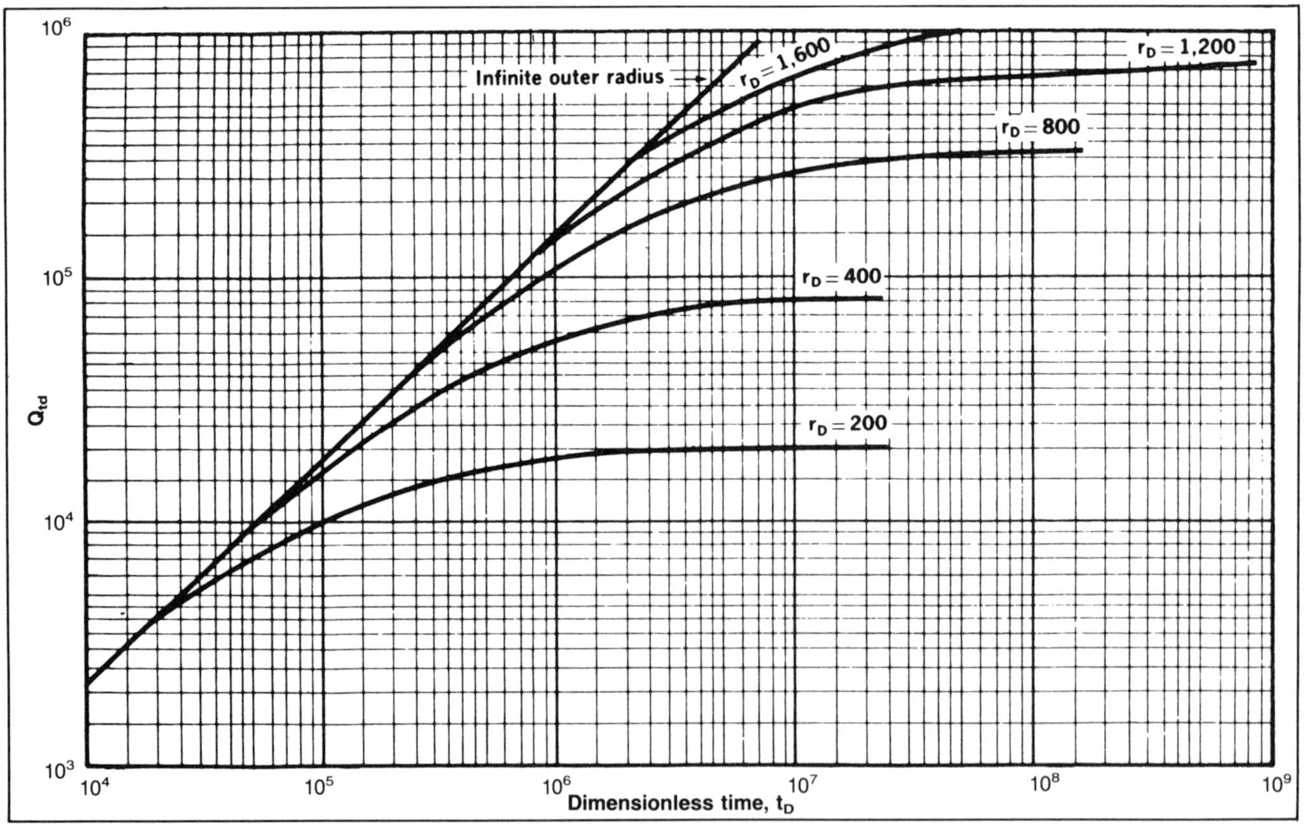

Fig. 3-6. Constant pressure functions. From Slider, Practical Reservoir Engineering Methods, copyright 1976, PennWell Books.

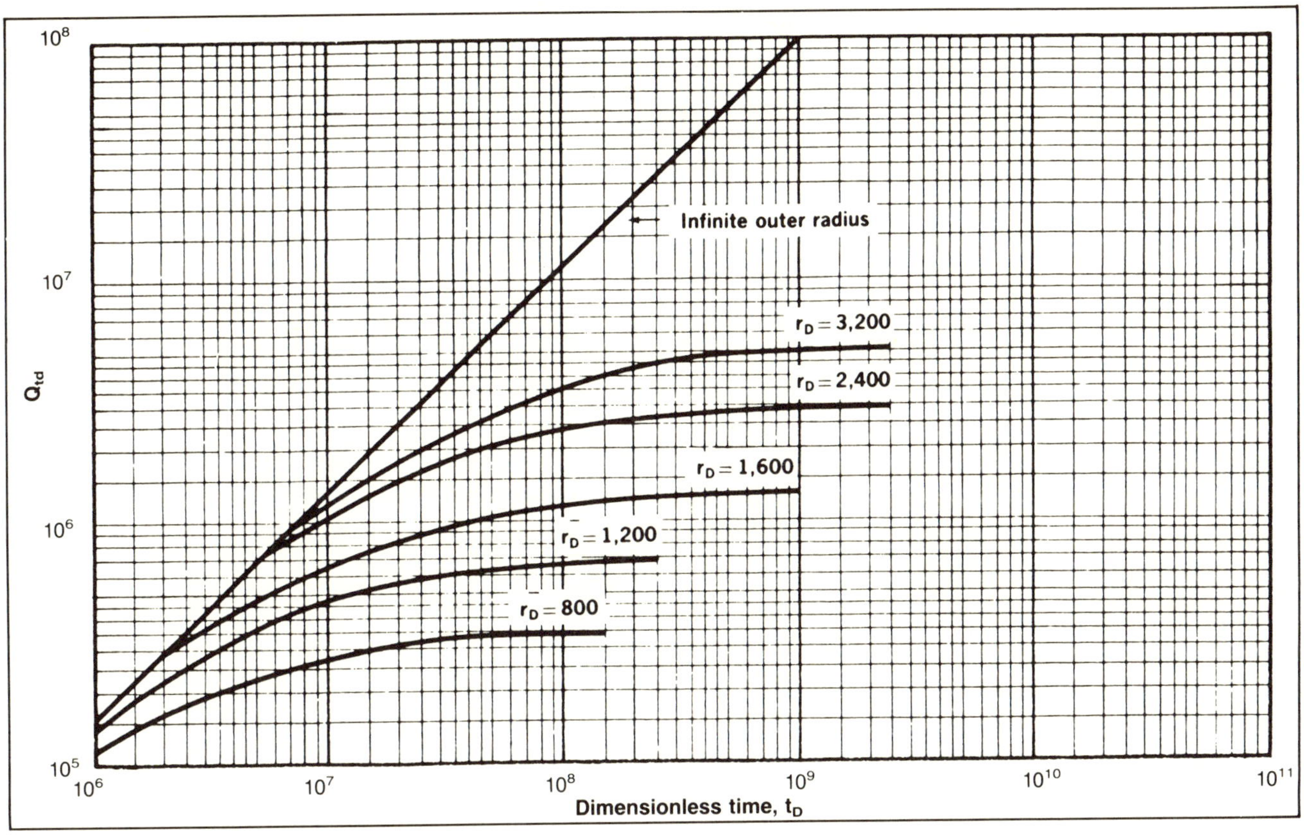

Fig. 3-7. Constant pressure functions. From Slider, Practical Reservoir Engineering Methods, copyright 1976, PennWell Books.

the actual case. The pressure behavior depends on the shape of the reservoir and the location of the well relative to the boundaries. The time to reach stabilized or pseudo-steady-state flow also depends on these factors.

The previous equations have been modified by Dietz[4] as follows. If the well is still infinite-acting,

$$\Delta p_D = 0.5 \left[\ln \left(\frac{A\, t_{DA}}{r_w^2} \right) + .809 \right], \qquad (3\text{-}26)$$

where A is the drainage area. For pseudo-steady-state,

$$\Delta p_D = 0.5 \left[\ln \left(\frac{A}{r_w^2 C_A} \right) + .809 \right] + 2\pi\, t_{DA}. \qquad (3\text{-}27)$$

Calculation of t_{DA} is based on the drainage area and is defined as

$$t_{DA} = \frac{2.64 \times 10^{-4}\, kt}{\phi\, \bar{\mu}\, \bar{C}\, A}. \qquad (3\text{-}28)$$

The shape factor C_A, as well as the value of t_{DA} required to reach pseudo-steady-state flow can be obtained from Table 3-2.

Example 3-3a:

Rework Example 3-3 if the well is located in the center of a rectangular shaped (1 × 4) drainage area containing 220 acres.
(9.58×10^6 ft^2).

Solution:

Check for pseudo-steady-state:

$$t_{DA} = \frac{2.64 \times 10^{-4}\,(20)(1800)}{0.15\,(0.0152)(0.00061)(9.58 \times 10^6)} = 0.71$$

Since $t_{DA} > 0.7$, the well is in pseudo-steady-state flow. From Table 3-2, $C_A = 5.38$ and

$$\Delta p_D = 0.5 \left[\left(\ln \frac{9.58 \times 10^6}{(0.4)^2\,(5.38)} \right) + .809 \right] + 2\pi(0.71)$$

$$\Delta p_D = 8.517 + 4.461 = 12.978$$

$$p_w^2 = p_i^2\,(1 - q_D\, \Delta\, p_D) = (2000)^2\,(1 - 0.0259(12.978))$$

$$p_w^2 = 2.655 \times 10^6 \qquad p_w = 1630 \text{ psia}$$

Rock Permeability

All of the equations for reservoir gas flow contain rock and fluid properties that must be known in order to be

TABLE 3-2

PSEUDO-STEADY STATE SHAPE FACTORS FOR VARIOUS RESERVOIRS

	$\ln C_A$	C_A	STABILIZED CONDITIONS FOR $t_{DA}>$		$\ln C_A$	C_A	STABILIZED CONDITIONS FOR $t_{DA}>$
IN BOUNDED RESERVOIRS							
	3.45	31.6	0.1	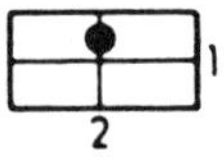	2.38	10.8	0.3
	3.43	30.9	0.1	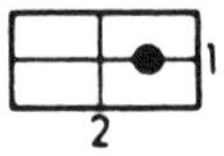	1.58	4.86	1.0
	3.45	31.6	0.1	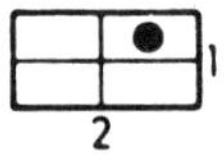	0.73	2.07	0.8
	3.32	27.6	0.2	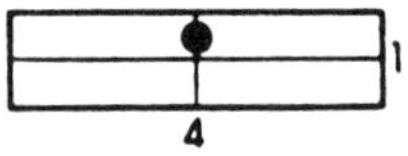	1.00	2.72	0.8
	3.30	27.1	0.2	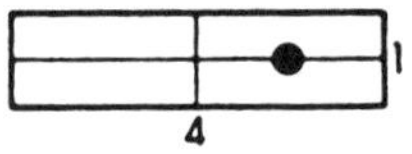	−1.46	0.232	2.5
	3.09	21.9	0.4	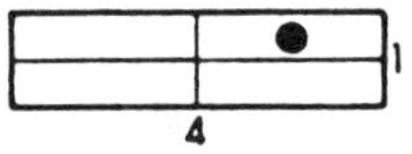	−2.16	0.115	3.0
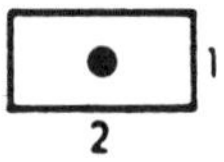	3.12	22.6	0.2		1.22	3.39	0.6
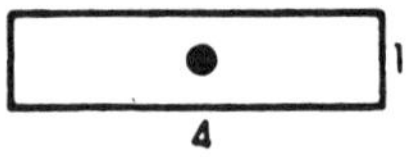	1.68	5.38	0.7		1.14	3.13	0.3
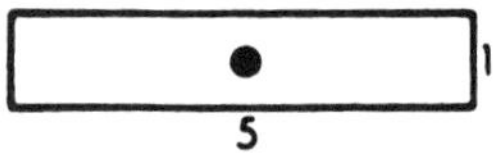	0.86	2.36	0.7	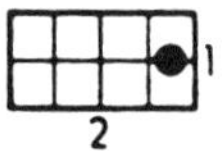	−0.50	0.607	1.0
	2.56	12.9	0.6	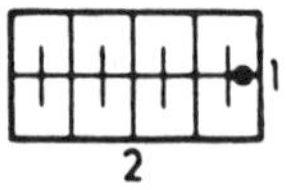	−2.20	0.111	1.2
	1.52	4.57	0.5	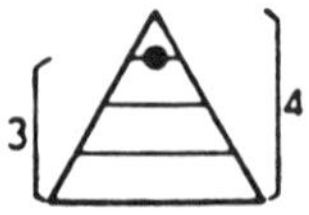	−2.32	0.098	0.9
				IN WATER-DRIVE RESERVOIRS			
					2.95	19.1	0.1
				IN RESERVOIRS OF UNKNOWN PRODUCTION CHARACTER			
					3.22	25	0.1

applied. The fluid properties were presented in Chapter 2. The only rock property involved in the equations is permeability. In many cases the flow equations will be used to determine permeability by measuring pressures and flow rates. This will be discussed in detail in a subsequent section.

A brief review of the behavior of effective or relative permeability when more than one fluid is present in the rock is given. In most dry gas reservoirs only gas will be flowing even though connate water will be present. In this case the permeability to gas will remain fairly constant over the life of the reservoir. If liquid condensation occurs in the reservoir the permeability to gas will decrease. Condensation of water vapor can also occur near the wellbore in some cases, which will also reduce the permeability to gas in this zone.

The absolute permeability of a rock is defined as a measure of the rock's ability to transmit a fluid when it is completely saturated with the flowing fluid. The effective permeability to a particular fluid is a measure of the ability of the rock to transmit that fluid in the presence of other fluids. The effective permeability to gas must be used in all of the flow equations presented earlier.

Permeability data are frequently presented as relative permeability. This is the ratio of effective to absolute permeability and therefore varies from zero to one.

Figure 3-8 presents typical oil and gas relative permeability as a function of liquid saturation. Similar data can be developed for any reservoir and fluid system by

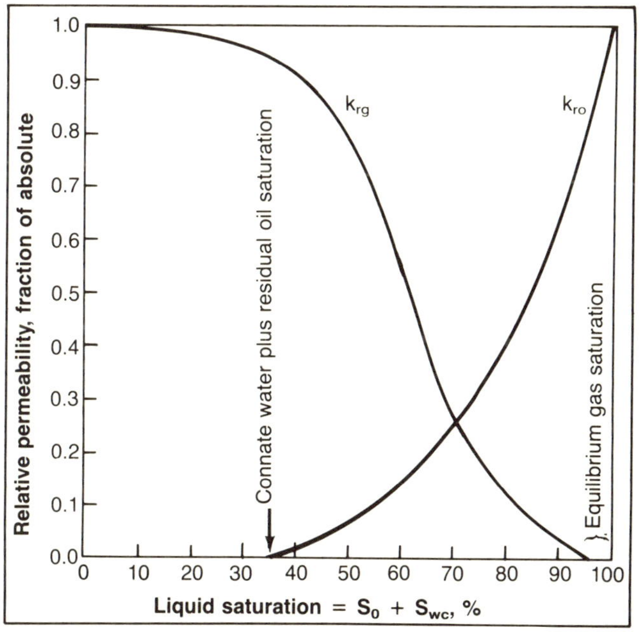

Fig. 3-8. Gas-oil relative permeability data. From Slider, Practical Reservoir Engineering Methods, copyright 1976, PennWell Books.

core analysis. This figure illustrates that the ability of the gas to flow is not seriously impaired until liquid saturation builds up to the point at which liquid begins to flow, about 35% in this case. It can also be seen that there is a gas saturation below which gas will not flow. This is the equilibrium or critical gas saturation and cannot be recovered.

Well Deliverability or Capacity

The stabilized flow capacity or deliverability of a gas well is required for planning the operation of any gas field. The flow capacity must be determined for different back pressures or flowing bottom-hole pressures at any time in the life of the reservoir and the change of flow capacity with average reservoir pressure change must be considered. The flow equations developed earlier are used in deliverability testing with some of the unknown parameters being evaluated empirically from well tests.

The most common method for determining gas well deliverability is called multipoint testing, in which a well is produced at several different rates (usually four) and from measured flow rates and well pressures, an inflow performance equation can be evaluated. There are basically two types of tests that can be conducted: the flow-after-flow test and the isochronal test. The isochronal test has been modified for very tight wells, but the modification cannot be theoretically justified.

The theory behind the tests is based on the pseudo-steady-state flow equation, Equation 3-11. Equation 3-11 may be written as follows:

$$\bar{p}_R^{\,2} - p_{wf}^{\,2} = Aq_{sc} + Bq_{sc}^{\,2} \tag{3-11}$$

where

$$A = \frac{1422\,T\,\bar{\mu}\,\bar{Z}}{kh}\left(\ln\left(\frac{.472\,r_e}{r_w}\right) + S\right), \text{ and}$$

$$B = \frac{1422\,T\,\bar{\mu}\,\bar{Z}}{kh}\,D.$$

It is sometimes convenient to establish a relationship between the two parameters that indicate the degree of turbulence occurring in a gas reservoir. These parameters are the velocity coefficient and the Turbulence Coefficient D. Equation 3-7 can be written for pseudo-steady-state flow as

$$\bar{p}_R^{\,2} - p_{wf}^{\,2} = \frac{1422\,T\,\bar{\mu}\,\bar{Z}}{kh}\left(\ln\frac{.472\,r_e}{r_w} + S\right)q_{sc} + \frac{3.161 \times 10^{-12}\,\gamma_g\,\bar{Z}\,T\,\beta}{r_w h^2}\,q_{sc}^{\,2} \tag{3-29}$$

This form of the equation includes the assumption that $r_e \gg r_w$. Equating the terms multiplying $q_{sc}^{\,2}$ in Equations 3-11 and 3-29 yields

$$B = \frac{1422\, T\, \bar{\mu}\, \bar{Z}}{kh} D = \frac{3.161 \times 10^{-12}\, \gamma_g\, \bar{Z}\, T}{r_w h^2} \beta$$

or

$$D = \frac{2.22 \times 10^{-15}\, \gamma_g\, k}{\bar{\mu}\, h\, r_w} \beta$$

Expressing β in terms of permeability from Equation 3-8, the above expression becomes

$$D = \frac{5.18 \times 10^{-5}\, \gamma_g}{\bar{\mu}\, h\, r_w\, k^{0.2}} \tag{3-30}$$

Equation 3-11 can also be written in the form

$$q_{sc} = C\,(\bar{p}_R^{\,2} - p_{wf}^{\,2})^n. \tag{3-31}$$

For wells in which turbulence is important the value of n approaches 0.5, while for wells in which turbulence is negligible, n approaches 1.0. In most cases the value of n obtained from well tests will fall between 0.5 and 1.0.

If values for the flow coefficient C and exponent n can be determined, the flow rate corresponding to any value of p_{wf} can be calculated and an inflow performance curve can be constructed. A parameter commonly used to characterize or compare gas wells is the flow rate that would occur if p_{wf} could be brought to zero. This is called the Absolute Open Flow Potential, or AOF.

Examination of Equation 3-31 reveals that a plot of $\Delta(p^2) = \bar{p}_R^{\,2} - p_{wf}^{\,2}$ versus q_{sc} on log-log scales should result in a straight line having a slope of $1/n$. At a value of $\Delta(p^2)$ equal to one, $C = q_{sc}$. This is made evident by taking the log of both sides of Equation 3-31.

$$\log(\bar{p}_R^{\,2} - p_{wf}^{\,2}) = \frac{1}{n}\log q_{sc} - \frac{1}{n}\log C$$

Once a value of n has been determined from the plot, the value of C can be calculated by using data from one of the tests that falls on the line. That is,

$$C = \frac{q_{sc}}{(\bar{p}_R^{\,2} - p_{wf}^{\,2})^n}.$$

Flow-After-Flow Tests

A flow-after-flow test starts from a shut-in condition. The well is opened on a particular choke size and is not disturbed until the flow rate and p_{wf} stabilize. This may require a considerable amount of time, depending on the permeability of the reservoir. A well is usually consid-

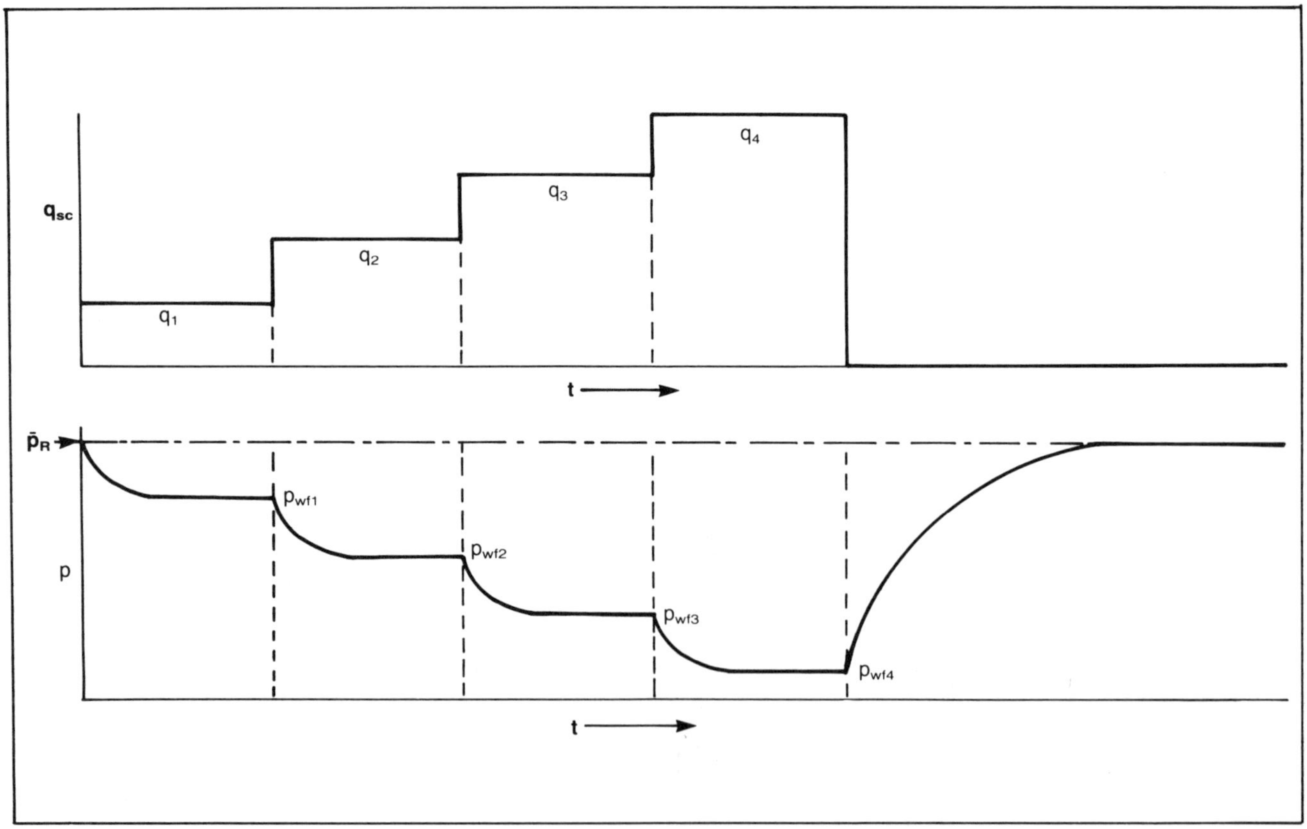

Fig. 3-9. Conventional test—flow rate and pressure diagrams.

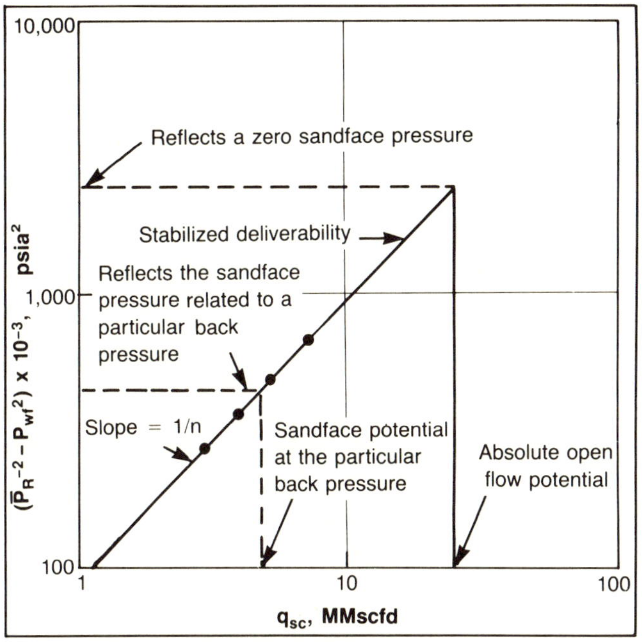

Fig. 3-10. Deliverability test plot.

ered to be stabilized if the pressure does not change over a 15-minute interval. Once stabilization is reached, q_{sc} and p_{wf} are measured, the rate is changed, and the procedure repeated for several flow rates, usually four. The behavior of flow rate and pressure with time is illustrated in Figure 3-9 for q_{sc} increasing in sequence. The tests may also be run in the reverse sequence. A plot of typical flow-after-flow data is shown in Figure 3-10.

Example 3-4:

A flow-after-flow test was performed on a well located in a low pressure reservoir in which the permeability was high. Using the following test data, determine:

1. The values of n and C for the deliverability equation.
2. The AOF.
3. The flow rate for p_{wf} = 160 psia.

Solution:

Test	*q_{sc}, Mscfd*	*p_{wf}, psia*	*$(\bar{p}_R^2 - p_{wf}^2) \times 10^{-3}$*
	0	201 ($\bar{p}_R$)	40.4 ($\bar{p}_R^2$)
1	2730	196	1.985
2	3970	195	2.376
3	4440	193	3.152
4	5550	190	4.301

A plot of q_{sc} versus $\Delta(p^2)$ is shown in Figure 3-11. From the plot it is apparent that tests 1 and 4 lie on the straight line and can thus be used to determine n.

$$n = \frac{\log q_1 - \log q_4}{\log \Delta (p^2)_1 - \log \Delta (p^2)_4}$$

$$n = \frac{\log(2730) - \log(5550)}{\log(1.985 \times 10^3) - \log(4.301 \times 10^3)} = 0.92.$$

Using Test 4 to calculate C:

$$C = \frac{q_{sc}}{(\bar{p}_R^2 - p_{wf}^2)^n} = \frac{5550}{(4.301 \times 10^3)^{0.92}} = 2.52 \frac{\text{Mscfd}}{\text{psia}^{1.84}}.$$

Therefore, the deliverability equation is

1. $q_{sc} = 2.52\,(\bar{p}_R^2 - p_{wf}^2)^{0.92}$
2. For $p_{wf} = 0$, $q_{sc} = 2.52\,(201^2 - 0^2)^{.92} = 43{,}579$ Mscfd
3. For p_{wf} = 160 psia,

$$q_{sc} = 2.52\,(201^2 - 160^2)^{.92} = 17{,}300 \text{ Mscfd}$$

A plot such as Figure 3-11 may be used directly to obtain the AOF and the well's inflow performance without calculating values for C and n. The AOF is determined by entering the ordinate at $\bar{p}_R^2$ and reading the AOF as 43.6 MMscfd. An inflow rate at any value of p_{wf} can be obtained by entering the ordinate at the appropriate value of $\Delta p^2 = \bar{p}_R^2 - p_{wf}^2$ and reading the rate from the abscissa.

Isochronal Testing

The isochronal, or equal time, test is based on the theory that at equal flow times the same volume of reservoir

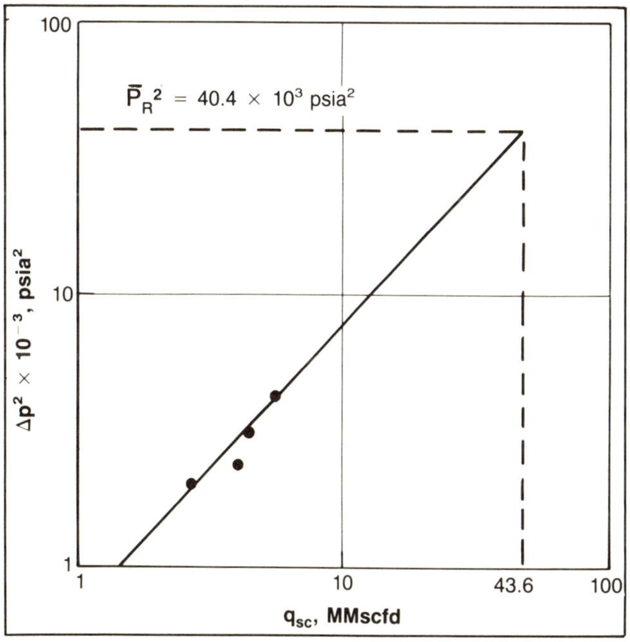

Fig. 3-11. q_{sc} versus $\Delta(p^2)$ plot.

is affected regardless of flow rate. The isochronal testing method, as introduced by Cullender[5] in 1955, has been modified to require even shorter testing times. The modified isochronal test is described in the next section.

The isochronal test was proposed as a means of determining deliverability in tight wells that require a long period of time to reach stabilization. At least one stabilized point is still required to evaluate the coefficient C in Equation 3-30. The procedure for conducting an isochronal test is:

1. Starting at a shut-in condition, open the well on a particular choke size for a period of time. Measure q_{sc} and p_{wf} at specific time periods for this choke size.
2. Shut the well in until the pressure returns to $\bar{p}_R$.
3. Open the well on a larger choke size and measure q_{sc} and p_{wf} at the same flowing time intervals as in Step 1. The total flow periods do not have to be of equal length but they usually are.
4. Shut the well in until the pressure returns to $\bar{p}_R$.
5. Repeat for several choke sizes, usually four.
6. On the last choke size, allow the well to flow until a stabilized condition is reached. This may require several hours or even days, but only one rate has to be flowed for the long period as compared to all the rates for flow-after-flow testing.

The behavior of the flow rate and pressure with time is illustrated in Figure 3-12 for increasing flow rates. The reverse order could also be used.

The test is analyzed by plotting $\bar{p}_R^2 - p_{wf}^2$ versus q_{sc} on log-log paper for each flow time at which the data were measured. This will produce one straight line for each flow time, the slopes of which will be equal. The slope allows determination of the exponent n, while the flow coefficient C can be determined using the stabilized or extended flow rate. A plot of typical isochronal flow data is shown in Figure 3-13.

Example 3-5:

An isochronal test was conducted on a well located in a reservoir that had an average pressure of 1952 psia. The well was flowed on four choke sizes, and the flow rate and flowing bottom-hole pressure were measured at 3 hr and 6 hr for each choke size. An extended test was conducted for a period of 72 hr at a rate of 6.0 MMscfd, at which time p_{wf} was measured at 1151 psia. Using the preceding data, find the following:

1. Stabilized deliverability equation.
2. *AOF*.
3. Generate an inflow performance curve.

Solution:

	$t = 3$ hr		$t = 6$ hr	
q_{sc}, Mscfd	p_{wf}	$\bar{p}_R^2 - p_{wf}^2 \times 10^{-3}$	p_{wf}	$\bar{p}_R^2 - p_{wf}^2 \times 10^{-3}$
2600	1793	597	1761	709
3300	1757	724	1657	1064
5000	1623	1177	1510	1530
6300	1505	1545	1320	2068
6000	Extended	flow, $t = 72$ hr	1151	2485

1. The slopes of both the 3 hr and 6 hr lines are apparently equal (Fig. 3-14). Using the first and last points on the 6 hr test to calculate n gives:

$$n = \frac{\log q_1 - \log q_4}{\log \Delta(p^2)_1 - \log \Delta(p^2)_4}$$

$$= \frac{\log(2600) - \log(6300)}{\log(709) - \log(2068)} = 0.83$$

Using the extended flow test to calculate C:

$$C = \frac{q_{sc}}{(\bar{p}_R^2 - p_{wf}^2)^n} = \frac{6000}{(2485 \times 10^3)^{0.83}} = 0.0295.$$

Therefore, the deliverability equation for q_{sc} in Mscfd is

$$q_{sc} = 0.0295(\bar{p}_R^2 - p_{wf}^2)^{0.83}$$

2. To calculate *AOF*, set $p_{wf} = 0$:

$$q_{sc} = 0.0295(1952^2 - 0)^{0.83} = 8551 \text{ Mscfd}$$

3. In order to generate an inflow performance curve, pick several values of p_{wf} and calculate the corresponding q_{sc}.

p_{wf}, psia	q_{sc}, Mscfd
1952	0
1800	1768
1400	4695
1000	6642
600	7875
200	8477
0	8551

The inflow performance curve is plotted in Figure 3-15.

If the log-log plot is used to determine the absolute open flow or the inflow performance, the line drawn through the stabilized test must be used.

Modified Isochronal Testing

The modified isochronal testing procedure was introduced so that even less flowing time is required for the

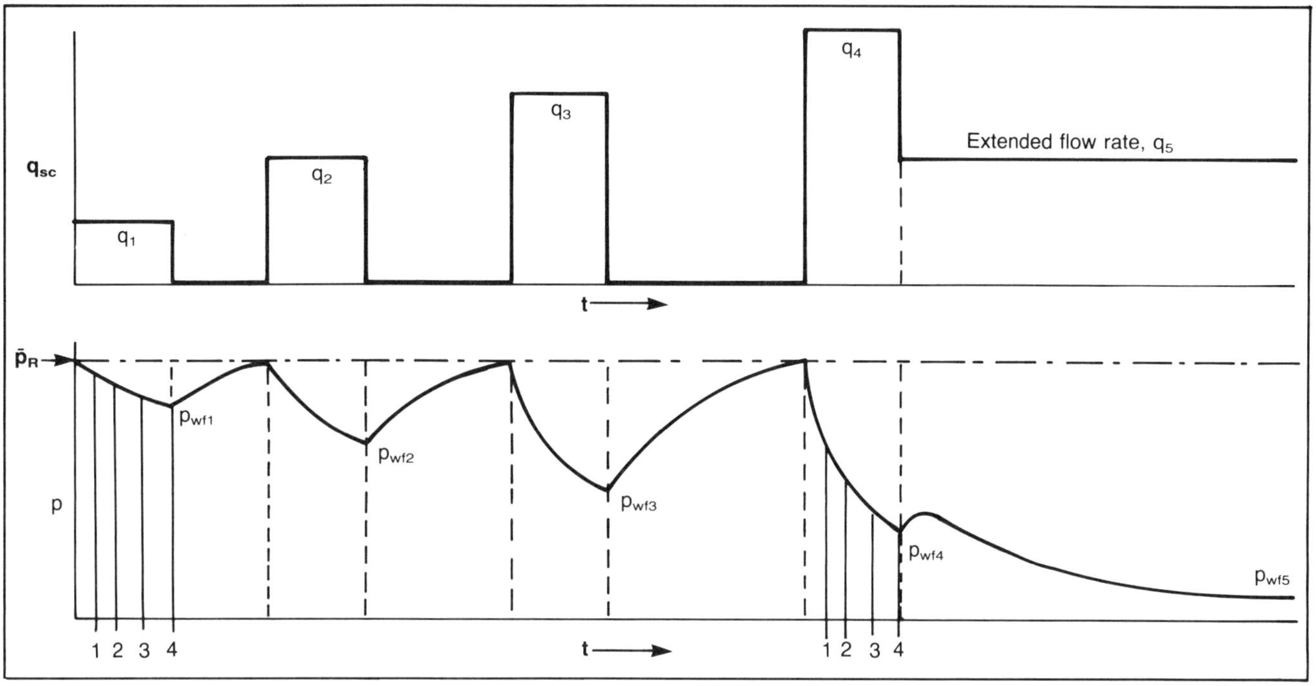

Fig. 3-12. Isochronal test.

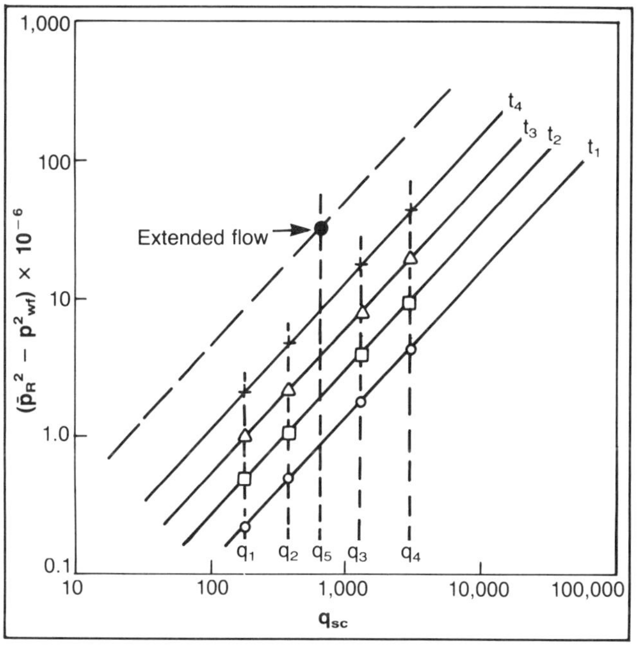

Fig. 3-13. Isochronal test data.

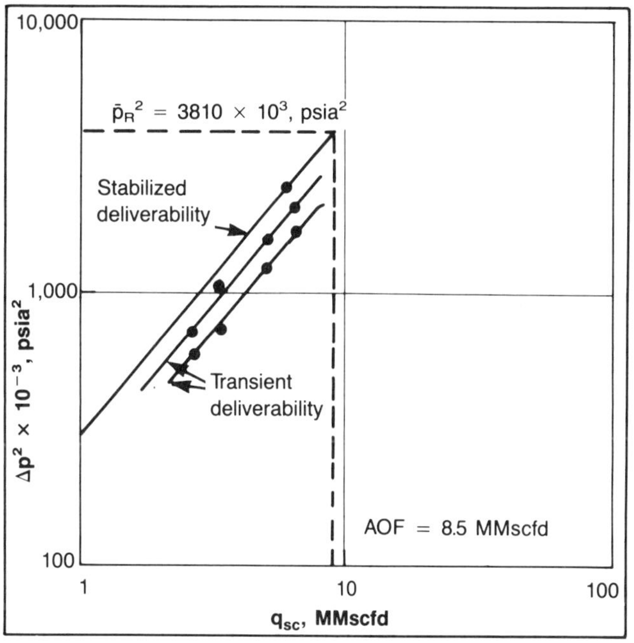

Fig. 3-14. Plot of Δp^2 versus q_{sc}-isochronal test.

well test. The procedure is very similar to the isochronal test, except that the shut-in period between each flow rate is not long enough to allow the well to return to the initial average reservoir pressure. In the modified method the well is shut-in for the same length of time that it was allowed to flow for each choke size. During this time the static well pressure will rebuild to some value, p_{ws}, which will be lower after each flow period. The flow rate-pressure behavior with time is shown in Figure 3-16. An extended flow period is still required to evaluate the flow coefficient, C. The analysis is the same as the regular isochronal test, with the exception that

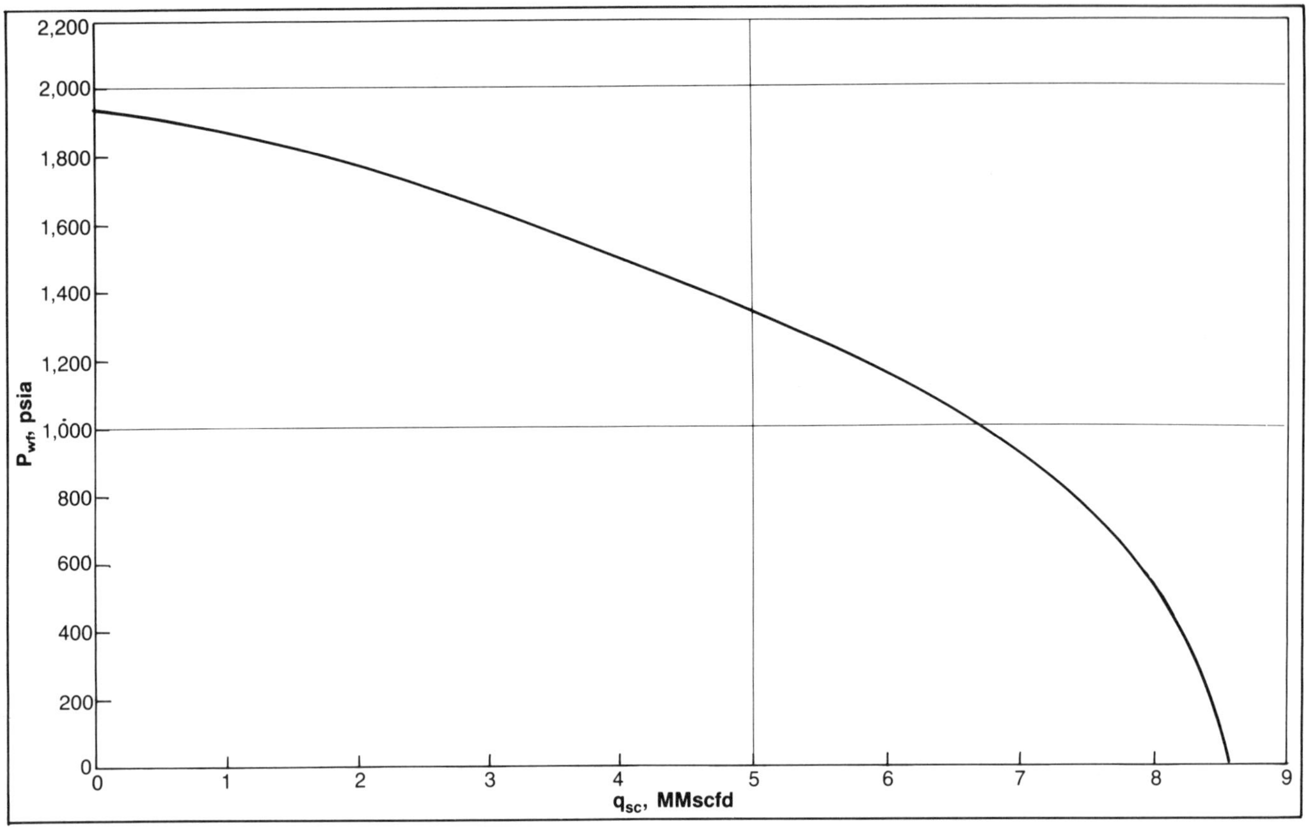

Fig. 3-15. Inflow performance curve, Example 3-5.

$$\Delta(p^2) = p_{wsi}^2 - p_{wfi}^2$$

is plotted versus q_{sc} to obtain a value for n. The value of C is calculated using the initial static or average reservoir pressure and the extended test values for p_{wf} and q_{sc}. Figure 3-17 illustrates a plot of typical modified isochronal test data.

Jones, Blount and Glaze Method

The method of plotting test data which was proposed by Jones, et al.,[7] can be applied to gas well testing to determine real or present time inflow performance relationships. The analysis procedure allows determination of turbulence or non-Darcy effects on completion efficiency irrespective of skin effect and laminar flow. The procedure also evaluates the laminar flow coefficient A, and if $k_g h$ is known, an estimate of skin effect can be made. The data required are either two or more stabilized flow tests or two or more Isochronal flow tests. At least one stabilized flow test is required to obtain a stabilized value of the laminar coefficient A. No transient tests are required to evaluate the completion efficiency if this method is applied. Jones, et al., also suggested methods to estimate the improvement in inflow performance which would result from reperforating a well to lengthen the completion interval and presented guidelines to determine if the turbulent effects were excessive. Equation 3-11 can be divided through by q_{sc} and written as

$$\frac{\bar{p}_R^2 - p_{wf}^2}{q_{sc}} = A + Bq_{sc}, \tag{3-11}$$

where A and B are the laminar and turbulent coefficients respectively and are defined in Equation 3-11. From Equation 3-11 it is apparent that a plot of $(\bar{p}_R^2 - p_{wf}^2)/q_{sc}$ versus q_{sc} on cartesian coordinates will yield a line that has a slope of B and an intercept of $A = \Delta(p^2)/q_{sc}$ as q_{sc} approaches zero. These plots apply to both linear and radial flow, but the definitions of A and B would depend on the type of flow.

In order to have some qualitative measure of the importance of the turbulent contribution to the total drawdown, Jones, et al., suggested comparison of the value of A calculated at the AOF of the well (A'), to the stabilized value of A. The value of A' can be calculated from $A' = A + B(AOF)$
where

$$AOF = \frac{-A + [A^2 + 4B\bar{p}_R^2]^{0.5}}{2B}. \tag{3-32}$$

Jones, et al., suggested that if the ratio of A' to A was

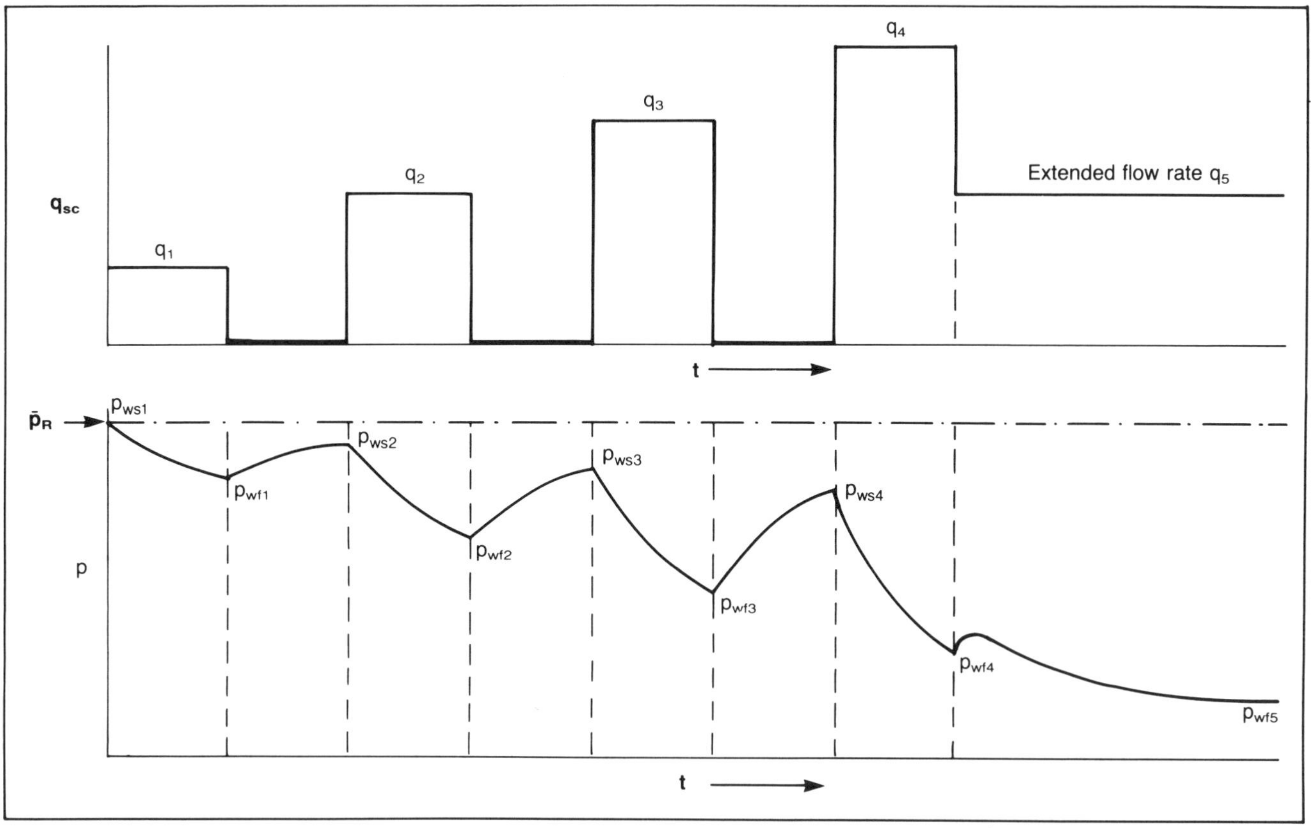

Fig. 3-16. Modified isochronal test.

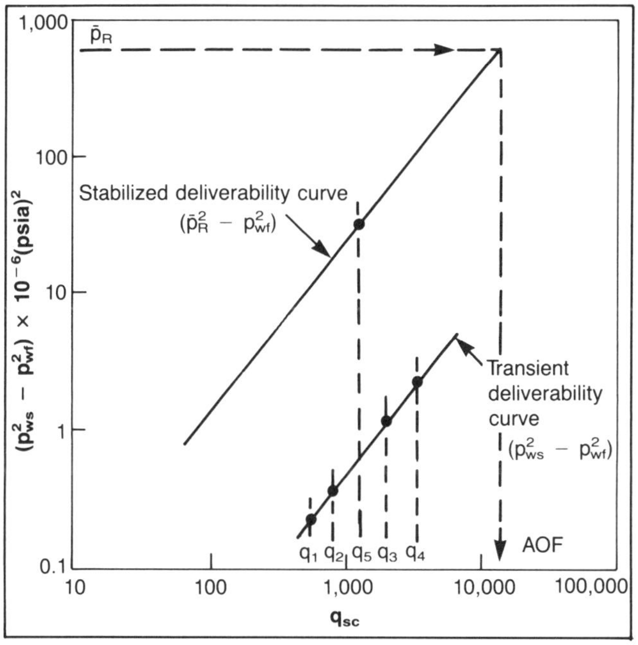

Fig. 3-17. Modified isochronal test.

greater than 2 or 3, then it is likely that some restriction in the completion exists. They also suggested that the formation thickness h used in the definition of B could be replaced by the length of the completed zone h_p, since most of the turbulent pressure drop occurs very near the wellbore. The effect of changing completion zone length on B and therefore on inflow performance can be estimated from

$$B_2 = B_1\left(\frac{h_{p1}}{h_{p2}}\right)^2$$

where

B_2 = turbulence multiplier after recompletion,
B_1 = turbulence multiplier before completion,
h_{p2} = new completion length, and
h_{p1} = old completion length.

Example 3-5a:

A four-point test was conducted on a gas well that had a perforated zone of 20 ft. Static reservoir pressure is 5250 psia. Using the Jones, et al., method, determine:

1. *A* and *B*
2. *AOF*
3. Ratio of *A'/A*
4. New *AOF* if the perforated interval is increased to 30 ft.

Test Data

Test No.	q_{sc}, *Mscfd*	p_{wf}, *psia*
1	9300	5130
2	6000	5190
3	5200	5203
4	3300	5225

Solution:

Test No.	q_{sc}, *Mscfd*	$(\bar{p}_R^2 - p_{wf}^2)/q_{sc}$, *psia²/Mscfd*
1	9300	133.9
2	6000	104.4
3	5200	94.5
4	3300	79.4

1. From a plot of this data, Fig. 3-17a, it is found that

$A = 48$ psia²/Mscfd

$B = 9.24 \times 10^{-3}$ psia²/Mscfd²

2. $$AOF = \frac{-48 + [48^2 + 4(9.24 \times 10^{-3})(5250)^2]^{0.5}}{2(9.24 \times 10^{-3})}$$

$AOF = 52{,}080$ Mscfd

3. $A' = 48 + 9.24 \times 10^{-3}(52080) = 529$

$A'/A = 529/48 = 11$

4. $B_2 = B_1(h_{p1}/h_{p2})^2$

$B_2 = 9.24 \times 10^{-3}(20/30)^2 = 4.1 \times 10^{-3}$

$$AOF_2 = \frac{-48 + [48^2 + 4(4.1 \times 10^{-3})(5250)^2]^{0.5}}{2(4.1 \times 10^{-3})}$$

$AOF_2 = 76{,}340$ Mscfd

The value of A' calculated in the previous example indicates a large degree of turbulence. The effect of increasing the perforated interval on the *AOF* is substantial. Further implications of the effects of well completion efficiency will be discussed in the section on Well Completion Effects.

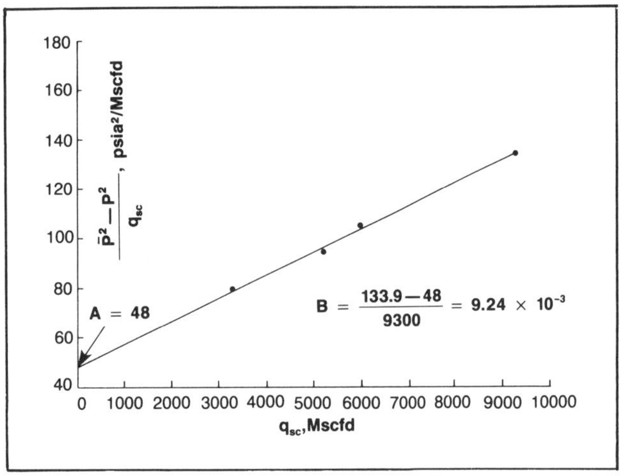

Fig. 3-17a. Jones et al., Method.

Laminar Inertia Turbulence (LIT) Analysis

All of the previously described methods for predicting gas well deliverability or inflow performance require at least one test conducted for a period long enough to reach stabilization. Equation 3-21 can be used to calculate the approximate time to stabilization as follows:

$$t_S = \frac{950\phi\bar{\mu}\bar{C}r_e^2}{k} = \frac{380\phi\bar{\mu}\bar{C}A}{k} \tag{3-33}$$

This can be a very long period of time in low permeability reservoirs, especially if a well is draining a large area.

Several methods have been proposed for obtaining a deliverability equation without a stabilized test.[6,7,8] Essentially the only difference in these methods is the method used to obtain the coefficients A and B in Equation 3-11. A method presented by Brar and Aziz[8] will be described in this section.

The pseudo-steady-state equation for gas flow is

$$\Delta(p^2) = \bar{p}_R^2 - p_{wf}^2 = Aq_{sc} + Bq_{sc}^2 \tag{3-11}$$

For unsteady-state flow, A varies with time and will be written as A_t. The equations can be written in terms of common logs rather than natural logs. For pseudo-steady-state

$$\Delta(p^2) = 2m\left[\log\left(\frac{.472r_e}{r_w}\right) + \frac{S}{2.303}\right]q_{sc} + 0.869\, mDq_{sc}^2 \tag{3-34}$$

where

$$m = \frac{1637T\bar{\mu}\bar{Z}}{kh} \tag{3-35}$$

For transient flow,

$$\Delta(p^2) = m\left[\log\left(\frac{kt}{\phi\bar{\mu}\bar{C}r_w^2}\right) - 3.23 + .869S\right]q_{sc} + 0.869mDq_{sc}^2 \tag{3-36}$$

Comparing Equations 3-34 and 3-36 to Equation 3-11 implies that

$$A = 2m\left[\log\left(\frac{0.472r_e}{r_w}\right) + \frac{S}{2.303}\right] \tag{3-37}$$

$$A_t = m\left[\log\left(\frac{kt}{\phi\bar{\mu}\bar{C}r_w^2}\right) - 3.23 + 0.869S\right] \tag{3-38}$$

and

$$B = 0.869mD \tag{3-39}$$

The object of the analysis is to be able to determine

the values of A and B for stabilized flow, then Equation 3-11 can be used to calculate inflow performance. The skin factor S, the turbulence coefficient D, and the permeability k can also be determined.

Equation 3-36 may be written as

$$\frac{\Delta(P^2)}{q_{sc}} = A_t + Bq_{sc}$$

where A_t and B are defined in Equations 3-38 and 3-39 respectively. The value of A_t will increase with time until stabilized flow is reached. A plot of $\Delta(p^2)/q_{sc}$ versus q_{sc} on cartesian coordinates will result in a series of straight, parallel lines having slopes equal to B and intercepts A_t equal to $\Delta(p^2)/q_{sc}$ for each flow time. The slopes and intercepts can also be determined using least squares analysis. Equation 3-38 can be expressed as

$$A_t = m\left[\log\frac{k}{\phi\bar{\mu}\bar{C}r_w^{\,2}} - 3.23 + .869S\right] + m\log t$$

Therefore a plot of A_t versus t on the semi-log scales will result in a straight line having a slope equal to m and an intercept at $t = 1$ hr ($\log 1 = 0$) equal to A_{t1}. The procedure for analyzing an Isochronal test is:

1. Determine A_t and B from transient tests for several flow times using plots of Equation 3-36 or least squares.
2. Plot A_t versus t on semi-log scales to determine m and A_{t1}.
3. Using the value of m, calculate k from Equation 3-35.
4. Solve Equation 3-38 for S using the values of m, k, and A_{t1}.
5. Calculate a stabilized value for A using Equation 3-37.
6. Using the value of B from Step 1, calculate D using Equation 3-39.
7. Calculate the stabilized well performance from Equation 3-11 using the stabilized values for A and B.

The method of least squares may be used to determine A_t and B from N transient flow tests.

$$A_t = \frac{\Sigma\frac{\Delta(p^2)}{q}\Sigma q^2 - \Sigma\Delta(p^2)\Sigma q}{N\Sigma q^2 - \Sigma q\Sigma q} \tag{3-40}$$

$$B = \frac{N\Sigma\Delta(p^2) - \Sigma\frac{\Delta(p^2)}{q}\Sigma q}{N\Sigma q^2 - \Sigma q\Sigma q} \tag{3-41}$$

Values for A_t and B will be obtained for each time at which p_{wf} was measured. The value of B should be constant, and Brar and Aziz suggest using the value of B obtained from the longest flow test as the representative value.

Example 3-6:

A modified isochronal test was conducted using four different flow rates, and the flowing bottom-hole pressure was measured at periods of 1, 2, 4, 6, and 8 hours. The test data are tabulated below.

h = 12 ft $\qquad \bar{p}_R$ = 922.6 psia
r_w = 0.23 ft $\qquad \bar{\mu}$ = 0.0116 cp
ϕ = 0.23 $\qquad \bar{Z}$ = 0.972
T = 582°R $\qquad \bar{C}$ = 0.00109 psia^{-1}
r_e = 2000 ft

	p_{ws}			
	922.6	*921.9*	*919.9*	*917.6*
	p_{wf}			
t	$q = 0.4746$	$q = 0.8797$	$q = 1.2716$	$q = 1.6589$
1.0	900.1	863.0	798.9	676.3
2.0	897.1	853.9	769.9	662.2
4.0	892.2	833.0	754.9	642.0
6.0	890.1	827.9	732.8	635.2
8.0	888.1	825.1	727.3	629.3

Use the test data to determine k, S, D, and the stabilized deliverability equation.

Solution:

Table 3-3 may be constructed to conveniently calculate the value for A_t and B. To further illustrate the procedure, some of the entries for t = 2 hr are calculated. For q_{sc} = 0.4746 MMscfd,

$$\Delta(p^2) = p_{ws}^{\,2} - p_{wf}^{\,2} = (922.6)^2 - (897.1)^2$$
$$= 46{,}402 \text{ psia}^2$$

$$\frac{\Delta(p^2)}{q} = \frac{46402}{0.4746} = 97{,}767\frac{\text{psia}^2}{\text{MMscfd}} = 97.767\frac{\text{Mpsia}^2}{\text{MMscfd}}$$

$$A_t = \frac{(677.583)(5.368) - (824.1)(4.2848)}{4(5.368) - (4.2848)(4.2848)}$$

$$= 34.11\frac{\text{Mpsia}^2}{\text{MMscfd}}$$

$$B = \frac{4(824.1) - (677.583)(4.2848)}{4(5.368) - (4.2848)^2} = 126.30\frac{\text{Mpsia}^2}{\text{MMscfd}^2}$$

The values calculated for A_t are plotted versus t on semi-log paper in Figure 3-18. The slope of the line is m = 58.7 Mpsia^2/MMscfd/cycle obtained by drawing a straight line through the last three points. The intercept at t = 1 hr can be read from the graph as 29.5 Mpsia^2/MMscfd.

TABLE 3-3

Flow rate	q	q^2	$t = 1.0$ Δp^2	$t = 1.0$ $\frac{\Delta p^2}{q}$	$t = 2.0$ Δp^2	$t = 2.0$ $\frac{\Delta p^2}{q}$	$t = 4.0$ Δp^2	$t = 4.0$ $\frac{\Delta p^2}{q}$	$t = 6.0$ Δp^2	$t = 6.0$ $\frac{\Delta p^2}{q}$	$t = 8.0$ Δp^2	$t = 8.0$ $\frac{\Delta p^2}{q}$
1	0.4746	0.2252	41.01	86.410	46.40	97.767	55.17	116.245	58.91	124.126	62.47	131.627
2	0.8797	0.7739	105.13	119.507	120.75	137.263	156.01	177.345	164.48	186.973	169.11	192.236
3	1.2716	1.6170	207.97	163.550	253.47	199.332	276.34	217.317	309.22	243.174	317.25	249.489
4	1.6589	2.7519	384.61	231.846	403.48	243.221	429.83	259.105	438.51	264.338	445.97	268.835
Σ	4.2848	5.3680	738.72	601.313	824.10	677.583	917.35	770.012	971.12	818.611	994.80	842.187
			A_t = 20.11, B = 121.57		A_t = 34.11, B = 126.30		A_t = 65.15, B = 118.89		A_t = 74.94, B = 121.09		A_t = 83.00, B = 119.07	

From Equation 3-35,

$$k = \frac{1637T\bar{\mu}\bar{Z}}{mh} = \frac{1637(528)(0.0116)(0.972)}{(58.7)(12)}$$

$k = 13.8$ *md*

Solving Equation 3-38 for S, using the value of A_t at $t = 1$ hr:

$$S = \left[\frac{A_{(1)}}{m} - \log\frac{k(1)}{\phi\bar{\mu}\bar{C}r_w^2} + 3.23\right]\left(\frac{1}{.869}\right)$$

$$S = \left[\frac{29.5}{58.7} - \log\left(\frac{13.8(1)}{.23(0.0116)(0.00109)(0.23)^2}\right) + 3.23\right](1.151)$$

$S = -4.22$

To obtain the turbulence coefficient, solve Equation 3-39 for D:

$$D = \frac{B}{0.869m} = \frac{119.07}{0.869(58.7)} = 2.23 \text{ MMscfd}^{-1}$$

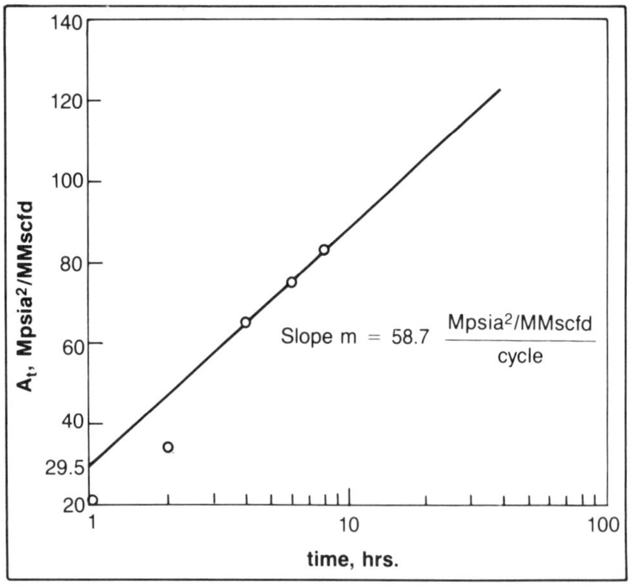

Fig. 3-18. A_t versus log t, Example 3-6.

Equation 3-34 may now be used to calculate A:

$$A = 2(58.7)\left[\log\left(\frac{.472(2000)}{0.23}\right) + \frac{(-4.22)}{2.303}\right]$$

$$A = 209 \frac{\text{psia}^2}{\text{Mscfd}}$$

The value chosen for B is $119.07 \frac{\text{Mpsia}^2}{\text{MMscfd}^2} = 0.1191 \frac{\text{psia}^2}{\text{Mscfd}^2}$

The stabilized flow equation for determining inflow performance is then $\bar{p}_R^2 - p_{wf}^2 = 209q_{sc} + 0.1191q_{sc}^2$ for p in psia and q_{sc} in Mscfd.

Equation 3-29 may be solved for q_{sc} to obtain

$$q_{sc} = \frac{-A + (A^2 + 4B(\bar{p}_R^2 - p_{wf}^2))^{0.5}}{2B} \tag{3-42}$$

For $A = 209$ and $B = 0.1191$

$$q_{sc} = \frac{-209 + (43681 + 0.4764(\bar{p}_R^2 - p_{wf}^2))^{0.5}}{0.2382}$$

The *AOF* for this well can be calculated using $p_{wf} = 0$ as

$$AOF = \frac{-209 + (43681 + 0.4764(922.6)^2)^{0.5}}{0.2382}$$

$AOF = 1936$ Mscfd

Factors Affecting Inflow Performance

Once a well has been tested and the deliverability or inflow performance equation established, it is sometimes desirable to be able to predict how changes in certain parameters will affect the inflow performance. These changes may be the result of reservoir depletion or time, or they may result from well workovers.

The effects of various changes can be estimated by reference to Equations 3-31 and 3-10.

$$q_{sc} = C(\bar{p}_R^2 - p_{wf}^2)^n \tag{3-31}$$

Comparing this with Equation 3-10, it can be seen that the effects of turbulence, Dq_{sc} are included in the exponent n, and the coefficient C contains several parameters subject to change.

$$C = \frac{703 \times 10^{-6}\, kh}{T\bar{\mu}\bar{Z}\left[\ln\left(\frac{.472 r_e}{r_w}\right) + S\right]} \tag{3-43}$$

The possible causes of changes in each parameter are discussed and modifications of C are suggested.

—The only factor that has an appreciable effect on *permeability to gas, k* is liquid saturation in the reservoir. As pressure declines from depletion, the remaining gas expands to keep S_g constant, unless retrograde condensation occurs or water influx is present. For dry gas reservoirs, the change in k with time can be considered negligible.

—In most cases the value of *formation thickness, h* can be considered constant. A possible exception is if the completion interval is changed by perforating a longer section. It is likely that the well would be re-tested at this time.

—*Reservoir temperature, T* will remain constant, except for possible small changes around the wellbore.

—*Gas Viscosity and Compressibility Factor,* $\bar{\mu}$ *and* $\bar{Z}$ are the parameters that are subject to the greatest change as $\bar{p}_R$ changes. The best method to handle these changes will be discussed in a later section on pseudo-pressure analysis. An approximation of the effect of changes in $\bar{p}_R$ on C can be made by modifying C as follows:

$$\frac{C_1}{C_2} = \frac{(\bar{\mu}\bar{Z})_2}{(\bar{\mu}\bar{Z})_1}. \tag{3-44}$$

—The *drainage radius,* r_e depends on the well spacing and can be considered constant once stabilized flow is reached.

—The *wellbore radius,* r_w can be considered to remain constant. It is possible that the effective wellbore radius can be changed by stimulation, but this can be accounted for in the skin factor.

—The *skin factor, S* can be changed by fracturing or acidizing a well. The well should be retested at this time to re-evaluate both C and n.

Example 3-7:

Generate future inflow performance curves for the well in Example 3-5 at values of average reservoir pressure of 1500 psia and 1000 psia. The following additional data are known:

$\gamma_g = 0.70$ Reservoir temperature = 200°F

Solution:

In order to correct for the changes in viscosity and compressibility factor, these will be calculated based on $\bar{p}_R$:

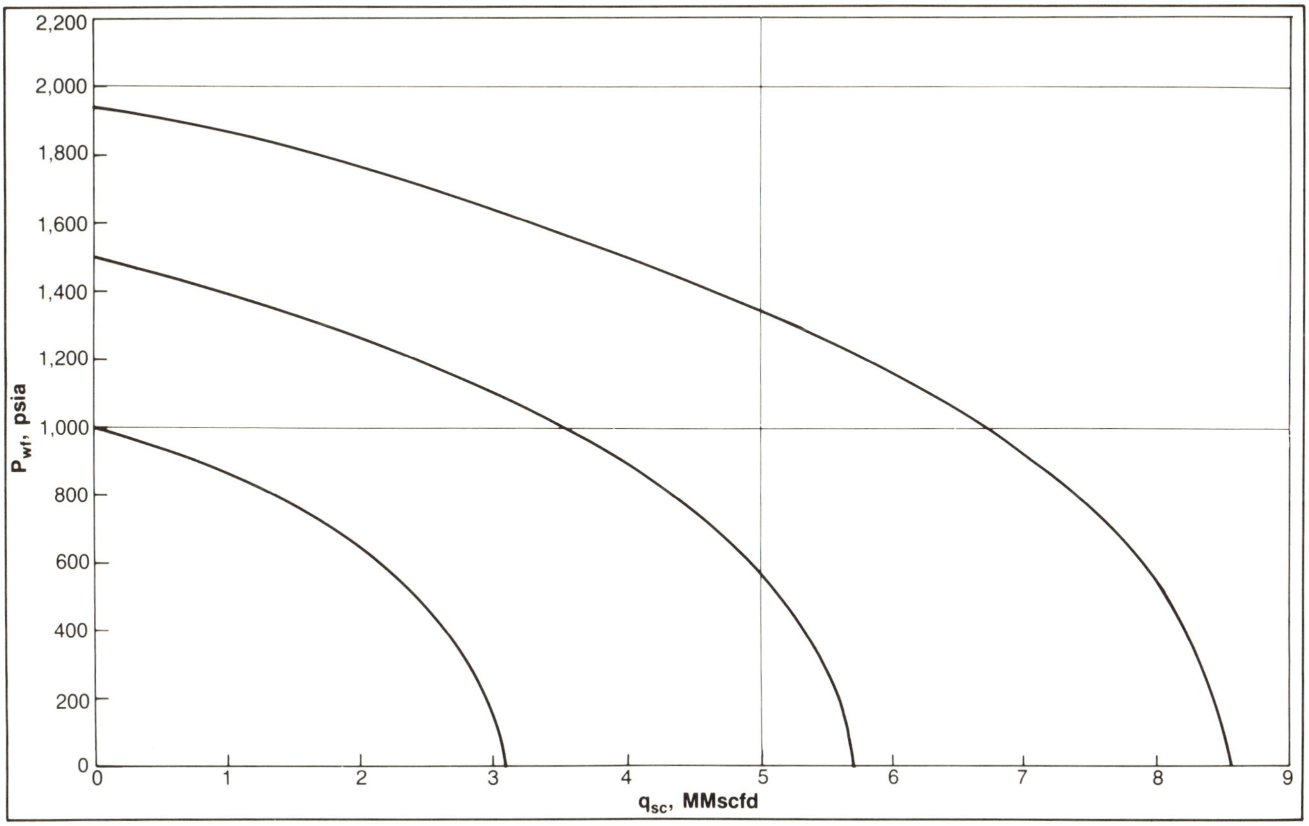

Fig. 3-19. Inflow performance curve, Example 3-7.

$\bar{p}_R$	μ	Z	μZ	C
1952	0.0156	0.86	0.0134	0.0295
1500	0.0148	0.88	0.0130	0.0304
1000	0.0134	0.91	0.0122	0.0324

For $\bar{p}_R$ = 1500 psia, $q_{sc} = 0.0304(1500^2 - p_{wf}^2)^{0.83}$
For $\bar{p}_R$ = 1000 psia, $q_{sc} = 0.0324(1000^2 - p_{wf}^2)^{0.83}$

	q_{sc}, Mscfd	
p_{wf}, psia	$\bar{p}_R$ = 1500	$\bar{p}_R$ = 1000
1500	0	—
1200	2437	—
1000	3494	0
600	4924	2136
300	5501	2861
0	5691	3094

The data are plotted in Figure 3-19.

Transient Testing

Methods have been presented for determining the stabilized deliverability or inflow performance of a gas well for use in planning equipment purchases and other field development procedures.

Much useful reservoir information can be obtained from various types of unsteady-state or transient gas well tests. Information that can be obtained from transient tests includes permeability k, skin factor S, turbulence coefficient D, and average reservoir pressure, $\bar{p}_R$. If a test is continued into the pseudo-steady-state flow regime, an estimate of reservoir size can be made. This is usually called a reservoir limit test.

The most common transient tests are drawdown tests and buildup tests. Essentially the same information can be obtained from each. The choice of which type of test to run depends on well and field conditions.

Any test that involves a change in flow rate is analyzed based on the principle of superposition. This principle, as it applies to well testing, is briefly described.

Principle of Superposition

The superposition principle in effect states that if a pressure disturbance is created in a reservoir, the disturbance continues to travel through the reservoir even though the source of the disturbance may change or cease. This means that in order to determine the pressure at a location as a function of time, all of the pressure disturbance effects must be added.

Superposition in Time. When the flow rate is changed in a well, the pressure disturbance caused by the previous flow rate continues to affect the reservoir. As an analogy, when the source of a noise is stopped, the sound waves already emitted do not stop. Consider the case in which a well is produced at some rate q_1 for a time t_1. The rate is then changed to q_2 and flow is continued. If the pressure is desired at some time t_2, the effects of both rates must be considered. The flow rate and pressure behavior are illustrated in Figure 3-20.

Adding the effect of the two rates flowing for their respective times gives:

$$p_i^2 - p_{wf}^2 = p_i^2 q_{D1} \Delta p_{D1} + p_i^2 q_{D2} \Delta p_{D2} \quad (3\text{-}45)$$

where

$$q_{D1} = \frac{1422 T \bar{Z} \bar{\mu}}{khp_i^2} q_1, \quad (3\text{-}46)$$

$$q_{D2} = \frac{1422 T \bar{Z} \bar{\mu}}{khp_i^2} (q_2 - q_1), \quad (3\text{-}47)$$

Δp_{D1} is based on total flowing time, $t_1 + t_2$, Δp_{D2} is based on flowing time $t - t_1$.

Superposition in Space. When more than one well is producing in a reservoir, the effects of both pressure disturbances must be added to calculate the total pressure effect at any point in the reservoir. This of course requires evaluation of pressure at points other than the well. This is the basis for interference testing involving two or more wells. Superposition in time and space can be applied simultaneously.

Pressure Drawdown Testing

Several important reservoir parameters can be determined by flowing a well at a constant rate and measuring flowing wellbore pressure as a function of time. This is called drawdown testing and it can utilize information obtained in both the transient and pseudo-steady-state flow regimes. If the flow extends to pseudo-steady-state, the test is referred to as a Reservoir Limit Test and can be used to estimate the reservoir pore volume. Both single rate and two rate tests are utilized, depending on the information required. Some of the reservoir parameters which may be obtained from drawdown testing are flow capacity kh, skin factor S, and turbulence coefficient D.

A drawdown test begins from a shut-in condition and a constant flow rate is maintained while pressure is measured constantly. The early time pressure data will be affected by wellbore storage and is usually used only to determine the beginning of the transient flow period. This can be identified as the beginning of the straight line segment of the plot of (Δp^2) versus time.

The equation for transient flow, Equation 3-20, may be written including formation damage and turbulence effects as

$$\Delta p_D = 0.5(\ln t_D + .809) + S' \quad (3\text{-}48)$$

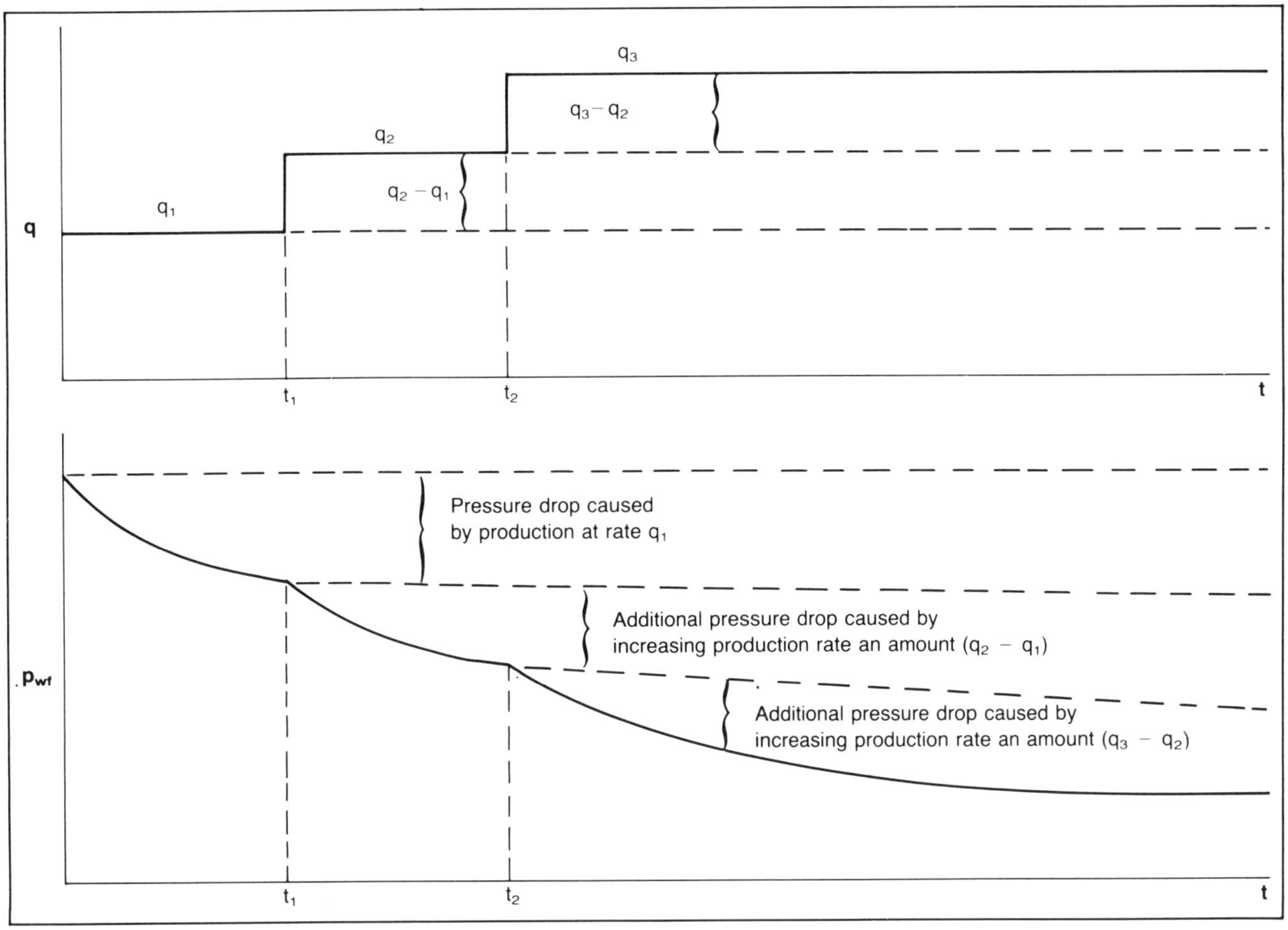

Fig. 3-20. Superposition. Courtesy the Energy Resources Conservation Board, Calgary.

where

$S' = S + Dq_{sc}$

S = actual well damage or improvement, such as clay swelling, or fractures, and may be positive or negative,

D = turbulence coefficient, which will always be positive.

In terms of real variables and common logs, Equation 3-48 becomes

$$p_i^2 - p_{wf}^2 = \frac{1637 q_{sc} T \bar{\mu} \bar{Z}}{kh} \left[\log t + \log \frac{k}{\phi \bar{\mu} \bar{C} r_w^2} - 3.23 + 0.869 S' \right], \quad (3\text{-}49)$$

where k is in millidarcys.

From this form of the equation it can be seen that a plot of $\Delta(p^2)$ vs. log t will give a straight line of slope m, where

$$m = \frac{1637 q_{sc} T \bar{\mu} \bar{Z}}{kh} \quad (3\text{-}50)$$

from which kh can be calculated. To obtain S', let t = 1 hr (log 1 = 0), then

$$p_i^2 - p_{1\,\text{hr}}^2 = m \left[\log \frac{k}{\phi \bar{\mu} \bar{C} r_w^2} - 3.23 + 0.869 S' \right], \quad (3\text{-}51)$$

where $p_{1\text{hr}}$ is obtained from an extrapolation of the linear segment of the plot. Solving for S' in Equation 3-51 gives

$$S' = 1.151 \left[\frac{p_i^2 - p_{1\,\text{hr}}^2}{m} - \log \frac{k}{\phi \bar{\mu} \bar{C} r_w^2} + 3.23 \right]. \quad (3\text{-}52)$$

Since S' is rate dependent, two single rate drawdown tests may be conducted to determine S and D. That is, from the two tests,

$$S'_1 = S + Dq_1$$

$$S'_2 = S + Dq_2.$$

The removable pressure drop due to actual damage can be calculated from

$$\Delta(p^2)_S = 0.869 mS, \quad (3\text{-}53)$$

and the rate dependent pressure drop from

$$\Delta(p^2)_D = 0.869mDq_{sc}. \qquad (3\text{-}54)$$

An estimate of the beginning and end of the transient flow period can be obtained from the dimensionless time once an estimate of k is obtained from the analysis. The approximate range of dimensionless time for transient flow is

$$0.03 \leq t_{De} \leq 0.25.$$

If the segment chosen for obtaining the straight line slope did not fall within this range, the data should be re-examined.

Example 3-8:

The following data apply to a well on which a drawdown test was conducted. Use the data to calculate k and S'.

$p_i = 3732$ psia $\quad T = 673°R \quad h = 20$ ft
$\phi = 0.10 \quad r_w = 0.29$ ft $\quad r_e = 2640$ ft
$\bar{\mu} = 0.021$ cp $\quad \gamma_g = 0.68 \quad \bar{Z} = 0.85$
$\bar{C} = 2.2 \times 10^{-4}$ psi^{-1} $\quad q_{sc} = 5.65$MMscfd

Solution:

t, hr	p_{wf}, psia	$\Delta(p^2) = p_i^2 - p_{wf}^2 \times 10^{-6}$, psia^2
1.60	3729	0.022
2.67	3546	1.354
3.20	3509	1.615
5.07	3491	1.741
6.13	3481	1.810
8.00	3433	2.142
15.20	3388	2.449
20.00	3366	2.598
30.13	3354	2.679
40.00	3342	2.759
60.27	3323	2.885
80.00	3315	2.939
100.27	3306	2.998
120.53	3295	3.071

From the graph of $\Delta(p^2)$ versus log t (Fig. 3-21) the slope is

$$m = \frac{3 \times 10^6 - 2.42 \times 10^6}{\log 100 - \log 10}$$

$$= 5.8 \times 10^5 \text{ psia}^2/\text{cycle, and}$$

$$kh = \frac{1637 q_{sc} T \bar{\mu} \bar{Z}}{m} = \frac{1637(5650)(673)(0.021)(0.85)}{5.8 \times 10^5},$$

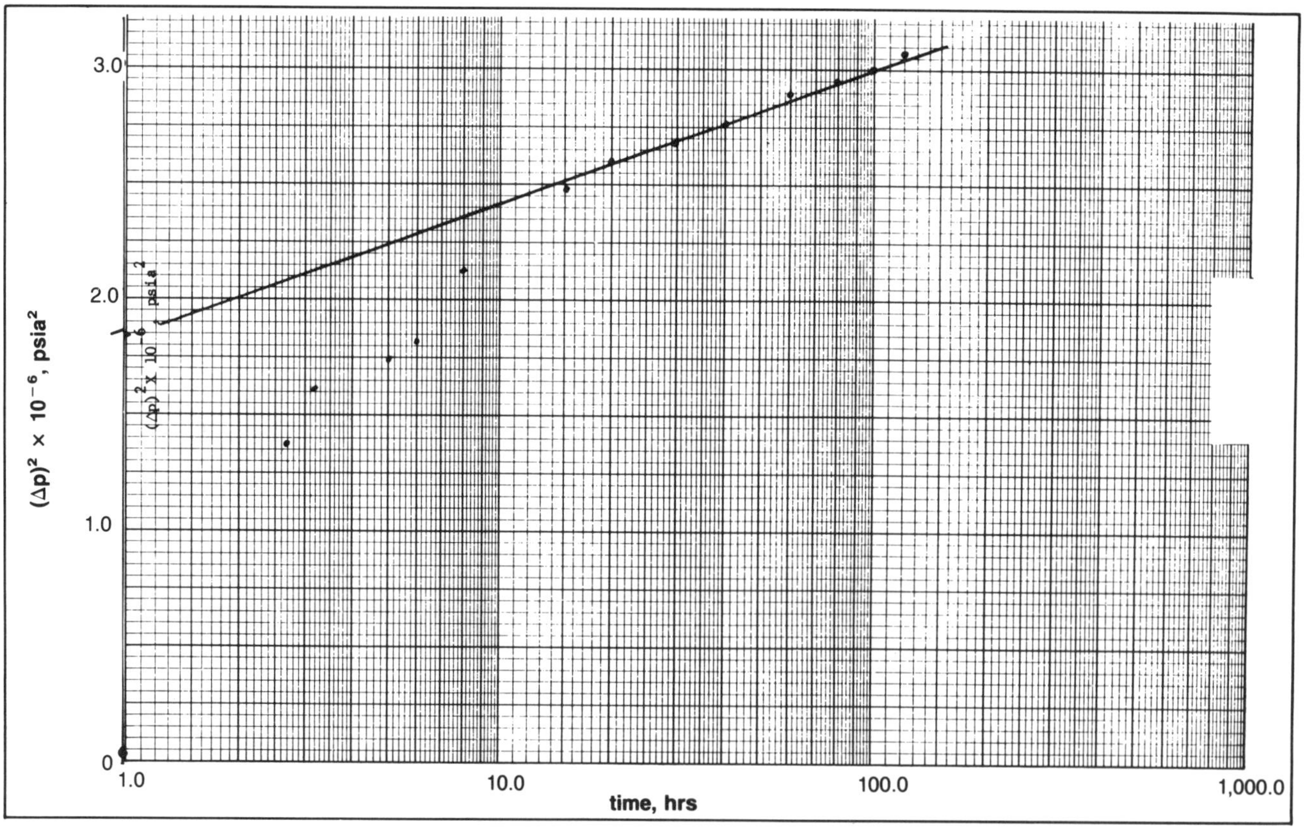

Fig. 3-21. Pressure drawdown test, Example 3-8.

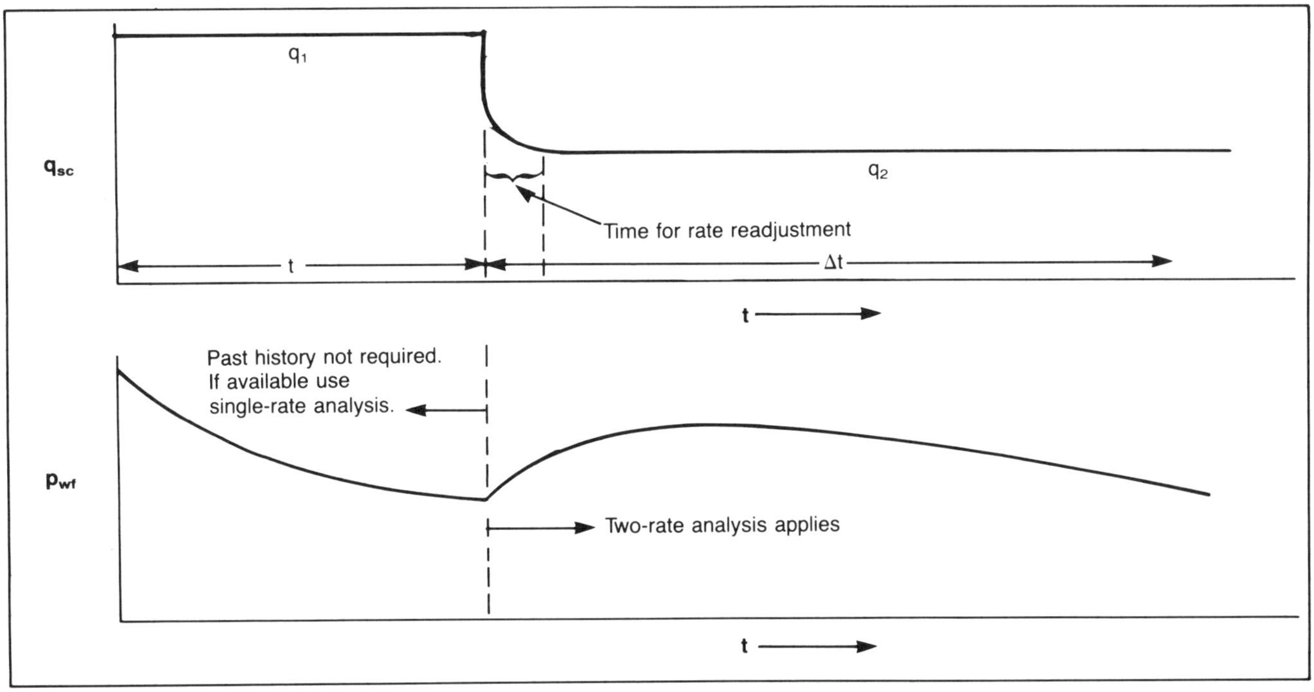

Fig. 3-22. Two-rate test—flow rate and pressure diagrams. Courtesy the Energy Resources Conservation Board, Calgary.

$$kh = 192 \ md\text{-ft},$$

$$k = \frac{192}{20} = 9.6 \ md.$$

To obtain p_{1hr}, extend the line to $t = 1$ hr and read

$$\Delta(p^2) = 1.84 \times 10^6 \text{ psia}^2$$

$$S' = 1.151 \left[\frac{1.84 \times 10^6}{0.58 \times 10^6} - \log \left(\frac{9.6}{0.1(0.021)(2.2 \times 10^{-4})(0.29)^2} \right) + 3.23 \right]$$

$$S' = 1.151(3.172 - 8.393 + 3.23) = -2.29$$

To estimate the extra pressure drop due to skin at the test flow rate,

$$\Delta(p^2)_{S'} = 0.869mS' = 0.869(5.8 \times 10^5)(-2.29)$$

$$\Delta(p^2)_{S'} = -1.154 \times 10^6 \text{ psia}^2$$

In order to check if the proper straight line segment was used to obtain the slope, the values of t corresponding to $t_{De} = 0.03$ and $t_{De} = 0.25$ are 38 and 318 hr, respectively, using $k = 9.6 \ md$. The points used on the graph to obtain the line fall between $t = 20$ hr and $t = 120$ hr.

Two-Rate Tests

The two-rate test consists of flowing the well at a rate q_1 for a period t and then changing the rate to q_2. The pressure and rate behavior are illustrated in Figure 3-22.

By applying superposition in time it can be shown that a plot of

$$\Delta(p^2) \text{ versus } \log \left(\frac{t + \Delta t}{\Delta t} \right) + \frac{q_2}{q_1} \log \Delta t$$

on cartesian coordinates will yield a straight line of slope

$$m = \frac{1637 \, q_1 \, T \, \bar{\mu} \, \bar{Z}}{kh}.$$

If the data from the first flow period are analyzed to obtain S_1', a value for S_2' can be calculated from

$$q_1 S_1' - q_2 S_2' = \frac{q_1 \, \Delta(p^2)_1 - \Delta(p^2)_0}{0.869m} - \frac{(q_1 - q_2)}{0.869} \left[\log \frac{k}{\phi \bar{\mu} \bar{c} r_w 2} - 3.23 \right], \tag{3-55}$$

where

$\Delta(p^2)_1$ is read at $\Delta t = 1$,

$\Delta(p^2)_0$ is read at $\Delta t = 0$.

Reservoir Limit Test

If a drawdown test is allowed to flow until the reservoir boundary is felt (pseudo-steady-state), the pressure behavior is governed by Equation 3-23 for circular reservoirs or by Equation 3-27 for noncircular reservoirs. By rearranging Equation 3-23 and putting it in terms of real variables, it can be shown that a plot of $\Delta(p^2)$ versus time on cartesian coordinates will yield a straight line of slope m. This can be used to estimate the in place volume of gas in the reservoir, G.

$$G = \frac{2.348\, q_{sc}\, T}{\mu\, C\, m}$$

where G is the gas volume in MMscf.

Pressure Buildup Testing

A pressure buildup test is the simplest test that can be run on a gas well. If the effects of wellbore storage can be determined, much useful information can be obtained. This information includes permeability, k, apparent skin factor, S', and average reservoir pressure, $\bar{p}_R$.

The test consists of flowing the well at a constant rate q_{sc}, for a period of time t, shutting the well in (at $\Delta t = 0$), and measuring wellbore pressure increase with shut-in time Δt.

The test was developed by Horner[9], and his method of analysis is generally considered best. Other methods include those of Miller, Dyes, and Hutchinson[10] and Muskat[15]. The method was extended to allow determination of average reservoir pressure for bounded reservoirs by Matthews, Brons, and Hazebroek (MBH)[11].

The theory behind the buildup test comes from superposition in time. In order to represent the shut-in condition, an injection rate of $-q_{sc}$ beginning at $\Delta t = 0$ is superposed on the flow rate q_{sc} that began at time $t = 0$. Writing Equation 3-48 for both q_{sc} and $-q_{sc}$ and adding the expressions results in

$$p_i^2 - p_{ws}^2 = \frac{1637\, q_{sc}\, T\, \bar{\mu}\, \bar{Z}}{kh} \log \frac{t + \Delta t}{\Delta t}. \tag{3-56}$$

From this expression it can be seen that a plot of p_{ws}^2 versus $\log((t + \Delta t)/\Delta t)$ will result in a straight line of slope $-m$, where

$$m = \frac{1637\, q_{sc}\, T\, \bar{\mu}\, \bar{Z}}{kh}, \tag{3-50}$$

from which kh or k can be determined.

Extrapolation of the line to an infinite shut in time Δt, or $(t + \Delta t)/\Delta t = 1$, results in a value for p_i^2 for an infinite reservoir. For a bounded reservoir this value is labeled p^{*2} and can be used to obtain $\bar{p}_R$, as described later.

The apparent skin factor can be determined by assuming that $(t/(t + \Delta t)) \approx 1$ at $\Delta t = 1$ hr and using the following equation

$$S' = 1.151 \left[\frac{p_{1hr}^2 - p_{wf}^2}{m} - \log \frac{k}{\phi \bar{\mu} \bar{C} r_w^2} + 3.23 \right], \tag{3-57}$$

where p_{1hr} is read from the extrapolated straight line at $\Delta t = 1$ hr, and p_{wf} is the flowing wellbore pressure at shut-in ($\Delta t = 0$). The permeability k is in millidarcys in this equation.

The value of $\bar{p}_R$ for finite or bounded reservoirs may be determined using the MBH method. The value obtained from extrapolating the line to infinite Δt, referred to as p^*, is neither the initial reservoir pressure p_i, nor the average reservoir pressure $\bar{p}_R$. It is a pseudo-pressure that will fall between these values. $\bar{p}_R$ can be obtained from p^* by using curves prepared by MBH for various shapes of drainage areas. A relationship between p^* and $\bar{p}_R$ is plotted as a function of dimensionless time based on drainage area, t_{DA}.

$$t_{DA} = \frac{2.64 \times 10^{-4}\, kt}{\phi\, \bar{\mu}\, \bar{C}\, A} \tag{3-28}$$

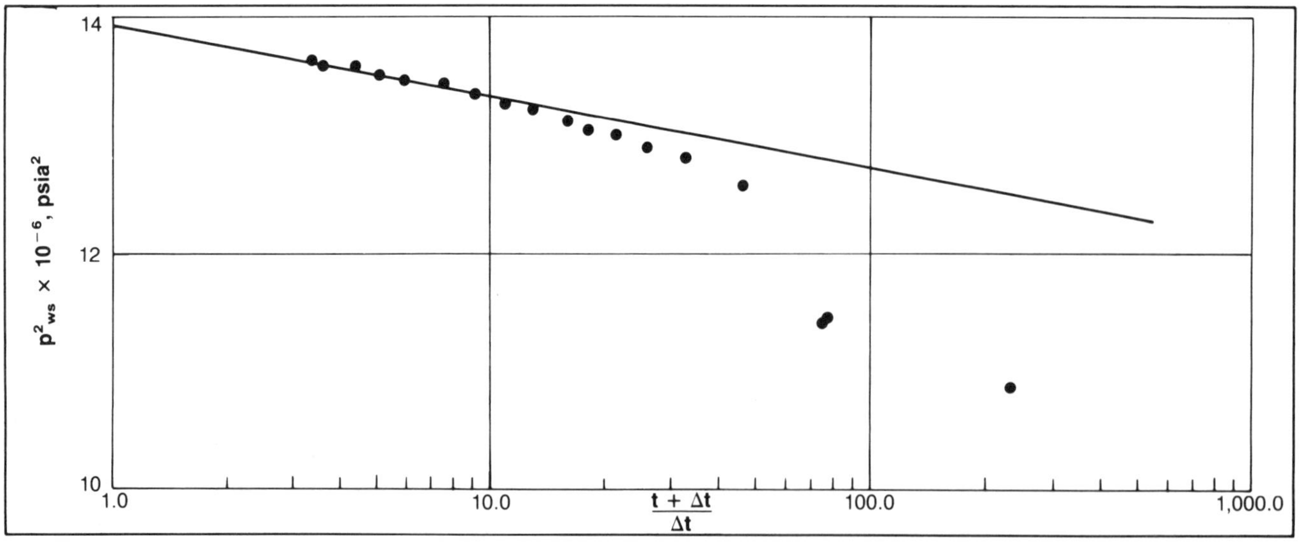

Fig. 3-23. Pressure buildup, Example 3-9.

where

t = producing time before shut-in, hrs, and
A = drainage area, ft^2.

The value obtained from the curves for the appropriate t_{DA} is

$$\frac{2.303(p^{*2} - \bar{p}_R^{\,2})}{m}, \quad (3.58)$$

from which a value of $\bar{p}_R$ can be obtained. An example MBH curve is illustrated in Figure 3-23. Curves for other conditions are in the appendix.

Example 3-9:

The well described in Example 3-8 was flowed at a rate of 5.65 MMscfd for a period of 120.5 hours and then shut-in for a buildup test. The flowing pressure at shut-in was 3295 psia. Calculate k, S', and $\bar{p}_R$ if the well is producing from the center of a square drainage area containing 22 × 10^6 sq. ft. The pressure versus time data are tabulated below.

Solution:

Δt, hr	p_{ws}, psia	$\frac{t + \Delta t}{\Delta t}$	$p_{ws}^2 \times 10^{-6}$, psia^2
0	3295	—	10.86
.53	3296	228.4	10.86
1.60	3385	76.3	11.46
2.67	3547	46.1	12.58
3.73	3573	33.3	12.77
4.80	3591	26.1	12.90
5.87	3605	21.5	13.00
6.93	3614	18.4	13.06
8.00	3623	16.1	13.13
9.87	3634	13.2	13.21
12.00	3644	11.0	13.28
14.67	3654	9.2	13.35
18.67	3664	7.5	13.42
24.53	3672	5.9	13.48
29.33	3676	5.1	13.51
35.73	3684	4.4	13.57
45.87	3688	3.6	13.60
49.87	3691	3.4	13.62

The plot of p_{ws}^2 versus $\log((t + \Delta t)/\Delta t)$ is shown in Figure 3-23.

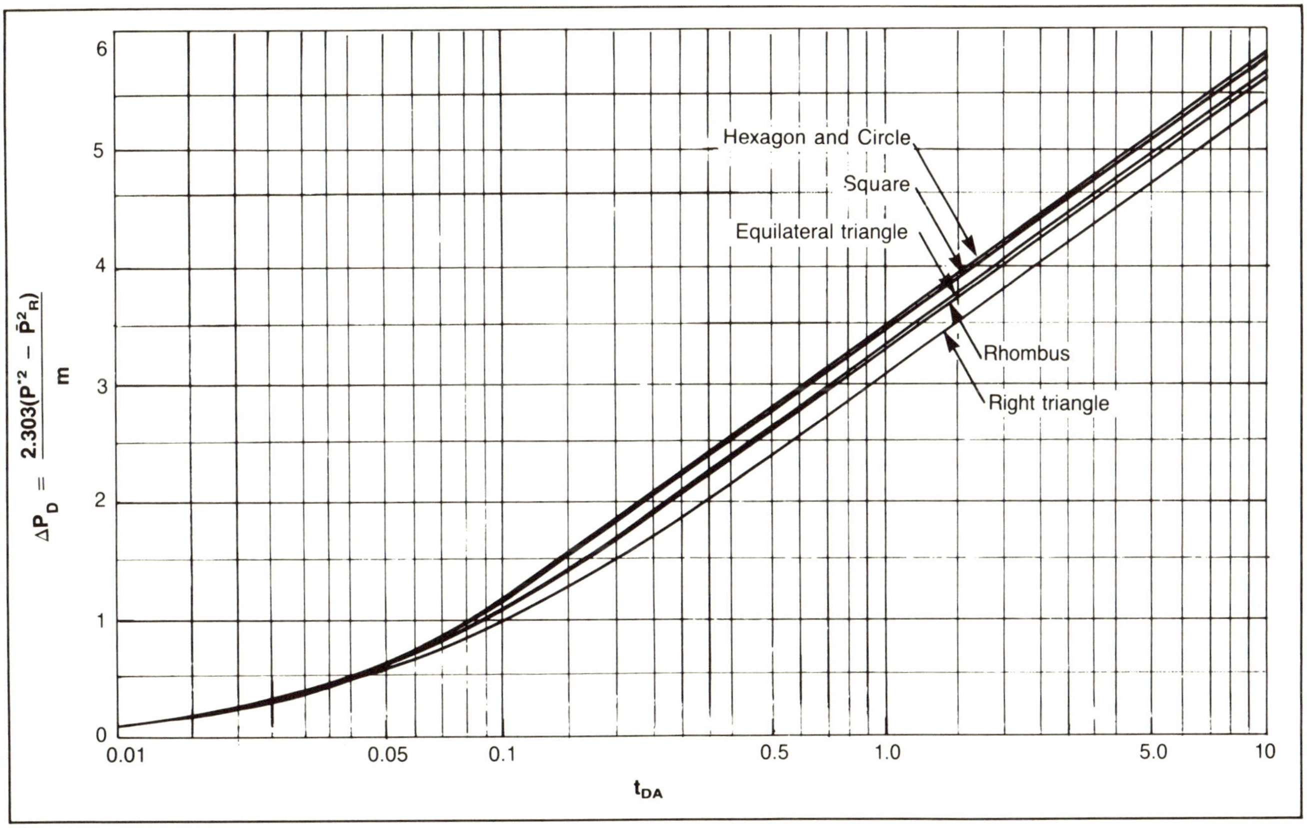

Fig. 3-24. MBH curves for well in center of area. Permission to publish by the Society of Petroleum Engineers of AIME. Copyright 1954 SPE-AIME.

The slope may be obtained by taking the change in p_{ws}^2 over 2 log cycles.

$$m = \frac{(13.92 - 12.72) \times 10^6}{\log 100 - \log 1} = 6 \times 10^5 \text{ psia}^2/\text{cycle}$$

$$kh = \frac{1637 q_{sc} T \bar{\mu} \bar{Z}}{m} = \frac{1637(5650)(673)(0.021)(0.85)}{6 \times 10^5}$$

$$kh = 185 \text{ md-ft} \qquad k = \frac{185}{20} = 9.3 \ md$$

At $\Delta t = 1$ hr, $\frac{t + \Delta t}{\Delta t} = 121.5$, $p_{1hr}{}^2 = 12.69 \times 10^6$

$$S' = 1.151\left[\frac{12.69 \times 10^6 - 3295^2}{6 \times 10^5} - \ldots \log\left(\frac{9.3}{(0.1)(0.021)(2.2 \times 10^{-4})(0.29)^2}\right) + 3.23\right]$$

$$S' = 1.151(3.055 - 8.379 + 3.23) = -2.41$$

The values obtained for k and S' agree with those obtained from the drawdown test, Example 3-8.

In order to estimate $\bar{p}_R$, obtain p^{*2} from the graph where $(t + \Delta t)/\Delta t = 1$ to be 13.92×10^6 psia^2. Using the value for k to calculate t_{DA} gives

$$t_{DA} = \frac{2.64 \times 10^{-4}(9.3)(120.5)}{(0.10)(0.021)(2.2 \times 10^{-4})(22 \times 10^6)} = 0.029$$

From Table 3-2, the well is still in unsteady-state flow if $t_{DA} < 0.1$, which is the case for this example. Therefore, $p^* = p_i = 3731$ psia.

Equation 3-11 may be used to predict the stabilized performance of a well by utilizing the results of a transient test. However, this requires isolation of the effects of actual formation damage or stimulation from those caused by non-Darcy or turbulent flow. That is, S' must be broken down into its components S and Dq_{sc}. If a transient test was conducted at only one flow rate, Equation 3-30 may be used to estimate D, and then S may be calculated from $S = S' - Dq_{sc}$.

Applying this procedure to Example 3-9 results in

$$D = \frac{5.18 \times 10^{-5}\,\gamma_g}{\bar{\mu}\, h\, r_w\, k^{0.2}} = \frac{5.18 \times 10^{-5}(.68)}{.021(20)(.29)(9.3)^{0.2}}$$

$$D = 1.85 \times 10^{-4} \text{ Mscfd}^{-1}$$

$$S = S' - Dq_{sc} = -2.41 - 1.85 \times 10^{-4}(5650)$$

$$S = -2.41 - 1.05 = -3.46$$

Real Gas Pseudo-Pressure Analysis

In deriving the equation for reservoir gas flow in terms of $\Delta(p^2)$, one step involved evaluation of the integral

$$2\int_{p_w}^{p_e} \frac{pdp}{\mu Z}. \tag{3-59}$$

Both μ and Z are functions of pressure and therefore should be included in the integration. In order to simplify the derivation it was assumed that μ and Z could be considered constant at reservoir temperature and average pressure in the drainage area, which resulted in $\bar{\mu}$ and $\bar{Z}$ appearing in all of the flow equations. For fairly low pressure reservoirs ($\bar{p}_R \leq 2500$ psia) this approach is adequate, but for some pressure ranges the μZ product is far from linear with pressure. A method to more rigorously evaluate the effects of μ and Z was introduced by Al-Hussainy in 1965[12]. He defined the integral in Equation 3-59 as the real gas pseudo-pressure, $m(p)$.

$$m(p) = \int_{p_m}^{p} 2\frac{p}{\mu Z}dp \tag{3-60}$$

All of the previously derived equations in this chapter can be modified to the pseudo-pressure analysis from the p^2 analysis by replacing the pressure squared terms with the pseudo-pressure terms and eliminating the $\bar{\mu}\bar{Z}$ product from the equation. As an example, Equation 3-11 becomes

$$m(\bar{p}_R) - m(p_w) = \frac{1422\, T\, q_{sc}}{kh}[\ln(0.472 r_e/r_w) + S + Dq_{sc}], \tag{3-61}$$

where $m(\bar{p}_R)$ is the value of the pseudo-pressure function at the pressure $\bar{p}_R$, and $m(p_w)$ is the value at pressure p_w. The definitions of some of the dimensionless variables defined in Table 3-1 will change if the $m(p)$ approach is used.

$$q_D = \frac{1422\, T\, q_{sc}}{kh\, m(p_i)},$$

$$p_D = \frac{m(p)}{m(p_i)\, q_D},$$

$$\Delta p_D = \frac{m(p_i) - m(p)}{m(p_i)\, q_D},$$

$$t_D = \frac{2.64 \times 10^{-4}\, kt}{\phi \mu_i C_i\, r^2},$$

where C_i and μ_i are evaluated at p_i, T. The $m(p)$ analysis may also be used in deliverability testing. This involves substitution of $\Delta m(p)$ for $\Delta(p^2)$ in the procedure to find A_t and B in the LIT analysis.

Use of the $m(p)$ approach requires generation of a table or graph of $m(p)$ versus p for the gas in question at reservoir temperature. This is easily accomplished by numerical integration of Equation 3-60 when values of μ and Z are available for the range of pressures required.

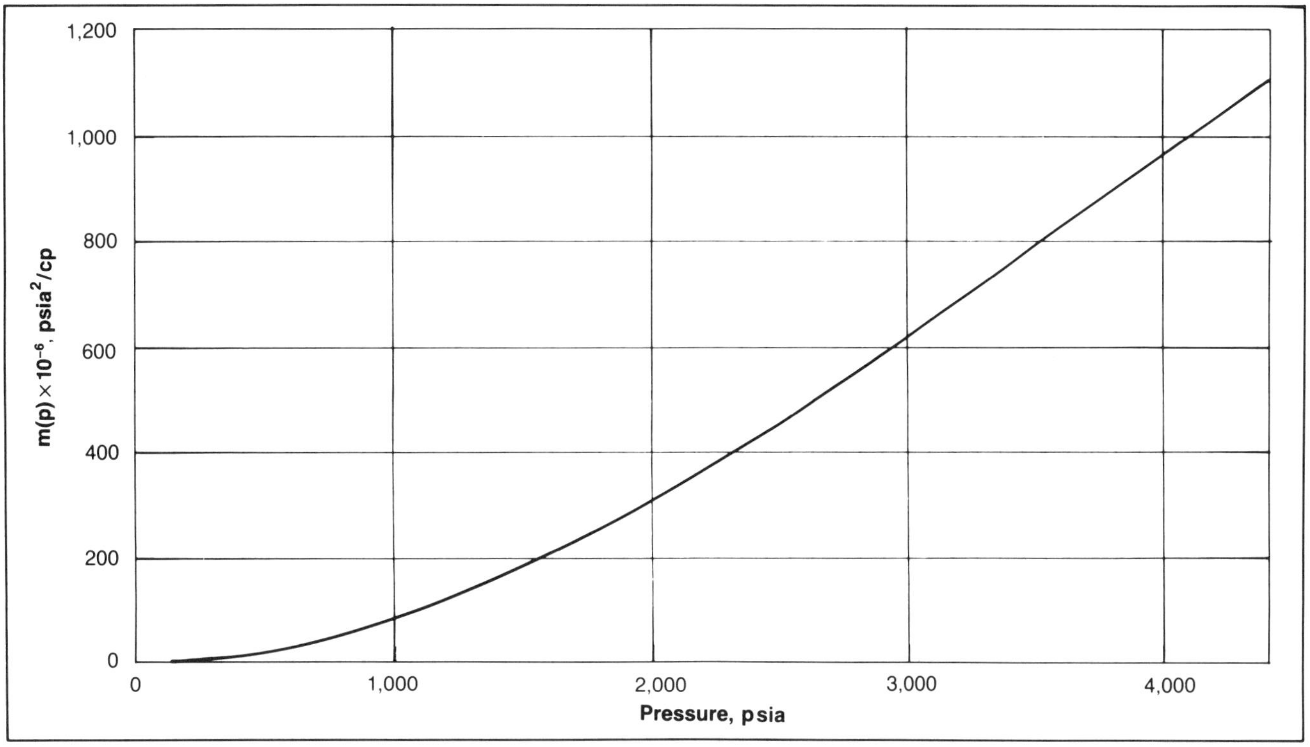

Fig. 3-25. Real gas pseudopressure, as a function of the actual pressure.

Correlations for μ and Z can be found in Chapter 2. A computer program for generating the $m(p)$ versus p values is given in the appendix. An example calculation of $m(p)$ data, as presented by Dake[13] is illustrated.

Example 3-10:

A reservoir existing at a temperature of 200°F contains a gas having a gravity of 0.85. Generate a relationship between $m(p)$ and p over the pressure range of 4400 to zero psia.

Solution:

The following table is used to tabulate the data, taking pressure increments of 400 psi for averaging. The results are plotted in Figure 3-25.

PVT data			*Numerical integration*				*Pseudo pressures*
p (psia)	μ (cp)	Z	$\frac{2p}{\mu Z}$	$\overline{\frac{2p}{\mu Z}}$	Δp	$\overline{\frac{2p}{\mu Z}} \times \Delta p$	$m(p) = \Sigma \overline{\frac{2p}{\mu Z}} \Delta p$ $(psia)^2/cp$
400	.01286	.937	66391	33196	400	13.278×10^6	13.278×10^6
800	.01390	.882	130508	98449	400	39.380×10^6	52.658×10^6
1200	.01530	.832	188537	159522	400	63.809×10^6	116.467×10^6
1600	.01680	.794	239894	214216	400	85.686×10^6	202.153×10^6
2000	.01840	.770	282326	261110	400	104.444×10^6	306.597×10^6
2400	.02010	.763	312983	297655	400	119.062×10^6	425.659×10^6
2800	.02170	.775	332986	322985	400	129.194×10^6	554.853×10^6
3200	.02340	.797	343167	338079	400	135.231×10^6	690.084×10^6
3600	.02500	.827	348247	345707	400	138.283×10^6	828.367×10^6
4000	.02660	.860	349711	348979	400	139.592×10^6	967.958×10^6
4400	.02831	.896	346924	348318	400	139.327×10^6	1107.285×10^6

Example 3-11:

Analyze the buildup test of Example 3-9 using the pseudo-pressure analysis to obtain k, S' and $\bar{p}_R$. The relationship between p and $m(p)$ is given in Figure 3-26.

Solution:

Values for $m(p)$ and $(t + \Delta t)/\Delta t$ are calculated and tabulated as follows.

Δt, hrs.	p_{ws}, psia	$\frac{t + \Delta t}{\Delta t}$	$m(p) \times 10^{-6}$, psia^2/cp
0.00	3295	—	709.8
0.53	3296	228.4	710.1
1.60	3385	76.3	742.7
2.67	3547	46.1	802.8
3.73	3573	33.3	812.6
4.80	3591	26.1	819.3
5.87	3605	21.5	824.6
6.93	3614	18.4	828.0
8.00	3623	16.1	831.4
9.87	3634	13.2	835.5
12.00	3644	11.0	839.3
14.67	3654	9.2	843.1
18.67	3664	7.5	846.8
24.53	3672	5.9	849.9
29.33	3676	5.1	851.4
35.73	3684	4.4	854.5
45.87	3688	3.6	855.9
49.87	3691	3.4	857.1

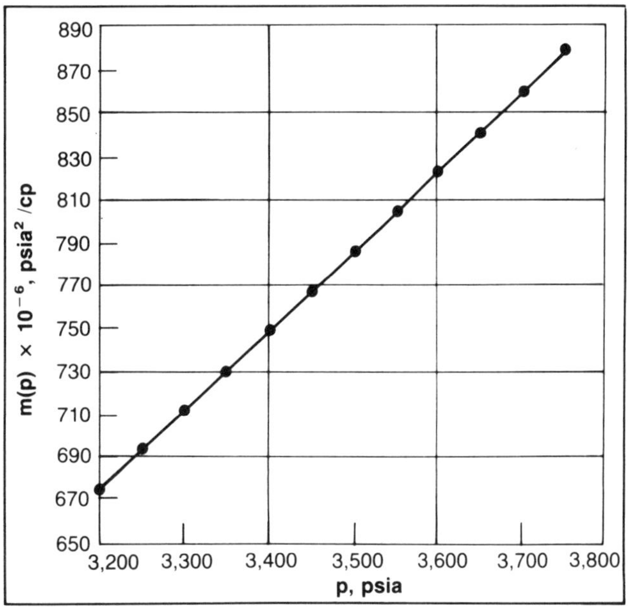

Fig. 3-26. m(p) versus p data, Example 3-11.

A plot of $m(p)$ versus $\log((t + \Delta t)/\Delta t)$ is shown in Figure 3-27. The slope of the straight line segment is:

$$m = \frac{(873.0 - 842.5) \times 10^6}{\log 10 - \log 1} = 30.5 \times 10^6$$

$$kh = \frac{1637\, q_{sc}\, T}{m} = \frac{1637(5650)(673)}{30.5 \times 10^6} = 204 \text{ md-ft}$$

$$k = \frac{204}{20} = 10.2\ md$$

At $\Delta t = 1$ hr, $\quad m(p_{1hr}) = 809 \times 10^6$

$$S' = 1.151 \left[\frac{m(p_{1hr}) - m(p_{wf})}{m} - \log \frac{k}{\phi\, \mu_i C_i\, r_w^2} + 3.23\right]$$

$$S' = 1.151 \left[\frac{(809.0 - 709.8) \times 10^6}{30.5 \times 10^6} - \log \ldots\right.$$

$$\left. \ldots \frac{10.2}{0.10(0.021)(2.2 \times 10^{-4})(0.29)^2} + 3.23\right]$$

$$S' = -2.2$$

These values of k and S' agree with those obtained in Example 3-9.

In order to obtain a value for $p^* = p_i\,(t_{DA} < 0.1)$, read $m(p^*)$ at $(t + \Delta t)/\Delta t = 1$ to be 873×10^6. This corresponds to a value of $p_i = 3730$ psia, which is equal to the result obtained previously.

It has generally been recommended in the literature that the pseudo-pressure analysis be used in preference to the pressure-squared analysis when $p_i > 2500$ psia. The only extra work involved in using the $m(p)$ method is generation of the $m(p)$ versus p data. This task has somehow delayed the acceptance of the $m(p)$ method among field engineers, and many prefer using the more familiar p^2 analysis. Some advantages of the $m(p)$ method are:

— It is more theoretically correct,
— It applies for all ranges of pressure,
— Iteration to obtain $\bar{p} = (p_i + p_{wf})/2$ for evaluating $\bar{\mu}$ $\bar{Z}$ is not necessary, and
— Once $m(p)$ versus p is available, it is simpler to use.

GAS RESERVES

All of the flow equations used in the previous sections of this chapter involved either initial reservoir pressure p_i, or average reservoir pressure $\bar{p}_R$. These pressures are functions of the gas in place in a reservoir and must be evaluated at various times in the reservoir life. The flow capacity or deliverability of the wells declines as gas is produced and $\bar{p}_R$ declines. Therefore, it is necessary to be able to predict $\bar{p}_R$ versus gas produced G_p.

In order to evaluate a gas reservoir the original gas in place G, must be determined and the gas recoverable at different values of $\bar{p}_R$ must be calculated. The decline in $\bar{p}_R$ with G_p can be calculated using the gas laws presented in Chapter 2 for a volumetric reservoir. If the volume occupied by gas changes because of water influx, it must be accounted for in the balance. This section will present methods to determine gas reservoir behavior using volumetric methods and material balance methods.

Reserve Estimates—Volumetric Method

The volumetric method for determining initial gas in place and reserves requires enough geologic data to determine reservoir pore volume and water saturation. Reservoir pressure is also required, but no production history is necessary. It is applied mainly in new fields for rough estimates.

The equation for calculating gas in place is

$$G = \frac{43560 A h \phi (1 - S_w)}{B_g}, \tag{3-62}$$

where

G = gas in place, scf,
A = area of reservoir, acres,
h = average reservoir thickness, ft,
ϕ = porosity,
S_w = water saturation, and
B_g = gas formation volume factor, ft³/scf.

This equation can be applied at both initial and aban-

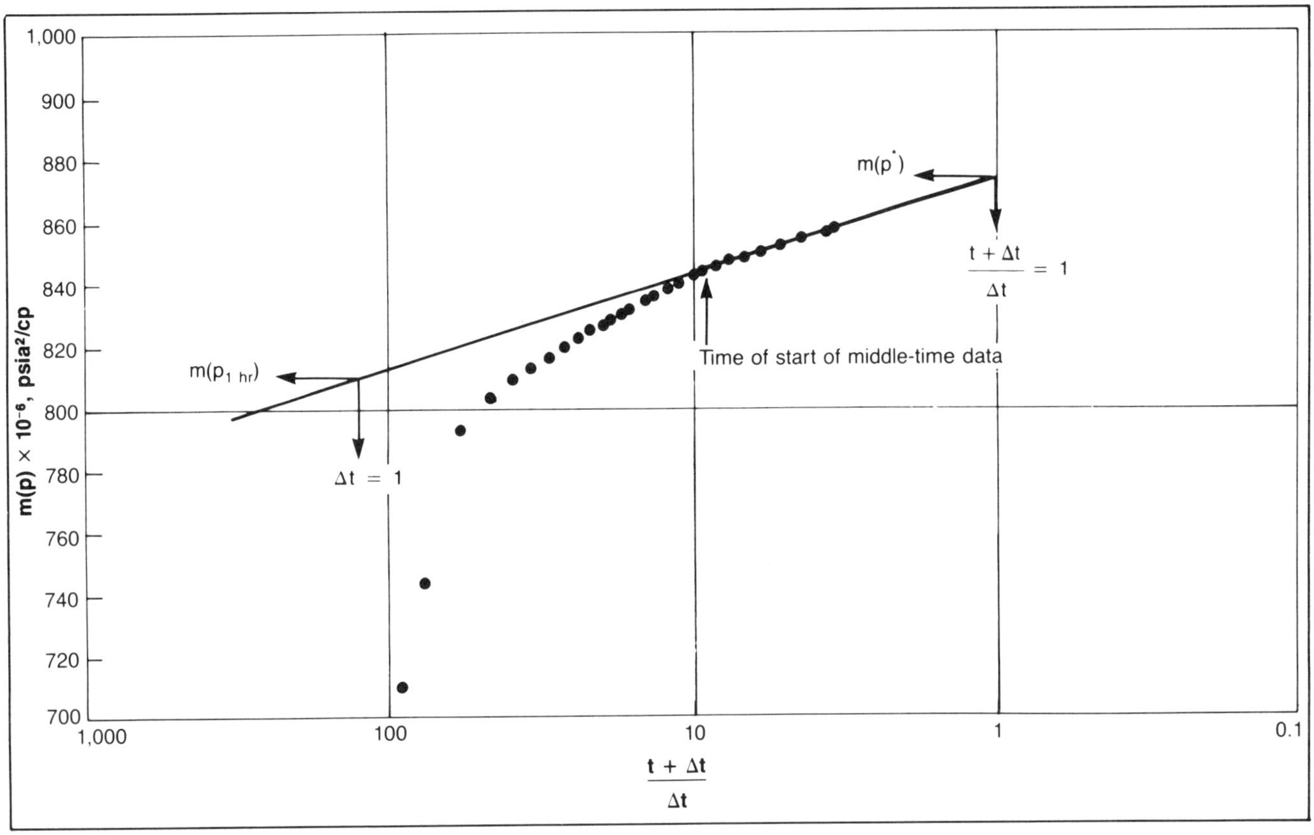

Fig. 3-27. Horner build-up plot for Example 3-11. Courtesy the Energy Resources Conservation Board, Calgary.

donment conditions in order to calculate the recoverable gas.

Gas produced = Initial gas − Remaining gas

or

$$G_p = 43560Ah\phi(1 - S_w)\left(\frac{1}{B_{gi}} - \frac{1}{B_{ga}}\right), \qquad (3\text{-}63)$$

where B_{ga} is evaluated at abandonment pressure. Application of the volumetric method assumes that the pore volume occupied by gas is constant. If water influx is occurring, A, h, and S_w will change.

Example 3-12:

A gas reservoir has the following characteristics. Calculate the gas recovered and the recovery factor at pressures of 1000 psia and 500 psia.

A = 2550 acres h = 50 ft ϕ = 0.20
S_w = 20% T = 186°F p_i = 2651 psia
γ_g = 0.70

p	Z	B_g
2651	0.83	0.0057
1000	0.90	0.0154
500	0.95	0.0346

Solution:

At p = 1000 psia

$$G_p = 43560(2550)(50)(0.2)(1 - 0.2) \cdot \left(\frac{1}{0.0057} - \frac{1}{0.0154}\right)$$

$$G_p = 8.886 \times 10^8(175.44 - 64.94)$$

$$= 9.818 \times 10^{10} \text{ scf}$$

$$E_g = 1 - \frac{B_{gi}}{B_{ga}} = 1 - \frac{0.0057}{0.0154} = 0.63 = 63\%$$

At p = 500

$$G_p = 8.886 \times 10^8\left(\frac{1}{0.0057} - \frac{1}{0.0346}\right)$$

$$G_p = 8.886 \times 10^8(175.44 - 28.90)$$

$$= 13.02 \times 10^{10} \text{ scf}$$

$$E_g = 1 - \frac{0.0057}{0.0346} = 0.835 = 83.5\%$$

The recovery factors for volumetric gas reservoirs will range from 80 to 90%. If a strong water drive is present, trapping of residual gas at higher pressures can reduce the recovery factor substantially, to the range of 50 to 80%.

Reserve Estimates—Material Balance Method

If enough production-pressure history is available for a gas reservoir, the initial gas in place G, the initial reservoir pressure, p_i, and the gas reserves can be calculated without knowing A, h, ϕ or S_w. This is accomplished by forming a mass or mole balance on the gas. That is,

Moles Produced = initial moles in place − remaining moles.

In equation form

$$n_p = n_i - n. \tag{3-64}$$

Applying the gas law, $pV = ZnRT$ gives

$$\frac{p_{sc}G_p}{T_{sc}Z_{sc}} = \frac{p_iV_i}{T_fZ_i} - \frac{pV_i}{T_fZ}, \tag{3-65}$$

where

T_f = formation temperature,
V_i = reservoir gas volume,
p_i = initial reservoir pressure, and
p = reservoir pressure after producing G_p scf.

The reservoir gas volume can be put in units of scf by use of B_{gi}. That is,

$$V_i = GB_{gi}. \tag{3-66}$$

Combining Equations 3-65 and 3-66 and solving for p/Z gives

$$\frac{p}{Z} = \frac{p_i}{Z_i} - \frac{T_f\,p_{sc}}{T_{sc}\,B_{gi}\,G}G_p, \tag{3-67}$$

from which it is obvious that a plot of p/Z versus G_p will produce a straight line of slope $(T_f\,p_{sc}/T_{sc}\,B_{gi}\,G)$ and intercept at $G_p = 0$ of p_i/Z_i. Thus, both G and p_i can be obtained graphically. Once these values are obtained, a value of gas recovered G_p, can be determined for any pressure. Equation 3-67 can also be expressed in terms of recovery factor as

$$\frac{p}{Z} = \frac{p_i}{Z_i}\left(1 - \frac{G_p}{G}\right). \tag{3-68}$$

Figure 3-28 illustrates typical plots of Equations 3-67 and 3-68.

Example 3-13:

The following production history was obtained from a volumetric gas reservoir. Plotting of p/Z versus G_p revealed that data points 1 and 4 fall on the best straight line through the data. Use these points to find G and p_i/Z_i. Also estimate the gas recovery if the reservoir pressure is drawn down to 300 psia. T_f = 200°F, γ_g = 0.9

	p, psia	Z	p/Z	G_p, Bcf
1	1885	0.767	2458	6.873
2	1620	0.787	2058	14.002
3	1205	0.828	1455	23.687
4	888	0.866	1025	31.009
5	645	0.900	717	36.207

Solution:

Using points 1 and 4, calculate the slope:

$$m = \frac{(p/Z)_1 - (p/Z)_4}{G_{p4} - G_{p1}} = \frac{2458 - 1025}{31.009 - 6.873} = 59.37\,\frac{\text{psia}}{\text{Bcf}}$$

To find p_i/Z_i, use Equation 3-67 and data point 1:

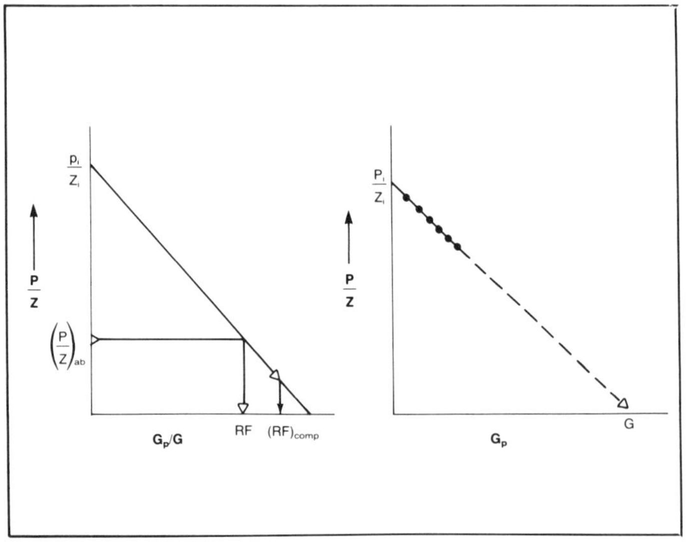

Fig. 3-28. Gas material balance plots.

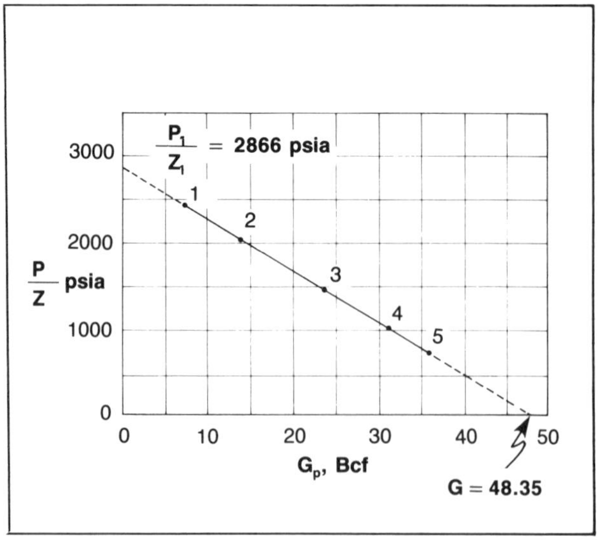

Fig. 3-28a. Example 3-13 Solution.

$$\frac{p_i}{Z_i} = \frac{p}{Z} + m\, G_p = 2458 + 59.37(6.873) = 2866 \text{ psia}$$

$$B_{gi} = 0.0283 \frac{T_f Z_i}{p_i} = \frac{0.0283(660)}{2866} = 0.0065 \text{ ft}^3/\text{scf}$$

$$G = \frac{T_f\, p_{sc}}{T_{sc}\, B_{gi}\, m} = \frac{p_i}{Z_i m} = \frac{2866}{59.37}$$

$$= 48.3 \text{ Bcf}$$

Recovery at $p = 300$ psia:

At $p = 300$ *psia*, $Z = 0.949$, $p/Z = 316$ psia

$$G_p = \left(\frac{p_i}{Z_i} - \frac{p}{Z}\right) \Big/ m = \frac{(2866 - 316)}{59.37} = 42.95 \text{ Bcf.}$$

The value of p_i can be obtained from p_i/Z_i by trial-and-error. Values of p_i are estimated; Z_i and p_i/Z_i are calculated until a value is determined that gives 2866 psia.

p_i	Z_i	p_i/Z_i
2200	.750	2933
2100	.752	2793
2150	.751	2863

Linear interpolation between 2200 and 2150 gives $p_i = 2151$ psia.

The above discussion applies to a reservoir in which the pore volume occupied by gas does not change. If water influx is occurring, V_i will be reduced by water influx as p declines.
Equation 3-67 then becomes

$$\frac{p}{Z} = \left(\frac{G\, B_{gi}}{G\, B_{gi} - W_e}\right)\frac{p_i}{Z_i} - \frac{T_f\, p_{sc}}{T_{sc}\,(G\, B_{gi} - W_e)}\, G_p. \tag{3-69}$$

The slope now includes a quantity that varies with G_p or time. This is the water influx W_e. Therefore, a plot of p/Z versus G_p will no longer be linear but will deviate upward with pressure decline depending on the strength of the water drive. This is illustrated in Figure 3-29.

If water influx is occurring, it can be calculated by using the constant pressure solution to the diffusivity equation, Equation 3-70. This involves the correct determination of certain water influx parameters. Havlena and Odeh[14] have described a procedure by which the material balance equation can be used to determine if these parameters have been correctly evaluated.

Water influx is a function of time, and can be expressed as

$$W_e = C\, \Sigma\, \Delta p\, Q_{tD}, \tag{3-70}$$

where

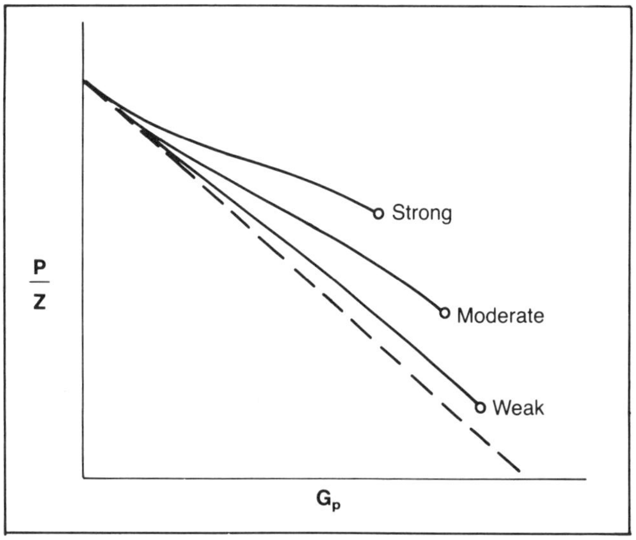

Fig. 3-29. Effect of water drive.

C = water influx constant,
$\Delta p = p_i - p_t$
Q_{tD} = dimensionless water influx.

The value of Q_{tD} depends on the ratio of the aquifer volume to the gas reservoir volume and the time that the reservoir has been produced. Values may be obtained from Figures 3-6 and 3-7.

Energy Plots

Many other graphical methods have been proposed for solving the material balance equation that are useful in detecting the presence of water influx. Equation 3-67 can be rearranged and using the definition of $B_{gi} = (p_{sc}\, Z_i\, T_f)/(p_i\, T_{sc})$, it can be written as

$$\frac{Z_i\, p}{p_i\, Z} = \frac{G\, B_{gi}}{G\, B_{gi} - W_e}\left[1 - \frac{G_p}{G}\right]. \tag{3-71}$$

For $W_e = 0$, this becomes

$$1 - \frac{Z_i\, p}{p_i\, Z} = \frac{G_p}{G}. \tag{3-72}$$

Taking the logarithm of both sides of Equation 3-72 yields

$$\log\left[1 - \frac{Z_i\, p}{p_i\, Z}\right] = \log G_p - \log G. \tag{3-73}$$

From Equation 3-73, it is obvious that a plot of $1 - Z_i p/p_i Z$ versus G_p on log-log coordinates will yield a straight line with a slope of one (45° angle). An extrapolation to one on the vertical axis ($p = 0$) yields a value for initial gas in place, G. The graphs obtained from this type of analysis have been referred to as Energy Plots. They have been found to be useful in detecting water influx early in the life of a reservoir. If W_e is not zero, the slope of the plot will be less than one, and will also decrease with time, since W_e increases with time. An increasing slope can only occur as a result of either gas leaking from the reservoir or bad data, since the increasing slope would imply that the gas occupied pore volume was increasing with time. This is also true for plots of p/Z versus G_p.

The detection of production problems by observing anomalies in these plots is discussed further in Chapter 9.

ABNORMALLY PRESSURED RESERVOIRS

In most gas reservoirs the gas is much more compressible than the rock and the connate water, which justifies ignoring the rock and water expansion in the material balance equation. However, in abnormally high pressured reservoirs, this may introduce some error in estimating gas reserves. Including rock and water compressibility, Equation 3-67 becomes

$$\frac{p}{Z} = \frac{\dfrac{p_i}{Z_i}\left[1 - \dfrac{G_p}{G}\right]}{1 - \left[\dfrac{(p_i - p)(C_w S_{wi} + C_f)}{1 - S_{wi}}\right]} \tag{3-74}$$

where

S_{wi} = connate water saturation,
C_w = connate water compressibility, and
C_f = formation compressibility.

An estimate of the effect of water and rock compressibility on errors in estimating gas-in-place from Equation 3-67 can be made by the following procedure:

1. Solve Equation 3-67 for $G_{cal.}$
2. Solve Equation 3-74 for G_{actual}
3. Substitute these expressions into the equation for percent error

$$\%\ \text{error} = \left[\frac{G_{actual} - G_{cal.}}{G_{actual}}\right] \times 100$$

This results in the following equation:

$$\%\ \text{error} = \frac{100\,(p_i - p)(C_w S_{wi} + C_f)\left[\dfrac{p_i Z}{Z_i p} - 1\right]}{1 - S_{wi}} \tag{3-75}$$

From Equation 3-75, it can be seen that the error in G obtained from a plot of p/Z versus G_p increases as C_w, C_f, and p_i increase. It has been found that for values of p_i less than 5000 psia and C_f less than 5×10^{-6} psi^{-1}, the error in neglecting the compressibilities was negligible. Further effects of abnormal reservoir pressure on reserve estimates are discussed in Chapter 9.

WELL COMPLETION EFFECTS

In many cases the inflow into a well is controlled more by the completion efficiency than by the actual reservoir characteristics.

There are basically three types of completions that may be made on a well, depending on the type of well, the well depth, and the type of reservoir or formation. In some cases the well is completed open-hole. That is, the casing is set at the top of the producing formation, and the formation is not exposed to cement. Also, no perforations are required. This type of completion is not nearly as common as it was several years ago, and most wells are now completed by cementing the casing through the producing formation.

The most widely used completion method is one in which the pipe is set through the formation, and cement is used to fill the annulus between the casing and the hole. This of course requires perforating the well to establish communication with the producing formation. This type of completion permits selection of the zones that are to be opened. The efficiency of the completion is highly dependent on the number of holes or perforations, the depth to which the perforation extends into the formation, the perforation pattern, and whether there is a positive pressure differential from the well to the formation or vice versa. Compaction of the formation immediately around the perforation can reduce the efficiency considerably.

In some reservoirs the lack of cementing material in the reservoir allows sand to be produced into the well. When completing wells in which the formation is incompetent or unconsolidated, a gravel pack completion scheme is frequently employed. In this type of completion a perforated or slotted liner or a screen liner is set inside the casing opposite the producing formation. The annulus between the casing and the liner is then filled with a sand that is more coarse than the formation sand. The size of the sand, or gravel, depends on the reservoir sand characteristics and on the type of gravel pack. The gravel pack sand also fills the perforation tunnels, and in some cases a zone is washed out behind the pipe, which is also filled with pack sand. Even though the pack sand is loosely packed and has a high permeability, non-Darcy or turbulent flow through the sand-filled perforation tunnels can cause a considerable pressure drop across the gravel pack. This pressure drop not only decreases inflow into the wellbore but also destroys the gravel pack if it is too large.

In order to calculate the extra pressure drop caused by the completion, the general inflow equations can be modified to include the completion efficiency for any type of completion. The equations for gas flow are given as:

$$q_{sc} = \frac{703 \times 10^{-6}\, k_g\, h(\bar{p}_R^2 - p_{wf}^2)}{\bar{\mu}_g\, \bar{Z}\, T[\ln(0.472 r_e/r_w) + S']}, \tag{3-10}$$

where

$$S' = S + Dq_{sc}$$

The value of S' can be obtained from a single transient test, but obtaining accurate values for S and D requires transient tests conducted at two different rates.

Equation 3-10 may be written in a different form, as shown previously.

$$\bar{p}_R^2 - p_{wf}^2 = Aq_{sc} + Bq_{sc}^2, \tag{3-11}$$

where A is the laminar coefficient, and B is the turbulence multiplier. Those coefficients may be written as composites of several terms that depend on the completion characteristics.

$$A = A_R + A_P + A_G \tag{3-76}$$

$$B = B_R + B_P + B_G \tag{3-77}$$

where

A_R = laminar reservoir component,
A_P = laminar perforation component,
A_G = laminar gravel pack component,
B_R = turbulent reservoir component,
B_P = turbulent perforation component, and
B_G = turbulent gravel pack component.

These components have different definitions for oil and gas flow. Only values of the overall coefficients A and B can be obtained from well tests; therefore equations for estimating the value of the components must be available if the effects of each are to be isolated.

Open-Hole Completions

The only effect of the completion on inflow performance of an open-hole completion will be caused by alteration of the reservoir permeability by damage or stimulation. The inflow equation becomes

$$\bar{p}_R^2 - p_{wf}^2 = A_R q_{sc} + B_R q_{sc}^2.$$

The laminar reservoir component includes the effect of Darcy or laminar flow in the reservoir plus any actual formation damage or stimulation. The defining equation is

$$A_R = \frac{1422\, \bar{\mu}_g\, \bar{Z}T}{k_{gR}\, h}[\ln(0.472 r_e/r_w) + S_d], \tag{3-78}$$

where

k_{gR} = unaltered reservoir permeability to gas, and
S_d = skin factor due to permeability alteration around the wellbore.

A value for S_d may be estimated from the following equation:

$$S_d = \left[\frac{k_R}{k_d} - 1\right] \ln(r_d/r_w), \tag{3-79}$$

where

k_R = reservoir permeability,
k_d = altered zone permeability,
r_w = wellbore radius, and
r_d = altered zone radius.

The actual calculation of an accurate value of S_d is difficult, because values of k_d and r_d must be estimated. If a value of S can be obtained from a transient test, this will be equal to S_d for an open hole completion.

The value of B_R will usually be low except for high rate gas wells. It may be calculated from

$$B_R = \frac{3.161 \times 10^{-12}\, \beta_R\, \gamma_g\, \bar{Z}T}{h^2 r_w}. \tag{3-80}$$

Values of the velocity coefficient β may be calculated from

$$\beta_R = \frac{2.33 \times 10^{10}}{k_R^{1.2}}. \tag{3-8}$$

A value for B_R can be calculated if a value of D is available from a transient test on an open hole completion. The units to be used in all of the equations presented in this chapter are the field units described previously.

Perforated Completions

The efficiency of a perforated completion depends on both the reservoir and perforation components in Equation 3-11. That is,

$$\bar{p}_R^2 - p_{wf}^2 = (A_R + A_P)q_{sc} + (B_R + B_P)q_{sc}^2.$$

The laminar perforation component includes the effects of the number and types of perforations and the effects of compaction around the perforations. These effects were discussed in detail by McLeod[16] and the discussion on perforated completions presented here is based largely on McLeod's work. The Equation is:

$$A_P = \frac{1422\, \bar{\mu}_g \bar{Z}T}{k_R\, h}(S_P + S_{dP}), \tag{3-81}$$

where

S_P = effect of flow converging into perforations, and

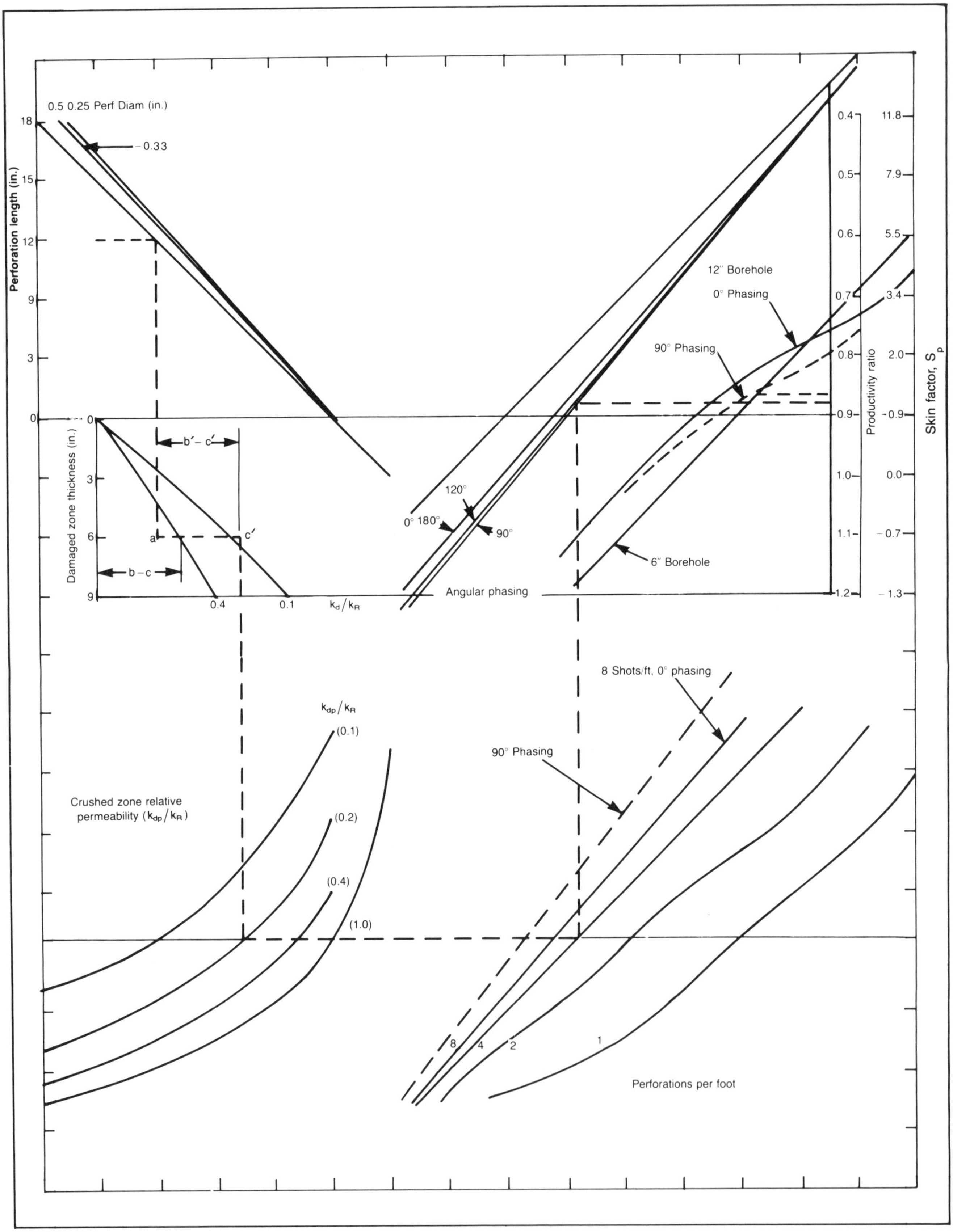

Fig. 3-30. Nomograph for productivity ratio. Permission to publish by the Society of Petroleum Engineers of AIME. Copyright 1981 SPE-AIME.

S_{dP} = effect of flow through the compacted and damaged zone around the perforation.

If sufficient data regarding the perforation are known, values for S_P and S_{dP} may be calculated. S_P is a function of perforating density, perforation length, perforation diameter, ratio of vertical to horizontal permeability, and damaged zone radius.

Values of S_p may be obtained from nomographs published by Hong[17] or Locke.[18] An equation for estimating S_P was given by Saidikowski.[19]

$$S_P = \left(\frac{h}{h_p} - 1\right)\left[\ln\left(\frac{h}{r_w}\left(\frac{k_R}{k_v}\right)^{0.5}\right) - 2\right] \tag{3-82}$$

where

h = total formation thickness,
h_p = perforated interval length,
k_R = reservoir permeability in the horizontal direction, and
k_v = vertical permeability.

The nomograph presented by Locke is shown in Figure 3-30.

McLeod derived an equation for calculating the effect of flow through the compacted zone as

$$S_{dP} = \left(\frac{h}{L_P N}\right)\left(\frac{k_R}{k_{dP}} - \frac{k_R}{k_d}\right) \ln(r_{dP}/r_P), \tag{3-83}$$

where

h = total formation thickness,
L_P = perforation length,
N = total number of perforations,
k_R = unaltered reservoir permeability,
k_d = altered reservoir permeability,
k_{dP} = compacted zone permeability,
r_P = perforation radius, and
r_{dP} = compacted zone radius

Figure 3-31, from Reference 16 shows a schematic of a perforated completion and the relationship among the various parameters in Equation 3-83.

The largest part of the pressure drop through a perforation is caused by turbulent or non-Darcy flow through the compacted zone. The equation for calculating this effect is

$$B_P = \frac{3.161 \times 10^{-12} \beta_{dP} \gamma_g \bar{Z} T}{r_P L_P^2 N^2}. \tag{3-84}$$

The value of the velocity coefficient must be calculated using the compacted zone permeability. The equation is

$$\beta_{dP} = \frac{2.33 \times 10^{10}}{k_{dP}^{1.2}}. \tag{3-85}$$

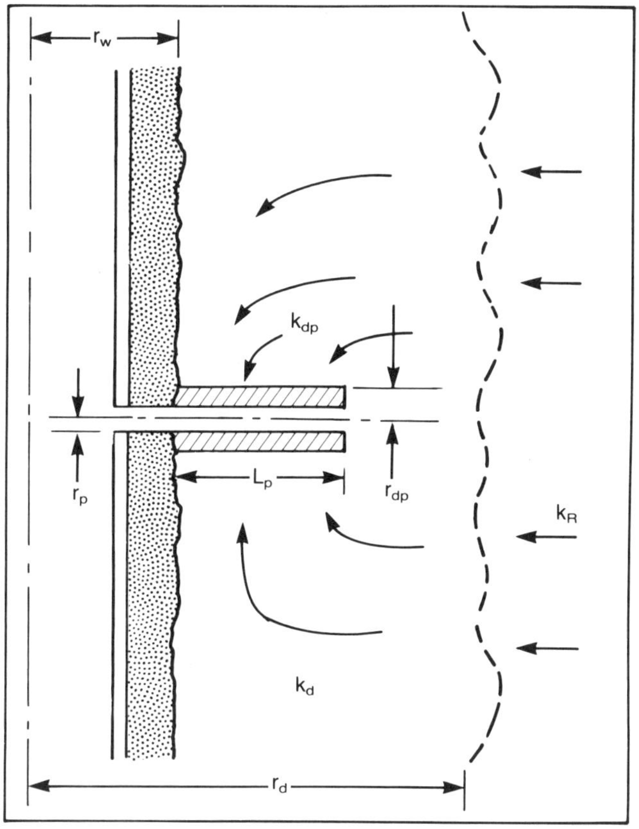

Fig. 3-31. Flow into a perforation.

There are several variables in the equations for perforated completions that are hard to determine. These include the altered zone permeability, the compacted zone permeability, the compacted zone radius, the perforation length, and the altered zone radius. Some of these parameters can be estimated from API-RP-43 test data published by the perforating companies. The following guidelines have been recommended by McLeod:

For wells perforated in mud

$$\frac{k_{dP}}{k_R} = \frac{k_c}{k} \tag{3-86}$$

For wells perforated in brine

$$\frac{k_{dP}}{k_d} = \frac{k_c}{k} \tag{3-87}$$

where k_c/k values are obtained from the API test data. Guidelines for estimating k_c/k when no tests are available were also presented by McLeod in Table 3-4.

McLeod also suggests that the compacted zone thickness is usually about 0.5 in. That is, $r_{dP} = r_P + 0.5$ if r_P is in inches. If no information is available regarding the altered zone radius, a value of $r_d = r_w + 1$ may be used, where r_w is given in ft.

TABLE 3-4
Perforating Parameter Guidelines

Fluid in Hole	Pressure Conditions	k_c/k
High solids mud	overbalance	0.01–0.03
Low solids mud	overbalance	0.02–0.04
Unfiltered brine	overbalance	0.04–0.06
Filtered brine	overbalance	0.08–0.16
Filtered brine	underbalance	0.15–0.25
Clean fluid	underbalance	0.30–0.50
Ideal fluid	underbalance	1.00

Perforated, Gravel-Packed Completions

The equation for a gravel-packed completion is

$$\bar{p}_R^2 - p_{wf}^2 = (A_R + A_P + A_G)q_{sc} + (B_R + B_P + B_G)q_{sc}^2. \quad (3\text{-}88)$$

For most gravel-packed wells the formation will have a high permeability because of the unconsolidated nature of the sand. This will also result in minimum damage from the compacted zone around the perforations. However, the effect of the linear flow through the perforation tunnel that is filled with pack sand can cause a significant non-Darcy flow pressure drop. The equations for A_G and B_G are

$$A_G = \frac{2844\,\bar{Z}T\bar{\mu}_g L}{k_G N r_P^2}, \quad (3\text{-}89)$$

$$B_G = \frac{1.263 \times 10^{-11}\,\beta_G \gamma_g \bar{Z}TL}{N^2 r_P^4}, \quad (3\text{-}90)$$

where

k_G = gravel permeability, md,
L = perforation tunnel length, ft, and

$$\beta_G = \frac{1.47 \times 10^7}{k_G^{0.55}}. \quad (3\text{-}91)$$

The following data, from Gurley,[20] may be used to estimate the gravel permeability based on its size.

Sieve Size	k_G, *md*
10–20	5.00×10^5
16–30	2.50×10^5
20–40	1.20×10^5
40–60	4.00×10^4

A schematic of a gravel packed completion is illustrated in Figure 3-32.

As illustrated in Figure 3-32, the tunnel length is defined as the radius of the hole minus the outside radius of the screen. In some cases it is defined as the hole radius minus the inside radius of the casing.

In analyzing gravel-packed completions it is sometimes convenient to break down the total pressure drawdown into two separate components. That is, the pressure drop in the reservoir and the pressure drop in the gravel pack. This can be expressed as

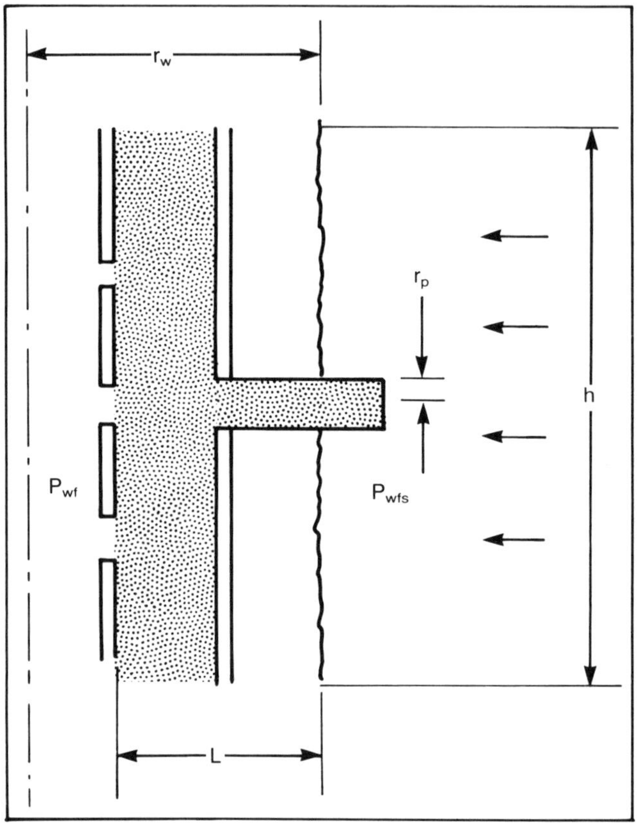

Fig. 3-32. Gravel-packed completion.

$$\bar{p}_R - p_{wf} = \bar{p}_R - p_{wfs} + (p_{wfs} - p_{wf}), \quad (3\text{-}92)$$

where p_{wfs} is the pressure existing at the sand face, as illustrated in Figure 3-32. Most operators agree that the pressure drop across the gravel pack, $p_{wfs} - p_{wf}$, should be below about 300 psi. The equations for the two pressure drops, ignoring A_P and B_P, may be written as

$$\bar{p}_R^2 - p_{wfs}^2 = A_R q_{sc} + B_R q_{sc}^2 \quad (3\text{-}93)$$

$$p_{wfs}^2 - p_{wf}^2 = A_G q_{sc} + B_G q_{sc}^2 \quad (3\text{-}94)$$

This type of analysis may also be applied to nongravel packed completions.

TIGHT GAS WELL ANALYSIS

Predicting the inflow performance for a very low permeability gas well can be difficult because of the necessity to obtain stabilized well tests. As was discussed previously, both the isochronal and modified isochronal tests require one stabilized production point for analysis.

Even though the LIT analysis can be used without a stabilized test, using this method to generate a stabilized IPR assumes that the well will eventually stabilize.

If a gas well is draining a fairly large area from a tight reservoir, the well may produce in the unsteady-state flow regime for most of its life. As an example, the stabilization time for a well draining a square or circular area can be estimated from

$$t_s = \frac{380\,\phi\,\bar{\mu}_g\,\bar{C}\,A}{k}. \tag{3-33}$$

Consider a well having the following characteristics:

$\phi = 10\%$, $\bar{\mu}_g = 0.011$ cp, $\bar{C} = 2.5 \times 10^{-4}$ psi^{-1}, $A = 640$ acres, $k = 0.05$ md

The stabilization time is approximately

$$t_s = \frac{380(0.1)(0.011)(2.5 \times 10^{-4})(640)(43{,}560)}{0.05}$$

$t_s = 58{,}266$ hours $= 6.65$ years.

This is of course an extreme case, and a well with permeability this low would require stimulation by a massive hydraulic fracture to be commercial.

A procedure for predicting the inflow performance of a fractured gas well as a function of time was presented by Agrawal, et al.,[21] in 1979. The method uses the constant well pressure solution to the diffusivity equation and utilizes computer generated, dimensionless type curves to evaluate the fracture parameters.

A procedure is described to analyze a drawdown test in which the wellbore flowing pressure is held constant and the decline of flow rate with time is measured. The following dimensionless variables are used in analyzing a well with a hydraulic fracture of length $X_e = 2X_f$. A schematic of the system geometry is shown in Figure 3-33.

The dimensionless time based on fracture half-length is

$$t_{DX} = \frac{0.000264\,k\,t}{\phi(\mu C)_i\,X_f^2}. \tag{3-95}$$

The dimensionless flow rate is expressed as

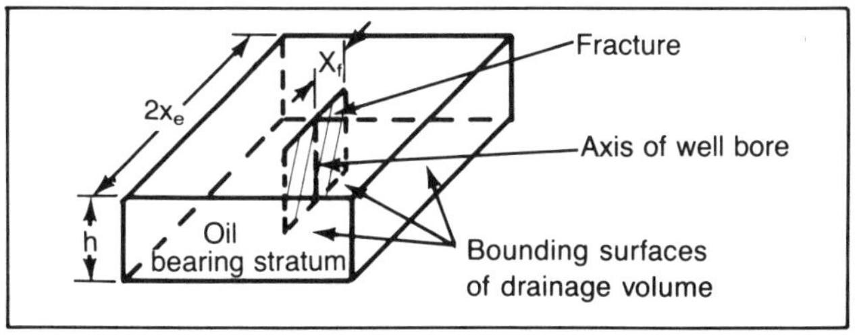

Fig. 3-33. Schematic view of a fractured well.

$$\frac{1}{q_D} = \frac{k\,h\,(p_i^2 - p_{wf}^2)}{1422\,q_{sc}\,\bar{\mu}\,\bar{Z}\,T}, \tag{3-96}$$

where

k = permeability, md,
t = flowing time, hours,
ϕ = porosity,
$(\mu C)_i$ = viscosity—compressibility product evaluated at p_i, T,
μ = viscosity, cp,
C = total compressibility, psi^{-1},
X_f = fracture half-length, ft,
h = formation thickness, ft,
p_i = initial reservoir pressure, psia,
p_{wf} = flowing wellbore pressure, psia,
q_{sc} = gas flow rate, Mscfd,
T = reservoir temperature, °R, and
Z = gas compressibility factor evaluated at T, $\bar{p} = (p_i + p_{wf})/2$.

A dimensionless fracture flow capacity was defined as

$$F_{CD} = \frac{k_f\,w}{k\,X_f} \tag{3-97}$$

where

k_f = fracture permeability, md,
k = formation permeability, md,
w = fracture width, ft, and
X_f = fracture half-length, ft.

Agrawal, et al., presented dimensionless curves of $1/q_D$ versus t_{DX} with F_{CD} as a parameter. An example curve is shown in Figure 3-34. Larger versions of this curve are available from the Society of Petroleum Engineers.

The following procedure may be used to calculate the production rate versus time and flowing wellbore pressure. A drawdown test is used to evaluate X_f, k and F_{CD}.

1. Plot $1/q_{sc}$ versus flowing time t on a log-log scale having the same scale as the type curve, Figure 3-34. It is convenient to use tracing paper to make this plot.
2. Shift the tracing plot in both the horizontal and vertical directions until a match is obtained between the measured data and one of the curves on the type curve. This determines the correct value of F_{CD}.
3. Select a match point at any convenient point on the graph. This establishes a correspondence between t and t_{DX} and between $1/q_{sc}$ and $1/q_D$.
4. Using the known values of t and q_D, calculate k and X_f:

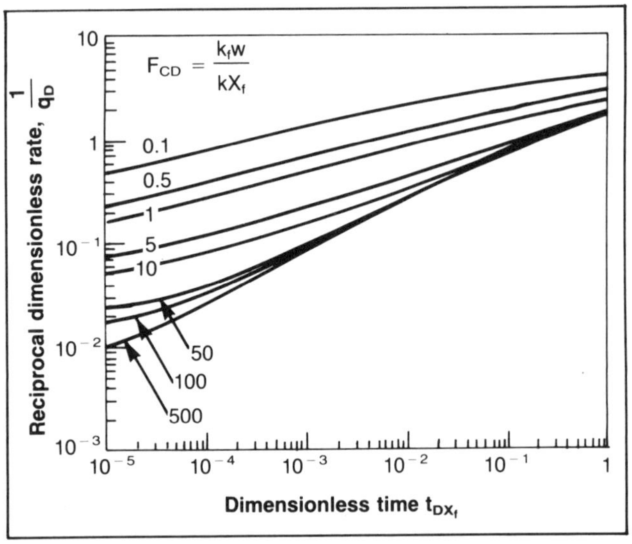

Fig. 3-34. Constant-pressure log-log type curves for finite flow-capacity vertical fractures. Permission to publish by the Society of Petroleum Engineers of AIME. Copyright 1979 SPE-AIME.

$$k = \frac{1422\, q_{sc}\, \bar{\mu}\, \bar{Z}\, T}{q_D\, h(p_i^2 - p_{wf}^2)}$$

$$X_f = \left[\frac{0.000264\, kt}{\phi\,(\mu C)_i\, t_{DX}}\right]^{0.5}$$

5. The actual fracture flow capacity can then be calculated from

$$k_f w = F_{CD}(kX_f).$$

To predict future performance:

6. Select a value for real time t and calculate t_{DX} using the values of k and X_f obtained in Step 4.
7. From Figure 3-34 determine the value of q_D for this t_{DX} using the correct F_{CD} curve. F_{CD} was determined in Step 2.
8. Calculate the actual flow rate corresponding to this time from

$$q_{sc} = \frac{q_D\, kh(p_i^2 - p_{wf}^2)}{1422\, \bar{\mu}\, \bar{Z}\, T}.$$

9. Repeat Steps 6 through 8 for various values of time t and p_{wf} to obtain inflow performance curves applicable at various times. The data may be plotted as shown in Figure 3-35 if the effect of changing p_{wf} values is to be considered. To observe the decline in flow rate with time at a constant bottom-hole flowing pressure, a plot such as shown in Figure 3-36 may be constructed.

The accuracy of the previously described method can be increased somewhat if formation permeability k is known from a pre-frac Horner type analysis. In this case, q_D can be calculated for each real flow rate q_{sc}, and a plot of $1/q_D$ versus t can be constructed and used to find a match for the proper F_{CD} value. The match would then be obtained by shifting the real data curve in the horizontal direction only.

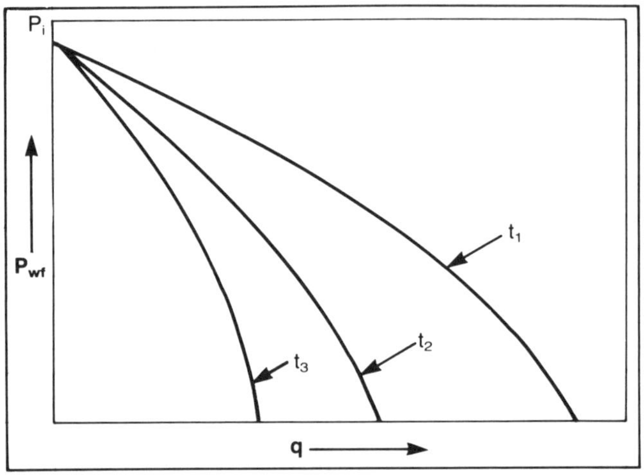

Fig. 3-35. Inflow performance for various times.

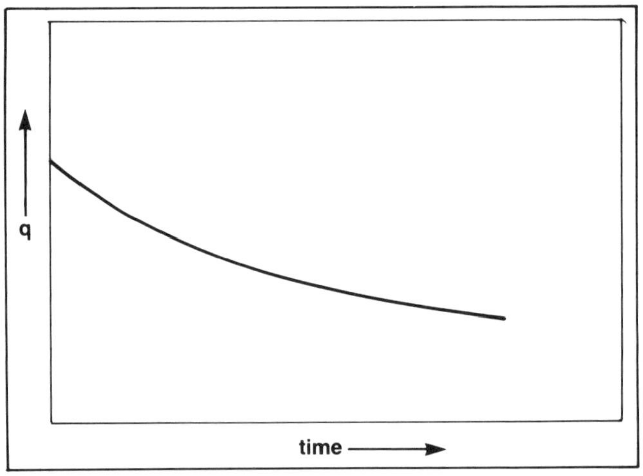

Fig. 3-36. Producing rate for constant p_{wf}.

The procedure outlined here is only one of many methods that may be used for analyzing fractured gas wells. A discussion of all of the methods is beyond the scope of this book.

GUIDELINES FOR GAS WELL TESTING

The accuracy of the results calculated using the previously presented procedures depends on the accuracy of the data used. Obtaining accurate field data can be accomplished only if the field personnel follow established procedures for data collection. These procedures, as presented in Reference 6, are summarized here.

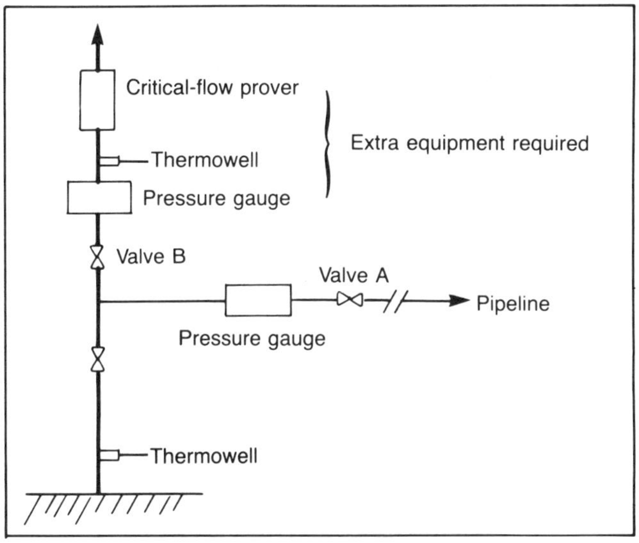

Fig. 3-37. Schematic flow diagram for measuring flow rates of production wells not equipped with flow-rate measurement equipment.

Testing Equipment

The two important factors that govern the selection of testing equipment are the nature of the produced fluids and the type of test being conducted. This section describes the essential features of various wellhead testing facilities that are necessitated by the presence of condensate, water, or acid gases in the natural gas being produced.

Sweet Dry Gas

The simplest configuration of wellhead testing facilities is required for a well producing a sweet dry gas. The testing equipment essentially consists of a flow rate measurement device, a shut-in and flowing pressure measurement device, a thermometer, gas sampling equipment, and the necessary fittings for connecting the equipment to the wellhead.

When the produced gas is being vented to the atmosphere, a commonly used flow rate measurement device is a critical flow prover that is attached to the top of the wellhead. Unless there are regulations to the contrary, the gas vented from the flow prover is not burned. A horizontal positioning of the flow prover should be avoided since high flow rates will set up a considerable torque that may cause the prover fittings or the wellhead to unwind.

If the well being tested is to be produced into a gas gathering system, the flow rate measurement is usually made with an orifice meter using a permanent or removable meter run.

In some instances, the production well that is to be tested does not have a permanently installed flow measurement device. To avoid interruption of flow or the disruption of stabilized flow in a well that has been producing for some time, a simple procedure illustrated by Figure 3-37 may be employed. As shown in this figure, if valve A is closed gradually, while valve B is being opened, maintaining a constant pressure in the flow string, the flow rate being measured by the flow prover will be the same as the production rate.

The desirability of constant flow rates during tests is very important. Figure 3-38 illustrates the wellhead rigging that may provide constant flow rates. Flow downstream from the flow prover is usually vented to the atmosphere. If, however, the produced gas must be flared, care must be taken to ensure that critical flow conditions are maintained in the flow prover. The theory and application of the metering devices are discussed in Chapter 7.

Sweet Wet Gas

The term "wet" is used to describe a natural gas containing heavier hydrocarbons that appear as a condensate

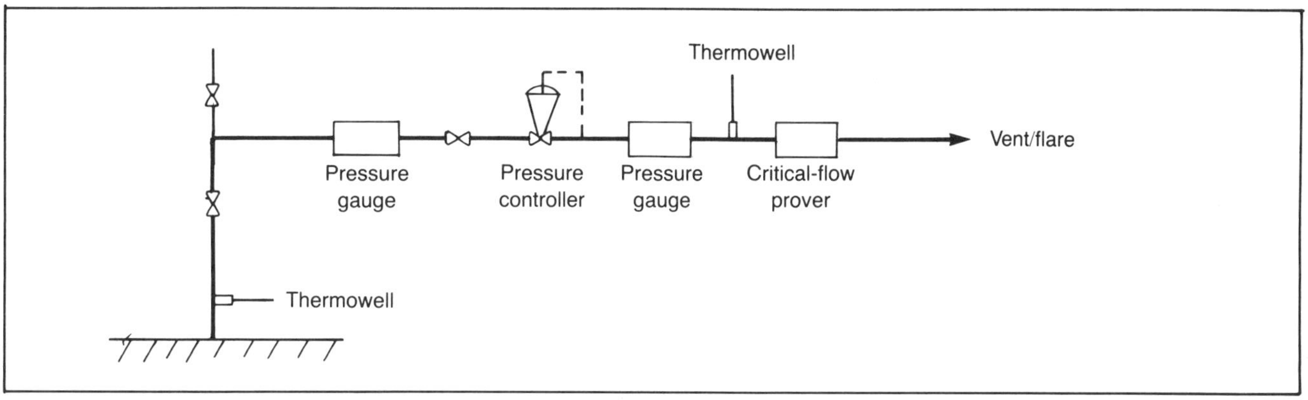

Fig. 3-38. Schematic flow diagram of wellhead rigging for constant-rate tests.

in the produced gas. In some instances, water may also be produced, but it is not included in the definition of a wet gas.

The presence of condensate in produced gas creates requirements for more complex testing facilities than those required for sweet, dry gas wells. A typical facility includes flow rate measurement devices, pressure measurement devices, thermometers, gas and condensate sampling equipment, line heaters, and separation facilities. Several stages of separation and a combination of measurements may be required for highly productive wells, but the most commonly used configurations involve either a single separator or two separators in series. These are illustrated by Figures 3-39 and 3-40, which are intended only as a guide in the selection of test equipment and do not represent the complete wellhead and separation facilities.

The requirement for line heaters is necessitated by the possibility of hydrate formation within the flow lines and testing equipment. Rather than using a line heater, glycol or alcohol may be injected into the gas stream to prevent the formation of hydrates.

Sour Gas

For testing sour gas wells, more elaborate facilities are required. In addition to the standard equipment, depending on whether the gas is dry or wet, a gas meter and a flow line to an appropriate flare stack are required. In addition, liquid seals may also be necessary to protect the gas meter and pressure measuring device from H_2S gas.

Flow Measuring

The accurate measurement of gas and liquid production rates is essential to the proper conduct and analysis of well tests. Correct sampling procedures are also necessary in order to obtain representative samples of the produced fluids and an accurate estimate of the constituents of the reservoir gas.

All flow measurement devices should be installed in accordance with recommended specifications since biased measurement errors can cause anomalous test results. Some of the more common biased errors are caused by insufficient pipe upstream and downstream of the meter, insufficient liquid retention time in the separators, inadequate liquid dumping cycles, incorrect meter coefficients or calibration factors, meter vibration, and other metering problems.

The most commonly used gas flow measurement devices are orifice meters and critical flow provers. Turbine and displacement meters are not as commonly used, but it is expected that with the advent of portable units their utility will increase.

When the condensate flow rate is being measured, turbine or positive displacement meters are usually used. The meter should be installed with sufficient straight pipe and should be located upstream of a snap-acting valve. This valve should be regulated so that there is sufficient retention time to ensure equilibrium of the gas and liquid in the separator.

In many instances, condensate is gaged in a stock tank. This method should be used only in instances where the input to the stock tank is from a low-pressure separator or when the flashed vapors are being collected. In the former case, care should be taken to ensure that the amount of flashed vapors is small enough to be neglected in recombination calculations.

Water flow rates may be measured with turbine or displacement meters, or gaged in a storage tank. If meters are being used, a snap-acting water dump valve should be used to ensure that there is sufficient flow to activate the meter. Since water, if present in gas, often presents a problem in the operation of gas wells and gas gathering systems, it is important to monitor any water production during a test. A knowledge of water production during tests is also important to the analysis and application of test results.

Pressure Measuring

The accurate measurement of static pressures and the pressures corresponding to flow rates measured during the flow periods of various tests is of great importance in gas well testing. Since interpretation of deliverability, drawdown and build-up test results must be based on the theory of flow in the reservoir, it follows that the important pressure in interpreting the tests is the reservoir sandface pressure (in the wellbore). Ideally, this pressure is measured directly through use of an accurate, carefully calibrated bottom-hole pressure gage. There are many types of such gages available today, all of which, when used properly, are quite adequate for accurate measurement of sandface pressures.

In some instances, due to mechanical difficulties, sour gases, or other reasons, it is not practical to use a bottom-hole gage. In such situations, wellhead pressures are measured and subsequently converted to reservoir sandface pressures by the methods described in Chapter 4. The highest possible accuracy in wellhead pressure measurement is important, and for best results these pressures should be taken with a dead-weight gage. This is a device that balances the force created by the well pressure with calibrated weights.

PROBLEMS IN GAS-WELL TESTING

There are several problems unique in testing gas wells that can result in erroneous results. These problems are

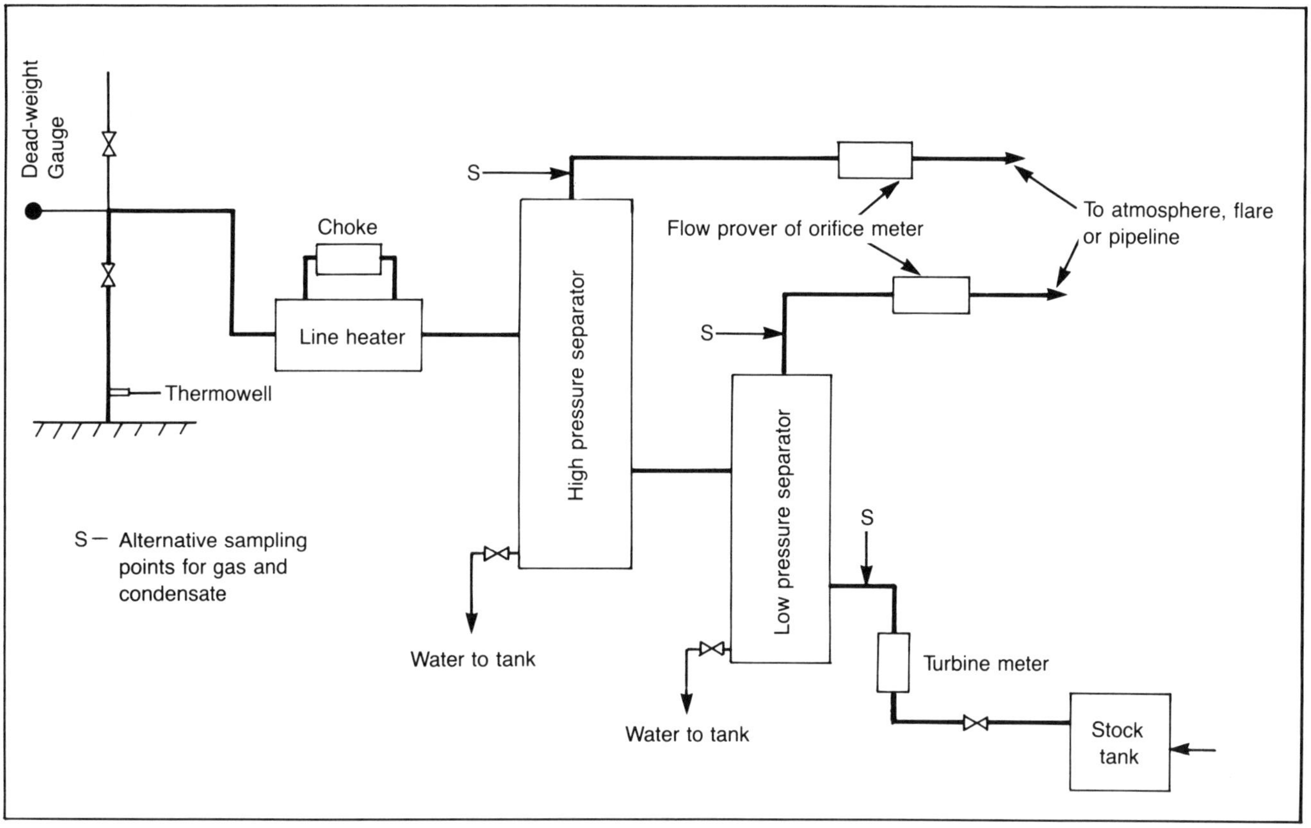

Fig. 3-39. Schematic flow diagram of surface well testing facilities for wet gas (two separators).

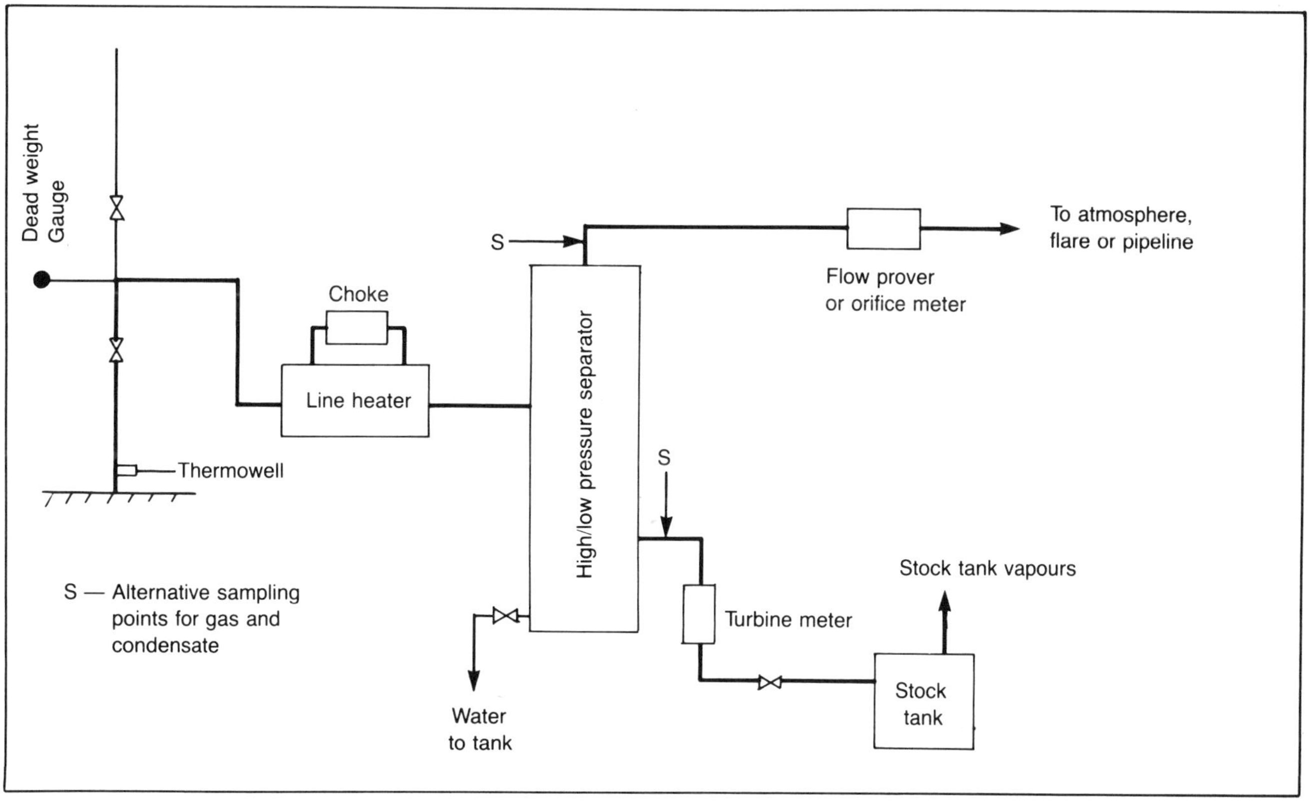

Fig. 3-40. Schematic flow diagram of surface well testing facilities for wet gas (single separator).

discussed from the standpoint of the types of errors that can result in the test data obtained.

Liquid Loading. The problem of liquid loading usually occurs when testing low productivity gas wells with high liquid-gas ratios. Wide variations of surface pressures may indicate liquid loading.

Hydrate Formation. The problem of hydrate formation occurs normally in high pressure gas wells. However, if adequate production equipment is available, this problem can be eliminated by maintaining the wellstream temperature above the hydrate formation temperature. Conditions promoting hydrate formation are discussed in Chapter 9.

Wet Gas Streams. Sometimes it is necessary to meter a gas stream at the wellhead. Such streams will often deposit liquid in the flow line downstream of the point where the orifice was installed. If the gas gravity is measured on a gas sample obtained at such a downstream point, its value will not represent the gravity of the gas that flowed through the orifice. In this instance, the measured gas gravity must be adjusted to give the gravity of the full stream. This can be accomplished by use of Equation 2-46.

Irregular Flow. One of the most frequent difficulties encountered in making accurate gas flow-rate measurements is the rapidly changing flow rates that accompany "slug flow," freeze-ups, and paraffin deposition in the wellhead choke. Because this problem is frequent and is a source of major error, it deserves special attention.

Adjusting the dampening screw on the differential meter will narrow the span of the differential variations. This should enable the operator to get a better differential reading. The differential pen should not be dampened to a point of no response.

It is a common experience that mercury meters, even though the wells flow steadily, exhibit substantial but relatively slow variations in differential reading during a 24 hour period because of atmospheric temperature changes. Temperature data may be obtained with a temperature recorder or periodically measured with a thermometer at the meter run. The variation in the flowing temperature factor is approximately 1 percent per 10°F of variation in temperature.

Sour (H_2S) Gas. Extreme caution should be exercised in handling gases and liquids containing hydrogen sulfide (H_2S). This hazardous substance is highly toxic and under certain concentration can cause illness and death. Special precautions should be taken when testing wells where hydrogen sulfide is present to be assured that exposure will not exceed the safe maximum allowable concentration for the work period required. Selfcontained breathing apparatus should be worn when hydrogen sulfide concentrations are present that might be injurious to health. Safety considerations are discussed in Chapter 9.

REPORTING DATA

As discussed previously, well test data are used for various purposes. Routine tests for production records may require only a minimum amount of data such as periodic fluid-rate measurements. On the other hand, special tests for equipment evaluation, well evaluation, reservoir evaluation, and state regulatory body requirements require additional data. As a result, a large amount of data may be required. The amount and type of data will vary from state to state in the case of state regulatory body requirements, and will depend on test objectives if needed for evaluation purposes. Therefore, rather than attempt to list data for a particular purpose, a complete list of well test data is given.

1. Fluid Measurements: gas rate for all streams, Mscf per day or MMscf per day; condensate rate, stock tank barrels per day; water rate, barrels per day; condensate gravity, API (at 60°F)
2. Pressure Measurements, psig: flowing wellhead pressure; shut-in wellhead pressure; flowing bottom-hole pressure (sometimes calculated); shut-in bottom-hole pressure (sometimes calculated); atmospheric pressure
3. Temperature Measurements, °F: bottom-hole temperature; wellhead temperature; stock-tank liquid temperature; atmospheric temperature
4. Choke Size, Inches
5. Separation conditions, psig and °F: number of separators; separator pressures and temperatures; stabilizer pressure and inlet and outlet temperatures
6. Time Data—Minutes, Hours, Days: duration of flowing test; length of time on test choke size before test period; shut-in time for pressure measurement

REFERENCES

1. Firoozabadi, A. and Katz, D. L.: "An Analysis of High Velocity Gas Flow Through Porous Media," *J. Pet. Tech.* (Feb., 1979) 221.
2. Van Everdingen, A. F. and Hurst, W.: "The Application of the LaPlace Transformation to Flow Problems in Reservoirs," (1949) *Trans.* AIME.
3. Aziz, K. and Flock, D. L.: "Unsteady-State Gas Flow–Use of Drawdown Data in the Prediction of Gas Well Behaviour," *J. Can. Pet. Tech. 2* (1).
4. Dietz, D. N.: "Determination of Average Reservoir Pressure From Build-Up Surveys," (1965) *Trans.* AIME.

5. Cullender, M. H.: "The Isochronal Method of Determining the Flow Characteristics of Gas Wells," (1955) *Trans.* AIME.
6. *Theory and Practice of the Testing of Gas Wells,* Third Edition, Alberta Energy Resources Conservation Board, (1975).
7. Jones, L. G., Blount, E. M., and Glaze, O. H.: "Use of Short Term Multiple Rate Flow Tests to Predict Performance of Wells Having Turbulence," paper SPE 6133 presented at the SPE 51st Annual Meeting, New Orleans, Oct. 3–6, 1976.
8. Brar, G. S. and Aziz, K.: "The Analysis of Modified Isochronal Tests to Predict the Stabilized Deliverability of Gas Wells Without Using Stabilized Flow Data," paper SPE 6134, presented at the SPE 51st Annual Meeting, New Orleans, Oct. 3–6, 1976.
9. Horner, D. R.: "Pressure Buildup in Wells," Third World Pet. Congress, Sect. II, 503–521.
10. Miller, C. C., Dyes, A. E., and Hutchinson, C. A.: "The Estimation of Permeability and Reservoir Pressures from Bottom Hole Pressure Build-up Characteristics," (1950) *Trans.* AIME.
11. Matthews, C. S., Brons, F., and Hazebroek, P.: "A Method for Determination of Average Pressure in a Bounded Reservoir," (1954) *Trans.* AIME.
12. Al-Hussainy, R. and Ramey, H. J.: "Application of Real Gas Flow Theory to Well Testing and Deliverability Forecasting," *J. Pet. Tech.* (May 1966).
13. Dake, L. P.: *Fundamentals of Reservoir Engineering,* Elsevier Scientific Pub. Co. (1978).
14. Havlena, D. and Odeh, A. S.: "The Material Balance as an Equation of a Straight Line," (1963) *Trans.* AIME.
15. Muskat, M.: *The Flow of Homogeneous Fluids Through Porous Media,* McGraw-Hill Book Co., New York (1937).
16. McLeod, H. O.: "The Effect of Perforating Conditions on Well Performance," *J. Pet. Tech.* (Jan., 1983).
17. Hong, K. C.: "Productivity of Perforated Completions in Formations With and Without Damage," *J. Pet. Tech.* (Aug., 1975).
18. Locke, S.: "An Advanced Method for Predicting the Productivity Ratio of a Perforated Well," *J. Pet. Tech.* (Dec., 1981).
19. Saidikowski, R. M.: "Numerical Simulations of the Combined Effects of Wellbore Damage and Partial Penetration," paper SPE 8204, (Sept., 1979).
20. Gurley, D. G., Copeland, C. T., and Hendrick, J. L.: "Design Plan and Execution of Gravel-Pack Completions," *J. Pet. Tech.* (Oct., 1977).
21. Agrawal, R. G., Carter, R. D., and Pollock, C. B.: "Evaluation and Performance Prediction of Low Permeability Gas Wells Stimulated by Massive Hydraulic Fracturing," *J. Pet. Tech.* (March, 1979).

4 Piping System Performance

CHAPTER 3 demonstrated that the ability of a gas reservoir to produce for a given set of reservoir conditions depends directly on the flowing bottom-hole pressure, p_{wf}. Figure 1-6 illustrates that p_{wf} depends on the separator pressure and the configuration of the piping system; that is,

$$p_{wf} = p_{sep} + \Delta p_{fl} + \Delta p_{ch} + \Delta p_{tub} + \Delta p_{rts} \quad (4\text{-}1)$$

where

p_{sep} = separator pressure,
Δp_{fl} = pressure drop in the flowline,
Δp_{ch} = pressure drop in the surface choke,
Δp_{tub} = pressure drop in the well tubing, and
Δp_{rts} = pressure drop in other restrictions, such as subsurface safety valves (SSSVS), valves and fittings, etc.

In order to determine the deliverability of the total well system, it is necessary to be able to calculate all of the pressure drops listed in Equation 4-1. This chapter will develop equations to make these calculations and demonstrate their application. The effects of liquids in the flow stream will be discussed, and the use of previously prepared pressure traverse curves for quick estimates will be demonstrated. Only steady-state, one-dimensional flow will be considered in this chapter.

BASIC FLOW EQUATION

The theoretical basis for most fluid flow equations is the general energy equation, an expression for the balance or conservation of energy between two points in a system. The energy equation is developed first, and using thermodynamic principles, is modified to a pressure gradient equation form.

The steady-state energy balance simply states that the energy of a fluid entering a control volume, plus any shaft work done on or by the fluid, plus any heat energy added to or taken from the fluid must equal the energy leaving the control volume. Figure 4-1 may be used to illustrate the control volume principle.

Considering a steady-state system, the energy balance may be written as

$$U'_1 + p_1V_1 + \frac{mv_1^2}{2g_c} + \frac{m\,g\,h_1}{g_c} + q' + W'_s = U'_2 + p_2\,V_2 + \frac{m\,v_2^2}{2\,g_c} + \frac{m\,g\,h_2}{g_c} \quad (4\text{-}2)$$

where

U' = internal energy,
pV = energy of expansion or compression,
$\frac{mv^2}{2g_c}$ = kinetic energy,
$\frac{m\,gh}{g_c}$ = potential energy,
q' = heat energy added to fluid, and
W'_s = work done on the fluid by the surroundings.

Dividing Equation 4-2 by m to obtain an energy per unit mass balance and writing in differential form gives:

$$dU = d\left(\frac{p}{\rho}\right) + \frac{vdv}{g_c} + \frac{g}{g_c}\,dh + dq + d\,W_s = 0. \quad (4\text{-}3)$$

This form of the energy balance equation is difficult to

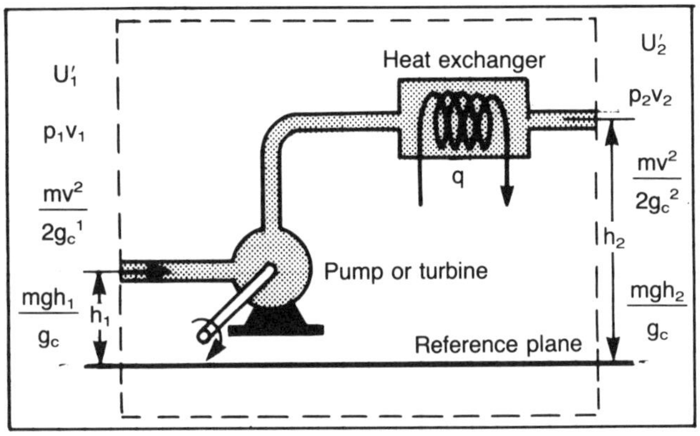

Fig. 4-1. Flow system control volume.

apply because of the internal energy term, so it is usually converted to a mechanical energy balance using well-known thermodynamic relations. From thermodynamics:

$$dU = dh - d\left(\frac{p}{\rho}\right), \tag{4-4}$$

and

$$dh = TdS + \frac{dp}{\rho},$$

or

$$dU = TdS + \frac{dp}{\rho} - d\left(\frac{p}{\rho}\right) \tag{4-5}$$

where

h = enthalpy,
S = entropy, and
T = temperature.

Substituting Equation 4-5 into Equation 4-3 and simplifying results in

$$TdS + \frac{dp}{\rho} + \frac{v\,dv}{g_c} + \frac{g}{g_c}dh + dq + dW_s = 0. \tag{4-6}$$

For an irreversible process, the Clausius inequality states that

$$dS \geq \frac{-dq}{T}, \text{ or}$$

$$TdS = -dq + dL_w,$$

where dL_w = losses due to irreversibilities, such as friction. Using this relationship and assuming no work is done on or by the fluid, Equation 4-6 becomes

$$\frac{dp}{\rho} + \frac{v\,dv}{g_c} + \frac{g}{g_c}dh + dL_w = 0 \tag{4-7}$$

If we consider a pipe inclined at some angle Θ to the horizontal, as in Figure 4-2, since $dh = dL \sin\Theta$

$$\frac{dp}{\rho} + \frac{v\,dv}{g_c} + \frac{g}{g_c}dL \sin\Theta + dL_w = 0.$$

Multiplying the equation by $\frac{\rho}{dL}$ gives

$$\frac{dp}{dL} + \frac{\rho\,v\,dv}{g_c\,dL} + \frac{g}{g_c}\rho \sin\Theta + \rho\frac{dL_w}{dL} = 0. \tag{4-8}$$

Equation 4-8 can be solved for pressure gradient, and if we consider a pressure *drop* as being positive in the direction of flow

$$\frac{dp}{dL} = \frac{g}{g_c}\rho \sin\Theta + \frac{\rho\,v\,dv}{g_c\,dL} + \left(\frac{dp}{dL}\right)_f, \tag{4-9}$$

where

$$\left(\frac{dp}{dL}\right)_f = \rho\frac{dL_w}{dL}$$

is the pressure gradient due to viscous shear or friction losses.

In horizontal pipe flow the energy losses or pressure drops are caused by change in kinetic energy and friction losses only. Since most of the viscous shear occurs at the pipe wall, the ratio of wall shear stress (τ_w) to kinetic energy per unit volume ($\rho\,v^2/2\,g_c$) reflects the relative importance of wall shear stress to the total losses. This ratio forms a dimensionless group and defines a friction factor.

$$f' = \frac{\tau_w}{\rho\,v^2/2g_c} = \frac{2\,\tau_w\,g_c}{\rho\,v^2} \tag{4-10}$$

To evaluate the wall shear stress, a force balance between pressure forces and wall shear stress can be formed. Referring to Figure 4-3,

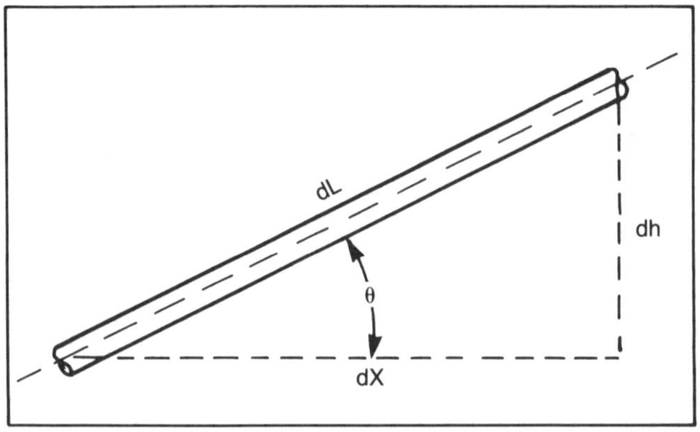

Fig. 4-2. Flow geometry.

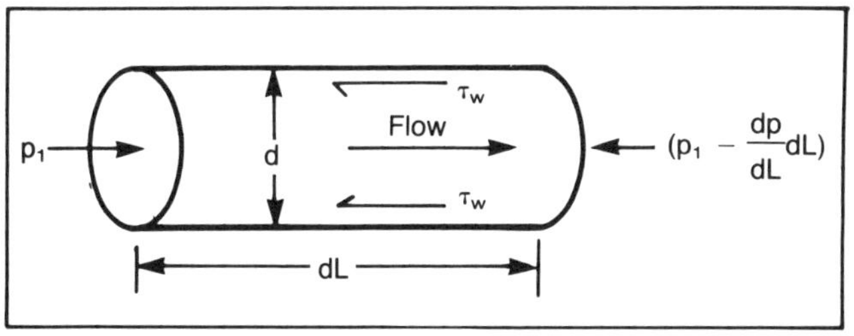

Fig. 4-3. Force balance.

$$\left[p_1 - \left(p_1 - \frac{dp}{dL}dL\right)\right]\frac{\pi d^2}{4} = \tau_w(\pi d)dL$$

$$\tau_w = \frac{d}{4}\left(\frac{dp}{dL}\right)_f. \tag{4-11}$$

Substituting Equation 4-11 into Equation 4-10 and solving for the pressure gradient due to friction gives

$$\left(\frac{dp}{dL}\right)_f = \frac{2f'\rho v^2}{g_c d},$$

which is the well-known Fanning equation. In terms of a Darcy-Weisbach or Moody friction factor, $f = 4f'$, and

$$\left(\frac{dp}{dL}\right)_f = \frac{f\rho v^2}{2g_c d}. \tag{4-12}$$

Laminar Single-Phase Flow

The friction factor for laminar flow can be determined analytically by combining Equation 4-12 with the Hagen-Poiseuille equation for laminar flow

$$v = \frac{d^2 g_c}{32\mu}\left(\frac{dp}{dL}\right)_f$$

or

$$\left(\frac{dp}{dL}\right)_f = \frac{32\mu v}{g_c d^2}.$$

Equating the expressions for frictional pressure gradient gives

$$\frac{32\mu v}{g_c d^2} = \frac{f\rho v^2}{2g_c d}$$

or

$$f = \frac{64\mu}{\rho v d} = \frac{64}{N_{Re}}.$$

The dimensionless group, $N_{Re} = \rho v d/\mu$ is the ratio of fluid momentum forces to viscous shear forces and is known as the Reynolds number. It is used as a parameter to distinguish between laminar and turbulent fluid flow. For engineering calculations, the dividing point between laminar and turbulent flow can be assumed to occur at a Reynolds number of 2100 for flow in a circular pipe. Using units of lbm/ft^3, ft/sec, ft and centipoise, the Reynolds number equation is

$$N_{Re} = \frac{1488\rho v d}{\mu}$$

Turbulent Single-Phase Flow

The ability to predict flow behavior under turbulent flow conditions is a direct result of extensive experimental studies of velocity profiles and pressure gradients. These studies have shown that both velocity profile and pressure gradient are very sensitive to characteristics of the pipe wall. A logical approach to defining friction factors is to begin with the simplest case, i.e., the smooth-wall pipe, proceed to the partially rough wall, and finally to the fully rough wall. Only the most accurate empirical equations available for friction factors are presented here.

Smooth-Wall Pipe. For smooth-wall pipes, several equations have been developed, each valid over different ranges of Reynolds numbers. The most commonly used equation—since it is explicit in f and also covers a wide range of Reynolds numbers ($3000 < N_{Re} < 3 \times 10^6$)—was presented by Drew, Koo, and McAdams[1] in 1932.

$$f = 0.0056 + 0.5N_{Re}^{-0.32} \tag{4-13}$$

An equation proposed by Blasius may be used for Reynolds numbers up to 100,000 for smooth pipes.

$$f = 0.316N_{Re}^{-0.25} \tag{4-14}$$

Rough-Wall Pipe. The inside wall of a pipe is not normally smooth, and in turbulent flow, the roughness can have a definite effect on the friction factor, and thus the pressure gradient. Wall roughness is a function of the pipe material, the method of manufacture, and the environment to which it has been exposed.

From a microscopic sense, wall roughness is not uniform. Individual protrusions, indentations, etc. vary in height, width, length, shape, and distribution. The absolute roughness of a pipe, ϵ, is the mean protruding height of relatively uniformly distributed and sized, tightly packed sand grains that would give the same pressure gradient behavior as the actual pipe.

Dimensional analysis suggests that the effect of roughness is not due to its absolute dimensions, but rather to its dimensions relative to the inside diameter of the pipe, ϵ/d. In turbulent flow, the effect of wall roughness has been found to be dependent on both the relative roughness and on the Reynolds number. If the laminar sublayer that exists within the boundary layer is thick enough, the behavior is similar to a smooth pipe. The sublayer thickness is directly related to the Reynolds number.

Nikuradse's[2] famous sand grain experiments formed the basis for friction factor data from rough pipes. His correlation for fully rough-wall pipe is still the best one available. The friction factor may be calculated explicitly from

$$\frac{1}{\sqrt{f}} = 1.74 - 2 \text{ Log} \left(\frac{2\,\epsilon}{d}\right). \tag{4-15}$$

The equation that is used as the basis for modern friction factor charts was proposed by Colebrook and White[3] in 1939.

$$\frac{1}{\sqrt{f}} = 1.74 - 2 \text{ Log} \left(\frac{2\,\epsilon}{d} + \frac{18.7}{N_{Re}\sqrt{f}}\right) \tag{4-16}$$

The friction factor cannot be extracted readily from the Colebrook equation. By rearranging the equation as follows, a trial-and-error procedure may be used to solve the equation for friction factor.

$$f_c = \left\{\frac{1}{1.74 - 2 \text{ Log}\left(\dfrac{2\epsilon}{d} + \dfrac{18.7}{N_{Re}\sqrt{f_g}}\right)}\right\}^2$$

Values of f_g are estimated and then f_c is calculated until f_g and f_c agree to an an acceptable tolerance. Using the Drew, Koo, and McAdams equation as an initial guess is recommended. After each unsuccessful iteration, the calculated value becomes the assumed value for the next iteration. Also, if more than one pressure loss calculation is to be made as in the case of the iterative procedures discussed in later sections, then the "converged" value of the previous calculation should be used for the initial guess in the next calculation. Convergence using this method is rapid, normally taking only two or three iterations. The variation of single-phase friction factor with Reynolds number and relative roughness is shown graphically in Figure 4-4. The Colebrook equation may be applied to flow problems in the smooth, transition, and fully rough zones of turbulent flow. For large values

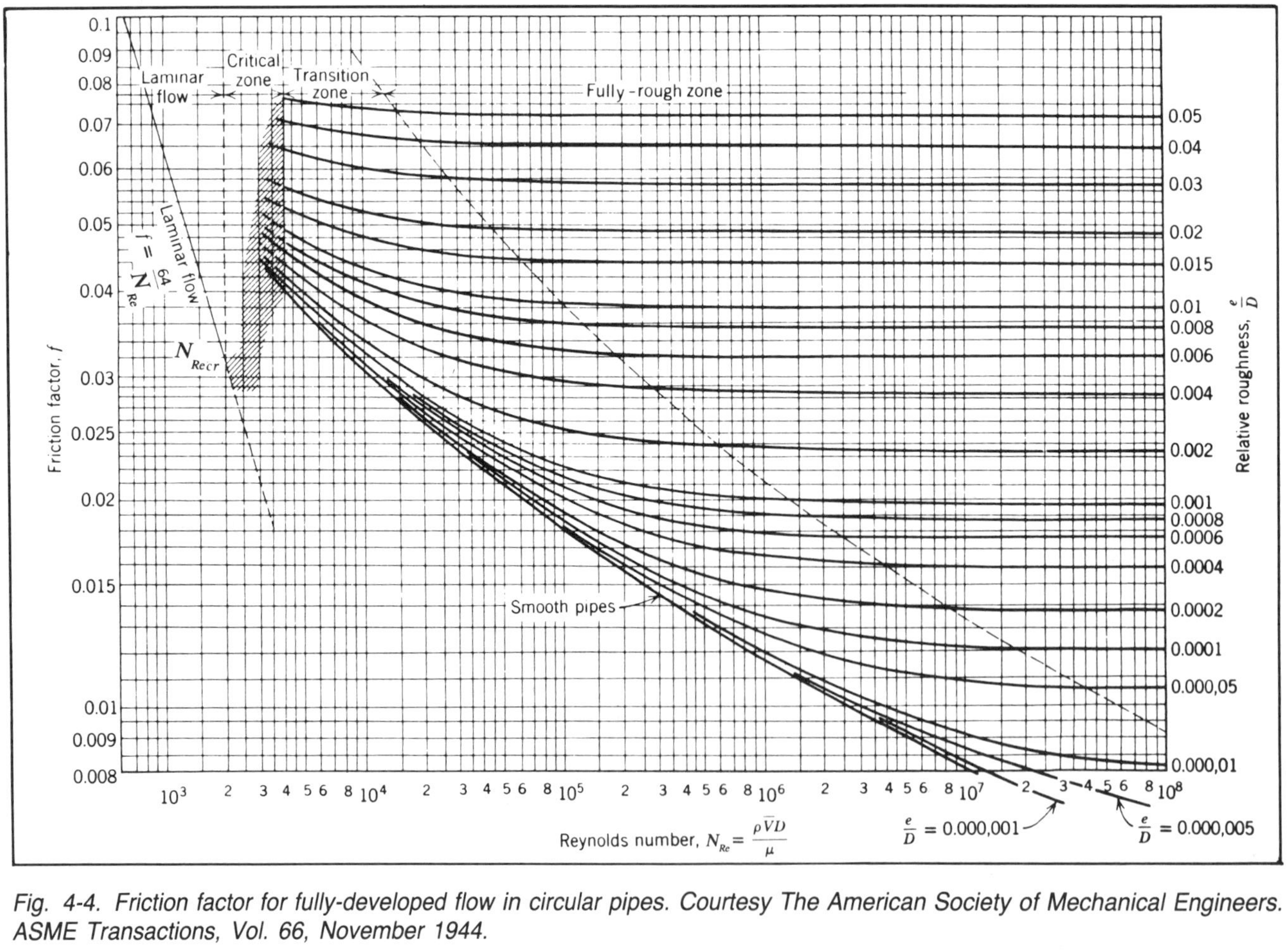

Fig. 4-4. Friction factor for fully-developed flow in circular pipes. Courtesy The American Society of Mechanical Engineers. ASME Transactions, Vol. 66, November 1944.

of Reynolds number, it degenerates to the Nikuradse equation.

An explicit friction factor equation was proposed by Jain[4] and compared in accuracy to the Colebrook equation. Jain found that for a range of relative roughness between 10^{-6} and 10^{-2} and a range of Reynolds number between 5×10^3 and 10^8 the errors were within $\pm 1.0\%$ when compared with the Colebrook equation. The equation gives a maximum error of 3% for Reynolds numbers as low as 2000. The equation is

$$\frac{1}{\sqrt{f}} = 1.14 - 2 \text{ Log}\left(\frac{\epsilon}{d} + \frac{21.25}{N_{Re}^{0.9}}\right). \qquad (4\text{-}17)$$

Equation 4-17 is recommended for all calculations requiring a friction factor determination for turbulent flow. It is much easier to use than Equation 4-16 and, since the value of ϵ will usually not be known to any high degree of accuracy, will give satisfactory results.

The determination of the value to use for pipe wall roughness in the friction factor equations is sometimes difficult. It is important to emphasize that ϵ is not a property that is physically measured. Rather, it is the sand grain roughness that would result in the same friction factor. The only way this can be evaluated is by comparison of the behavior of a normal pipe with one that is sand-roughened. Moody has done this, and his results, given in Figure 4-5, are still the accepted values. These values should not be considered inviolate and could change significantly by such things as paraffin deposition, erosion, or corrosion. Thus, if measured pressure gradients are available, a friction factor and Reynolds number can be calculated, and an effective ϵ/d obtained from the Moody diagram. This value of ϵ/d should then be used for future predictions until updated again. If no information on roughness is available, a value of $\epsilon = 0.0006$ ft is recommended for tubing and line pipe that has been in service for some time.

Example 4-1:

A liquid of specific gravity 0.82 and viscosity of 3 cp (.003 kg/m-sec) flows in a 4 in. (101.6 mm) diameter pipe at a velocity of 30 ft/sec (9.14 m/sec). The pipe material is new commercial steel. Calculate the friction factor using both the Colebrook equation and the Jain equation.

Solution:

From Figure 4-5, for commerical steel, $\epsilon/d = 0.00045$
Colebrook Solution: Use the Drew, Koo and McAdams equation for a first guess.

$$N_{Re} = \rho\, v\, d/\mu = (820)(9.14)(.1016)/.003 = 253{,}824$$

$$f_g = 0.0056 + 0.5\, N_{Re}^{-0.32} = 0.0056 + 0.5\,(253{,}824)^{-0.32}$$

$$f_g = 0.015$$

$$f_c = \left[1.74 - 2 \text{ Log}\left(\frac{2\epsilon}{d} + \frac{18.7}{N_{Re}\sqrt{f_g}}\right)\right]^{-2}$$

$$f_c = \left[1.74 - 2 \text{ Log}\left(2(.00045) + \frac{18.7}{253{,}824\sqrt{.015}}\right)\right]^{-2}$$

$$f_c = 0.0183$$

This value is not close enough to f_g; therefore another trial is required using $f_g = 0.0183$.

$$f_c = \left[1.74 - 2 \text{ Log}\left(2(.00045) + \frac{18.7}{253{,}824\sqrt{.0183}}\right)\right]^{-2}$$

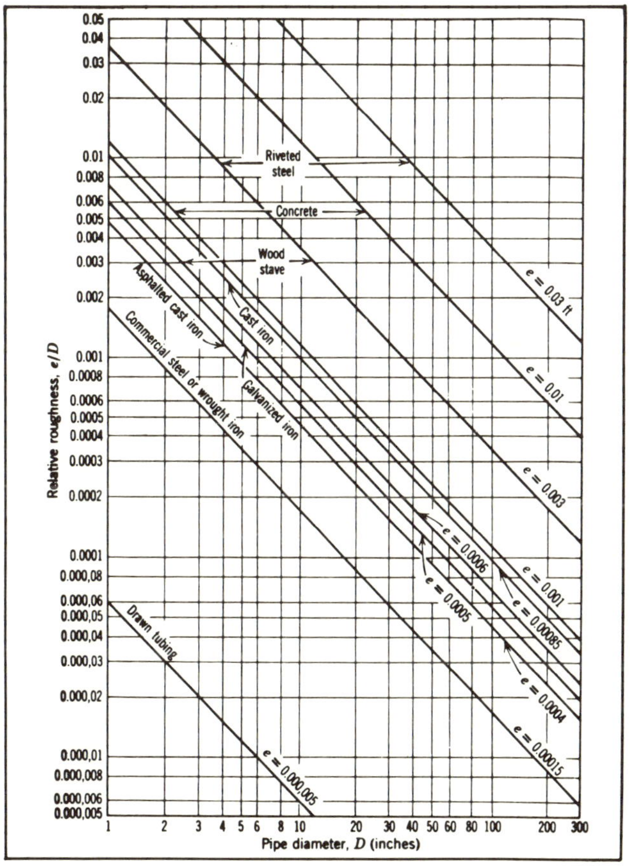

Fig. 4-5. Relative roughness values for pipes of common engineering materials. Courtesy The American Society of Mechanical Engineers. ASME Transactions, Vol. 66, November 1944.

$f_c = 0.0182$

A third trial using $f_g = 0.0182$ gives $f_c = 0.0182$.
Jain Solution:

$$f = \left[1.14 - 2\,\mathrm{Log}\left(\frac{\epsilon}{d} + \frac{21.25}{N_{Re}^{0.9}}\right)\right]^{-2}$$

$$f = \left[1.14 - 2\,\mathrm{Log}\left(0.00045 + \frac{21.25}{(253{,}824)^{0.9}}\right)\right]^{-2}$$

$f = 0.0183$

Combining Equations 4-9 and 4-12, the pressure gradient equation, which is applicable to any fluid at any pipe inclination angle, becomes

$$\frac{dp}{dL} = \frac{g}{g_c}\rho \sin \Theta + \frac{f\rho v^2}{2g_c d} + \frac{\rho\, v\, dv}{g_c\, dL}, \tag{4-18}$$

where the friction factor, f, is a function of Reynolds number and pipe roughness. This relationship is shown in the Moody diagram (Fig. 4-4). The total pressure gradient can be considered to be composed of three distinct components; that is,

$$\frac{dp}{dL} = \left(\frac{dp}{dL}\right)_{el} + \left(\frac{dp}{dL}\right)_f + \left(\frac{dp}{dL}\right)_{acc}, \tag{4-19}$$

where

$$\left(\frac{dp}{dL}\right)_{el} = \frac{g}{g_c}\rho \sin \Theta$$

is the component due to potential energy or elevation change. It is also referred to as the hydrostatic component since it is the only component that would apply at conditions of no flow.

$$\left(\frac{dp}{dL}\right)_f = \frac{f\rho v^2}{2g_c d}$$

is the component due to friction losses.

$$\left(\frac{dp}{dL}\right)_{acc} = \frac{\rho\, v\, dv}{g_c\, dL}$$

is the component due to kinetic energy change or convective acceleration. Equation 4-18 applies for any fluid in steady-state, one-dimensional flow for which f, ρ, and v can be defined.

The elevation change or hydrostatic component is zero for horizontal flow only. It applies for compressible or incompressible, steady-state or transient flow in both vertical and inclined pipes. For downward flow the sine of the angle is negative, and the hydrostatic pressure increases in the direction of flow.

The friction loss component applies for any type of flow at any pipe angle. It always causes a drop of pressure in the direction of flow. In laminar flow the friction losses are linearly proportional to the fluid velocity. In turbulent flow the friction losses are proportional to v^n, where $1.7 \le n \le 2$.

The kinetic energy change or acceleration component is zero for constant area, incompressible flow. For any flow condition in which a velocity change occurs, such as compressible flow, a pressure drop will occur in the direction of the velocity increase.

Although single-phase flow has been studied extensively, it still requires an empirically determined friction factor for turbulent flow calculations. The dependence of this friction factor on pipe roughness, which must usually be estimated, makes the calculated pressure gradients subject to considerable error.

Equation 4-18 is a differential equation and must be integrated in order to apply it to calculate pressure drop as a function of flow rate or velocity and pipe diameter. It must be combined with a continuity equation and an equation of state to express velocity and density in terms of pressure. The following sections describe various assumptions made in integrating the equation for application.

If a computer is available, the equation can be integrated numerically by dividing the pipe into small increments and evaluating the gas or fluid properties at average pressure and temperature in the increment. If small enough increments are taken, the accuracy will be very good. A procedure for calculating a pressure traverse in a pipe using this method is outlined below, and a flow chart is presented in Figure 4-6.

1. Starting with the known pressure, p_1, at location L_1 select a length increment, ΔL.
2. Estimate a pressure increment, Δp, corresponding to length increment, ΔL.
3. Calculate the average pressure and, for nonisothermal cases, the average temperature in the increment.
4. From laboratory data or empirical correlations, determine the necessary fluid and *PVT* properties at conditions of average pressure and temperature (ρ_g, v_g, μ_g).
5. Calculate the pressure gradient, dp/dL, in the increment at average conditions of pressure, temperature, and pipe inclination, using Equation 4-18.
6. Calculate the pressure increment corresponding to the selected length increment, $\Delta p = \Delta L\,(dp/dL)$.
7. Compare the estimated and calculated values of Δp obtained in steps 2 and 6. If they are not sufficiently close, estimate a new pressure increment and return to step 3. Repeat steps 3 through 7 until the estimated and calculated values are sufficiently close.
8. Set $L = L_1 + \Sigma\Delta L$ and $p = p_1 + \Sigma\Delta p$.

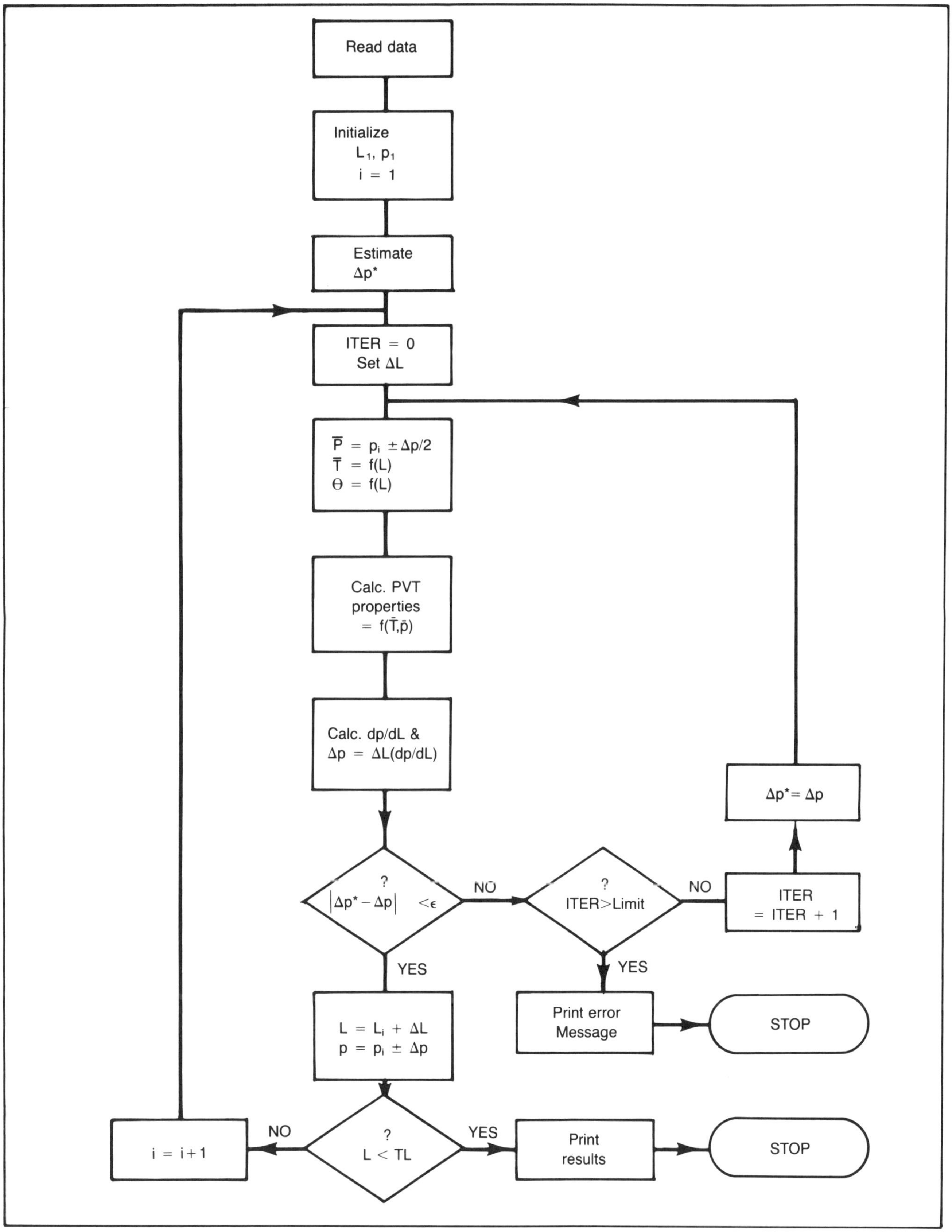

Fig. 4-6. Flow chart for calculating a pressure traverse (incrementing on pressure).

9. If $\Sigma\Delta L$ is less than the total conduit length, return to step 2.

Using this procedure the length increments can be selected so that their sum is exactly equal to the total conduit length and interpolation is not required in the last step.

FLOW IN WELLS

Several methods are available for calculating static and flowing pressure drop in gas wells. The most widely used method is that of Cullender and Smith[5]. All of the methods begin with Equation 4-18, with modifications for flow geometry. In most cases the acceleration gradient is ignored. Since it is frequently necessary to calculate the static bottom-hole pressure in a gas well, this procedure will be presented first.

Static Bottom-Hole Pressure

For a vertical ($\Theta = 90°$, $\sin \Theta = 1$), shut-in ($v = o$) gas well, Equation 4-18 becomes

$$\frac{dp}{dh} = \frac{g\,\rho_g}{g_c}, \tag{4-20}$$

where

$$\rho_g = \frac{pM}{ZRT}.$$

Combining this with Equation (4-20),

$$\frac{dp}{p} = \frac{g\,M\,dh}{g_c Z\,R\,T}. \tag{4-21}$$

Average Pressure and Temperature Method. If Z is evaluated at average pressure and temperature in the increment,

$$\int_{p_{ts}}^{p_{ws}} \frac{\mathrm{dp}}{\mathrm{p}} = \frac{g\,M}{g_c R\,\bar{Z}\,\bar{T}} \int_o^H dh,$$

from which

$$p_{ws} = p_{ts}\,\mathrm{EXP}\left(\frac{g\,M\,H}{g_c R\,\bar{Z}\,\bar{T}}\right), \tag{4-22}$$

This equation holds for any consistent set of units. For conventional field units,

$$p_{ws} = p_{ts}\,\mathrm{EXP}\,[(0.01875\,\gamma_g H)/(\bar{T}\bar{Z})] \tag{4-23}$$

where

p_{ws} = static or shut-in *BHP*, psia,
p_{ts} = static tubing pressure, psia,
γ_g = gas gravity (air = 1),
H = well depth, ft,
$\bar{T}$ = average temperature in the tubing, °R, and
$\bar{Z}$ = gas compressibility factor evaluated at $\bar{T}$, $\bar{p} = (p_{ws} + p_{ts})/2$.

Evaluation of $\bar{Z}$ makes the calculation iterative, and the procedure outlined previously can be used.

Example 4-2:

Using the following data, calculate p_{ws} with Equation 4-23.

H = 10,000 ft, $\gamma_g = 0.6$, p_{ts} = 4000 psia,
T_s = 70°F = 530°R, T_f = 220°F = 680°R

Solution:

A good first guess for p_{ws} can be obtained from

$$p^*_{ws} = p_{ts}\,(1 + 2.5 \times 10^{-5}\,H)$$

$$p^*_{ws} = 4000(1 + 2.5 \times 10^{-5}\,(10000)) = 5000 \text{ psia}$$

$$\bar{T} = \frac{T_s + T_f}{2} = \frac{530 + 680}{2} = 605°\text{R}$$

$$\bar{p} = \frac{p_{ts} + p^*_{ws}}{2} = \frac{4000 + 5000}{2} = 4500 \text{ psia}$$

$$Z = 0.932$$

$$p_{ws} = 4000\,\mathrm{EXP}\,[(0.01875)(.6)(10000)/(605)Z]$$

$$p_{ws} = 4000\,\mathrm{EXP}\,[0.18595/Z] = 4000\,\mathrm{EXP}\,[0.18595/0.932]$$

$$p_{ws} = 4883$$

This is not close enough to the estimated value of 5000 psia. Set the calculated value of p_{ws} as the next estimated value and continue until convergence is reached.

p^*_{ws}	$\bar{p}$	$\bar{Z}$	p_{ws}
5000	4500	.932	4883
4883	4442	.928	4887
4887	4444	.928	4887

The calculation could also be made by estimating an initial value for Z and comparing calculated and estimated values until convergence on Z is obtained.

Cullender and Smith Method. The method presented by Cullender and Smith[5] takes into account the variation of temperature with depth and the variation of Z with pressure and temperature. From Equation 4-21,

$$\int_{p_{ts}}^{p_{ws}} \frac{TZ}{p}\,dp = \frac{M}{R}\int_o^H dh = \frac{MH}{R} = 0.01875\,\gamma_g\,H.$$

The integral is written in short notation as

$$\int_{p_{ts}}^{p_{ws}} \frac{TZ}{p}\, dp = \int_{p_{ts}}^{p_{ws}} I\, dp = 0.01875\, \gamma_g H$$

Using a series expansion, the value of the integral is approximated by

$$2 \int I\, dp = (p_{ms} - p_{ts})(I_{ms} + I_{ts}) + (p_{ws} - p_{ms})(I_{ws} + I_{ms}), \quad \text{(4-24)}$$

where

p_{ms} = pressure at mid-point of well, $H/2$,
I_{ms} = I evaluated at p_{ms}, $\bar{T}$,
I_{ts} = I evaluated at p_{ts}, T_s,
I_{ws} = I evaluated at p_{ws}, T_f.

The calculation procedure consists of dividing the well into two equal segments of length, $H/2$, finding the pressure p_{ms} at $H/2$ and using this value to calculate p_{ws}. I_{ts} can be evaluated from known surface conditions; that is,

$$p_{ms} = p_{ts} + \frac{.01875\, \gamma_g H}{I_{ms} + I_{ts}},$$

$$p_{ws} = p_{ms} + \frac{.01875\, \gamma_g H}{I_{ms} + I_{ws}}.$$

Example 4-3:

Work Example 4-2 using the Cullender and Smith method.

Solution:

Temperature at any depth $h = 70 + \frac{220 - 70}{10000} h = 70 + .015h$

Calculate I_{ts}:

At $T = 70$, $p = 4000$, $Z = .84$

$$I_{ts} = \frac{TZ}{p} = \frac{530(.84)}{4000} = 0.1113$$

Estimate p_{ms}:

$$p^*_{ms} = p_{ts}\,(1 + 2.5 \times 10^{-5}\, H/2)$$

$$= 4000\,(1 + 2.5 \times 10^{-5}\,(5000)) = 4500 \text{ psia}$$

$$\bar{T} = 70 + .015(5000) = 145$$

$$Z = 0.93$$

Calculate I_{ms}:

$$I_{ms} = \frac{\bar{T}Z}{p_{ms}} = \frac{605\,(.93)}{4500} = 0.1250$$

Calculate p_{ms}:

$$p_{ms} = p_{ts} + \frac{0.01875\, \gamma_g H}{I_{ms} + I_{ts}}$$

$$= 4000 + \frac{0.01875(.6)(10000)}{0.1250 + 0.1113}$$

$$p_{ms} = 4000 + 476 = 4476$$

This is not close enough to the estimated value of 4500 psia, therefore set $p^*_{ms} = 4476$ and repeat.

At $\bar{T} = 145$, $p^*_{ms} = 4476$, $Z = .93$

$$I_{ms} = \frac{605\,(.93)}{4476} = 0.1257$$

$p_{ms} = 4000 + 475 = 4475$, which is close enough.

Estimate p_{ws}:

$$p^*_{ws} = p_{ms}\,(1 + 2.5 \times 10^{-5}\, H/2)$$

$$= 4475\,[1 + 5000(2.5 \times 10^{-5})]$$

$p^*_{ws} = 5034$ psia, $T = 220°F$, $Z = 1.006$

Calculate I_{ts}:

$$I_{ts} = \frac{680(1.006)}{5034} = 0.1359$$

$$p_{ws} = p_{ms} + \frac{0.01875\, \gamma_g H}{I_{ms} + I_{ws}}$$

$$= 4475 + \frac{112.5}{0.1257 + 0.1359} = 4905 \text{ psia}$$

For the second trial, $Z = 0.998$

$$I_{ts} = \frac{680(.998)}{4905} = 0.1384$$

$$p_{ws} = 4475 + 426 = 4901 \text{ psia}$$

This compares to 4887 calculated using the average pressure and temperature method.

Flowing Bottom-Hole Pressure

For a flowing well the velocity is not zero, and ignoring acceleration, Equation 4-18 becomes, for a well inclined at an angle $\emptyset$ from the vertical,

$$\frac{dp}{dL} = \frac{g}{g_c}\, \rho \cos \emptyset + \frac{f \rho v^2}{2 g_c d}. \quad \text{(4-25)}$$

Several methods have been presented for integrating Equation 4-25 depending on the assumptions made for handling temperature and Z-factor. Only the average pressure and temperature and Cullender and Smith methods will be discussed.

Average Pressure and Temperature Method. Substituting the expression for gas density in terms of p, T, and Z into Equation 4-25 results in

$$\frac{dp}{dL} = \frac{pM}{ZRT}\left(\cos \emptyset + \frac{f v^2}{2 g_c d}\right). \tag{4-26}$$

Integration of Equation 4-26 assuming an average temperature in the flow string and evaluating Z at average conditions of pressure and temperature gives

$$p_{wf}^2 = p_{tf}^2 \text{ EXP } (S) + \frac{25\,\gamma_g\,q^2\,\bar{T}\,\bar{Z}\,f\,(MD)\,(\text{EXP } (S) - 1)}{S\,d^5}, \tag{4-27}$$

where

p = psia,
$S = 0.0375\ \gamma_g\ (TVD)/\bar{T}\bar{Z}$,
MD = measured depth, ft,
TVD = true vertical depth, ft,
$\bar{T}$ = °R,
q = MMscfd,
d = inches, and
$f = f\,(N_{Re}, \epsilon/d)$ (Jain or Colebrook equation).

The solution procedure is the same as for a shut-in well except for evaluation of the friction factor, which requires calculating a Reynolds number and estimating pipe roughness. Iteration is required since Z must be evaluated at $\bar{p} = (p_{tf} + p_{wf})/2$.

Dividing the well into several length increments and using the procedure described earlier will give more accurate results. Actually, any of the methods will give identical results if the well is divided into short enough increments.

Convergence is sometimes obtained faster if iteration is performed on the Z-factor rather than the unknown pressure. The procedure for this method is:

1. Estimate Z^* (A good first estimate is 0.9).
2. Calculate the unknown pressure using Equation 4-27 with $Z = Z^*$.
3. Calculate the average pressure, $\bar{p} = (p_{tf} + p_{wf})/2$.
4. Evaluate Z at $\bar{p}$ and $\bar{T}$.
5. Compare Z and Z^*. If not close enough, set $Z^* = Z$ and go to Step 2. Repeat until $abs(Z - Z^*)/Z < 0.001$ or any other tolerance preferred. When the tolerance is met, the pressure calculated in Step 2 is the correct value.

Example 4-3a:

Use the average pressure and temperature method to calculate the flowing bottom-hole pressure for the following directional well:

$\gamma_g = 0.75$, $MD = 10{,}000$ ft, $TVD = 7{,}000$ ft
$T_s = 110$°F, $T_f = 245$°F, $p_{tf} = 2000$ psia,
$q_{sc} = 4.915$ MMscfd, $d = 2.441$ in., $\epsilon = 0.0006$ in.,
$\bar{\mu} = 0.012$ cp

Solution:

In terms of mass flow rate, the Reynolds number is

$$N_{Re} = \frac{C\,\gamma_g\,q_{sc}}{\mu\,d} \tag{4-28}$$

where

Variable	Units: Field	Units: SI
q_{sc} = gas flow rate	MMscfd	MM m³/day
γ_g = gas gravity	—	—
μ = gas viscosity	cp	kg/m-sec
d = pipe inside diameter	in.	m
C = constant	20011	17.96

$$N_{Re} = \frac{20011\,\gamma_g q_{sc}}{\bar{\mu}\,d} = \frac{20011(0.75)(4.915)}{0.012(2.441)} = 2.518 \times 10^6$$

From Equation 4-17, $f = 0.015$

(1) Estimate $Z^* = 0.9$

$$S = \frac{0.0375(0.75)(7000)}{638\,Z^*} = \frac{0.3086}{Z^*}$$

(2) $p_{wf}^2 = (2000)^2 \text{ EXP } (0.3086/Z^*) +$

$$\frac{25(.75)(4.915)^2(638)\,Z^*\,(.015)\ \text{(TERM CONTINUED)}}{\frac{0.3086}{Z^*}(2.441)^5}\ \frac{(10{,}000)[\text{EXP } (0.3086/Z^*) - 1]}{}$$

$$p_{wf}^2 = 4 \times 10^6 \text{ EXP } (0.3086/Z^*) + 1.621 \times 10^6\,(Z^*)^2\,(\text{EXP}(.3086/Z^*) - 1)$$

For $Z^* = 0.9$, $p_{wf}^2 = 5.636 \times 10^6 + 536{,}966$

$p_{wf}^2 = 6.173 \times 10^6$, $p_{wf} = 2485$ psia

(3) $\bar{p} = (p_{tf} + p_{wf})/2 = \dfrac{2000 + 2485}{2} = 2242$ psia

(4) At $p = 2242$ psia and $T = 178$°F, $Z = 0.806$

(5) $\dfrac{abs(Z - Z^*)}{Z} = \dfrac{0.9 - 0.806}{0.806} = 0.117$,

which is too large.

(2)′ For $Z^* = 0.806$, $p_{wf}^2 = 5.866 \times 10^6 + 491{,}187$

$p_{wf}^2 = 6.357 \times 10^6$, $p_{wf} = 2521$ psia

(3)′ $\bar{p} = \dfrac{2000 + 2521}{2} = 2261$ psia

(4)′ At $p = 2261$ psia and $T = 178$°F, $Z = 0.805$

(5)′ $\frac{\text{abs}(Z - Z^*)}{Z} = \frac{\text{abs}(.805 - .806)}{.805} = 0.001$, which is close enough.

Therefore, $p_{wf} = 2521$ psia.

Cullender and Smith Method. Derivation of the Cullender and Smith method for flowing wells begins with Equation 4-26. The following substitutions are made for velocity:

$$v = \frac{q}{A},$$

$$q = q_{sc}\frac{p_{sc}\,T\,Z}{T_{sc}\,pZ_{sc}},$$

which gives

$$\frac{dp}{dL} = \frac{pM\cos\phi}{ZRT} + \frac{MTZ\,p_{sc}^{\,2}\,f\,q_{sc}^{\,2}}{Rp\,T_{sc}^{\,2}\,2\,g_c d\,A^2}$$

or

$$\frac{p}{ZT}\frac{dp}{dh} = \frac{M}{R}\left[\left(\frac{p}{ZT}\right)^2\cos\phi + C\right],$$

where

$$C = \frac{8\,p_{sc}^{\,2}\,q_{sc}^{\,2}\,f}{T_{sc}^{\,2}\,g_c\,\pi^2\,d^5}$$

which is constant for a given flow rate in a particular pipe size. Separating the variables gives:

$$\int_{p_{tf}}^{p_{wf}} \frac{\frac{p}{ZT}\,dp}{\left(\frac{p}{ZT}\right)^2\cos\phi + C} = \frac{M}{R}\int_o^{MD} dL, \tag{4-29}$$

which is applicable for any consistent set of units. Substituting field units and integrating the right-hand side of Equation 4-29 gives:

$$\int_{p_{tf}}^{p_{wf}} \frac{\frac{p}{ZT}\,dp}{0.001\left(\frac{p}{ZT}\right)^2\frac{TVD}{MD} + F^2} = 18.75\,\gamma_g\,MD, \tag{4-30}$$

where

$$F^2 = \frac{0.667\,f\,q_{sc}^{\,2}}{d^5}, \tag{4-31}$$

and $\frac{TVD}{MD} = \cos\phi.$

Writing Equation 4-30 in short notation and dividing the well into two increments of length $H/2$ gives:
Upper half of well:

$$18.75\,\gamma_g(MD) = (p_{mf} - p_{tf})(I_{mf} + I_{tf}),$$

Lower half of well:

$$18.75\,\gamma_g(MD) = (p_{wf} - p_{mf})(I_{wf} + I_{mf}),$$

where

$$I = \frac{\frac{p}{TZ}}{0.001\left(\frac{p}{TZ}\right)^2\frac{TVD}{MD} + F^2} \tag{4-32}$$

The solution procedure is similar to that for the static case, but is more involved because of the more complicated definition of I. For practical purposes, F can be considered a constant since the only variable in the Reynolds number used in evaluating f is gas viscosity. Viscosity is a function of pressure, but for simplification of the calculations it can be evaluated at $\bar{T}$ and the known pressure.

Example 4-4:

The following data pertain to a flowing gas well. Use the Cullender and Smith method to calculate the flowing bottom-hole pressure.

$\gamma_g = 0.75$, $H = 10{,}000$ ft, $T_s = 110°F$,
$T_f = 245°F$, $p_{tf} = 2000$ psia, $q_{sc} = 4.915$ MMscfd,
$d = 2.441$ in., $\epsilon = 0.0006$ in., $\bar{\mu} = 0.012$ cp
$\phi = 0°$

Solution:

Calculate f and F^2:

$$N_{Re} = \frac{20011(0.75)(4.915)}{0.012(2.441)} = 2.518 \times 10^6$$

From Equation 4-17: $f = 0.015$

$$F^2 = \frac{0.667\,(0.015)(4.915)^2}{(2.441)^5} = 0.00279$$

Calculate I_{tf}:

At $p = 2000$ psia, $T = 110°F$, $Z = 0.71$

$$\frac{p}{TZ} = \frac{2000}{(570)(.71)} = 4.942$$

$$I_{tf} = \frac{4.942}{(0.001)(4.942)^2 + 0.00279} = 181.60$$

Estimate p^*_{mf}: (First Trial)

$p^*_{mf} = 2000\,(1 + 2.5 \times 10^{-5}(5000)) = 2250$ psia

Calculate I_{mf}:

At $p = 2250$, $T = 110 + 67.5 = 178$, $Z = 0.797$

$$\frac{p}{TZ} = \frac{2250}{(638)(.797)} = 4.425$$

$$I_{mf} = \frac{4.425}{(0.001)(4.425)^2 + 0.00279} = 197.81$$

Calculate p_{mf}:

$$p_{mf} = p_{tf} + \frac{18.75\ \gamma_g H}{I_{mf} + I_{tf}} = 2000 + \frac{18.75(.75)(10000)}{197.81 + 181.60}$$

$p_{mf} = 2000 + 371 = 2371$ (not close enough to p^*_{mf})

Calculate I_{mf}: (Second Trial)

At $p = 2371$, $T = 178$, $Z = 0.796$

$$\frac{p}{TZ} = \frac{2371}{638(.796)} = 4.669$$

$$I_{mf} = \frac{4.669}{0.001(4.669)^2 + 0.00279} = 189.88$$

Calculate p_{mf}:

$$p_{mf} = 2000 + \frac{140625}{189.88 + 181.60} = 2379$$

Calculate I_{mf}: (Third Trial)

At $p = 2379$, $T = 178$, $Z = 0.796$

$$\frac{p}{TZ} = \frac{2379}{638(.796)} = 4.684$$

$$I_{mf} = \frac{4.684}{0.001(4.684)^2 + 0.00279} = 189.41$$

Calculate p_{mf}:

$$p_{mf} = 2000 + \frac{140625}{189.41 + 181.60} = 2379 \text{ psia}$$

Therefore, the pressure at the mid-point of the well is 2379 psia. The value of p_{wf} is now calculated.

Estimate p^*_{wf}:

$p^*_{wf} = 2379\ (1 + 2.5 \times 10^{-5}(5000)) = 2676$

Calculate I_{wf}:

At $p = 2676$, $T = 245$, $Z = 0.867$

$$\frac{p}{TZ} = \frac{2676}{705(0.867)} = 4.378$$

$$I_{wf} = \frac{4.378}{(0.001)(4.378)^2 + 0.00279} = 199.39$$

Calculate p_{wf}: (First Trial)

$$p_{wf} = p_{mf} + \frac{140625}{199.39 + 189.41} = 2379 + 362 = 2741$$

Calculate I_{wf}: (Second Trial)

At $p = 2741$, $T = 245$, $Z = 0.868$

$$\frac{p}{TZ} = \frac{2741}{705(0.868)} = 4.479$$

$$I_{wf} = \frac{4.479}{(0.001)(4.479)^2 + 0.00279} = 196.00$$

Calculate p_{wf}:

$$p_{wf} = 2379 + \frac{140625}{196.00 + 189.41} = 2744 \text{ psia}$$

This is close enough to the previously calculated value of 2741 psia. Therefore, the flowing bottom-hole pressure is 2744 psia.

Annular Flow. Many gas wells are dually completed, and one zone may produce through the tubing-casing annulus. This presents no problem in calculating static pressures, and either of the previously described methods can be used. For flowing wells either the Average Pressure and Temperature or the Cullender and Smith method may be used if the Hydraulic Radius concept is employed. The only modifications necessary are in calculating the effective diameter and the Reynolds number. It can be shown that the correct effective diameter is

$$d_h = d_c - d_t \tag{4-33}$$

where

d_h = effective diameter,
d_c = casing inside diameter,
d_t = tubing outside diameter.

There are no published data on values of roughness for an annulus. It would seem that a somewhat rougher surface than that for a round pipe would be applicable since the flow must pass the tubing collars. If one measurement of pressure drop and flow rate can be made, roughness can be back-calculated, and this value can be used for other flow rates.

The Cullender and Smith method can also be used to estimate pressure drop occurring during gas injection by using a negative value for F^2 in Equation 4-30.

FLOW IN PIPELINES

For most practical cases of gas flow in pipelines, the line can be considered horizontal and the hydrostatic or elevation component, as well as the acceleration component, can be dropped from the general equation, Equation 4-18. This is not true if the line is transporting liquids, as will be discussed later. The general equation is

$$\frac{dp}{dX} = \frac{f\rho\, v^2}{2g_c d} = \frac{pMf\,v^2}{ZRT\,2g_c d} \tag{4-34}$$

Many solutions to Equation 4-34 have been proposed over the years. The difference in most of the solutions results from the method used to handle the friction factor, f, and the gas compressibility factor, Z. In most cases temperature is assumed constant and Z is evaluated at average pressure in the line. This requires an iterative solution if one of the pressures is unknown. Integration of Equation 4-34 over some distance, L, between upstream pressure p_1 and downstream pressure p_2 results in

$$p_1^2 - p_2^2 = \frac{25\,\gamma_g\, q^2\, \bar{T}\,\bar{Z}\, f\, L}{d^5}, \tag{4-35}$$

where

p = psia,
L = ft,
$\bar{T}$ = °R
q = MMscfd (14.7 psia, 60°F)
d = inches, and
$f = f\,(N_{Re}, \epsilon/d)$ (Moody diagram or Jain eq.).

Equation 4-35 was derived using base or standard conditions as 14.7 psia and 60°F. It can be put in a more general form by leaving the base conditions in the equation as variables. It is frequently advantageous to express Equation 4-35 in terms of flow rate; that is

$$q = \frac{C\,T_b}{p_b}\left[\frac{p_1^2 - p_2^2}{\gamma_g f \bar{T}\bar{Z} L}\right]^{0.5} d^{2.5}. \tag{4-36}$$

The value of C depends on the units used in the equation. Table 4-1 gives the value of C for various combinations of units.

A so-called efficiency factor is sometimes used in Equation 4-36 to account for the fact that actual pipelines frequently deliver less gas than is calculated. The efficiency factor usually varies between about 0.7 and 0.92 and is usually obtained from "experience."

TABLE 4-1
Value of *C* for Various Units

p	*T*	*d*	*L*	*q*	*C*
psia	°R	in.	mi	scfd	77.54
psia	°R	in.	ft	scfd	5634
psia	°R	in.	ft	MMscfd	5.634×10^{-3}
kPa	°K	*m*	*m*	m^3/d	1.149×10^6

Example 4-5:

A pipeline is to deliver 320 MMscfd of gas to a downstream pressure of 600 psia. Using the following data, calculate the required upstream pressure.

$\bar{T}$ = 45°F = 505°R, γ_g = 0.67, d = 25.375 in.,
L = 100 miles = 528,000 ft, ϵ = 0.0006 in.

Solution:

Use Equation 4-35. p_1 must be estimated to determine $\bar{p}$ and $\bar{Z}$. Estimate $\Delta p/\Delta L$ = 0.0005 psi/ft. Then p^*_1 = p_2 + 0.0005 (528,000) = 600 + 264 = 864 psia.

$$\bar{p} = \frac{p^*_1 + p_2}{2} = \frac{864 + 600}{2} = 732 \text{ psia}$$

At p = 732 psia, T = 45°F, Z = 0.844,

μ = 0.012 cp
Determine f:

$$N_{Re} = \frac{20011\,\gamma_g\, q}{\mu\, d} = \frac{20011\,(0.67)(320)}{0.012\,(25.375)} = 14.1 \times 10^6$$

From Equation 4-17 (Jain Eq.), f = 0.0097

Calculate p_1:

$$p_1^2 = p_2^2 + \frac{25\,\gamma_g\, q^2\, \bar{T}\bar{Z} fL}{d^5}$$

$$p_1^2 = (600)^2 + \frac{25(0.67)(320)^2(505)(Z)(0.0097)(528{,}000)}{(25.375)^5}$$

p_1^2 = 360,000 + 421,680 Z = 360,000 + 355,898

p_1^2 = 715,898 $\quad p_1$ = 846 psia

This is not close enough to the estimated value of 864. Set p^*_1 = 846 and re-calculate p_1:
Second Trial:

$$\bar{p} = \frac{600 + 846}{2} = 723 \text{ psia}$$

Z = 0.846

p_1^2 = 360,000 + 421,680 (.846) = 716,567

p_1 = 847 psia

Since two successive values of p_1 are approximately equal, the iterative solution has converged. The required upstream pressure is 847 psia.

Example 4-6:

Calculate the flow capacity of the following pipeline.

p_1 = 4140 kPa, p_2 = 2760 kPa, L = 30,000 m,
d = 0.152 m, ϵ = 4.572×10^{-5} m, γ_g = 0.75
T_b = 288°K, p_b = 101.4 kPa, $\bar{T}$ = 294°K,
μ = 1.2×10^{-5} kg/m-sec, $\bar{Z}$ = 0.889

Solution:

The solution will be trial-and-error or iterative since friction factor is required in Equation 4-36. This requires a value of Reynolds number, which requires flow rate, the unknown. Rather than estimate flow rate, it is easier to estimate friction factor. The procedure is as follows:

1. Estimate f^*
2. Calculate q using Equation 4-36
3. Calculate $N_{Re} = f(q)$
4. Calculate f using the Moody diagram or Equation 4-17
5. Compare f and f^*. If not close, set $f^* = f$ and go to Step 2. Continue until $f = f^*$.

1. Estimate $f^* = 0.01$, a reasonable value for pipelines.

2. $$q = \frac{1.149 \times 10^6\,(288)}{101.4}\left[\frac{(4140)^2 - (2760)^2)}{0.75\,(f)\,(294)(0.889)(30{,}000)}\right]^{0.5}(0.152)^{2.5}$$

$$q = 31{,}783/f^{0.5} = 31{,}783/(0.01)^{0.5}$$

$$= 317{,}830 \text{ m}^3/d$$

3. $$N_{Re} = \frac{17.96\,\gamma_g q}{\mu d} = \frac{17.96(0.75)\,q}{(1.2 \times 10^{-5})(0.152)}$$

$$= 7.385 \times 10^6\,q$$

$$N_{Re} = 7.385 \times 10^6\,(0.31783) = 2.347 \times 10^6$$

4. $$\frac{\epsilon}{d} = \frac{4.572 \times 10^{-5}}{0.152} = 0.0003$$

$$f = \left[1.14 - 2\log\left(0.0003 + \frac{21.25}{(2.347 \times 10^6)^{0.9}}\right)\right]^{-2}$$

$$= 0.015$$

5. Comparing f and f^*, $0.015 \neq 0.010$, therefore, set $f^* = 0.015$.

2.′ $$q = \frac{31783}{(0.015)^{0.5}} = 259{,}507$$

3.′ $N_{Re} = 7.385 \times 10^6\,(0.259{,}507) = 1.916 \times 10^6$

4.′ $f = 0.015$

5.′ $f = f^*$, therefore $q = 259{,}507 \text{ m}^3/\text{day}$.

The previous examples illustrate that if either the pressure drop or flow rate is unknown, the solution is iterative. If the unknown is diameter, the solution will also be iterative since diameter is necessary to evaluate the friction factor. This fact has prompted several investigators to substitute a specific equation for f into the general flow equation to make the solution for either q or d noniterative or explicit. The friction factor specified may be either diameter dependent only or Reynolds number dependent. None includes the dependence on pipe roughness.

The expressions incorporated into Equation 4-35 for f in several of the most popular pipeline equations are listed below.

Equation	f
Panhandle A	$\frac{0.085}{N_{Re}^{0.147}}$
Panhandle B	$\frac{0.015}{N_{Re}^{0.183}}$
IGT[6]	$\frac{0.187}{N_{Re}^{0.2}}$
Weymouth	$\frac{0.032}{d^{1/3}}$

Using these relationships for friction factor in Equation 4-36, the general form of the pipeline flow equation without f becomes

$$q = a_1 E\left(\frac{T_b}{p_b}\right)^{a_2}\left[\frac{p_1^2 - p_2^2}{\bar{T}\bar{Z}L}\right]^{a_3}\left(\frac{1}{\gamma_g}\right)^{a_4} d^{\,a_5} \tag{4-37}$$

where E is the efficiency factor, and the values of the a_i constants used in the various equations are tabulated below.

Equation	a_1	a_2	a_3	a_4	a_5
Panhandle A	435.87	1.0788	0.5394	0.4604	2.618
Panhandle B	737.00	1.0200	0.5100	0.4900	2.530
IGT	337.90	1.1110	0.5560	0.4000	2.667
Weymouth	433.50	1.0000	0.5000	0.5000	2.667

The units to be used in Equation 4-37 are:

q = cu ft/day measured at T_b, p_b,
T = °R,
p = psia,
L = miles, and
d = inches.

Example 4-7:

Using the following data, calculate the flow capacity of the pipeline using the Weymouth equation, the Panhandle B equation, and the Panhandle A equation.

$p_1 = 847$ psia, $p_2 = 600$ psia, $d = 25.375$ in.,
$L = 100$ miles, $\gamma_g = 0.67$, $\bar{T} = 505$°R, $\bar{Z} = 0.846$
$T_b = 520$°R, $p_b = 14.7$ psia, $E = 1.0$

Solution:

$$\frac{p_1^2 - p_2^2}{\bar{T}\bar{Z}L} = \frac{847^2 - 600^2}{505\,(0.846)(100)} = 8.366$$

$$\frac{T_b}{p_b} = \frac{520}{14.7} = 35.374$$

Weymouth:

$$q = 433.5\ (35.374)^{1.0}\ (8.366)^{0.5} \left(\frac{1}{0.67}\right)^{0.5} (25.375)^{2.667}$$

q = 301,610,000 cu ft/day = 301.6 MMscfd

Panhandle B:

$$q = 737.0\ (35.374)^{1.02}\ (8.366)^{0.51} \left(\frac{1}{0.67}\right)^{0.49} (25.375)^{2.53}$$

q = 359,732,857 cu ft/day = 359.7 MMscfd

Panhandle A:

$$q = 435.87\ (35.374)^{1.0788}\ (8.366)^{0.5394} \left(\frac{1}{.67}\right)^{0.4604}$$

$$(25.375)^{2.618}$$

q = 364,247,375 cu ft/day = 364.2 MMscfd

The flow capacities using the various equations vary considerably and can be compared to the results of Example 4-5, which showed that the line would deliver 320 MMscfd at the specified pressures. If it is assumed that this is the correct value for q, efficiency factors of 1.061, 0.890, and 0.879 would have to be applied to the Weymouth, Panhandle B, and Panhandle A equations respectively to make the calculated results agree.

The choice of which equation to use can be difficult. It is generally assumed that the Panhandle A equation is more applicable at Reynolds numbers in the transition region, and the B equation is more accurate in the fully turbulent region. If pipe roughness data are available, the value of friction factor can be determined and Equation 4-36 should be used.

All of the pipeline equations considered to this point have been applicable to horizontal or near-horizontal pipes only. That is, all of the calculated pressure drop is due to friction. If the gas changes elevation, an elevation component will apply even though it is very small for most pipelines. However, the elevation or hydrostatic component may be included in the general flow equation as follows:

$$q = \frac{C T_b}{p_b}\left[\frac{(p_1^2 - p_2^2 e^s)S}{\gamma_g f \bar{T} \bar{Z} (e^s - 1)L}\right]^{0.5} d^{2.5}, \tag{4-38}$$

where

e = natural log base = 2.718,

$s = \dfrac{0.0375\ \gamma_g H}{\bar{Z}\bar{T}}$, and

H = elevation change, ft.

This method of handling the elevation change assumes that the rate of elevation change with pipeline length is constant. However, if this is not the case, very little error will result because of this assumption unless the average pressure in the line is very high, resulting in a large gas density effect.

Pipelines in Series

If a pipeline or gathering line consists of sections of different diameter pipe, the flow capacity of the entire pipeline can be calculated by first determining an equivalent length of a line of some arbitrary diameter that would have the same flow capacity as the system. The general flow equation or any of the specialized equations can then be used to determine the flow capacity. By using the fact that for a series pipeline

$$\Delta p_T = \Delta p_1 + \Delta p_2 + \Delta p_3 + \text{---} = \sum_{i=1}^{N} \Delta p_i,$$

it can be shown that

$$\frac{fL_e}{d^5} = \sum_{i=1}^{N} \frac{f_i L_i}{d_i^5}, \tag{4-39}$$

where N = number of pipe segments. If the specialized equations are used, the friction factor is not required and the equation becomes

$$\frac{(L_e)^{2a_3}}{(d)^{2a_5}} = \sum_{i=1}^{N} \frac{(L_i)^{2a_3}}{(d_i)^{2a_5}}, \tag{4-40}$$

where a_3 and a_5 are obtained from the previous table. For example, if the Weymouth equation is used, since a_3 = 0.5 and a_5 = 2.667, Equation 4-40 becomes

$$L_e = d^{5.333} \sum_{i=1}^{N} \frac{L_i}{d_i^{5.333}}. \tag{4-41}$$

Pipelines in Parallel

For pipes in parallel the total flow rate is the sum of the rates in the individual pipes, or

$$q_T = q_1 + q_2 + \ldots + q_N = \sum_{i=1}^{N} q_i. \tag{4-42}$$

If the lengths of the individual pipes are the same, the total flow capacity is calculated from

$$q_T = C \sum_{i=1}^{N} \frac{d_i^{2.5}}{f_i^{0.5}}. \tag{4-43}$$

When applied to the specialized equations, Equation 4-43 becomes

$$q_T = C' \sum_{i=1}^{N} d_i^{a_5}. \tag{4-44}$$

The diameter d_o of a single line having the same flow capacity as N parallel lines of diameters d_i is

$$d_o = \left[\sum_{i=1}^{N} d_i^{a_5} \right]^{1/a_5} . \tag{4-45}$$

In many cases only part of an existing pipeline will be paralleled or "looped" in order to increase its flow capacity. Equations 4-41, 4-44, and 4-45 can be combined to give

$$q_{\text{new}} = \frac{q_{\text{old}}}{\left[1 + Y \left(\frac{1}{(1 + W)^2} - 1 \right) \right]^{0.5}}, \tag{4-46}$$

where

q_{new} = new flow capacity after looping,
q_{old} = flow capacity before looping,
Y = fraction of the original line that was paralleled starting at the outlet,
$W = \left(\frac{d_2}{d_1} \right)^{2.5} \left(\frac{f_2}{f_1} \right)^{0.5}$,
d_1 = diameter of original line,
d_2 = diameter of parallel line,
f_1 = friction factor for original line, and
f_2 = friction factor for parallel line.

Using the specialized equations, the expression for W becomes

$$W = \left(\frac{d_2}{d_1} \right)^{a_5} . \tag{4-47}$$

Equation 4-46 can easily be solved for either W or Y to determine either the necessary line size for the parallel line or the length of line to be looped.

EFFECTS OF LIQUIDS

The equations presented in the previous sections were derived for calculating the relationship between flow rate and pressure drop for dry gases. There are many cases in gas production operations in which some liquid will be traveling in the pipe along with the gas. These include gas wells producing some condensate or water and pipelines in which condensation may occur or water may form. The presence of these liquids greatly increases the pressure drop for a given gas rate and reduces the efficiency of gathering systems.

If the liquid loading is low, the increased pressure drop in the tubing can be handled by adjusting the gas gravity used in the vertical gas flow equations. This will not suffice for pipeline flow because the liquids will accumulate in the low sections in the line. It then becomes necessary to apply a two-phase pressure drop method to design the piping system. Methods for use in both wells and pipelines will be discussed in this section.

Well Performance

The two-phase flow problem in flowing wells can be handled by either fluid gravity adjustment or by applying a two-phase flow correlation. There are many such correlations available, but only the Hagedorn and Brown[7] method will be discussed in detail for vertical flow. The Beggs and Brill[8] method, which will be used for pipeline flow, can also be used for well performance, but it has been found to sometimes overpredict pressure drop in vertical flow. Other well-known vertical, two-phase flow correlations are those of Poettmann and Carpenter[9], Orkiszewski[10], and Duns and Ros[11].

Gravity Adjustment. The gravity adjustment procedure consists of adjusting the gas gravity to a mixture gravity to account for the added density due to the liquid and then using one of the equations presented earlier to calculate pressure drop in a well. Either Equation 4-27 or the Cullender and Smith method can be used. The mixture gravity is given by

$$\gamma_{\text{m}} = \frac{\gamma_g + 4591\ \gamma_L/R}{1 + 1123/R}, \tag{4-48}$$

where

γ_m = adjusted fluid gravity (air = 1),
γ_g = dry gas gravity,
γ_L = liquid specific gravity, and
R = producing gas-liquid ratio, scf/STB.

The gravity adjustment method may be used with confidence for wells producing at high gas-liquid ratios. If a well is producing at a gas-liquid ratio of less than approximately 10000 scf/STB, the two-phase correlations should be used. A gas-liquid ratio of 10000 scf/STB expressed in terms of liquid loading is 100 STB/MMscf.

Hagedorn and Brown Method. The Hagedorn and Brown method, ignoring acceleration, requires solution of Equation 4-25 for each increment that the well is divided into.

$$\frac{dp}{dh} = \frac{g}{g_c} \rho_m \cos \emptyset + \frac{f \rho_f {v_m}^2}{2 g_c d} \tag{4-49}$$

Empirical correlations are presented for determining the mixture density, ρ_m and the friction factor, f. The parameters in Equation 4-49 are defined by

$\rho_m = \rho_L H_L + \rho_g (1 - H_L)$,
ρ_L = liquid density,

ρ_g = gas density,
H_L = liquid holdup (fraction of pipe occupied by liquid),
$\emptyset$ = angle of well segment from vertical,
$v_m = v_{sL} + v_{sg}$,
v_{sL} = superficial liquid velocity $= q_L/A_p$,
v_{sg} = superficial gas velocity $= q_g/A_p$,
A_p = area of flow string $= \pi d^2/4$,
d = flow string ID,
$\rho_f = \rho_n^2/\rho_m$
$\rho_n = \rho_L\lambda + \rho_g(1 - \lambda)$, and
$\lambda = v_{sL}/v_m$.

The friction factor is calculated using the Jain equation or found from the Moody diagram using the pipe relative roughness and the following Reynolds number:

$$N_{Rem} = \frac{\rho_n v_m d}{\mu_m}, \tag{4-50}$$

where

$\mu_m = \mu_L^{H_L} \mu_g^{(1-H_L)}$,
μ_L = liquid viscosity, and
μ_g = gas viscosity.

Determination of H_L requires the use of three empirical correlations. These are presented in Figures 4-7, 4-8, and 4-9. In order to determine H_L from these figures, the following dimensionless numbers must be evaluated from known data:

$$N_{Lv} = v_{sL}(\rho_L/g\sigma)^{.25},$$

$$N_{gv} = v_{sg}(\rho_L/g\sigma)^{.25},$$

$$N_d = d(\rho_L g/\sigma)^{.5}, \text{ and}$$

$$N_L = \mu_L(g/\rho_L\sigma^3)^{.25},$$

where σ = gas-liquid surface tension. These equations are valid for any consistent set of units. For field units the equations are

$$N_{Lv} = 1.938\, v_{sL}(\rho_L/\sigma)^{.25},$$

$$N_{gv} = 1.938\, v_{sg}(\rho_L/\sigma)^{.25},$$

$$N_d = 120.872\, d(\rho_L/\sigma)^{.5}, \text{ and}$$

$$N_L = 0.15726\, \mu_L(1.0/\rho_L\sigma^3)^{.25},$$

where

v_{sL}, v_{sg} = ft/sec,
ρ_L = lbm/cu ft,
σ = dynes/cm,
d = ft, and
μ_L = centipoise.

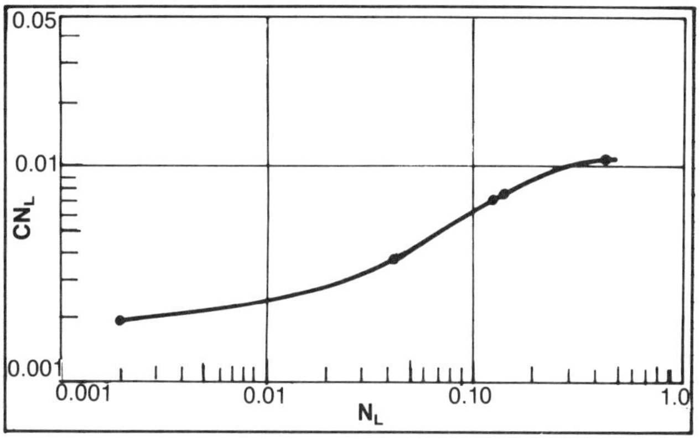

Fig. 4-7. Correlation for viscosity number coefficient C. Permission to publish by the Society of Petroleum Engineers of AIME. Copyright 1965 SPE-AIME.

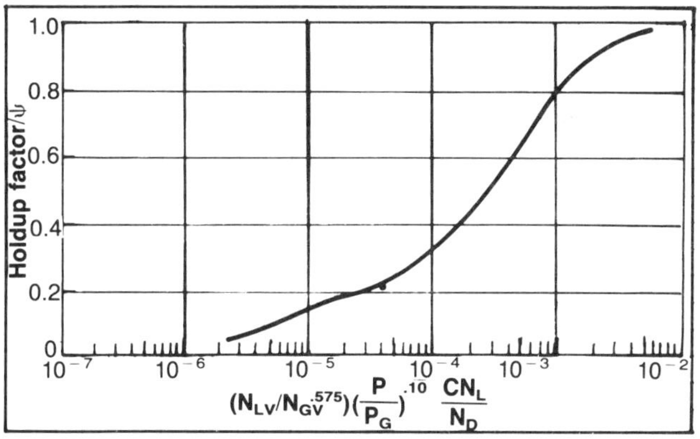

Fig. 4-8. Holdup-factor correlation. Permission to publish by the Society of Petroleum Engineers of AIME. Copyright 1965 SPE-AIME.

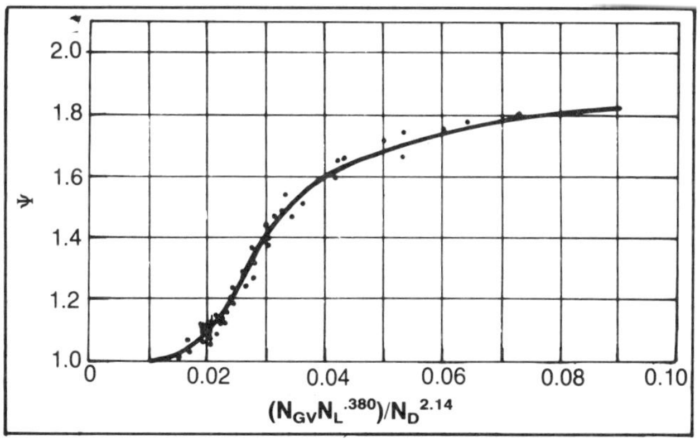

Fig. 4-9. Correlation for secondary correction factor. Permission to publish by the Society of Petroleum Engineers of AIME. Copyright 1965 SPE-AIME.

The procedure for finding H_L is:

1. Calculate N_L
2. Find CN_L from Figure 4-7
3. Calculate
   $$\bar{X}_H = \frac{N_{Lv}(CN_L)p^{0.1}}{N_d N_{gv}^{0.575} p_a^{0.1}},$$
   where p_a = base pressure (14.7 psia)
4. Find
   $$\frac{H_L}{\psi} \text{ from Figure 4-8}$$
5. Calculate
   $$\bar{X}_\psi = \frac{N_{gv} N_L^{0.38}}{N_d^{2.14}}$$
6. Find ψ from Figure 4-9
7. Calculate $H_L = \psi(H_L/\psi)$

A constraint on liquid holdup is that $H_L \geqslant \lambda$.

Once H_L is determined, N_{Re} and thus f can be calculated. The pressure gradient can then be calculated. This is Step 5 in the procedure for calculating a pressure traverse, which was presented earlier. All of the fluid properties and velocities used in the above equations are evaluated at the average pressure and temperature in the tubing increment.

Example 4-8:

During the calculation of a pressure traverse in a gas well producing liquid, the following conditions were determined at the average pressure and temperature in the pipe increment:

p = 1500 psia	v_{sg} = 30 ft/sec
T = 180°F	v_{sL} = 5 ft/sec
d = 2.992 in.	ϵ = 0.0006 ft
μ_g = 0.012 cp	ρ_L = 50 lbm/cu ft
μ_L = 0.45 cp	ρ_g = 8 lbm/cu ft
σ = 25 dynes/cm	

Using the Hagedorn and Brown method, determine the pressure gradient.

Solution:

Before finding H_L and f, some preliminary calculations are made:

$$v_m = v_{sL} + v_{sg} = 5 + 30 = 35 \text{ ft/sec}$$

$$\lambda = 5/35 = 0.143$$

$$\rho_n = 50(0.143) + 8(1 - 0.143) = 14 \text{ lb/ft}^3$$

$$A_p = \pi d^2/4 = 0.7854(0.249)^2 = 0.0487 \text{ ft}^2$$

$$\rho_L/\sigma = 50/25 = 2$$

$$N_{Lv} = 1.938(5)(2)^{.25} = 11.52$$

$$N_{gv} = 1.938(30)(2)^{.25} = 69.14$$

$$N_d = 120.872(0.249)(2)^{.5} = 42.56$$

$$N_L = 0.15726(0.45)[1/(50)(25)^3]^{.25} = 0.0024$$

Determine H_L:

1. $N_L = 0.0024$
2. From Figure 4-7, $CN_L = 0.002$
3. $\bar{X}_H = \dfrac{11.52(0.002)(1500)^{0.1}}{42.56(69.14)^{0.575}(14.7)^{0.1}} = 7.53 \times 10^{-5}$
4. From Figure 4-8, $H_L/\psi = 0.29$
5. $\bar{X}_\psi = \dfrac{69.14(0.0024)^{0.38}}{(42.56)^{2.14}} = 2.28 \times 10^{-3}$
6. From Figure 4-9, $\psi = 1.0$
7. $H_L = 1.0(0.29) = 0.29$

$$\rho_m = 50(0.29) + 8(1 - 0.29) = 20.18 \text{ lb/ft}^3$$

$$\rho_f = (14)^2/20.18 = 9.71 \text{ lb/ft}^3$$

$$\mu_m = (0.45)^{0.29}(0.012)^{(1-.29)} = 0.034 \text{ cp}$$

$$N_{Rem} = \frac{1488(14)(35)(0.249)}{0.034} = 5.29 \times 10^6$$

$$\frac{\epsilon}{d} = \frac{0.0072}{2.992} = 0.0024$$

From Figure 4-4 or Equation 4-17,

$$f = 0.025$$

$$\frac{dp}{dh} = 20.18 + \frac{0.025(9.71)(35)^2}{2(32.2)(0.249)} = 20.18 + 18.54$$

$$\frac{dp}{dh} = 38.72 \text{ lb/ft}^3 = 0.269 \text{ psi/ft}$$

A FORTRAN computer subroutine for the Hagedorn and Brown method is included in the appendix.

Pipeline Performance

Two methods will be presented for handling pipelines in which both liquid and gas are flowing. The Flanigan[12] method accounts for the added pressure drop caused by lifting the liquid up the hills in a hilly-terrain pipeline and ignores any pressure recovery in the downhill sections. The angle of the hill is of no consequence in this method.

The Beggs and Brill correlation can be used to account for the hydrostatic or elevation pressure drop and is applicable to downward two-phase flow such as might occur in offshore gathering lines.

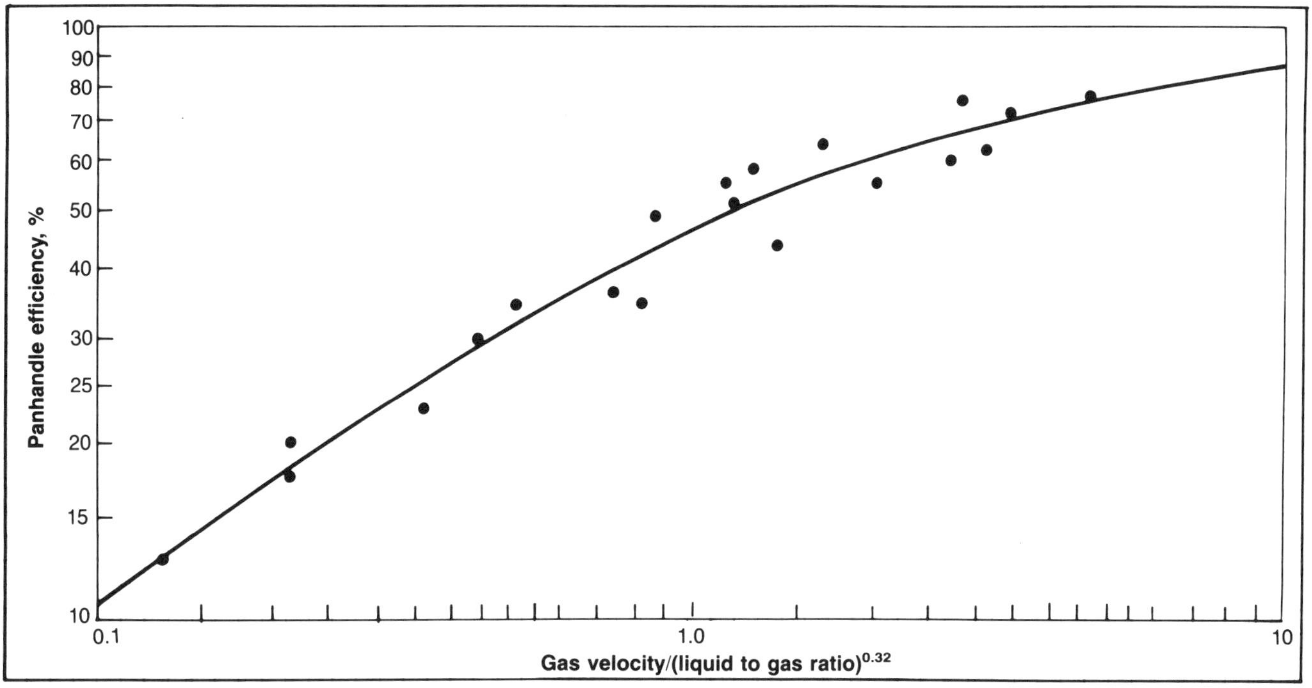

Fig. 4-10. Flanigan efficiency factor. Courtesy PennWell Publishing Company.

Flanigan Method. Flanigan proposed using the Panhandle A equation to calculate the pressure drop due to friction, based on the gas flow rate. A correlation for efficiency factor as a function of superficial gas velocity and liquid loading is presented in Figure 4-10. The gas velocity and liquid to gas ratio are in ft/sec and bbls/MMscf, respectively.

The additional pressure drop due to hills is calculated from

$$\Delta p_{e1} = \frac{\rho_L H_F \Sigma h}{144} \tag{4-51}$$

where

Δp_{e1} = pressure drop due to hills, psi,
ρ_L = liquid phase density, lb/ft^3,
H_F = holdup factor, and
Σh = sum of the vertical heights of the hills.

The holdup factor is a function of gas superficial velocity and is calculated from

$$H_F = \frac{1}{1 + 0.3264\, v_{sg}^{\;1.006}}. \tag{4-52}$$

The superficial gas velocity used in Figure 4-10 and Equation 4-52 must be calculated at the average pressure and temperature in the line. This requires an iterative solution since either p_1 or p_2 is unknown and $\bar{p} = 0.5(p_1 + p_2)$. If either q or d is unknown, they must be estimated before v_{sg} can be calculated.

Beggs and Brill Method. The Beggs and Brill method requires the determination of the flow pattern that would exist in the pipeline if the pipe were horizontal. Different equations are used to calculate liquid holdup for each flow pattern. The flow patterns defined are shown in Figure 4-11.

Determination of the correct flow pattern requires calculating several dimensionless numbers, including a two-phase Froude number.

The following variables are used to determine which flow pattern would exist if the pipe were in a horizontal position. This flow pattern is a correlating parameter and gives no information about the actual flow pattern unless the pipe is horizontal.

$$N_{FR} = \frac{v_m^{\;2}}{g\,d}$$

$$\lambda_L = \frac{v_{sL}}{v_m}$$

$$L_1 = 316\,\lambda_L^{\;0.302}$$

$$L_2 = 0.0009252\lambda_L^{\;-2.4684}$$

$$L_3 = 0.10\,\lambda_L^{\;-1.4516}$$

$$L_4 = 0.5\,\lambda_L^{\;-6.738}$$

The horizontal flow pattern limits are:

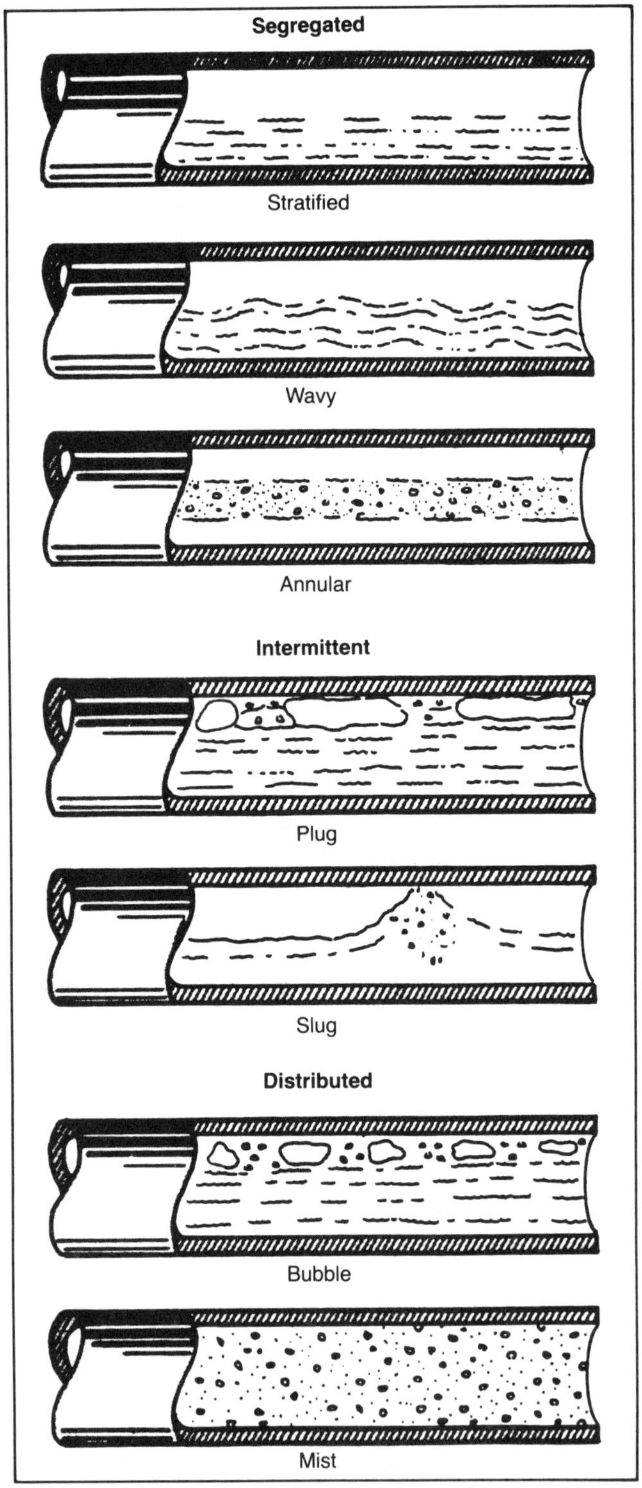

Fig. 4-11. Horizontal flow patterns.

Segregated:

Limits: $\lambda_L < 0.01$ and $N_{FR} < L_1$

or $\lambda_L \geq 0.01$ and $N_{FR} < L_2$

Transition:

Limits: $\lambda_L \geq 0.01$ and $L_2 < N_{FR} \leq L_3$

Intermittent:

Limits: $0.01 \leq \lambda_L < 0.4$ and $L_3 < N_{FR} \leq L_1$

or $\lambda_L \geq 0.4$ and $L_3 < N_{FR} \leq L_4$

Distributed:

Limits: $\lambda_L < 0.4$ and $N_{FR} \geq L_1$

or $\lambda_L \geq 0.4$ and $N_{FR} > L_4$

When the flow falls in the transition region, the liquid holdup must be calculated using both the segregated and intermittent equations and interpolated using the following weighting factors.

$$H_L \text{ (transition)} = A \times H_L \text{ (segregated)} + B \times H_L \text{ (intermittent)},$$

where

$$A = \frac{L_3 - N_{FR}}{L_3 - L_2}, \quad \text{and}$$

$$B = 1 - A.$$

The same equations are used to calculate liquid holdup for all flow patterns. The coefficients and exponents used in the equations are different for each flow pattern.

The liquid holdup depends on flow pattern and is calculated from

$$H_{L(\phi)} = H_{L(0)}\psi, \tag{4-53}$$

where $H_{L(0)}$ is the holdup that would exist at the same flow and pressure conditions in a horizontal pipe. It is calculated from:

$$H_{L(0)} = \frac{a\,\lambda_L^{\,b}}{N_{FR}^{\,c}}, \tag{4-54}$$

where a, b, and c are determined for each flow pattern from Table 4-2.

The value calculated for $H_{L(0)}$ is constrained by

$$H_{L(0)} \geqslant \lambda_L.$$

The factor for correcting the holdup for the effect of pipe inclination is given by:

$$\psi = 1 + C\,[\sin(1.8\phi) - 0.333 \sin^3(1.8\phi)],$$

where ϕ is the actual angle of the pipe from horizontal, and

$$C = (1 - \lambda_L) \ln(\alpha \lambda_L^{\,e} N_{LV}^{\,f} N_{FR}^{\,g}),$$

TABLE 4-2

Flow Pattern	a	b	c
Segregated	0.98	0.4846	0.0868
Intermittent	0.845	0.5351	0.0173
Distributed	1.065	0.5824	0.0609

TABLE 4-3

Flow Pattern	α	e	f	g
Segregated uphill	0.011	−3.768	3.539	−1.614
Intermittent uphill	2.96	0.305	−0.4473	0.0978
Distributed uphill	No correction		$C = 0$ $\psi = 1$	$H_1 = f(\phi)$
All flow patterns downhill	4.70	−0.3692	0.1244	−0.5056

where α, *e*, *f*, and *g* are determined for each flow condition from Table 4-3.

The value for C must be positive, and if a negative value is calculated, C is set equal to zero.

Once $H_{L(\phi)}$ is determined, the two-phase density is calculated from

$$\rho_s = \rho_L H_L + \rho_g H_g,$$

where $H_g = 1 - H_L$.

The pressure gradient due to elevation change is then

$$\left(\frac{dp}{dZ}\right)_{el} = \frac{g}{g_c} \rho_s \sin \phi. \quad (4\text{-}55)$$

The pressure gradient due to friction is

$$\left(\frac{dp}{dZ}\right)_f = \frac{f_{tp} \rho_n v_m^2}{2 g_c d}, \quad (4\text{-}56)$$

where:

$$\rho_n = \rho_L \lambda_L + \rho_g \lambda_g$$

$$f_{tp} = f_n \frac{f_{tp}}{f_n}$$

The no-slip friction factor f_n is determined from the Moody diagram or from Equation 4-17 using the following Reynolds number:

$$N_{Re} = \frac{\rho_n v_m d}{\mu_n},$$

where

$$\mu_n = \mu_L \lambda_L + \mu_g \lambda_g.$$

The ratio of the two-phase to no-slip friction factor is calculated from:

$$\frac{f_{tp}}{f_n} = e^s, \quad (4\text{-}57)$$

where

$$S = [\ln (y)]/\{-0.0523 + 3.182 \ln (y) - 0.8725 [\ln (y)]^2 + 0.01853 [\ln(y)]^4\}, \quad (4\text{-}58)$$

and

$$y = \frac{\lambda_L}{[H_{L(\phi)}]^2}.$$

The value of *S* becomes unbounded at a point in the interval $1 < y < 1.2$; for *y* in this interval, the function *S* is calculated from

$$S = \ln (2.2y - 1.2).$$

Although the acceleration pressure gradient is very small except for high velocity flow, it should be included for more accurate results.

$$\left(\frac{dp}{dZ}\right)_{acc} = \frac{\rho_s v_m v_{sg}}{g_c p} \frac{dp}{dZ}. \quad (4\text{-}59)$$

If an acceleration term is defined as

$$E_k = \frac{\rho_s v_m v_{sg}}{g_c p},$$

the total pressure gradient can be calculated from

$$\frac{dp}{dZ} = \frac{\left(\frac{dp}{dZ}\right)_{el} + \left(\frac{dp}{dZ}\right)_f}{1 - E_k} \quad (4\text{-}60)$$

Example 4-9:

Using the following data for a hilly terrain pipeline, calculate the outlet pressure using the Beggs and Brill method.

$q'_o = 7140$ STB/*D* p_1(inlet) = 425 psia
$q'_g = 25.7$ MMcf/*D* $\bar{T} = 90°F$, $\gamma_g = 0.70$
$d = 12$ in. $\gamma_o = 0.83 = 40°API$

Divide the pipeline into two sections. Section 1 rises 300 ft in one mile. Section 2 drops 300 ft in 3000 ft.

Solution:

Section 1

1. Estimate Δp and calculate $\bar{p}$

   $\Delta p^* = 30$ psi, $\bar{p} = 425 - 30/2 = 410$ psia

2. From fluid property correlations, at 410 psia and 90°F:

   R_s = 96 scf/STB σ_o = 19.6 dyne/cm
   B_o = 1.047 z = .925
   μ_o = 2.4 cp μ_g = 0.0105 cp

3. Calculate flow rates and densities

$$\rho_o = \frac{350 \gamma_o + 0.0764 R_s \gamma_g}{5.615 B_o}$$

$$= \frac{350 (.83) + .0764 (96)(.7)}{5.615 (1.047)}$$

$$\rho_o = 50.29 \text{ lbm/cu ft}$$

$$\rho_g = \frac{2.7 p \gamma_g}{Z T} = \frac{2.7(410)(.7)}{.925(550)} = 1.52 \text{ lbm/cu ft}$$

$$q_o = 6.49 \times 10^{-5}\, q'_o B_o = 6.49 \times 10^{-5}\,(7140)(1.047) = 0.485 \text{ ft}^3/\text{sec}$$

$$q_g = \frac{3.27 \times 10^{-7} Z(q'_g - q'_o R_S)T}{P}$$

$$q_g = \frac{3.27 \times 10^{-7}(.925)[25.7 \times 10^6 - 7140(96)](550)}{410}$$

$$q_g = 10.1 \text{ ft}^3/\text{sec}$$

4. Calculate the in-situ superficial velocities

$$v_{sL} = q_L/A = .485/.785(1)^2 = 0.617 \text{ ft/sec}$$
$$v_{sg} = q_g/A = 10.1/.785 = 12.87 \text{ ft/sec}$$
$$v_m = v_{sL} + v_{sg} = .617 + 12.87 = 13.49 \text{ ft/sec}$$

5. Determine the flow pattern

$$\lambda_L = \frac{v_{sL}}{v_m} = \frac{.617}{13.49} = 0.0457$$

$$N_{FR} = v_m^2/gd = (13.49)^2/(32.2)(1) = 5.65$$

$$N_{Lv} = 1.938\, v_{sL}\left(\frac{\rho_L}{\sigma_L}\right)^{.25} = 1.51$$

$$L_1 = 316\, \lambda_L^{0.302} = 124$$
$$L_2 = 0.0009252\, \lambda_L^{-2.4684} = 1.86$$
$$L_3 = 0.10\, \lambda_L^{-1.4516} = 8.82$$

Since $\lambda_L > .01$ and $L_2 < N_{FR} \leq L_3$,
Flow pattern is *transition*, therefore interpolation is required.

6. Calculate the liquid holdup
   a. *Segregated*

$$H_{L(0)} = \frac{.98\, \lambda_L^{.4846}}{N_{FR}^{.0868}} = .189$$

$$C = (1 - \lambda_L) \ln(.011\, \lambda_L^{-3.768} N_{Lv}^{3.539} N_{FR}^{-1.614}) = 5.516$$

$$\emptyset = \arcsin(300/5280) = 3.257°$$

$$\psi = 1 + C\,[\sin(1.8\,\emptyset) - \sin^3(1.8\,\emptyset)/3] = 1.56$$

$$H_{L(\emptyset)} \text{ (segregated)} = H_{L(0)}\, \psi = .189\,(1.56) = .295$$

   b. *Intermittent*

$$H_{L(0)} = \frac{.845\, \lambda_L^{.5351}}{N_{FR}^{.0173}} = .157$$

$$C = (1 - \lambda_L) \ln(2.96\, \lambda_L^{.305} N_{Lv}^{-.4437} N_{FR}^{.0978}) = .1246$$

$$\psi = 1 + C(.1015) = 1.013$$

$$H_{L(\emptyset)} \text{ (intermittent)} = .157(1.013) = .159$$

$$A = \frac{L_3 - N_{FR}}{L_3 - L_2} = \frac{8.82 - 5.65}{8.82 - 1.86} = .455$$

$$B = 1 - A = .545$$

$$H_L \text{ (transition)} = A \times H_L \text{ (seg.)} + B \times H_L \text{ (int.)}$$
$$= .455\,(.295) + .545\,(.159) = .221$$

7. Calculate the actual and no-slip densities

$$\rho_s = \rho_L H_L + \rho_g H_g = 50.29\,(.221) + 1.52\,(.779) = 12.3 \text{ lbm/cu ft}$$

$$\rho_n = \rho_L \lambda_L + \rho_g \lambda_g = 50.29\,(.0457) + 1.52\,(.9543) = 3.75 \text{ lbm/cu ft}$$

8. Calculate the friction factor

$$N_{Ren} = \frac{1488\, \rho_n v_m d}{\mu_L \lambda_L + \mu_g \lambda_g} = \frac{1488\,(3.75)(13.49)(1)}{2.4\,(.0457) + .0105\,(.9543)} = 6.29 \times 10^5$$

From Equation 4-17

$$f_n = .0126$$

$$y = \frac{\lambda_L}{H_{L(\emptyset)}^2} = \frac{.0457}{(0.221)^2} = .936$$

$$X = \ln y = -.066$$

$$S = X/[-0.0523 + 3.182\,X - 0.8725\,X^2 + 0.01853\,X^4] = 0.248$$

$$f_{tp} = f_n \text{ EXP }(S) = .0126\,(1.28) = 0.016$$

9. Calculate the pressure gradient

$$\frac{dp}{dL} = \frac{\frac{g}{g_c}\rho_s \sin\emptyset + \frac{f_{tp}\rho_n v_m^2}{2 g_c d}}{1 - \frac{\rho_s v_m v_{sg}}{g_c p}}$$

$$= \frac{12.3\,(.057) + \frac{.016\,(3.75)(13.49)^2}{64.4\,(1)}}{1 - \frac{12.3\,(13.49)(12.87)}{32.2\,(410)(144)}}$$

$$= \frac{0.70 + 0.17}{0.999} = 0.871 \text{ psf/ft} = 0.0061 \text{ psi/ft}$$

10. Calculate the pressure drop

$$\Delta p = \frac{dp}{dL}\Delta L = .0061\,(5280) = 32.2 \text{ psi}$$

For hand calculations this is close enough to the estimated Δp^* of 30 psi. If a computer program were being used, the calculated Δp would be taken as the new estimated Δp^* and the procedure would be repeated until the error was smaller. The pressure at the end of section 1 is 425 − 32.2 = 392.8 psia. The same procedure must be followed to calculate the pressure at the end of section 2.

Section 2

1. Estimate $\Delta p^* = 0$, $\bar{p} = 392.8$ psia

2. $R_s = 90 \qquad \mu_g = 0.0105$ cp

   $B_o = 1.046 \qquad \sigma_o = 19.6$ dyne/cm

   $\mu_o = 2.4$ cp $\qquad z = .925$

3. $$\rho_o = \frac{350(.83) + .0764(90)(.7)}{5.615(1.046)} = 50.28 \text{ lbm/cu ft}$$

   $$\rho_g = \frac{2.7(392.8)(.7)}{.925(550)} = 1.46 \text{ lbm/cu ft}$$

   $$q_o = \frac{.485(1.046)}{1.047} = .484 \text{ ft}^3\text{/sec}$$

   $$q_g = \frac{3.27 \times 10^{-7}(.925)}{392.8} \times \frac{[25.7 \times 10^6 - 7140(90)]550}{392.8} = 10.61 \text{ ft}^3\text{/sec}$$

4. $v_{sL} = .484/.785 = 0.616$ ft/sec

   $v_{sg} = 10.61/.785 = 13.52$ ft/sec

   $v_m = .616 + 13.52 = 14.14$ ft/sec

5. $\lambda_L = .616/14.14 = .04356$

   $N_{FR} = 14.14^2/(32.2) = 6.21$

   $$N_{Lv} = \frac{1.51(.616)}{.617} = 1.51$$

   $L_1 = 316(.04356)^{.302} = 122.7$

   $L_2 = 2.11 \quad L_3 = 9.45$

   Since $\lambda_L > .01$ and $L_2 < N_{FR} < L_3$, flow pattern is *transition.*

6. a. *Segregated*

   $$H_{L(0)} = \frac{.98(.04356)^{.4846}}{(6.21)^{.0868}} = .183$$

   $$C = (1 - \lambda_L) \ln (4.7\, \lambda_L^{-.3692} N_{Lv}^{.1244} N_{FR}^{-.5056}) = 1.75$$

   $\emptyset = -\arcsin(300/3000) = -5.74°$

   $\psi = 1 + C\ (\sin(1.8\emptyset) - \sin^3(1.8\,\emptyset)/3) = 1 - 1.75(.177) = .69$

   $H_{L(\emptyset)} = .183(.69) = .126$

   b. *Intermittent*

   $$H_{L(0)} = \frac{.845(.04356)^{.5351}}{(6.21)^{.0173}} = .153$$

   $C = 1.75, \quad \psi = .69$

   $H_{L(\emptyset)} = .153(.69) = .106$

   $$A = \frac{9.45 - 6.21}{9.45 - 2.11} = .44 \quad B = .56$$

   H_L (transition) = .44(.126) + .56(.106) = .115

7. $\rho_s = 50.28(.115) + 1.46(1 - .115) = 7.07$ lbm/cu ft

   $\rho_n = 50.28(.04356) + 1.46(1 - .04356) = 3.59$ lbm/cu ft

8. $$N_{Ren} = \frac{1488(3.59)(14.14)}{2.4(.04356) + .0105(.9564)} = 6.59 \times 10^5$$

   $$f_n = .0125 \qquad y = \frac{.04356}{(.115)^2} = 3.29$$

   $X = \ln y = 1.192$

   $S = .319$

   $f_{tp} = .0125 \text{ EXP } (.319) = .017$

9. $$\frac{dp}{dL} = \frac{7.07(-.10) + \dfrac{.017(3.59)(14.14)^2}{64.4(1)}}{.999} = \frac{-.707 + .189}{.999}$$

   $= -.518$ psf/ft $= -.0036$ psi/ft

10. $\Delta p = (-.0036)(3000) = -10.8$ psi

The estimated Δp^* was zero. Iteration through two more trials gives a pressure drop of −8.6 psi. The pressure at the outlet end of the pipe is then 425 − 32.2 + 8.6 = 401.4 psia.

GAS FLOW THROUGH RESTRICTIONS

There are several locations in the gas production system where the gas must pass through relatively short restrictions. Examples of these restrictions are perforations, subsurface safety valves, and surface chokes. The flow may be either critical or subcritical. In critical flow the velocity of the gas through the restriction is equal to the velocity of sound in the gas. Since pressure disturbances travel at the velocity of sound, a disturbance in the pressure downstream of the restriction cannot affect the upstream pressure or the flow rate.

In sub-critical flow the flow rate depends on both the upstream and downstream pressures. Surface chokes are usually sized so that flow will be critical, whereas in subsurface safety valves the flow is subcritical. Flow through well perforations will also be subcritical. Equations for both critical and subcritical flow are given in this section. Flow through well perforations was discussed in Chapter 3.

A general equation for flow through restrictions can be obtained by combining the Bernoulli equation with an equation of state and assuming that there are no irreversible or friction losses taking place. An empirical discharge coefficient is included to account for the simplifying assumptions used in deriving the equation. The following equation may be used for both critical (sonic) or subcritical (subsonic) flow. Tables 4-4 and 4-5 give values for the constants in the equation for various systems of units.

$$q_{sc} = \frac{C_n (p_1)(d)^2}{\sqrt{\gamma_g (T_1) Z_1}} \sqrt{\left(\frac{k}{k-1}\right)\left[\left(\frac{p_2}{p_1}\right)^{2/k} - \left(\frac{p_2}{p_1}\right)^{k+1/k}\right]}, \quad (4\text{-}61)$$

where

$$C_n = \frac{C_s (C_d) T_{sc}}{P_{sc}},$$

q_{sc} — volumetric gas flow rate,
C_n — coefficient based on system of units, discharge coefficient and standard conditions,
d — ID of bore opening to gas flow,
γ_g — gas specific gravity (air = 1.0), dimensionless,
k — ratio of specific heats = C_p/C_v, dimensionless,
p_1 — upstream pressure, absolute units,
p_2 — downstream pressure, absolute units,
T_1 — upstream temperature, absolute units,
Z_1 — compressibility factor at p_1 and T_1, dimensionless,
C_s — coefficient based on system of units,
C_d — discharge coefficient (empirical), dimensionless,
T_{sc} — standard temperature base, absolute units,
p_{sc} — standard pressure base, absolute units, and
R_{pc} — critical pressure ratio, dimensionless.

TABLE 4-4
Coefficients and Units for Equation 4-61

Symbol	English System	Metric System	SI Metric System
q_{sc}	Mscf/d	m^3/d	m^3/d
d	in.	mm	mm
p abs	psia	kg_f/cm^2	kPa
T abs	°R	°K	°K
C_s	27.611	1.6259	1.6259

TABLE 4-5
Coefficient for Equation 4-61

System of Units	C_d	p_{sc} abs	T_{sc} abs	C_n
English	0.865	14.696 psi	491.68°R	799.06
	0.865	14.696 psi	519.68°R	844.57
Metric	0.865	1.0332 kg/cm²	273.16°K	371.83
	0.865	1.0332 kg/cm²	288.72°K	393.01
SI Metric	0.865	101.325 kPa	273.16°K	3.7915
	0.865	101.325 kPa	288.72°K	4.0075

Values of k can be obtained from:

$$k = \frac{C_p}{C_v} = \frac{MC_p}{MC_p - 1.987}$$

where

M = molecular weight, lbm/mole,
C_p = specific heat, BTU/lbm − °R

The pressure ratio at which flow becomes critical depends on the k value for the flowing gas and is given by

$$R_{pc} = \left(\frac{2}{k+1}\right)^{k/k-1}$$

In calculating the values for C_n given in Table 4-5, a discharge coefficient of 0.865 was used. The discharge coefficient actually depends on the Reynolds number, the ratio of the diameter of the pipe to the diameter of the restriction, and the geometry of the restriction.

Example 4-10:

Using the following data, find the flow rate through the choke for:

a) p_2 = 2837 kPa, and
b) p_2 = 1420 kPa

d = 10 mm, γ_g = 0.69, k = 1.25, C_d = 0.865
p_{sc} = 101.325 kPa T_{sc} = 288.72°K,
T_1 = 333°K, p_1 = 3546 kPa, Z_1 = 0.93

Solution:

From Table 4-5, C_n = 4.0075

$$R_{pc} = \left(\frac{2}{k+1}\right)^{k/k-1} = \left(\frac{2}{1.25+1}\right)^{1.25/.25} = 0.555$$

(a) $\frac{p_2}{p_1} = \frac{2837}{3546}$ = 0.80, therefore flow is subcritical.

Using Equation 4-61:

$$q_{sc} = \frac{4.0075(3546)(10)^2}{[(0.69)(333)(.93)]^{.5}} \left(5[(.8)^{1.6} - (.8)^{1.8}]\right)^{0.5}$$

$$q_{sc} = 97{,}213\ (0.391) = 38{,}000\ m^3/d$$

(b) $\frac{p_2}{p_1} = \frac{1420}{3546} = 0.4$, therefore flow is critical.

$$q_{sc} = 97{,}213\ \left(5[(.555)^{1.6} - (.555)^{1.8}]\right)^{0.5}$$

$$q_{sc} = 97{,}213\ (0.465) = 45{,}235\ m^3/d$$

Equation 4-61 has been modified for particular types of wellhead chokes. An equation which is used for the types of chokes manufactured by the Thornhill-Craver company is given below. This equation applies for 6 in. long chokes with rounded entrances operating in critical flow. The equation is

$$q_{sc} = \frac{605.4\ A\ p_1 C_d}{(T\ \gamma_g)^{0.5}} \tag{4-61a}$$

where

q_{sc} = flow rate, Mscfd,
A = area of choke opening, in.^{2},
p_1 = upstream pressure, psia,
C_d = discharge coefficient, usually = 0.82,
T = upstream temperature, °R, and
γ_g = gas specific gravity.

Example 4-11:

Recalculate the flow rate in Example 4-10b using Equation 4-61a.

d = 10 mm = 0.394 in. p_1 = 3546 kPa = 514 psia
T_1 = 333°K = 600°R γ_g = 0.69

Solution:

$$A = .7854(0.394)^2 = 0.122\ \text{in.}^2$$

$$q_{sc} = \frac{605.4(0.122)(514)(0.82)}{[(600)(0.69)]^{0.5}} = 1530\ \text{Mscfd}$$

$$= 43{,}337\ m^3/d$$

An equation for calculating the pressure drop across a subsurface safety valve operating in subcritical flow was presented by the API[13] in 1974. For the English System of units given in Table 4-4, the equation is

$$p_1 - p_2 = \frac{2.7\ \gamma_g\, p_1}{Z_1\, T_1}(1 - \beta^4) \left[\frac{6.23 \times 10^{-4}\, Z_1\, T_1\, q_{sc}}{p_1\, d^2\, C_d\, Y}\right]^2, \tag{4-62}$$

where

$\beta = d/d_p$,
d_p = pipe diameter, inches,
C_d = discharge coefficient (API suggests using 0.9), and
Y = expansion factor,

$$Y = 1 - [0.41 + 0.35\beta^4]\left(\frac{p_1 - p_2}{k\, p_1}\right).$$

The solution of Equation 4-62 is iterative since Y is a function of $\Delta p = p_1 - p_2$. The value for Y ranges from about 0.67 to 1.0. For quick estimates of Δp, a value of 0.85 can be used.

Example 4-12:

A subsurface safety valve having a bean diameter of 1.0 in. is installed in a gas well equipped with 3.5 in. tubing (2.992 in. ID). The well is flowing at a rate of 20 MMscfd. Calculate the pressure drop across the SSSV if the pressure upstream of the SSSV is 2000 psia. The temperature is 180°F. Assume C_d = 0.9, Y = 0.85. Gas gravity is 0.70.

Solution:

$$\beta = \frac{d}{d_p} = \frac{1.0}{2.992} = 0.334 \quad Z_1 = 0.84$$

$$\Delta p = \frac{2.7(0.7)(2000)}{0.84(640)}(1 - (.334)^4) \cdot \left[\frac{6.23 \times 10^{-4}(.84)(640)(20000)}{(2000)(1.0)^2(.9)(.85)}\right]^2$$

$$\Delta p = 6.944\ (4.378)^2 = 134\ \text{psi}$$

USE OF PRESSURE TRAVERSE CURVES

Several equations were presented earlier for calculating flowing bottom-hole pressures in gas wells. Solution of these equations is iterative and, unless a computer is available, can require a considerable amount of time to solve. Estimates in the field can be made by using previously prepared pressure traverse curves, which can be calculated using conditions relative to a specific field. These curves can be prepared using either Equation 4-27 or the Cullender and Smith method. If small length increments are used in the calculations, Equation 4-27 is sufficiently accurate. Samples of flowing gas pressure traverse curves prepared using Cullender and Smith (see Figs. 4-12 through 4-16) are included for illustration. An example problem illustrates the application procedure for determining flowing bottom-hole pressure. It should be pointed out that when using the prepared traverse curves, one has no control over gas gravity, flowing temperature, gas viscosity, or pipe roughness. These curves were taken from Brown.[19]

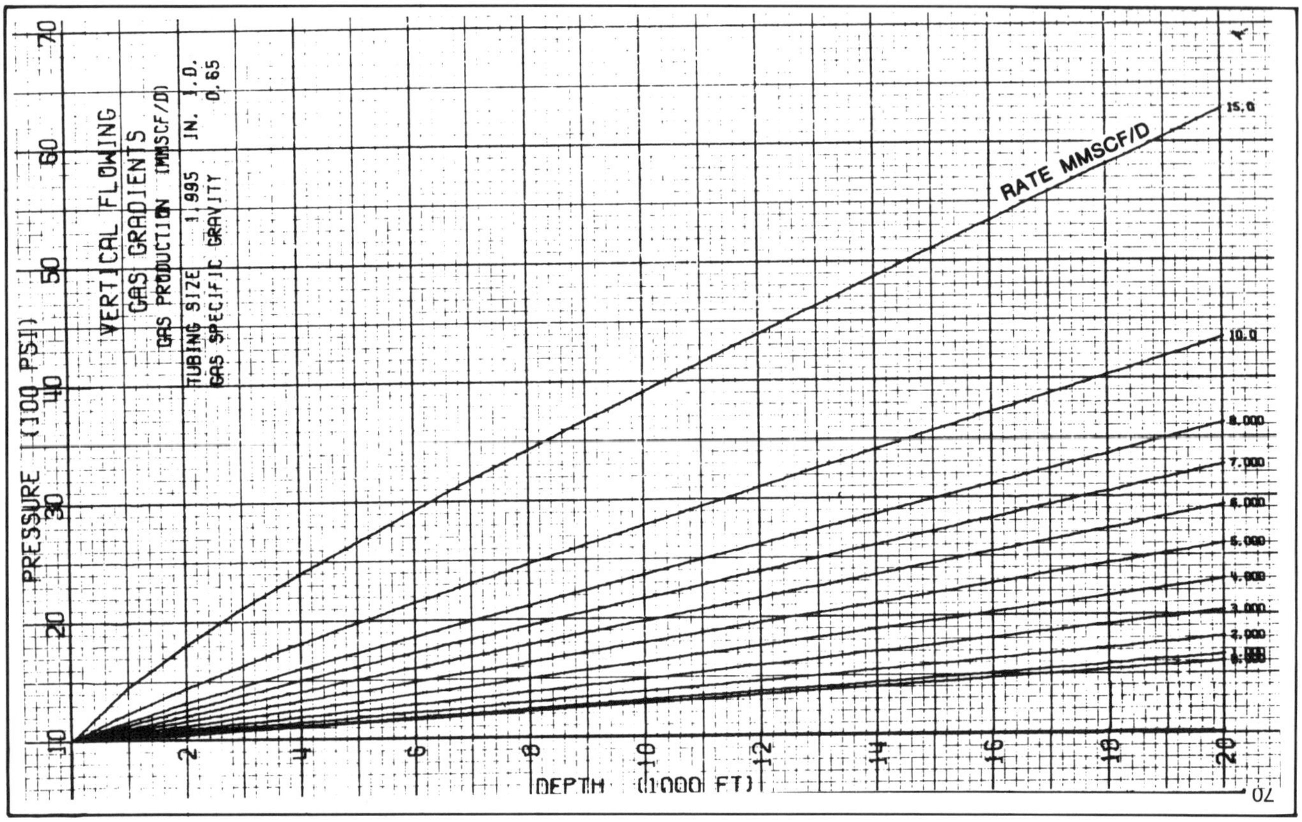

Fig. 4-12. Vertical flowing gas gradients.

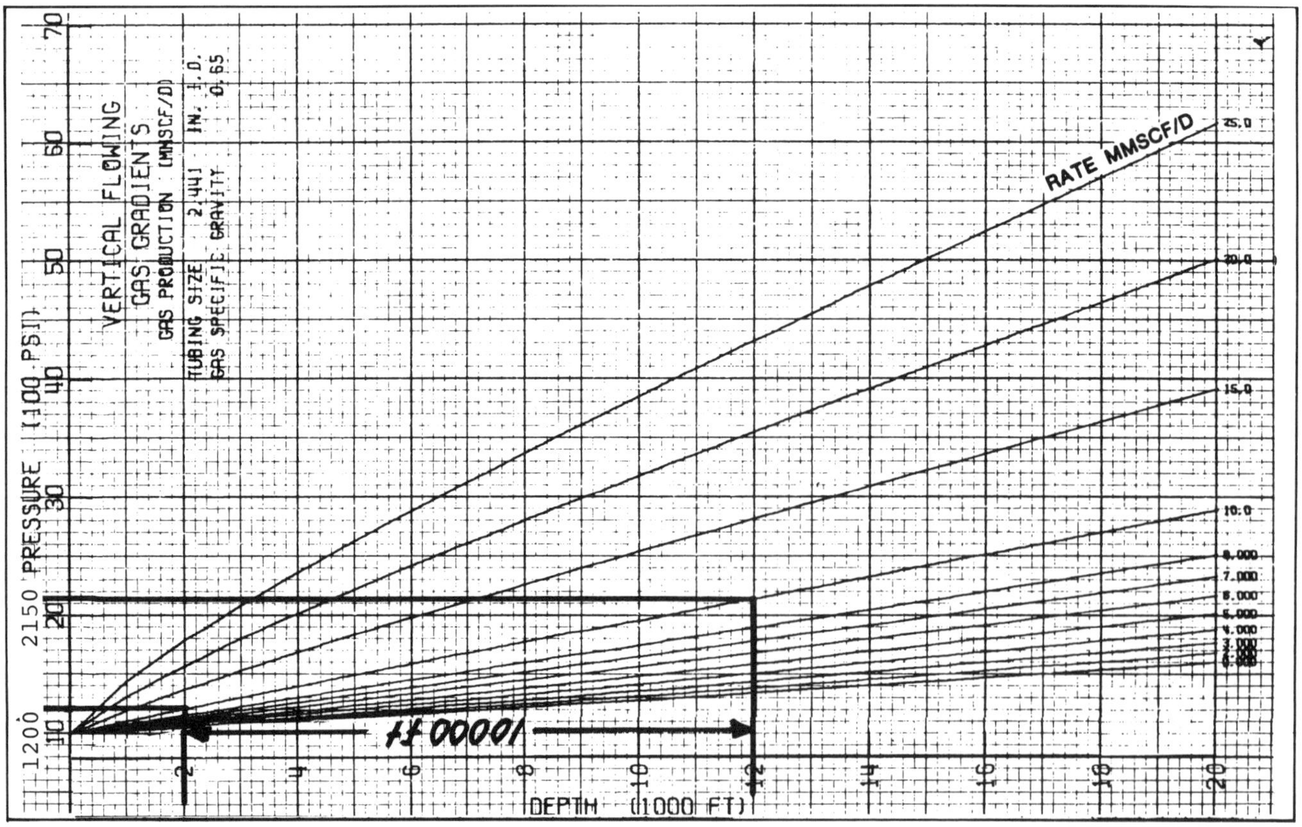

Fig. 4-13. Vertical flowing gas gradients.

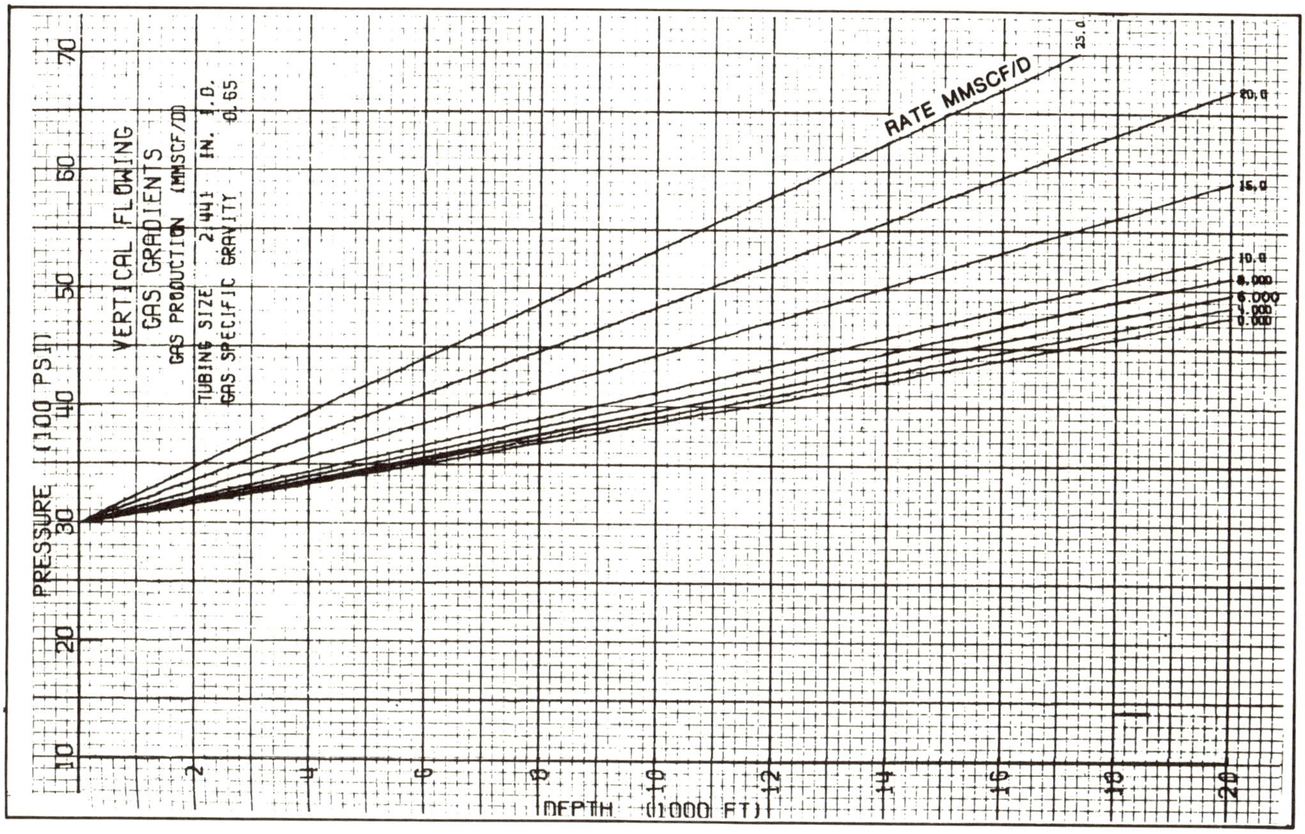

Fig. 4-14. Vertical flowing gas gradients.

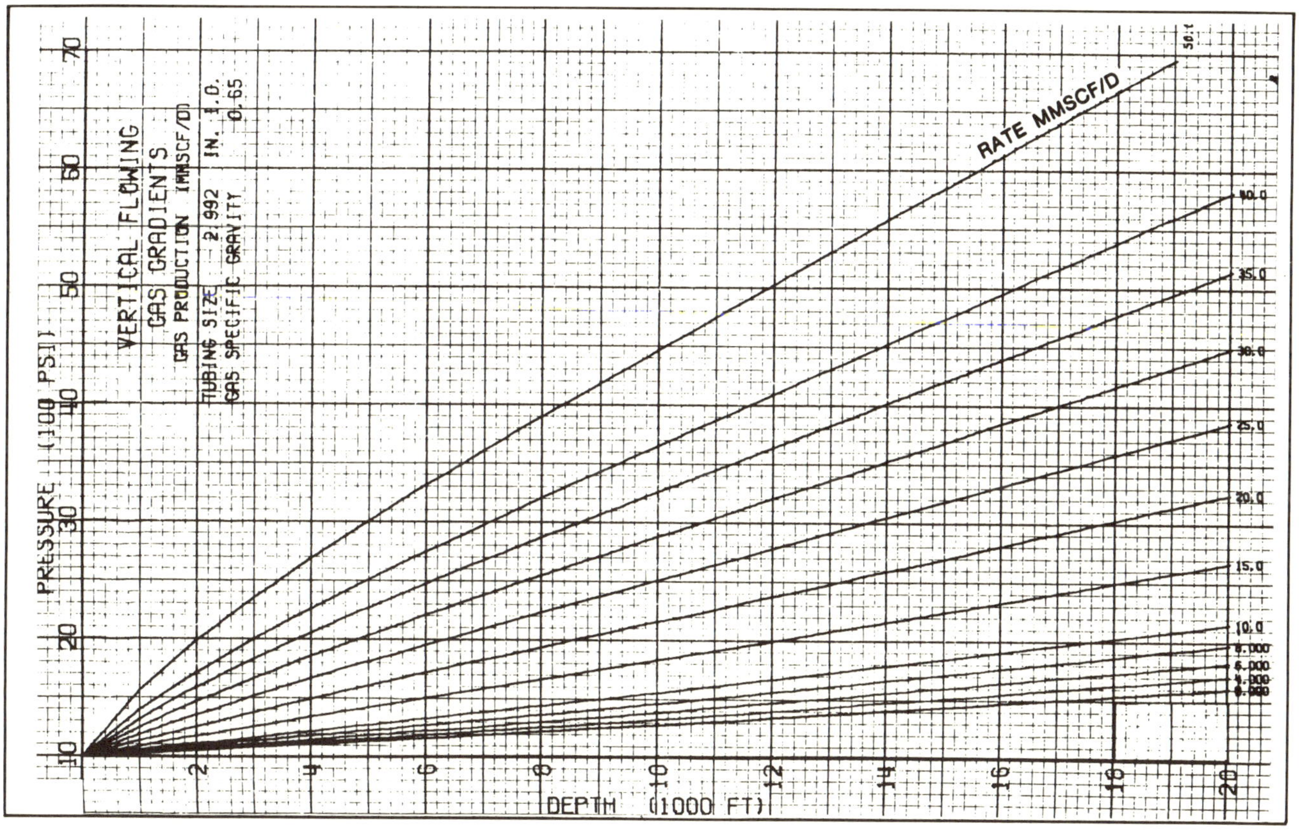

Fig. 4-15. Vertical flowing gas gradients.

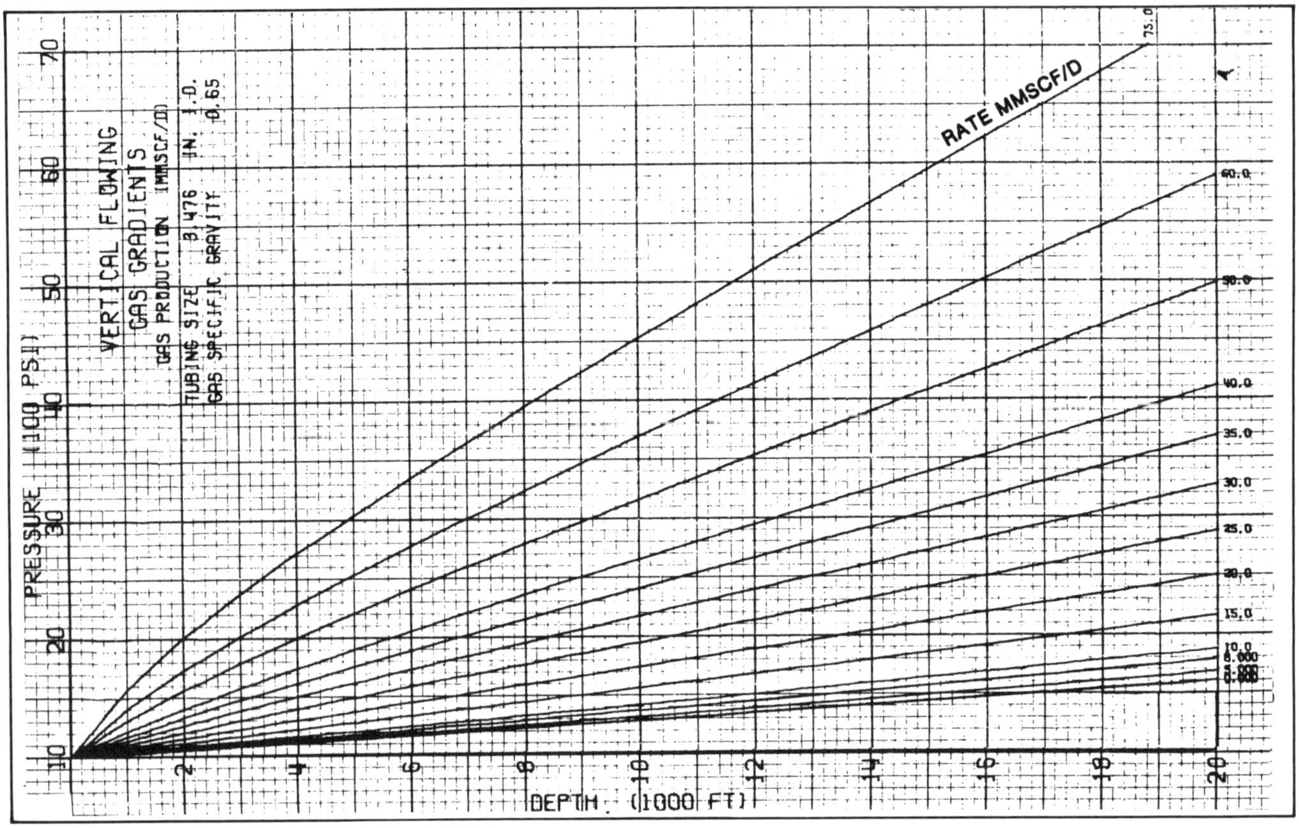

Fig. 4-16. Vertical flowing gas gradients.

Example 4-13:

Using the traverse curves, estimate the flowing bottom-hole pressure for the well producing under the following conditions:

H = 10,000 ft, p_{tf} = 1200 psig, d = 2.441 in., q_{sc} = 9.8 MMscfd

---

Solution:

Select Figure 4-13 as the curve that closely matches the given conditions. The following procedure will yield an estimate of p_{wf}:

1. Starting at the known pressure of 1200 psig on the pressure axis, draw a vertical line to intersect the 10 MMscfd traverse line.
2. Draw a horizontal line to the depth axis. This locates the equivalent depth of 2000 ft, which represents the wellhead.
3. Add the well depth to the equivalent depth found in Step 2 to get 10,000 + 2000 = 12,000 ft.
4. Draw a horizontal line from 12,000 ft to intersect the 10 MMscfd line.
5. From this intersection, draw a vertical line to the pressure axis and read p_{wf} = 2150 psig.

LIQUID REMOVAL FROM GAS WELLS

One of the most common factors that can reduce the deliverability of a gas well is the increased flowing well-bore pressure caused by the accumulation of liquids in the well bore. Liquid accumulation can occur in wells that have never produced large quantities of liquid.

When the velocity of the fluid in the tubing is too low to lift the liquids to the surface the liquid accumulates and begins to build up in the bottom of the well. The increased hydrostatic head acting on the formation further reduces the flow rate and thus the velocity of the fluid. This process continues until the well dies or begins to flow intermittently. The source of these liquids is either condensation of hydrocarbons or water from the gas or water entering from the formation.

Several methods have been used to either keep the liquid from accumulating or to remove the liquid as it accumulates. Some of these methods are discussed in this section.

Minimum Flow Rate for Continuous Liquid Removal

A model for calculating the minimum gas velocity for removing liquid droplets from wells was presented by

Turner, et al. in 1969[14]. The model was based on the fact that a freely falling particle in a fluid will reach a terminal velocity when the drag forces are equal to the gravitational forces. The terminal velocity is a function of the size, shape, and density of the particle and the density and viscosity of the fluid in which it is falling. The interfacial tension between the two fluids also influences this velocity since the droplet will shatter if the surface tension of the liquid is too small.

By writing a force balance on a droplet suspended in gas, the following equation was obtained:

$$v_t = \frac{1.3\,\sigma^{1/4}\,(\rho_L - \rho_g)^{1/4}}{C_d^{\,1/4}\,\rho_g^{\,1/2}} \tag{4-63}$$

where

v_t = terminal velocity of the droplet, ft/sec,
σ = interfacial tension, lbf/ft,
ρ_L = liquid density, lbm/ft^3,
ρ_g = gas density, lbm/ft^3, and
C_d = drag coefficient.

The equation was adjusted using experimental data to evaluate the drag coefficient, and when surface tension is expressed in dynes/cm, Equation 4-63 becomes

$$v_t = 20.4\,\frac{\sigma^{1/4}\,(\rho_L - \rho_g)^{1/4}}{\rho_g^{\,1/2}}. \tag{4-64}$$

Turner, et al. also presented simplified versions of Equation 4-64 for field use that are easier to use since only the minimum pressure in the flow string is required. Equations were presented for use when the liquid is either water or condensate, which necessitate assuming average values for the fluid properties. The assumptions used were

Liquid	*σ, dynes/cm*	*ρ, lbm/ft³*
Water	60	67
Condensate	20	45

It was also assumed that the gas density could be put in terms of pressure if a gas gravity of 0.6 and a temperature of 120°F were used. This resulted in separate equations for water and condensate.

$$v_g\,(\text{water}) = \frac{5.62(67 - 0.0031p)^{1/4}}{(0.0031p)^{1/2}} \tag{4-65}$$

$$v_g\,(\text{condensate}) = \frac{4.02(45 - 0.0031p)^{1/4}}{(0.0031p)^{1/2}} \tag{4-66}$$

where the velocities are in ft/sec and the pressures are in psia. The minimum volumetric flow rate for a particular velocity and pipe size may be calculated from

$$q_{sc\,(\text{min.})} = \frac{3.06 v_g A p}{T Z}, \tag{4-67}$$

where

$q_{sc\,(\text{min.})}$ = minimum flow rate for continuous liquid removal, MMscfd,
v_g = gas velocity, ft/sec,
A = conduit area, ft^2,
T = flowing temperature, °R,
Z = gas compressibility factor evaluated at T and the pressure used to calculate v_g, and
p = wellhead pressure, psia.

Equations 4-65 and 4-66 were derived using wellhead flowing pressures since this was the only pressure measured in the field tests. Actually, the minimum velocity in a gas well will occur at the point of highest pressure, that is, at the bottom of the well. Using the bottom-hole flowing pressure to determine $q_{sc\,(\text{min.})}$ will introduce a safety factor in the design.

A nomograph of Equation 4-67 is presented in Figure 4-17. The product qZ is obtained from the nomograph, and to obtain $q_{sc\,(\text{min.})}$, the Z factor must be calculated at surface conditions.

Example 4-14:

A gas well is producing against a fixed wellhead pressure of 1150 psia through 3-1/2 in. (2.992 ID) tubing. Wellhead temperature is 140°F and gas gravity is 0.70. Calculate the minimum flow rate required to keep this well unloaded if it produces salt water along with the gas.

Solution:

$$v_g = \frac{5.62(67 - 0.0031\,p)^{0.25}}{(0.0031\,p)^{0.5}}$$

$$v_g = \frac{5.62\,(67 - 0.0031\,(1150))^{0.25}}{(0.0031\,(1150))^{0.5}} = 8.4 \text{ ft/sec}$$

at p = 1150, T = 140°F, γ_g = 0.7, the Z-factor is 0.86.

$$A = \frac{\pi}{4} d^2 = .7854\,(2.992/12)^2 = 0.0488 \text{ ft}^2$$

$$q_{sc\,(\text{min.})} = \frac{3.06\,p\,v_g\,A}{T Z} = \frac{3.06(1150)(8.4)(0.0488)}{(600)(0.86)}$$

$$q_{sc\,(\text{min.})} = 2.8 \text{ MMscfd}$$

Turner, et al., observed that their method was applicable for liquid loading up to 130 bbl/MMscf. If both water and condensate are present, the equation for water should be used.

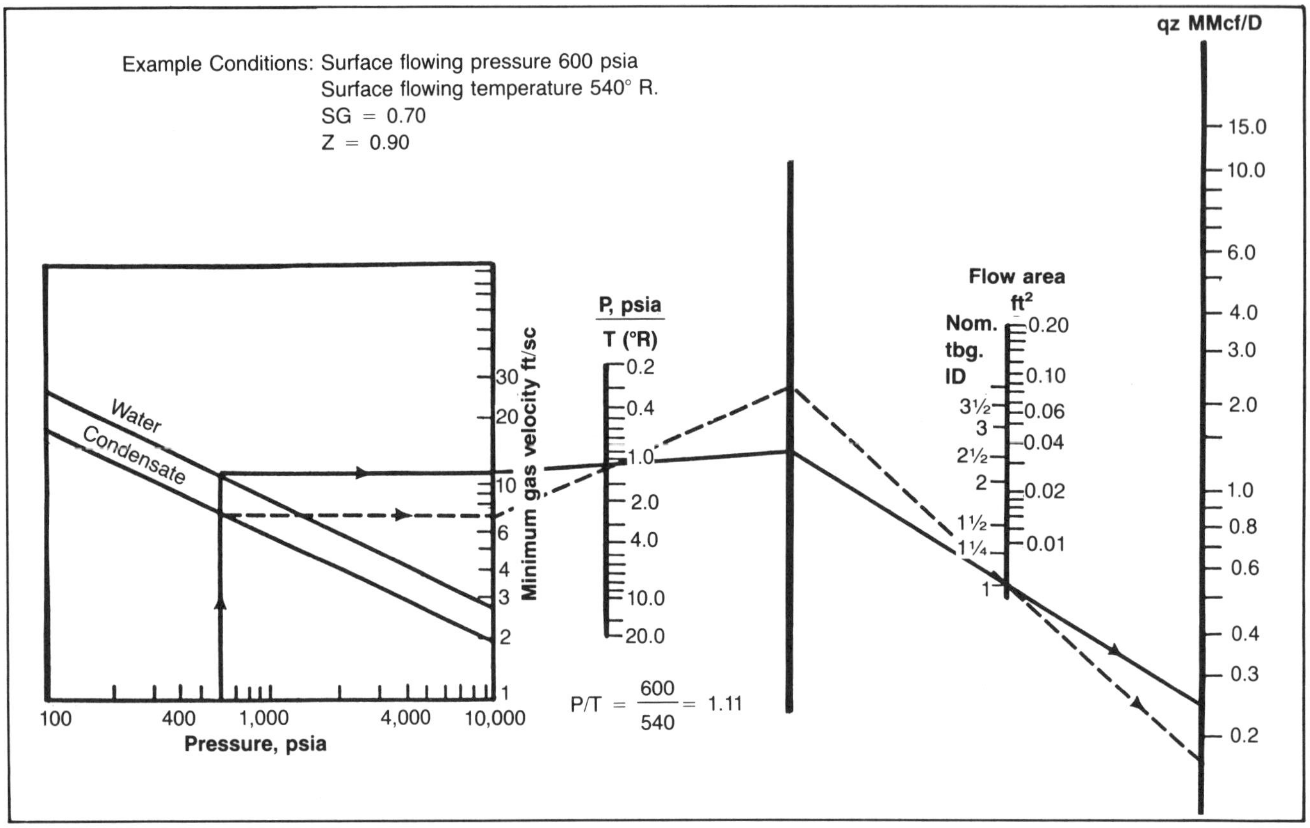

Fig. 4-17. Nomograph for calculating gas rate required to lift liquids through tubing of various sizes (after Turner, et al.). Permission to publish by the Society of Petroleum Engineers of AIME. Copyright 1969 SPE-AIME.

Liquid Removal Methods

Various methods for removing liquids from gas wells, sometimes called dewatering, have been used in the past. Descriptions of the methods and the degree of success of the various methods are described in papers by Hutlas and Granberry[15] and by Libson and Henry[16]. Libson and Henry found that the minimum surface velocity for keeping wells unloaded in the Intermediate Shelf Area of southwest Texas was 1000 ft/min or about 17 ft/sec. Methods for liquid removal include pumping units, gas lift, plunger lift, intermittent flow with flow controllers, small tubing installation, and soap injection.

Beam Pumping Units. Pumping units may be used to pump the liquids up the tubing, allowing the gas to be produced through the annulus. An advantage of using pumping units is that they do not depend on gas velocity for lift and can be used to deplete the field to a very low pressure. It is desirable to have the tubing set as close as possible to the bottom perforations or even below the perforations. A liquid cushion above the pump helps prevent gas from entering the pump.

On low liquid-rate wells, difficulties arise because a beam pumping unit cannot be adjusted for very low rates. The wells are sometimes pumped intermittently.

Plunger Lift. A plunger installed in a gas well producing liquid acts as an interface between the gas stored in the annulus and the liquid accumulated in the tubing above the plunger. A schematic of a plunger lift well is shown in Figure 4-18. The well remains shut in for a period of time and is then opened to allow the plunger, pushed by the annulus gas, to unload the liquids.

The automatic opening of the well may be accomplished by a motor valve on the flowline. The motor valve may be operated by a clock or by monitoring the flow rate of the well. If a clock is used the time cycles are adjusted by trial-and-error to find the optimum.

Surface flow controllers may be used that permit the well to flow until the gas velocity drops to some critical value. The well is then shut in for a time period that still must be determined by trial. The advantage of the flow controller is that it allows the well to flow for the maximum length of time before shut-in.

Small Tubing. Smaller tubing may be installed in gas wells as the flow rate decreases in order to maintain the

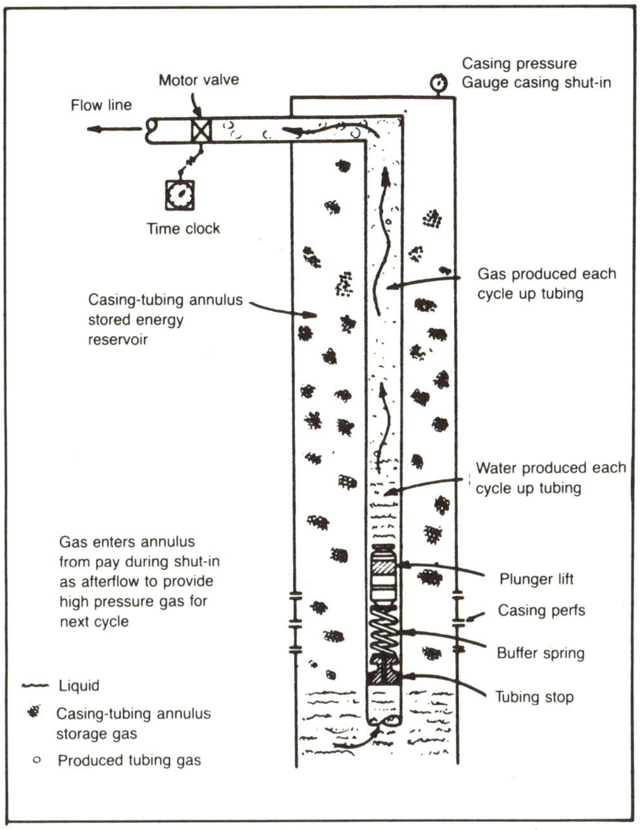

Fig. 4-18. Plunger lift operations.

velocity above a critical value. In some cases a smaller string may be run inside the existing tubing string, such as running 1 in. inside 2-7/8 in. tubing. This method works best on low volume wells in which friction loss is not severe.

Gas-Lift. A recent development in removing liquid from gas wells is the combination of a liquid diverter and gas-lift system. This system lifts liquid up the tubing string and produces gas through the annulus.

A liquid diverter is a device that opens when a predetermined head of liquid is accumulated above it, allowing the liquid to enter the tubing from the annulus. The tubing is open to atmospheric pressure at the surface. As liquid is diverted to the tubing, it accumulates until the gas-lift valve opens to lift the liquid slug to the surface.

The gas-lift valve may be actuated by the liquid head in the tubing if it is a so-called fluid operated valve. Various other means have been proposed to control the opening and closing of the gas-lift valve. Some of these are described by Hutlas and Granberry[15].

Soap Injection. Injection of surfactants or foaming agents into the annulus with a chemical pump and timer has produced successful results in some wells. By the reduction of the surface tension, water can be unloaded continuously in a foamed state. Soap injection has not been as successful in wells producing condensate because of the difficulty involved in obtaining a surfactant that will foam condensate.

EROSIONAL VELOCITY

When fluid flows through a pipe at high velocities, erosion of the pipe can occur. This is especially true for high capacity gas flow in which the in-situ velocity may exceed 60 to 70 ft/sec. Erosion is not as much of a problem in oil wells, although some high gas-liquid ratio wells may be subject to erosion.

The velocity at which erosion begins to occur cannot be determined exactly, and if solid particles such as sand are in the fluid, erosion may occur at relatively low velocities.

The velocity at which erosion may occur has been related to the density of the fluid by the following equation.

$$v_e = \frac{C}{\rho^{0.5}} \tag{4-68}$$

where

v_e = erosional velocity, ft/sec,
ρ = fluid density, lbm/ft^3, and
C = a constant that ranges between 75 and 150.

A good value for C has been found to be about 100. If C is set equal to 100, and the gas equation of state is used to express density, Equation 4-68 becomes

$$v_e = \frac{100}{\left[\frac{29\, p\, \gamma_g}{ZRT}\right]^{0.5}},$$

where p, T, and Z are determined at the conditions at which the velocity is to be determined.

The equation may be expressed in terms of gas flow rate at standard conditions by

$$q_e = 1.86 \times 10^5 A \left(\frac{p}{ZT\,\gamma_g}\right)^{0.5}, \tag{4-69}$$

where

q_e = erosional flow rate, Mscfd,
A = area of the pipe, ft^2,
p = lowest pressure in the pipe, psia,
T = temperature at point where p is determined, °R,
Z = gas compressibility factor at p, T, and
γ_g = gas gravity.

Example 4-15:

A gas well is producing through 2-7/8 in tubing at a wellhead pressure of 800 psia. The wellhead temperature is 140° F and gas gravity is 0.65. Determine the maximum rate at which this well can produce without exceeding the erosional velocity.

Solution:

$$A = \frac{\pi}{4} d^2 = 0.7854(2.441/12)^2 = 0.032 \text{ ft}^2$$

At $p = 800$, $T = 140°F$ the value of Z is 0.91

$$q_e = 1.86 \times 10^5(0.032)(800/(.91)(600)(.65))^{0.5}$$

$$q_e = 8936 \text{ Mscfd} = 8.9 \text{ MMscfd}$$

PREDICTING FLOWING TEMPERATURES

All of the correlations presented previously require a value of fluid temperature in order to calculate the required fluid property or pressure drop. The flowing temperature profile in a gas well or an oil well is usually assumed to be linear between the surface temperature and the bottom-hole temperature. A linear temperature profile is also frequently assumed for surface flowline calculations. The linear assumption for well flow will usually not introduce significant errors if a good value for surface flowing temperature can be obtained. The heat loss from a fluid in a pipe is a function of the mass flow rate in the pipe, and will therefore change with a change in producing rate.

An algorithm for coupling pressure and heat loss calculations requires an iterative solution because the overall heat transfer coefficient and the enthalpy change depend on pressure. If some average heat transfer coefficient can be determined, an approximate temperature profile can be calculated independent of the pressure loss calculations. This will of course be less accurate, but in many cases the amount of data available will not be sufficient to perform the more accurate calculation.

Flowing Temperatures in Wells

An equation for temperature in a well as a function of location, L, as derived by Ramey[17], can be written as

$$T_L = T_1 - G_T[L - A(1 - \text{EXP}(-L/A))], \quad (4\text{-}70)$$

where

T_1 = temperature at fluid entry ($L = 0$),
T_L = temperature at location L,
G_T = geothermal gradient,
A = relaxation distance $wC_p/\pi dU$,
w = mass flow rate, $= \rho_{sc}q_{sc}$,
C_p = specific heat of the flowing fluid,
d = pipe diameter,
U = overall heat transfer coefficient, and
L = distance from fluid entry.

When the equation is written in this form it assumes that the fluid and surroundings temperature are equal at the inlet to the pipe. This will be the case for flowing wells where T_1 is the reservoir temperature. Also included is the assumption that the heat loss is independent of time. This assumption limits application of Equation 4-70 to wells that have been producing for a considerable length of time.

When multiphase flow is occurring in a well, the variables involved in evaluating the relaxation distance, A, are very difficult to determine, especially the overall heat transfer coefficient U. In view of this fact, Shiu and Beggs[18] developed an empirical method to estimate A based on measured temperature profiles from 270 wells. Using the measured temperatures, T_L at various locations, L, a value of A for each test was calculated from Equation 4-70. An equation to estimate A was then developed as a function of data that will usually be known. The equation is

$$A = C_1\, w^{C_2}\, \rho_L{}^{C_3}\, d^{C_4}(API)^{C_5}\, \gamma_g{}^{C_6}, \quad (4\text{-}71)$$

where

A = relaxation distance, ft,
w = total mass flow rate, lbm/sec,
ρ_L = liquid density at standard cond., lbm/ft^3,
d = pipe ID, in.,
API = oil gravity, °API,
γ_g = gas gravity (air = 1),
C_1 = 0.0149,
C_2 = 0.5253,
C_3 = 2.9303,
C_4 = −0.2904,
C_5 = 0.2608, and
C_6 = 4.4146.

Equation 4-71 is applicable for flowing oil wells only, although a similar approach could be used for gas wells.

Flowing Temperatures in Pipelines

In order to calculate a temperature profile in a pipeline it is usually assumed that the temperature of the surroundings is constant. Modification of Equation 4-70 to account for this results in

$$T_L = T_S + (T_I - T_S)\ \text{EXP}\ (-L/A), \quad (4\text{-}72)$$

where T_S is the surroundings temperature, and the other variables are defined in Equation 4-70.

For flow of gases the Joule-Thomson effect may be

included, but since this effect depends on pressure, an iterative solution is required. The more rigorous equation is

$$T_L = T_S + \mu A(dP/dL) + [T_1 - T_S - \mu A(dP/dL)]\, EXP\,(-L/A), \quad (4\text{-}73)$$

where

μ = Joule-Thomson coefficient, and
dP/dL = pressure gradient at L.

If measured values of temperature and flow rate are available for one flow condition, a value for the group of terms $C_p/\pi U$ may be calculated. The value of this group can be assumed constant for a particular well or even for an entire field if necessary. The procedure is:

1. Using measured values of T and L, calculate A
2. Using the measured flow rate q_{sc}, calculate $C' = \dfrac{C_p}{\pi U} = \dfrac{Ad}{q_{sc}\rho_{sc}}$
3. For other conditions of q_{sc} and/or d, calculate $A = \dfrac{C'\, q_{sc}\rho_{sc}}{d}$

This procedure can be used for both wells and pipelines.

REFERENCES

1. Drew, T. B., Koo, E. C., and McAdams, W. H.: *Trans. Am. Inst. Chem. Engrs.*, 28, 56 (1930).
2. Nikuradse, J.: *Forschungsheft*, p. 301, (1933).
3. Colebrook, C. F.: *J. Inst. Civil Engrs.*, Vol. 11, p. 133, (1938).
4. Jain, A. K.: "An Accurate Explicit Equation for Friction Factor," *J. Hydraulics Div. ASCE*, 102, No. HY5, May, 1976.
5. Cullender, M.H. and Smith, R. V.: "Practical Solution of Gas Flow Equations for Wells and Pipelines with Large Temperature Gradients," *Trans.* AIME 207, (1956).
6. IGT Report No. 10: "Steady Flow in Gas Pipelines," PRC Project NB-13.
7. Hagedorn, A. R. and Brown, K. E.: "Experimental Study of Pressure Gradients Occurring During Continuous Two-Phase Flow in Small Diameter Vertical Conduits," *J. Pet. Tech.*, (April 1965), 475–484.
8. Beggs, H. D. and Brill, J. P.: "A Study of Two-Phase Flow in Inclined Pipes," *J. Pet. Tech.* (May 1973) 607–617.
9. Poettman, F. H. and Carpenter, P. G.: "The Multiphase Flow of Gas, Oil and Water Through Vertical Flow Strings with Application to the Design of Gas-Lift Installations," *Drill. and Prod. Prac.*, API (1952) 257–317.
10. Orkiszewski, J.: "Predicting Two-Phase Pressure Drops in Vertical Pipes," *J. Pet. Tech.* (June 1967) 829–838.
11. Duns, H., Jr. and Ros, N. C. J.: "Vertical Flow of Gas and Liquid Mixtures in Wells," *Proc.*, 6th World Pet. Congress (1963), 451.
12. Flanigan, O.: "Effect of Uphill Flow on Pressure Drop in Design of Two-Phase Gathering Systems," *Oil and Gas J.* (Mar. 10, 1958).
13. API 14B: "Users Manual for API 14B Subsurface Controlled Subsurface Safety Valve Sizing Computer Program," API, Washington, D.C., June, (1974).
14. Turner, R. G., Hubbard, M. G., and Dukler, A. E.: "Analysis and Prediction of Minimum Flow Rate for the Continuous Removal of Liquids From Gas Wells," *J. Pet. Tech.* (Nov. 1969), 1475–1482.
15. Hutlas, E. J. and Granberry, W. R.: "A Practical Approach to Removing Gas Well Liquids," *J. Pet. Tech.* (Aug. 1972), 916–922.
16. Libson, T. M., and Henry, J. R.: "Case Histories: Identification of and Remedial Action for Liquid Loading In Gas Wells—Intermediate Shelf Gas Play." *J. Pet. Tech.* (April 1980), 685–693.
17. Ramey, H. J.,: "Wellbore Heat Transmission," *J. Pet. Tech.* (April 1962).
18. Shiu, K. C. and Beggs, H. D.: "Predicting Temperatures in Flowing Oil Wells," J. Energy Res. Tech., (March 1980); *Trans.* AIME.
19. Brown, K.E.: *The Technoloygy of Articifial Lift Methods*, Vol. III, PennWell Publishing Co., (1981).

5 Gas Compression

THERE are many cases in gas production operations in which the pressure of a gas must be raised to a higher value. As the pressure in a gas reservoir depletes, it will eventually reach a point where it will no longer overcome all the pressure losses in the system and the pressure of the line into which the gas is being delivered. It is then necessary to add a compressor to the system to supplement the reservoir energy. In this type of application, the suction or intake pressure, and possibly the volume compressed, will change with time even though the discharge pressure may remain constant. Compressors have been used to lower the wellhead pressure below atmospheric, that is, to pull a vacuum on the well in order to obtain maximum rates.

Compressors are also used to overcome the losses incurred in the long distance transportation of natural gas through transmission lines. This may require large capacity machines operating at essentially constant conditions.

The reinjection of gas for pressure maintenance or cycling requires compression of produced gas to a high pressure to move sufficient volumes into the reservoir. Injection of gas into storage fields also requires a compressor capable of operating under a wide range of conditions.

The engineer is concerned with essentially two types of compressor design problems: (1) determination of the power required to compress a certain volume of gas from some given intake pressure to a given discharge pressure and (2) estimation of the capacity of an existing compressor under required pressure increase conditions. It is also frequently necessary to calculate the temperature increase occurring in the gas as it is compressed.

TYPES OF COMPRESSORS

The principal types of compressors are positive-displacement, or intermittent flow, units and continuous flow units. Positive-displacement units are those in which successive volumes of gas are confined within a closed space and elevated to a higher pressure. Continuous flow units are those in which a rapidly rotating element accelerates the gas as it passes through the element, converting the velocity head into pressure, partially in the rotating element and partially in stationary diffusers or blades. The principal compressor types are defined below (see Fig. 5-1).

—*Reciprocating compressors* are positive-displacement machines in which the compressing and displacing element is a piston having a reciprocating motion within a cylinder.

—*Rotary positive-displacement compressors* are machines in which compression and displacement are effected by the positive action of rotating elements.

—*Sliding-vane compressors* are rotary positive-displacement machines in which axial vanes slide radially in a rotor eccentrically mounted in a cylindrical casing. Gas trapped between vanes is compressed and displaced.

—*Liquid-piston compressors* are rotary positive-displacement machines in which water or other liquid is used as the piston to compress and displace the gas handled.

—*Two-impeller straight-lobe compressors* are rotary positive-displacement machines in which two straight mating lobed impellers trap gas and carry it from intake to discharge. There is no internal compression.

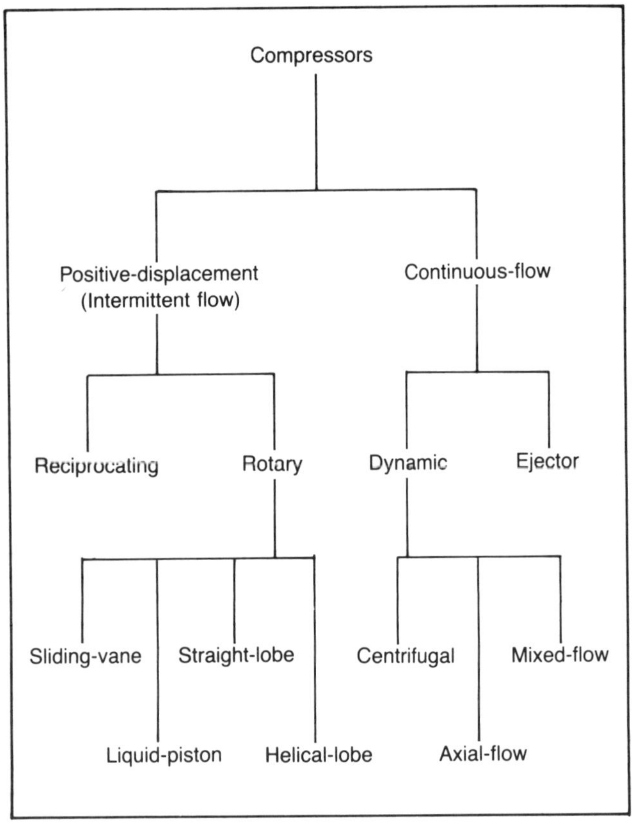

Fig. 5-1. Chart of principal compressor types. From Compressed Air and Gas Data, copyright 1980. Courtesy Ingersoll-Rand.[4]

—*Helical- or spiral-lobe compressors* are rotary positive-displacement machines in which two intermeshing rotors, each with a helical form, compress and displace the gas.

—*Centrifugal compressors* are dynamic machines in which one or more rotating impellers, usually shrouded on the sides, accelerate the gas. Main gas flow is radial.

—*Axial compressors* are dynamic machines in which gas acceleration is obtained by the action of the bladed rotor shrouded on the blade ends. Main gas flow is axial.

—*Mixed-flow compressors* are dynamic machines with an impeller form combining some characteristics of both the centrifugal and axial types.

—*Ejectors* are devices that use a high velocity gas or steam jet to entrain the inflowing gas, then convert the velocity of the mixture to pressure in a diffuser.

Every compressor is made up of one or more basic elements. A single element, or a group of elements in parallel, comprises a single-stage compressor.

Many compression problems involve conditions beyond the practical capability of a single compression stage. Too great a compression ratio (absolute discharge pressure divided by absolute intake pressure) may cause excessive discharge temperature or other design problems. It therefore may become necessary to combine elements or groups of elements in series to form a multistage unit, in which there will be two or more steps of compression. The gas is frequently cooled between stages to reduce the temperature and volume entering the following stage.

Note that each stage is an individual basic compressor within itself. It is sized to operate in series with one or more additional basic compressors, and even though they may all operate from one power source, each is still a separate compressor. The following simplified discussions show the principles of operation of each principal type of compressor.

Positive Displacement Compressors

The basic reciprocating compression element is a single cylinder compressing on only one side of the piston (single-acting). A unit compressing on both sides of the piston (double-acting) consists of two basic single-acting elements operating in parallel in one casting.

The reciprocating compressor uses automatic spring-loaded valves that open only when the proper differential pressure exists across the valve. Inlet valves open when the pressure in the cylinder is slightly below the intake pressure. Discharge valves open when the pressure in the cylinder is slightly above the discharge pressure.

Figure 5-2, diagram A, shows the basic element with the cylinder full of low pressure gas. On the theoretical *pV* diagram, point 1 is the start of compression and both valves are closed.

Diagram B shows the compression stroke, the piston having moved to the left, reducing the original volume of gas with an accompanying rise in pressure. Both valves remain closed. The *pV* diagram shows compression from point 1 to point 2, and that the pressure inside the cylinder has reached that in the receiver.

Diagram C shows the piston completing the delivery stroke. The discharge valve opened just beyond point 2. Compressed gas is flowing out through the discharge valve to the receiver.

After the piston reaches point 3, the discharge valve will close, leaving the clearance space filled with gas at discharge pressure. During the expansion stroke, diagram D, both the inlet and discharge valves remain closed, and gas trapped in the clearance space increases in volume, causing a reduction in pressure. This continues as the piston moves to the right until the cylinder pressure drops below the inlet pressure at point 4. The inlet valve will now open and gas will flow into the cylinder until the end of the reverse stroke at point 1. This is the intake or suction stroke, illustrated by diagram E. At point 1 on the *pV* diagram, the inlet valve will close, and the

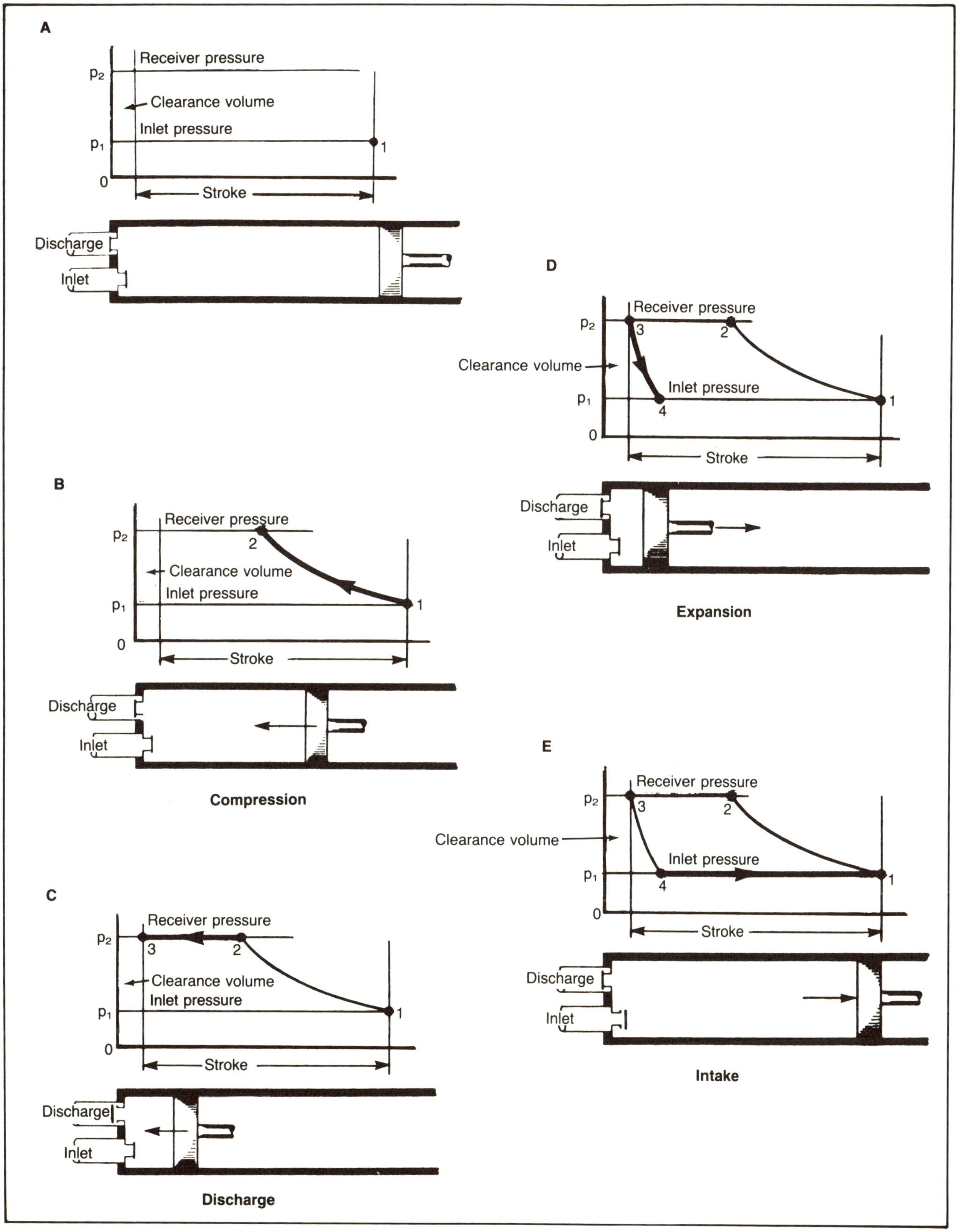

Fig. 5-2. The various steps in a reciprocating compressor cycle. From Compressed Air and Gas Data, copyright 1980. Courtesy Ingersoll-Rand.[4]

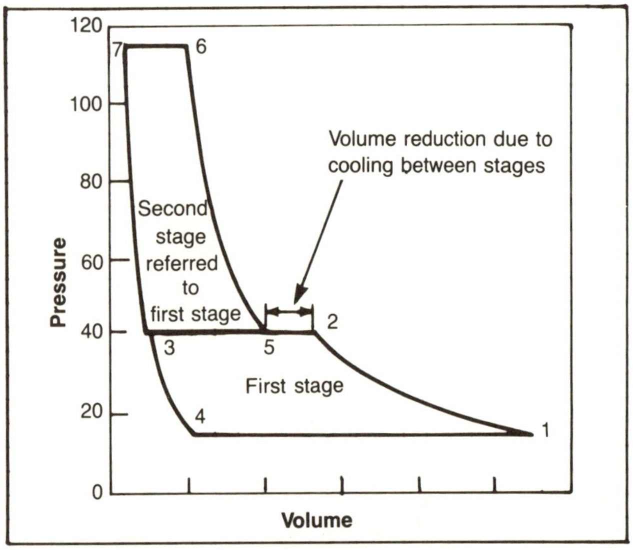

Fig. 5-3. Combined theoretical indicator card for a two-stage, two-element 100 psig positive-displacement compressor. From Compressed Air and Gas Data, copyright 1980. Courtesy Ingersoll-Rand.[4]

cycle will repeat on the next revolution of the crank.

In a simple two-stage reciprocating compressor, the cylinders are proportioned according to the total compression ratio, the second stage being smaller because the gas, having already been partially compressed and cooled, occupies less volume than at the first stage inlet. Looking at the *pV* diagram (Fig. 5-3), the conditions before starting compression are points 1 and 5 for the first and second stages, respectively, after compression, points 2 and 6, and after delivery, 3 and 7. Expansion of gas trapped in the clearance spaces as the pistons reverse brings the pressures and volumes to points 4 and 8, and on the intake stroke the cylinders are again filled at points 1 and 5, and the cycle is set for repetition. Multiple staging of any positive displacement compressor follows this pattern.

The rotary sliding-vane compressor has as its basic element the cylindrical casing with its heads and rotor assembly. When running at design pressure, the theoretical *pV* diagram is identical to the reciprocator. There is one difference of importance, however. The sliding-vane machine has no valves. The times in the cycle when

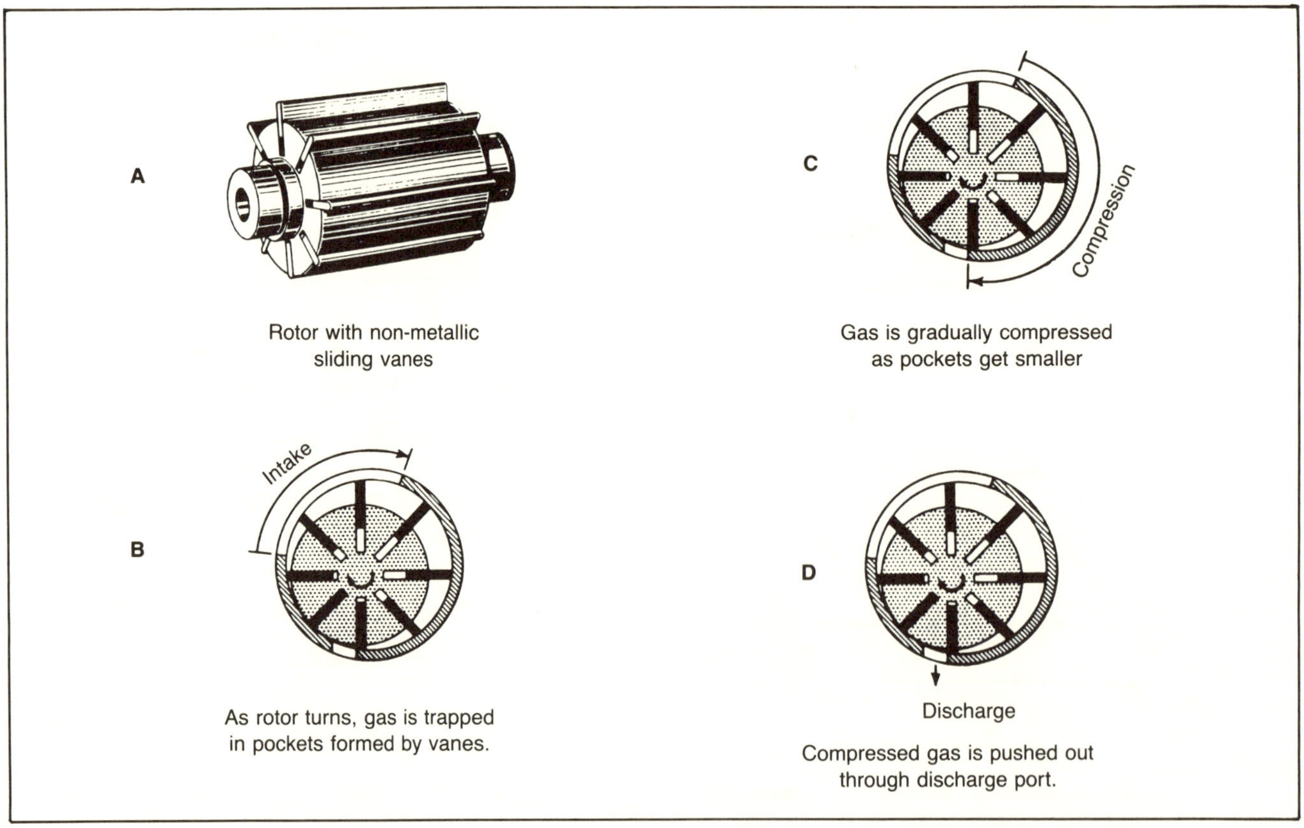

Fig. 5-4. The steps in compression for a sliding-vane rotary compressor. From Compressed Air and Gas Data, copyright 1980. Courtesy Ingersoll-Rand.[4]

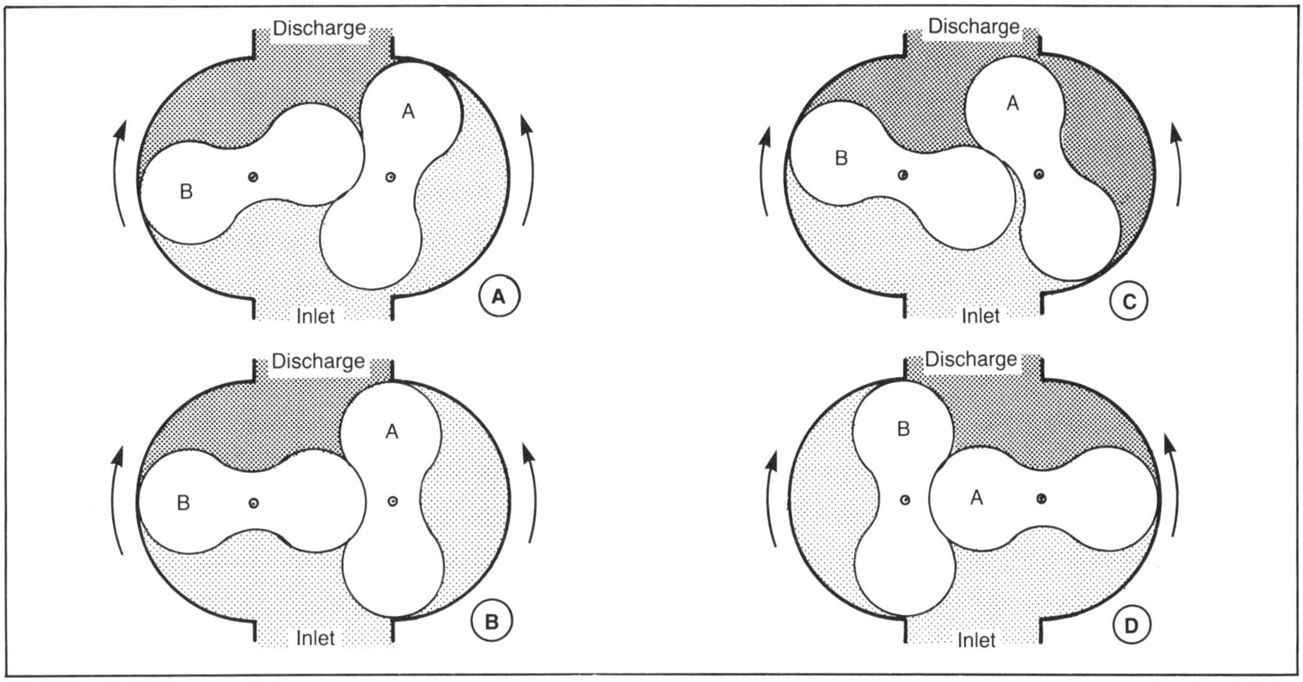

Fig. 5-5. The operating cycle of a two-impeller straight-lobe rotary compressor. From Compressed Air and Gas Data, copyright 1980. Courtesy Ingersoll-Rand.[4]

the inlet and discharge are open are determined by the location of ports over which the vanes pass (Fig. 5-4).

The pocket volume decreases as the rotor turns and the gas is compressed. Compression continues until the discharge port is uncovered by the leading vane of each pocket. This point must be preset or built-in when the unit is manufactured. Thus, the compressor always compresses the gas to design pressure, regardless of the pressure in the receiver into which it is discharging.

A two-impeller straight-lobe positive-displacement compressor element consists of a casing containing duplicate symmetrical rotors or impellers usually having a figure eight cross section. These intermesh, are kept in phase by external timing gears, and rotate in opposite directions.

There is no compression or reduction of gas volume during the turning of the rotors. The rotors merely move the gas from the inlet to the discharge. Compression is by backflow into the casing from the discharge line at the time the discharge port is uncovered. Displacement of the compressed gas into the discharge system then takes place. There is no contact between the impellers or between impellers and casing. Sealing is by close clearances and lubrication is not required within the gas chamber. One impeller is driven directly while the other is driven through phasing gears. The operation can be visualized from the diagrams of Figure 5-5. Light shading shows gas at inlet pressure. Dark shading shows gas at discharge pressure.

There are several other types of positive displacement compressors available, as indicated in Figure 5-1. Most of these are seldom used in the gas industry and are not described.

Dynamic Compressors

Compression in any dynamic compressor depends on the transfer of energy from a rotating set of blades to the gas. The rotor accomplishes this energy transfer by changing the momentum and pressure of the gas. The momentum (related to kinetic energy) then is converted into useful pressure energy by slowing the gas down in a stationary diffuser or another set of blades.

The centrifugal designation is used when the gas flow is radial, and the energy transfer is predominantly due to a change in the centrifugal forces acting on the gas.

The axial designation is used when the gas flow is parallel to the compressor shaft. Energy transfer is caused by the action of a number of rows of blades on a rotor, each row followed by a fixed row fastened to the casing.

The centrifugal compressor has an impeller with radial or backward slanted vanes usually between two shrouds. The gas is forced through the impeller by the mechanical action of the rapidly rotating impeller vanes. The velocity generated is converted into pressure. Figure 5-6 illustrates a single-stage centrifugal compressor with radial vanes. This utilizes a radial diffuser and a volute gas collector ending in a volute diffuser.

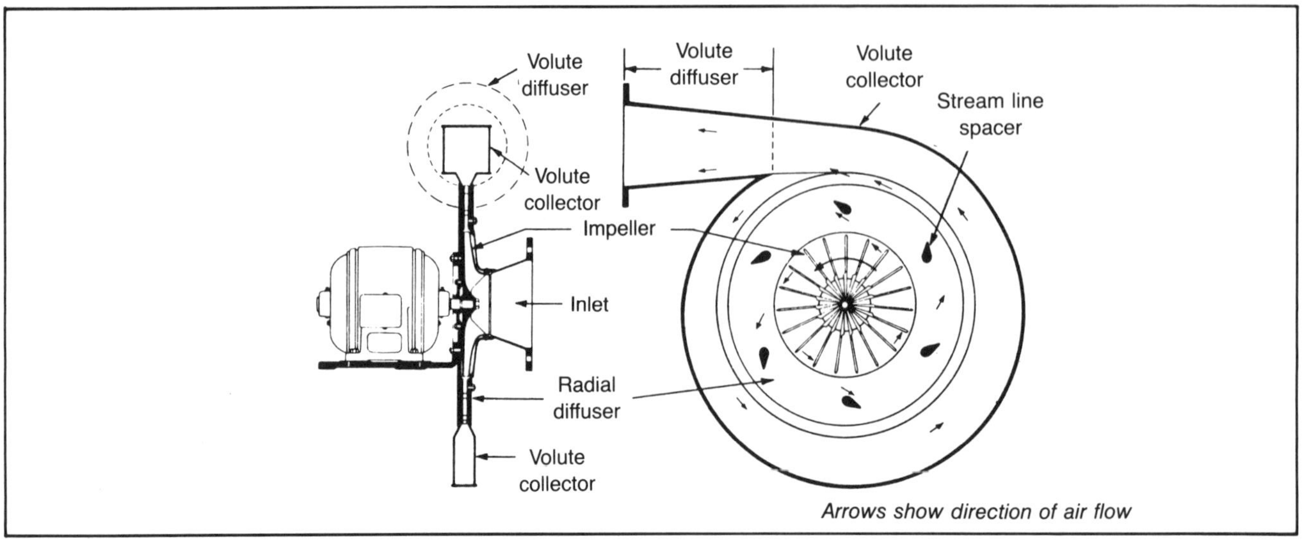

Fig. 5-6. Typical overhung impeller single-stage centrifugal compressor. From Compressed Air and Gas Data, copyright 1980. Courtesy Ingersoll-Rand.[4]

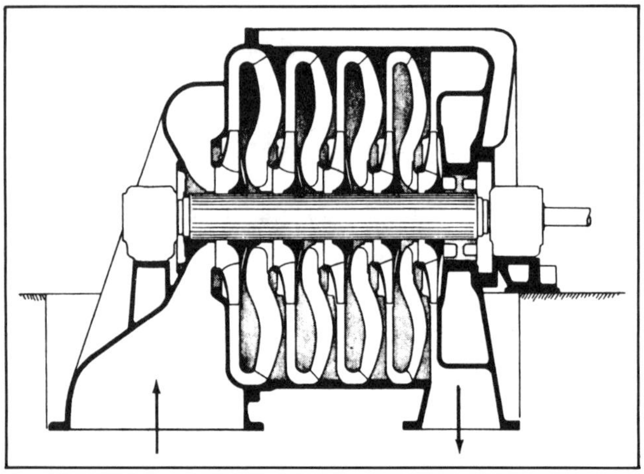

Fig. 5-7. Cross-section of a typical multistage uncooled centrifugal compressor. From Compressed Air and Gas Data, copyright 1980. Courtesy Ingersoll-Rand.[4]

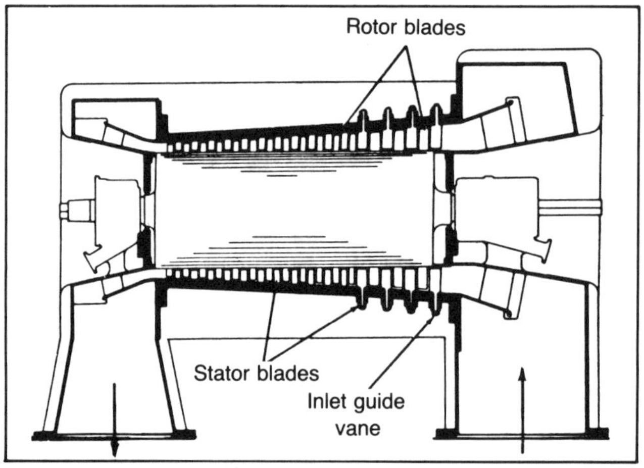

Fig. 5-8. Cross-section of a typical axial-flow dynamic compressor. From Compressed Air and Gas Data, copyright 1980. Courtesy Ingersoll-Rand.[4]

Multistage centrifugal compressors utilize two or more impellers arranged for series flow, each with a radial diffuser and return channel separating impellers. The number of impellers per casing is dependent upon many factors, but usually eight to ten is the limit. Figure 5-7 shows a section of a typical uncooled multistage compressor.

An axial-flow dynamic compressor is shown in Figure 5-8. It is essentially a large-capacity high-speed machine with characteristics quite different from the centrifugal. Each stage consists of two rows of blades, one row rotating and the next row stationary. The rotor blades impart velocity and pressure to the gas as the rotor turns, the velocity being converted to pressure in the stationary blades. Frequently about half the pressure rise is generated in the rotor blades and half in the stator. Figure 5-8 shows a multistage unit. Gas flow is predominantly in an axial direction, there being no appreciable vortex action.

Ejector Compressors

An ejector consists of a relatively high-pressure motive steam or gas nozzle discharging a high-velocity jet across a suction chamber into a venturi-shaped diffuser. The gas, whose pressure is to be increased, is entrained by the jet in the suction chamber. The mixture at this point has high velocity and is at the pressure of the induced gas. Compression takes place as velocity energy is transformed into pressure inside the diffuser.

Ejectors are principally used to compress from pressures below atmospheric to a discharge pressure close to atmospheric. They may, however, involve compression from a near atmospheric intake to some higher level.

A vacuum ejector, using steam as motive fluid and compressing air, is shown in Figure 5-9. Pressure and velocity changes are indicated for various sections of the device. Temperature changes follow the pressure curve closely.

Ejectors have no moving parts. They can handle liquid carry-over without physical damage although they should not be exposed to a steady flow of liquid.

COMPRESSOR DESIGN

The principal compressor design problems are the determination of compressor capacity and the determination of power requirements. In making the calculations for these designs, various simplifying assumptions can be made. There are three methods of design: (1) analytical expressions, (2) Mollier diagrams, and (3) quick estimates. The best method in any given situation depends on both the degree of accuracy required and the data available. Detailed design procedures are given for reciprocating and centrifugal compressors only.

Design Methods

Two basic compression cycles or processes are applicable to both positive displacement and dynamic compressors. Although neither of the two basic processes is commerically attainable, they are useful as a basis for calculations and comparisons.

Isothermal compression occurs when the temperature

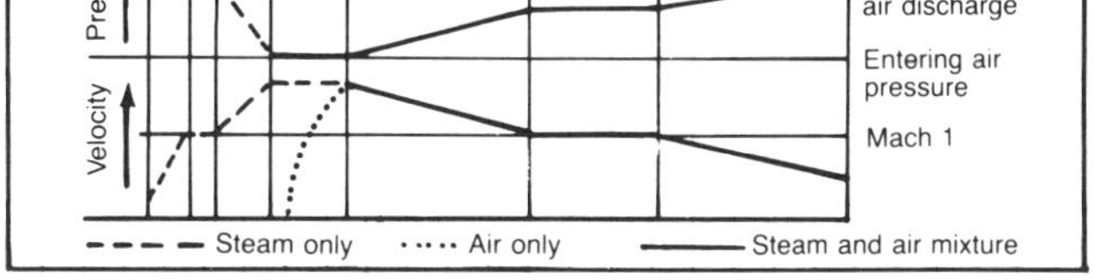

Fig. 5-9. A diagram of the pressure and velocity variations within a steam jet ejector handling air. From Compressed Air and Gas Data, copyright 1980. Courtesy Ingersoll-Rand.[4]

is kept constant as the pressure increases. This requires continuous removal of the heat of compression. The compression process follows the formula

$$p_1V_1 = p_2V_2 = \text{Constant}. \tag{5-1}$$

However, it is never commercially possible to remove the heat of compression as rapidly as it is generated.

Adiabatic compression is obtained when there is no heat added to or removed from the gas during compression. The compression process is expressed by Equation 5-2.

$$p_1V_1^k = p_2V_2^k. \tag{5-2}$$

The factor k is the ratio of specific heat at constant pressure to the specific heat at constant volume. Determination of k for a gas was discussed in Chapter 4.

Adiabatic compression is likewise never exactly obtained, since with some types of units there may be heat loss during part of the cycle and heat gain during another part. With other types of compressors there may be a definite heat gain. Nevertheless, the adiabatic cycle is rather closely approached with most positive-displacement units and therefore they are usually designed using the adiabatic cycle.

Dynamic units generally are designed based on the polytropic cycle where the pV relationship is

$$p_1V_1^n = p_2V_2^n. \tag{5-3}$$

The exponent n is experimentally determined for a given type of machine and may be lower or higher than the adiabatic exponent k. In positive displacement and internally cooled dynamic compressors n is usually less than k. In uncooled dynamic units it is usually higher than k due to internal gas friction. Although n actually changes during compression, an average or effective value, calculated from experimental information, is used.

Thermodynamically, the isentropic or adiabatic process is reversible, while the polytropic process is irreversible. Also, all compressors operate on a theoretical steady-flow process.

Although the exponent n is seldom required, the quantity $(n - 1)/n$ is frequently needed. This can be obtained from the following equation, although it is necessary that the polytropic efficiency (η_p) be known (or approximated) from a prior test. The k value of any gas or gas mixture is either known or can be calculated.

$$\frac{n-1}{n} = \frac{k-1}{k\eta_p}, \tag{5-4}$$

Either n or $(n - 1)/n$ can also be experimentally calculated from test data if inlet and discharge pressures and temperatures are known. The following formula may be used:

$$\frac{T_2}{T_1} = \left(\frac{p_2}{p_1}\right)^{(n-1)/n} = r^{(n-1)/n}. \quad (5\text{-}5)$$

This equation may also be used to estimate discharge temperatures when n or $(n - 1)/n$ is known.

It is obvious that k and n can have quite different values. There has been a tendency in the past to use these symbols interchangeably to represent the ratio of specific heats. This is incorrect, and the difference between them should be carefully observed. In the absence of experimental data, an approximation of the polytropic compression efficiency can be obtained from Figure 5-10.

Figure 5-11 shows the theoretical zero clearance isothermal and adiabatic cycles on a pV basis for a compression ratio of four. The area ADEF represents the work required when operating on the isothermal basis, and ABEF, the work required when operating on the adiabatic basis. Obviously, the isothermal area is considerably less than the adiabatic and would be the desired cycle for greatest compression economy.

In addition to the isothermal and adiabatic compression curves shown in Figure 5-11, the dotted lines show typical polytropic curves for a water-cooled reciprocating cylinder (AC) and for a noncooled dynamic unit (AC').

All basic compressor elements, regardless of type, have certain limiting operating conditions. When any limitation is involved it becomes necessary to multistage the compression process. Each stage will utilize at least one basic element designed to operate in series with the other elements of the machine.

The limitations vary with the type of compressor, but the most important include:

1. Discharge temperature—all types

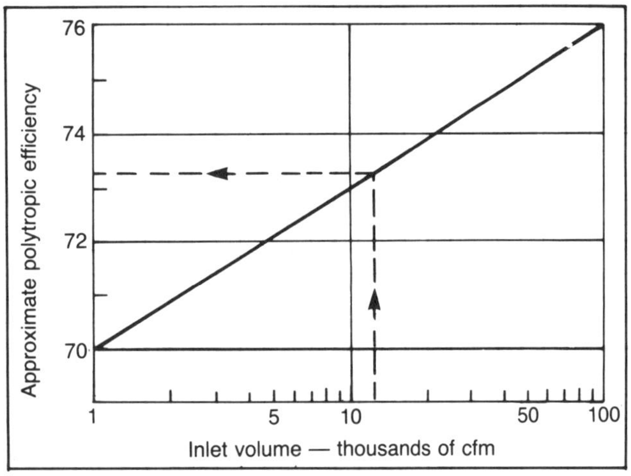

Fig. 5-10. Approximate polytropic compression efficiency of a dynamic compressor versus inlet capacity. From Compressed Air and Gas Data, copyright 1980. Courtesy Ingersoll-Rand.[4]

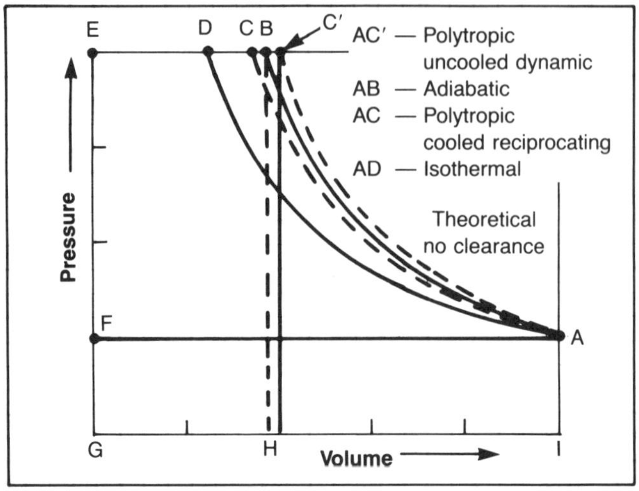

Fig. 5-11. Theoretical indicator card showing various gas compression processes. From Compressed Air and Gas Data, copyright 1980. Courtesy Ingersoll-Rand.[4]

2. Pressure rise (or differential)—dynamic units and most positive-displacement types
3. Compression ratio—dynamic units
4. Effect of clearance—reciprocating units, and
5. Desirability of saving power.

A reciprocating compressor usually requires a separate cylinder for each stage, with intercooling of the gas between stages. In a reciprocating unit, all cylinders are commonly combined into one unit assembly and driven from a single crankshaft.

It was previously noted that the isothermal cycle is more economical of power. Cooling the gas after partial compression to a temperature equal to original intake temperature obviously will reduce the power required in the second stage.

Dynamic compressors would, in most cases, have several staging elements in the same casing. There would normally be no intercooling between those stages contained in a given casing although the internal diffusers or diaphragms are sometimes water-cooled. The gas can be brought out to external exchangers for intercooling at the end of several stages of compression. Frequently, large units have two casings (separate machines) in series to cover the entire compression range, each machine equipped with several stages and with external intercooling between the machines. They are usually coupled in tandem to one driver. The extent to which intercooling is used depends largely on power cost.

Reciprocating Compressors

As stated earlier, the capacity and power required are the two principal factors that must be determined in designing compressors. Temperature increase and cooling requirements are also frequently required. Several methods for calculating required power are presented in this section. Also, the effect of changing conditions, such as clearance, specific heat ratio, and compressibility are discussed.

Capacity. The amount of gas a given compressor can pump depends on the actual displacement volume of the intake cylinder and the volumetric efficiency. Since all of the gas is not discharged from the cylinder because of the clearance between the piston and the cylinder, the total volume of the cylinder is not available for new gas. The capacity of a compressor is calculated from

$$q = \frac{\pi d^2 L\, S\, E_v}{4}, \tag{5-6}$$

where

q = flow capacity,
d = piston diameter,
L = stroke length,
S = compressor speed, and
E_v = volumetric efficiency.

On double-acting compressors the volume occupied by the rod must be accounted for. The volumetric efficiency is calculated from

$$E_v = 1 - A - C\left[\frac{Z_1\, r^{1/k}}{Z_2} - 1\right], \tag{5-7}$$

where

A = factor to allow for leakage, friction, etc., usually between 0.03 and 0.06,
C = clearance, which varies from 0.04 to 0.16,
Z_1 = gas compressibility factor at suction conditions,
Z_2 = gas compressibility factor at discharge conditions,
r = compression ratio, p_2/p_1,
p_1 = suction pressure, and
p_2 = discharge pressure.

Equations 5-6 and 5-7 are applicable for any consistent set of units.

Example 5-1:

A single acting reciprocating compressor having a piston diameter of 4 in. and a stroke length of 6 in. is to compress gas from 100 psig and 100°F to 400 psig. Calculate the flow capacity in scfd if the compressor runs at a speed of 500 rpm. Other data are:

$k = 1.3, \quad C = 6\%, \quad A = 5\%$

Solution:

$$r = p_2/p_1 = 414.7/114.7 = 3.62$$

$$T_2 = T_1\, r^{(k-1)/k} = (100 + 460)(3.62)^{.23} = 753°\text{R}$$

$$Z_1 = 0.98, \quad Z_2 = 0.95, \quad Z_1/Z_2 = 1.03$$

$$E_v = 1 - .05 - .06\,[1.03(3.62)^{.77} - 1] = 0.85$$

$$q = \pi(4/12)^2(6/12)(500)(.85)/4 = 18.54 \text{ ft}^3/\text{min}$$

$$q = 18.54\,\frac{\text{ft}^3}{\text{min}} \times \frac{1440 \text{ min}}{\text{day}} = 26704 \text{ ft}^3/\text{day}$$

$$q_{sc} = \frac{q}{B_g} \qquad B_g = \frac{.0283\,ZT}{p} = \frac{.0283(.98)(560)}{114.7}$$

$$B_g = 0.135 \text{ ft}^3/\text{scf}$$

$$q_{sc} = \frac{26704}{0.135} = 1.97 \times 10^5 \text{ scfd}$$

Power Requirement

The power requirement of any compressor is the prime basis for sizing the driver and for selection and design of compressor components. The actual power requirement is related to a theoretical cycle through a compression efficiency that has been determined by tests on prior machines. Compression efficiency is the ratio of the theoretical to the actual gas horsepower, and as used by the industry, does not include mechanical friction losses. These are added later either through the use of a mechanical efficiency or by adding actual mechanical losses previously determined. The mechanical efficiencies of positive-displacement compressors range from 88 to 95%, depending upon the size and type of unit. Dynamic units also have certain relatively small hydraulic losses that are often disregarded for estimating purposes. Positive-displacement machines are designed based on the adiabatic cycle while dynamic units generally are designed using the polytropic cycle.

In calculating horsepower, the compressibility factor, Z, must be considered since its influence is considerable with many gases, particularly at high pressure. The power required may be determined using analytic equations or Mollier diagrams. Charts for quick estimates are also available. The equations are derived by starting with the general energy equation. Neglecting potential and kinetic energy changes and friction losses, the equation becomes

$$w = \int_{p_1}^{p_2} V\,dp. \tag{5-8}$$

To integrate this equation a relationship between V and p must be used. For adiabatic flow, pV^k = constant, and upon integrating, adjusting for gas compressibility and units, the following equation can be obtained:

$$w = \frac{3.027\, p_{sc} T_1\, k}{T_{sc}(k-1)} (r^{Z_1(k-1)/k} - 1), \qquad (5\text{-}9)$$

where

w = power required, Hp/MMscfd,
p_{sc} = pressure at standard conditions, psia,
T_{sc} = temperature at standard conditions, °R, and
T_1 = suction temperature, °R.

Example 5-2:

Calculate the horsepower required to compress 1 MMscfd of a 0.6 gravity gas from 100 psia and 80°F to 1600 psia using the adiabatic equation. Assume that a two stage compressor is used, and that the gas is cooled back to 80°F between stages. Also find the final discharge temperature.

Other data are:

$k = 1.28$, $p_{sc} = 14.65$, $T_{sc} = 60°F$

Solution:

The total compression ratio is 1600/100 = 16. Therefore, use a value of $r = 4$ for each stage.

First Stage

$p_1 = 100$, $p_2 = 400$, $Z_1 = 0.985$,

$$\frac{Z_1(k-1)}{k} = 0.22$$

$$w = \frac{3.027(14.65)(540)(1.28)}{520(.28)} [(4)^{0.22} - 1]$$

$w_1 = 210.5(0.36) = 75.8$ Hp/MMscfd

The gas temperature after compression from 100 to 400 psia is

$T_2 = T_1\, r^{(k-1)/k} = 540(4)^{.22} = 733°R = 273°F$

The gas is cooled to 80°F before entering the second stage.

Second Stage

$p_1 = 400$ psia, $p_2 = 1600$ psia, $Z_1 = 0.94$,

$$\frac{Z_1(k-1)}{k} = 0.21$$

$w_2 = 210.5(4^{.21} - 1) = 71.6$ Hp/MMscfd

$T_2 = T_{final} = 273°F$

The total power required is

$Hp = q(w_1 + w_2) = 1$ MMscfd (75.8 + 71.6) Hp/MMscfd

$Hp = 147.4$

In order to size the prime mover for the compressor a compression efficiency must be estimated. Figure 5-12 can be used for determining efficiency as a function of

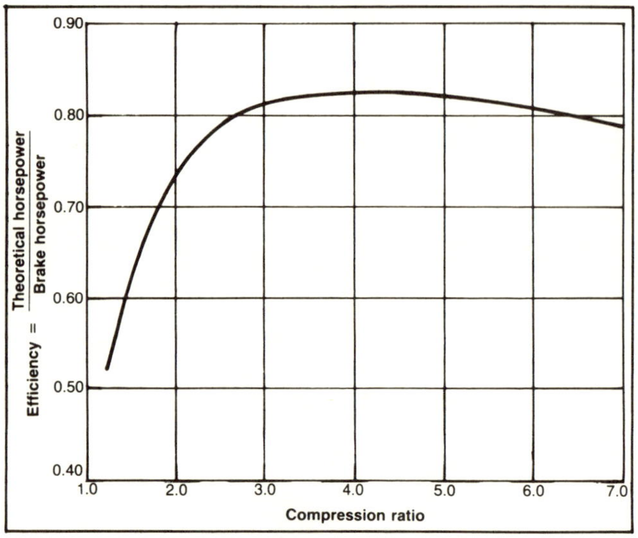

Fig. 5-12. Compressor efficiency. From Compressed Air and Gas Data, copyright 1980. Courtesy Ingersoll-Rand.[4]

the compression ratio. From Figure 5-12, for $r = 4$, $E = 83\%$. Therefore, the actual brake horsepower required is

$$bhp = \frac{147.4}{0.83} = 177.6.$$

Power requirement and cooling requirements can be determined using plots of enthalpy versus entropy (Mollier diagrams) if a plot is available for the gas that is to be compressed. When cooling the gas, which occurs at constant pressure, the change in enthalpy can be used to determine the amount of heat to be removed. From thermodynamics,

$$\Delta H = Q - w, \qquad (5\text{-}10)$$

where

ΔH = change in enthalpy,
Q = heat transferred during compression, and
w = work.

For adiabatic compression, $Q = 0$ and

$$-w = \Delta H. \qquad (5\text{-}11)$$

In units of horsepower per million standard cubic feet, Equation 5-11 becomes

$$w = 0.0432\, \Delta H, \qquad (5\text{-}12)$$

where

w = power required, Hp/MMscfd, and
ΔH = enthalpy change, BTU/lb-mole.

When cooling the gas, $w = 0$ and therefore

$$Q = \Delta H. \qquad (5\text{-}13)$$

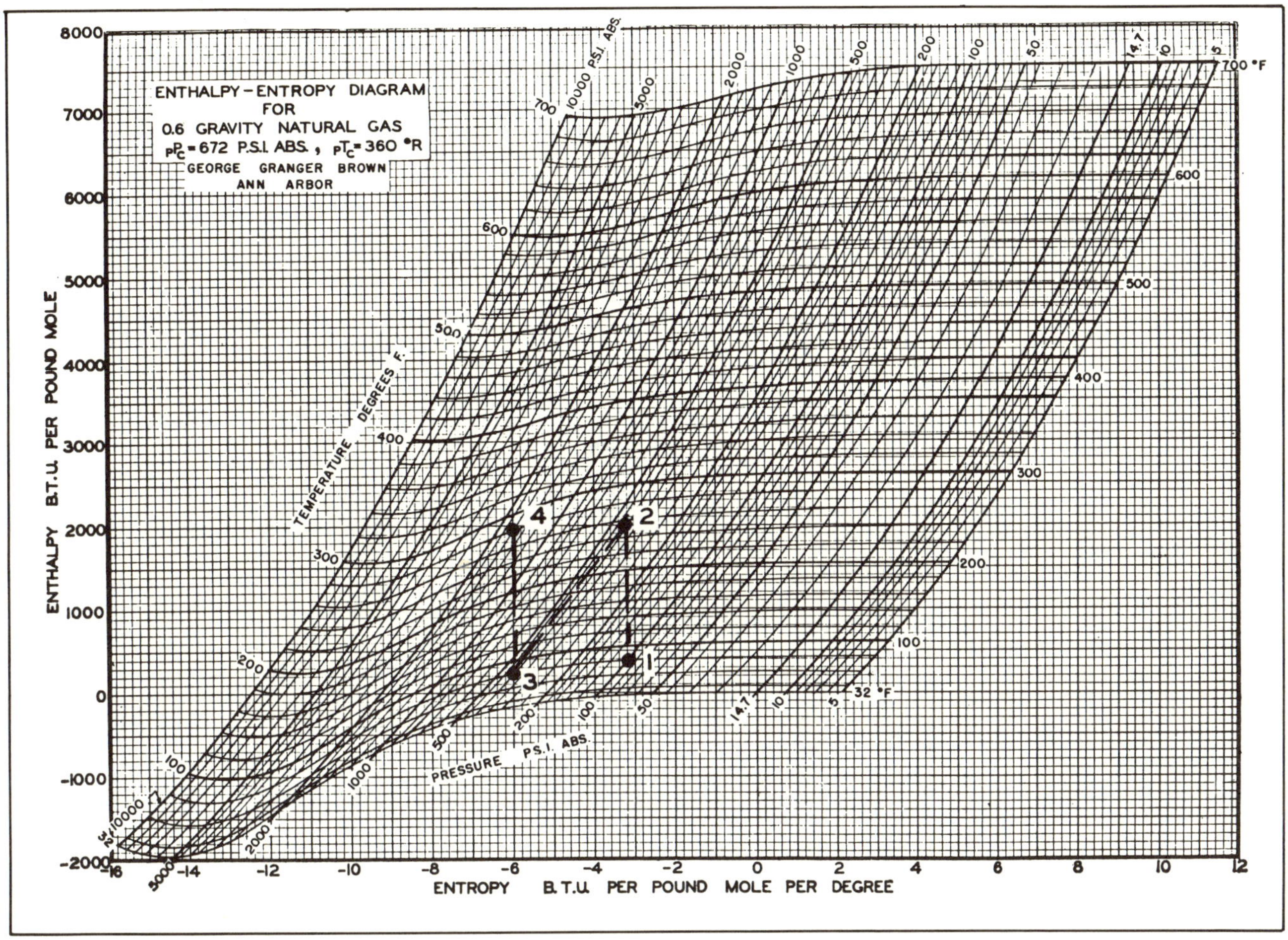

Fig. 5-13. Enthalpy-entropy diagram for 0.6 gravity natural gas. Permission to publish by the Society of Petroleum Engineers of AIME. Copyright 1945 SPE-AIME.[2]

Example 5-3:

Use the Mollier diagram method to solve Example 5-2. Also calculate the heat to be removed between stages to cool the gas to 80°F.

Solution:

Select Figure 5-13 to represent the gas.

First Stage

The first stage compresses the gas from 100 psia to 400 psia at constant entropy. Enter Figure 5-13 at $p_1 = 100$ and $T_1 = 80°F$ and read $H_1 = 380$ BTU/lb-mole. Proceed vertically ($\Delta S = 0$) to $p_2 = 400$ psia and read $H_2 = 1990$, $T_2 = 260°F$. Follow the 400 psia line to $T_3 = 80°F$ and read $H_3 = 220$. The amount of cooling required is then

$$Q = \Delta H_{23} = H_2 - H_3 = 1990 - 220$$
$$= 1770 \text{ BTU/lb-mole,}$$

or

$$Q = 1770 \frac{\text{BTU}}{\text{lb-mole}} \times \frac{\text{lb-mole}}{379 \text{ scf}} = 4.67 \frac{\text{BTU}}{\text{scf}}$$

The power required for stage 1 is

$$w = 0.0432\ \Delta H_{12} = 0.0432(H_2 - H_1)$$

$$w = 0.0432(1990 - 380) = 70 \text{ Hp/MMscfd}$$

Second Stage

From point 3 where $p = 400$, $T = 80$ and $H = 220$, proceed vertically to the final pressure of 1600 psia and read $T_4 = 274°F$, $H_4 = 1920$ BTU/lb-mole. The power required for the second stage is then

$$w = 0.0432(H_4 - H_3) = 0.0432(1920 - 220)$$
$$= 73.4 \text{ Hp.}$$

The total horsepower required is then

$$Hp_{total} = q(w_1 + w_2) = 1 \text{ MMscfd}(70 + 73.4) = 143.4$$

This compares well with the value of 147.4 Hp obtained using the equations. The Mollier diagram solution is considered to be more accurate and should preferably be used. Mollier diagrams for other gases are included in the appendix.

For making rough estimates of power requirement the

following equation can be used:

$$bhp = 22\, r\, N\, q_{sc}\, F, \qquad (5\text{-}14)$$

where

bhp = brake horsepower,
r = compression ratio per stage,
N = number of stages,
q_{sc} = flow rate in MMscfd, and
F = correction for pressure drop between stages, depending on N.

N	F
1	1.00
2	1.08
3	1.10

Graphical solutions to Equation 5-9 have been prepared for specific conditions of compression efficiency, standard conditions, and gas gravity. Examples are shown

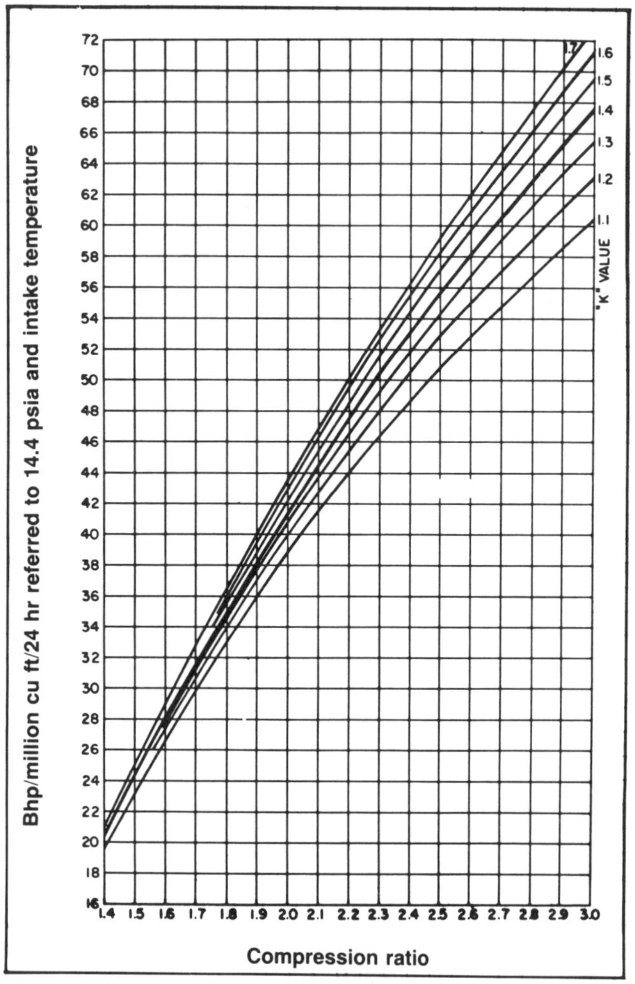

Fig. 5-14. Compressor power requirements. From Compressed Air and Gas Data, copyright 1980. Courtesy Ingersoll-Rand.[4]

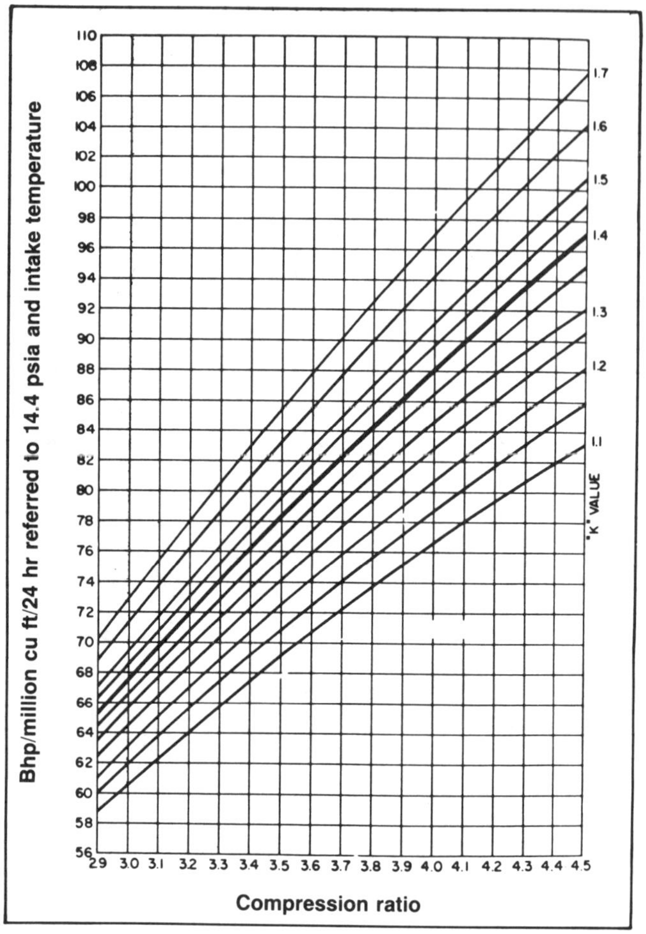

Fig. 5-15. Compressor power requirements. From Compressed Air and Gas Data, copyright 1980. Courtesy Ingersoll-Rand.[4]

in Figures 5-14 through 5-16.[3] Correction to other gas gravities and for low suction pressure are obtained from Figures 5-17 and 5-18 respectively.

Example 5-4:

Solve Example 5-2 using both Equations 5-14 and the graphs presented in Figures 5-14 through 5-16.

q_{sc} = 1 MMscfd, p_1 = 100 psia, p_2 = 1600 psia
k = 1.28, γ_g = 0.6

Solution:

Using Equation 5-14,

$bhp = 22(4)(2)(1)(1.08) = 190$ Hp

Using Figure 5-15, for $r = 4$ and $k = 1.28$, read $bhp/$MMscfd = 84. For 2 stages this will be 2(84) = 168 Hp. There is no correction for low suction pressure or gas gravity since r is greater than 3.4, the limit of the correction graphs.

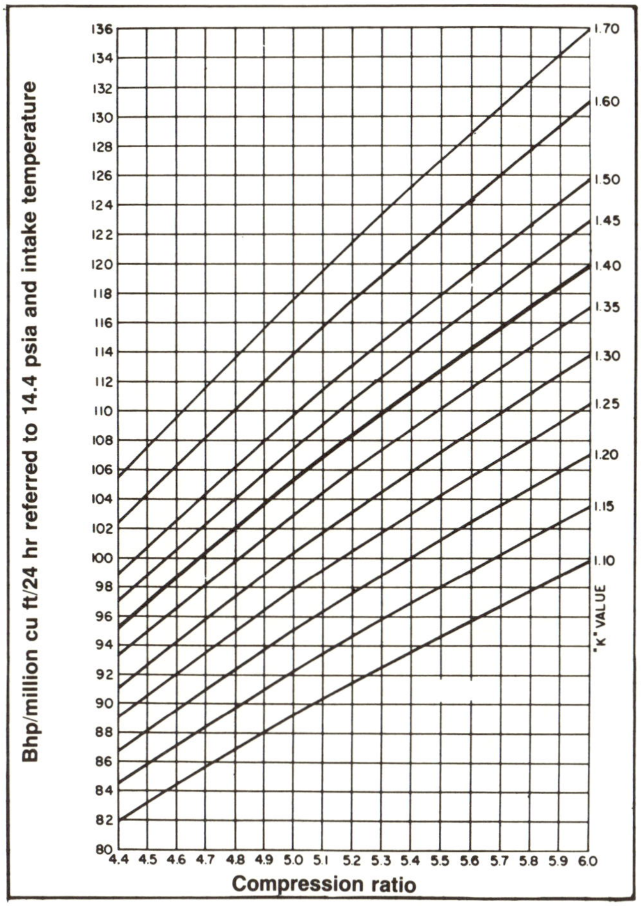

Fig. 5-16. Compressor power requirements. From Compressed Air and Gas Data, copyright 1980. Courtesy Ingersoll-Rand.[4]

Multistaging

It has been previously stated that there are practical limits to the amount of compression permissible within a single stage or element beyond which two or more steps or stages must be used with intercooling between stages in most cases.

Theoretically minimum power with perfect intercooling and no pressure loss between stages is obtained by making the ratio of compression the same in all stages. The following formula may be used to determine the optimum compression ratio per stage:

$$r_s = \sqrt[s]{r_t} = (r_t)^{1/s} \tag{5-15}$$

where

r_s is the theoretically best compression ratio per stage,

r_t is the overall compression ratio $\left(\dfrac{p_{\text{final}}}{p_{\text{initial}}}\right)$, and

s is the number of stages.

Even though this method results in minimum power, the energy required will vary only a fraction of one percent for rather large variations in the actual compression ratios for individual stages. Designers take advantage of this fact for economic and engineering reasons.

Each stage is best designed as a separate compressor, the capacity of each stage being separately calculated from the first stage real intake volume, corrected to the actual pressure and temperature conditions existing at the higher stage cylinder inlet, allowing for pressure drop through cooler and piping. Allowances must also be made for any reduction in the moisture content if there is condensation between stages in an intercooler. The theoretical power per stage can then be calculated and the total horsepower obtained. A compression ratio per stage greater than four is seldom used because of heating and rod loading problems.

Effect of Clearance

Although cylinder clearance reduces the volumetric efficiency of a compressor (Eq. 5-7), it cannot be completely eliminated. Clearance may be altered for capacity control. Figures 5-19 and 5-20 show the effect of clear-

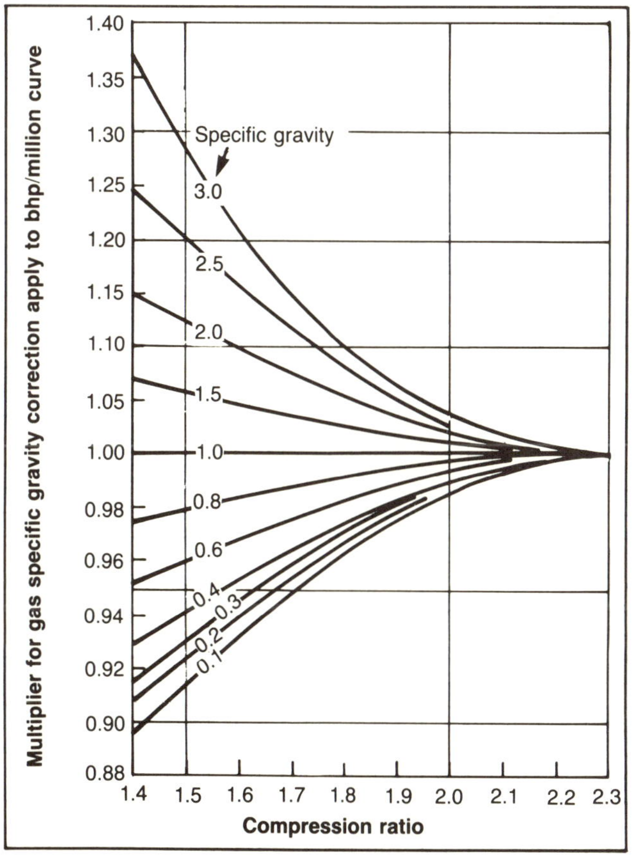

Fig. 5-17. Specific gravity correction. From Compressed Air and Gas Data, copyright 1980. Courtesy Ingersoll-Rand.[4]

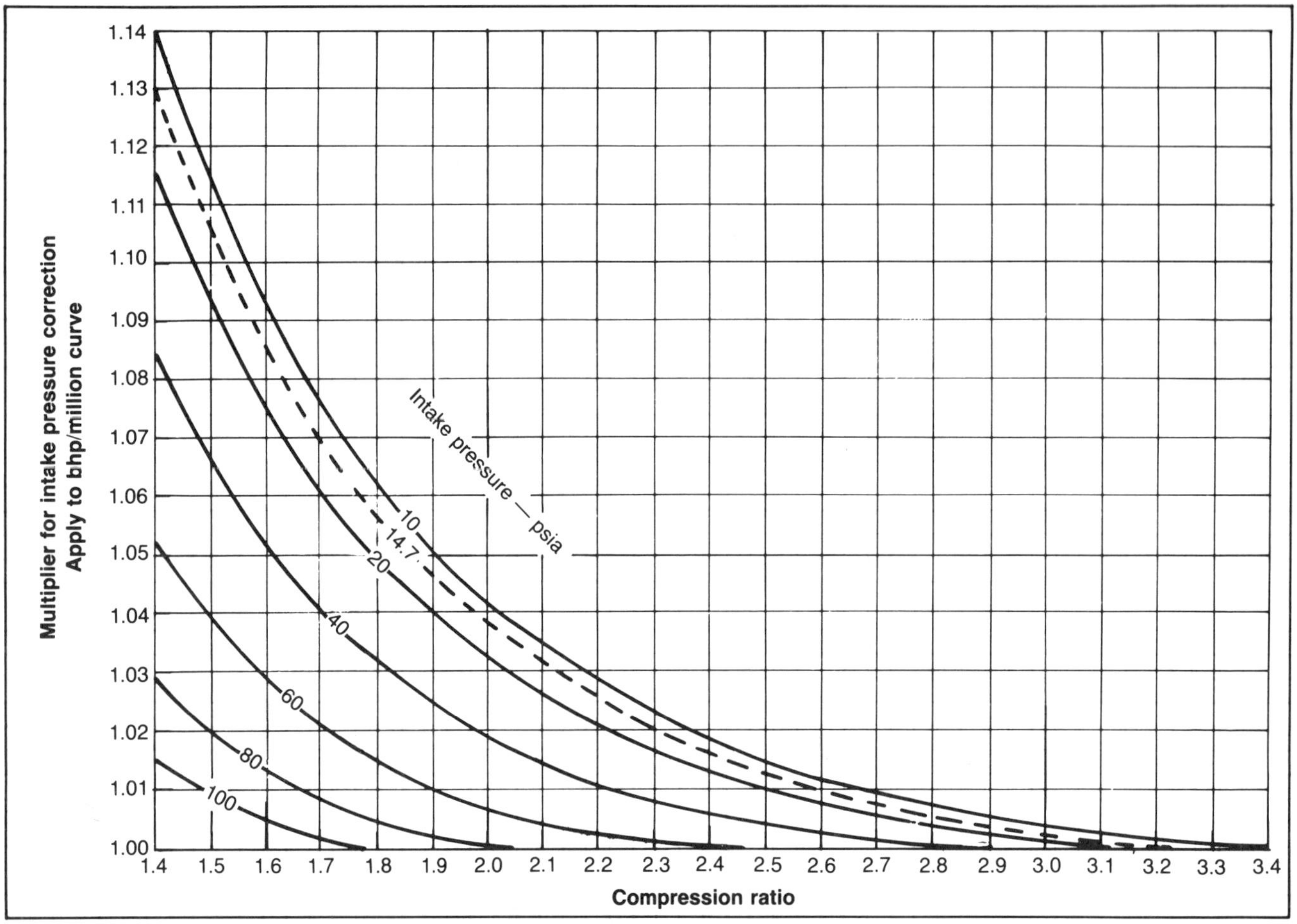

Fig. 5-18. Suction pressure correction. From Compressed Air and Gas Data, copyright 1980. Courtesy Ingersoll-Rand.[4]

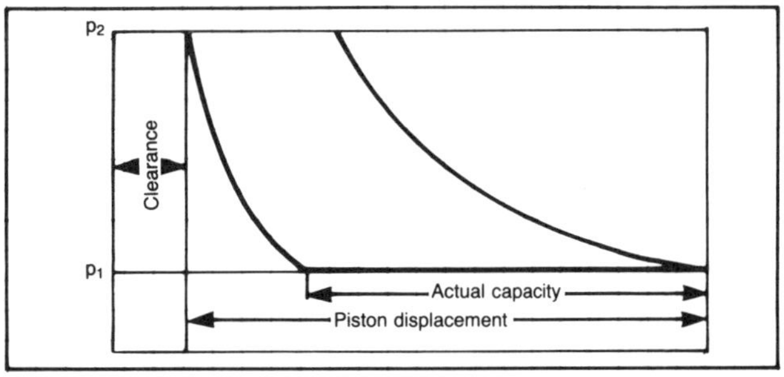

Fig. 5-19. Effect of clearance on the capacity of a reciprocating compressor. From Compressed Air and Gas Data, copyright 1980. Courtesy Ingersoll-Rand.[4]

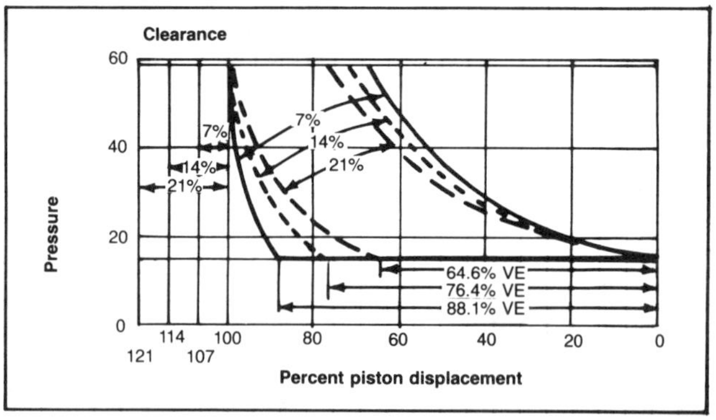

Fig. 5-20. Quantitative effect of various cylinder clearances on the volumetric efficiency at constant compression ratio. From Compressed Air and Gas Data, copyright 1980. Courtesy Ingersoll-Rand.[4]

ance by means of a *pV* diagram. Figure 5-21 illustrates how the clearance may be varied to alter volumetric efficiency and therefore capacity. Actual clearance control is accomplished by adding pockets or bottles to the cylinder, which may be open or closed, or by adjustable clearance valves.

Effect of Specific Heat Ratio

Figure 5-22 shows the effect of specific heat ratio, $k = C_p/C_v$, on volumetric efficiency for a fixed compres-

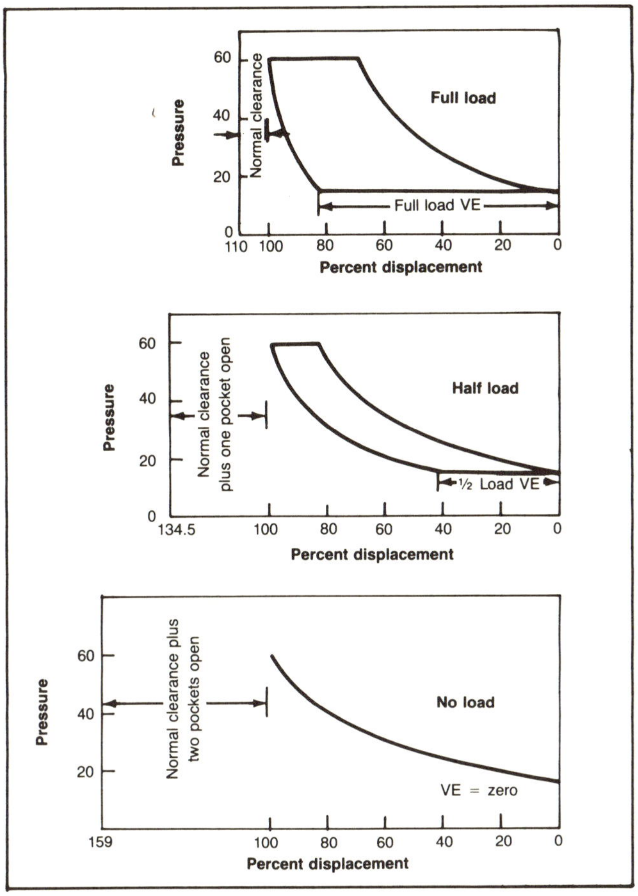

Fig. 5-21. Effect of adding clearance for capacity purposes. From Compressed Air and Gas Data, copyright 1980. Courtesy Ingersoll-Rand.[4]

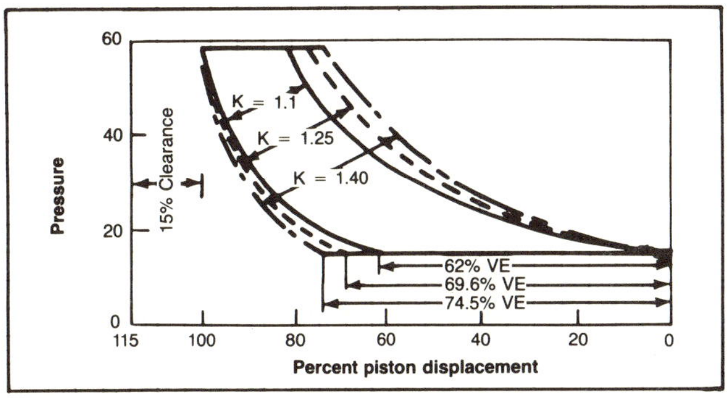

Fig. 5-22. Effect of various specific heat ratios on the volumetric efficiency of a given cylinder. From Compressed Air and Gas Data, copyright 1980. Courtesy Ingersoll-Rand.[4]

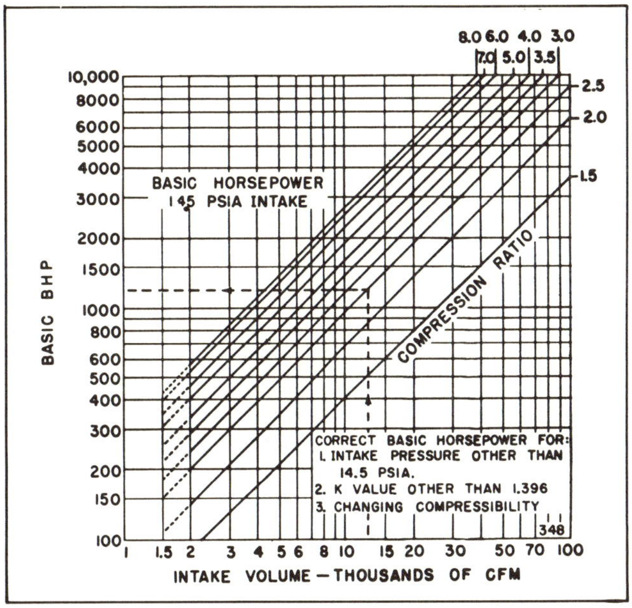

Fig. 5-23. Basic horsepower. From Compressed Air and Gas Data, copyright 1980. Courtesy Ingersoll-Rand.[4]

sion ratio and clearance. The lower the value of k, the lower the volumetric efficiency.

CENTRIFUGAL COMPRESSORS

The performance of a centrifugal compressor is usually estimated based on the polytropic process. The same equation (Eq. 5-9) given for a reciprocating compressor can be used if k is replaced with n, where

$$\frac{n-1}{n} = \frac{k-1}{k\,\eta_p}. \tag{5-4}$$

The concept of polytropic head is frequently used in designing centrifugal machines since the head developed is independent of the fluid being moved. The polytropic head is the work in energy per unit mass of the gas. It can be calculated from

$$H_p = \frac{R\,T_1\,n\,\bar{Z}}{M(n-1)}\left(r^{(n-1)/n} - 1\right), \tag{5-16}$$

where

H_p = polytropic head,
R = gas constant,
T_1 = gas inlet temperature,
$\bar{Z}$ = average compressibility factor, $(Z_1 + Z_2)/2$, and
M = gas molecular weight.

The actual power required then depends on the amount of gas being moved, or the mass flow rate, w. In terms of horsepower,

$$\text{Hp} = \frac{w\,H_p}{33000\,\eta_p}, \tag{5-17}$$

where

Hp = power required, horsepower,
w = mass flow rate, lb_m/min,
H_p = polytropic head, ft-lb/lb_m, and
η_p = polytropic efficiency.

In determining the brake horsepower required, mechanical losses must be added. These will usually vary between 10 and 50 Hp, depending on the machine. The power required can also be obtained using a Mollier diagram, as outlined previously in the discussion of reciprocating compressors.

Example 5-5:

Using the following data, estimate the horsepower required for a centrifugal compressor to compress 50 MMscfd of a 0.6 gravity gas.

p_1 = 100 psia T_1 = 80°F
p_2 = 400 psia k = 1.28
p_{sc} = 14.65 psia T_{sc} = 60°F
Z_1 = 0.988 B_{gl} = 0.131 ft³/scf

Solution:

The actual inlet volume is

$$q = q_{sc} B_g = 50 \times 10^6 \frac{\text{scf}}{\text{day}} \times 0.131 \frac{\text{ft}^3}{\text{scf}} \times \frac{\text{day}}{1440 \text{ min}}$$

$$= 4550 \frac{\text{ft}^3}{\text{min}}.$$

From Figure 5-10, η_p = 0.72

$$\frac{n-1}{n} = \frac{k-1}{k\eta_p} = \frac{1.28-1}{1.28(0.72)} = 0.3$$

$$T_2 = T_1 r^{(n-1)/n} = 540\,(4)^{0.3} = 818°\text{R}$$

$$Z_2 = 0.991 \quad \bar{Z} = \frac{0.988 + 0.991}{2} = 0.99$$

$$M = 28.96\,\gamma_g = 28.96(.6) = 17.38 \text{ lbm/lb-mole}$$

$$H_p = \frac{R\,T_1\,n\,\bar{Z}}{M(n-1)}(r^{(n-1)/n} - 1)$$

$$H_p = \frac{1545(540)(.99)}{17.38(0.3)}((4)^{0.3} - 1) = 81{,}700 \frac{\text{ft-lb}}{\text{lbm}}$$

$$w = 50 \times 10^6 \frac{\text{scf}}{\text{day}} \frac{17.38 \text{ lbm}}{379 \text{ scf}} \frac{1 \text{ day}}{1440 \text{ min}} = 1592 \frac{\text{lbm}}{\text{min}}$$

$$\text{Hp} = \frac{w H_p}{33{,}000\,\eta_p} = \frac{1592(81{,}700)}{(33{,}000)(0.72)} = 5474 \text{ Hp}$$

Estimates of the horsepower, number of stages, and compressor speed can be made by referring to Figures 5-23 through 5-26. The horsepower required for a cen-

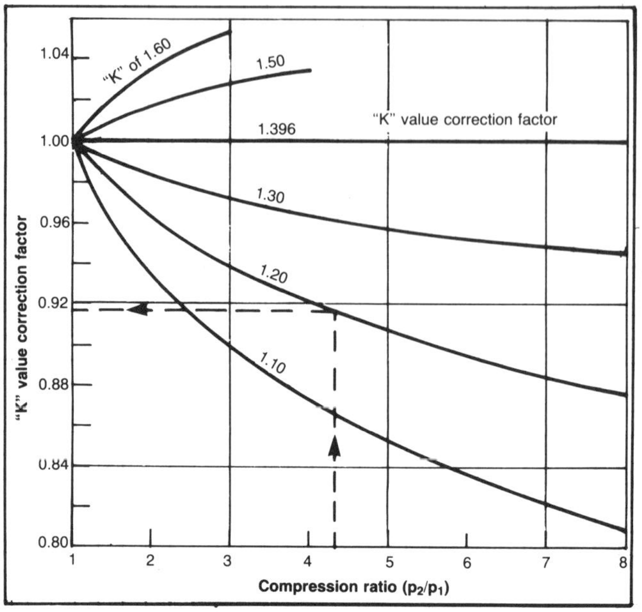

Fig. 5-24. K value correction. From Compressed Air and Gas Data, copyright 1980. Courtesy Ingersoll-Rand.[4]

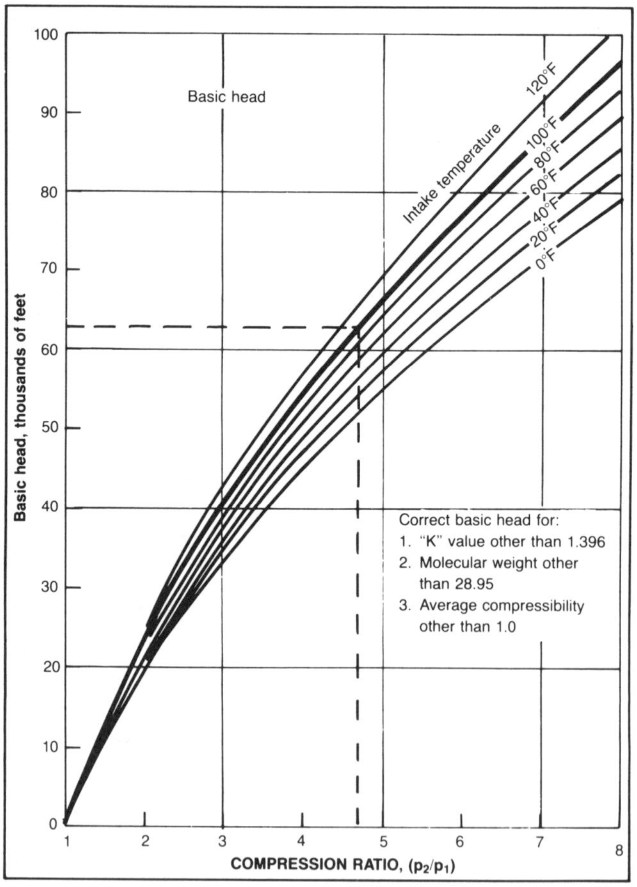

Fig. 5-25. Basic head. From Compressed Air and Gas Data, copyright 1980. Courtesy Ingersoll-Rand.[4]

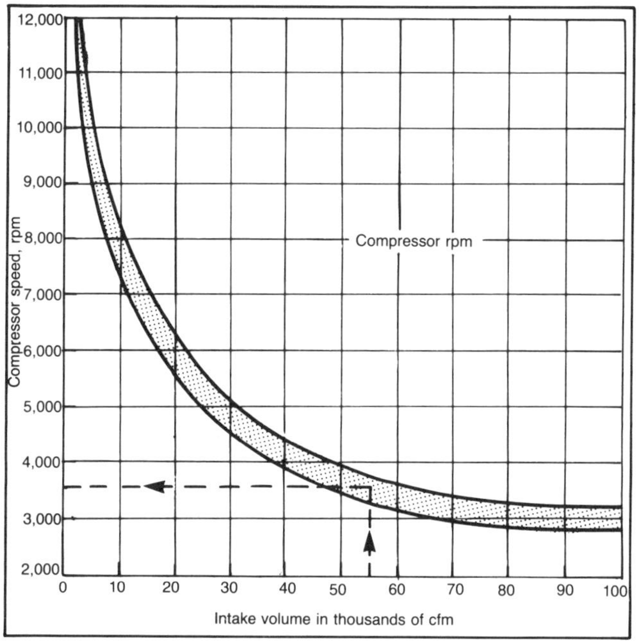

Fig. 5-26. Compressor speed. From Compressed Air and Gas Data, copyright 1980. Courtesy Ingersoll-Rand.[4]

trifugal compressor is obtained from:

$$\text{Hp} = (\text{BASIC BHP})\,\frac{p_1 \bar{Z}}{14.5\, Z_1}\,(K_{\text{corr}})$$

where

(BASIC BHP) is obtained from Figure 5-23,
(K_{corr}) is obtained from Figure 5-24,
p_1 = suction pressure, psia,
$\bar{Z}$ = average Z factor, and
Z_1 = Z factor at p_1, T_1

The number of stages required for a centrifugal compressor may be estimated from:

$$N = \frac{(\text{BASIC HEAD})(K_{\text{corr}})(28.95)\bar{Z}}{M(9500)}$$

where

N = number of stages,
(BASIC HEAD) is obtained from Figure 5-25,
(K_{corr}) is obtained from Figure 5-24,
$\bar{Z}$ = average Z factor, and
M = molecular weight of the gas.

Example 5-6:

Re-work Example 5-5 using Figure 5-23 and Figure 5-24. Also, estimate the number of stages required and the approximate compressor speed using Figure 5-25 and Figure 5-26.

Solution:

From Figure 5-23, for an intake volume of 4550 ft^3/min, the basic bhp ≈ 720 for a compression ratio of 4. From Figure 5-24, for a k value of 1.28 the k value correction factor is 0.956 for r = 4. Therefore,

$$\text{bhp} = \frac{720(0.956)(100)(0.99)}{14.5(0.988)} = 4757 \text{ HP}.$$

In order to estimate the number of stages required, from Figure 5-25, for r = 4 and T_1 = 80°F, Basic Head ≈53,000 ft.
The number of stages is then

$$N = \frac{(53000)(0.956)(28.95)(0.99)}{17.38(9500)} = 8.9$$

This indicates that 9 stages will be required.
The compressor speed is estimated by entering Figure 5-26 at an intake volume of 4550 ft^3/min and reading the speed as 9800 rpm.

REFERENCES

1. Ridgeway, R. S.: CNGA Meeting, Los Angeles, Ca., Mar. 1, 1945.
2. Brown, G. G.: "A Series of Enthalpy-Entropy Charts for Natural Gases," *Trans.*, AIME (1945) 160.
3. NGPSA Data Book, 9th Edition, Tulsa, Ok (1972).
4. *Compressed Air and Gas Data,* Ingersoll-Rand Co., Woodcliff Lake, N.J.

6 Total System Analysis

THE object of any gas production operation is to move the gas from some point in the reservoir to the sales line. In order to accomplish this, the gas must pass through many areas of pressure drop, or if a compressor is used, pressure is gained or increased. The restrictions to flow may include some or all of the following, as illustrated in Figure 6-1.

1. Porous medium
2. Gravel pack or perforations
3. Bottom-hole choke
4. Well tubing
5. Subsurface safety valve (SSSV)
6. Surface choke
7. Well flowline
8. Separator pressure
9. Flowline from compressor to sales line
10. Sales line pressure

Procedures to calculate the pressure drop or resistance to flow in the porous medium and gravel pack are given in Chapter 3. Pressure drops in all of the other restrictions may be calculated using the material in Chapter 4. The selection of a compressor to create the necessary pressure gain in the system can be made using information given in Chapter 5.

Although all of these separate components of the production system can be analyzed independently, in order to determine the performance of the well they must be combined in a total system or nodal analysis. This is most easily accomplished by dividing the total system into two separate subsystems and determining the effects of changes made in one or both of the subsystems on the well performance. Selection of a location or node at which to divide the system depends on the purpose of the analysis. It is usually more convenient to select the division point at or as close as possible to the part of the system being analyzed. For example, if the effect of subsurface safety valve size is being analyzed, the system is divided at the SSSV. Figure 6-2 illustrates some of the most commonly used division points or nodes.

The analysis is accomplished by determining and plotting flow rate versus pressure existing at the division point or node for each of the subsystems. These are plotted as separate and independent curves on the same graph, and since there can be only one pressure existing at the division point, the intersection of the two subsystem curves gives the total system performance or flow rate, which will satisfy the requirement that the flow into the node must equal the flow out of the node. If the component being analyzed must be sized according to the flow rate and pressure change across the component, such as a SSSV or compressor, it is often convenient to plot pressure difference between the two subsystems versus flow rate. The system analysis procedures will be illustrated in this chapter in the form of example calculations.

TUBING AND FLOWLINE SIZE EFFECT

The size of piping installed in a well can have a large effect on the flow capacity or deliverability of the well. In some cases it can be the controlling factor and can cause the well to produce at a low rate even though the reservoir may be capable of producing much more gas.

Constant Wellhead Pressure

The simplest case will be considered first, that is, the case of a constant wellhead pressure. This case might

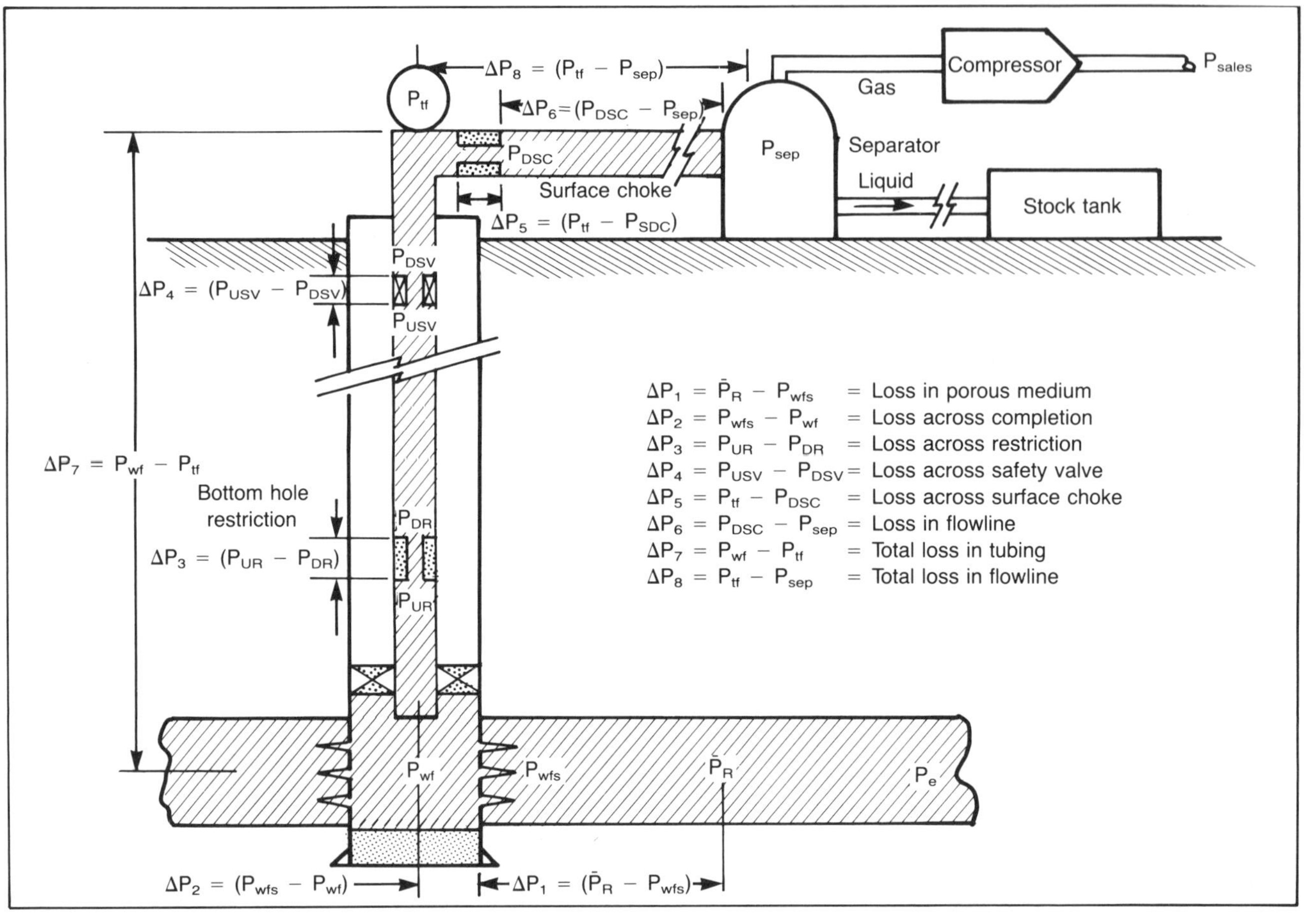

Fig. 6-1. Possible pressure losses in complete system.

occur if the well has a very short flowline between the wellhead and separator. In this case the well is divided at the bottom of the hole, node 6. The expressions for the inflow and outflow are.

$$\text{Inflow:} \quad \bar{p}_R - \Delta p_{\text{res}} = p_{wf}$$

$$\text{Outflow:} \quad p_{tf} + \Delta p_{\text{tub}} = p_{wf}$$

The solution procedure is:

1. Assume various values of p_{wf} and determine q_{sc} from inflow performance methods.
2. Plot p_{wf} versus q_{sc}.
3. Assume various flow rates, and starting at the fixed wellhead pressure, calculate p_{wf} for each q_{sc} using Equation 4-27 or the Cullender and Smith method Eq. 4-29.
4. Plot p_{wf} versus q_{sc} on the same graph as in Step 2. The intersection of the curves gives the flow capacity and p_{wf} for this particular tubing size.

To determine the effect of other tubing sizes, Steps 3 and 4 can be repeated for various tubing sizes. The effect of p_{tf} can also be determined by repeating for various wellhead pressures. For this case, the two subsystems are: (1) the reservoir and (2) the tubing plus the wellhead pressure.

Example 6-1:

A stabilized deliverability test was conducted on a gas well to obtain reservoir inflow performance data. Determine the flow capacity of the well for both 1.995 in. and 2.441 ID tubing if p_{tf} is constant at 1000 psia. Other data are:

$n = 0.83$ $\quad C = 0.0295$ Mscfd/psia^2

$\bar{p}_R = 1952$ psia $\quad H = 10000$ ft

Solution:

The equation for generating the reservoir inflow performance curve is

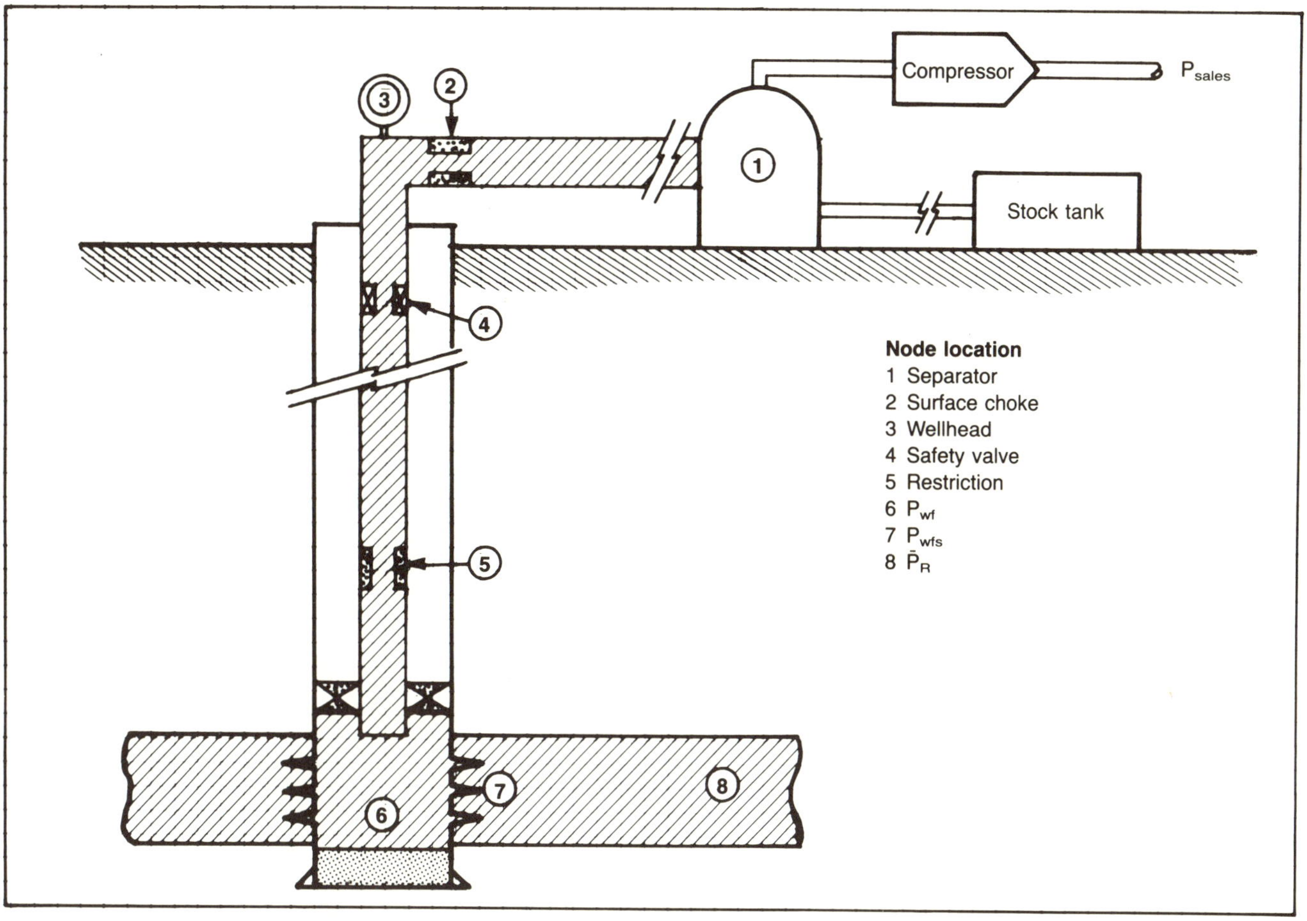

Fig. 6-2. Location of various nodes.

$$q_{sc} = C(\bar{p}_R^2 - p_{wf}^2)^n = 0.0295\ (1952^2 - p_{wf}^2)^{0.83}.$$

(This is the same well as was analyzed in Example 3-5)

1. Assume various values of p_{wf} and calculate q_{sc}:

Inflow	
p_{wf}, psia	q_{sc}, Mscfd
1952	0
1800	1768
1400	4695
1000	6642
600	7875
200	8477
0	8551

2. Plot p_{wf} versus q_{sc} (Fig. 6-3)
3. Assume various flow rates and find the value of p_{wf} required to overcome the pressure drop in the tubing and the wellhead pressure for each flow rate. This is required for each tubing size. For illustration purposes, the values of p_{wf} are obtained from Figures 4-12 and 4-13. For more accuracy, the well flow equations should be used.

	Outflow	
	p_{wf}, psia	
q_{sc}, Mscfd	d = 1.995	d = 2.441
1000	1300	1290
2000	1370	1300
3000	1500	1370
4000	1620	1400
5000	1800	1580
6000	1970	1620

4. Plot p_{wf} versus q_{sc} for both tubing strings on Figure 6-3. The only value of p_{wf} that will satisfy both subsystems is the intersection of the inflow curve with the tubing performance curves. This occurs at the following values:

Tubing ID	p_{wf}, psia	q_{sc}, Mscfd
1.995	1560	3500
2.441	1440	4350

Thus, by installing the larger tubing, the well's flow

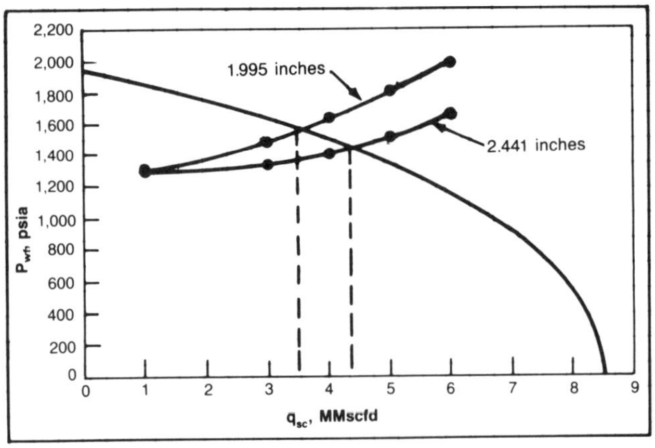

Fig. 6-3. Solution to Example 6-1.

capacity can be increased by 850 Mscfd or about a 24% increase.

Variable Wellhead Pressure

If a well is equipped with a flowline of considerable length, the size of the flowline can affect the flow capacity of the system. When flowline size effect is being considered, it is convenient to divide the system at the wellhead into two subsystems: (1) the reservoir plus the tubing and, (2) the flowline plus the separator pressure.

The inflow and outflow expressions are:

$$\text{Inflow:} \quad \bar{p}_R - \Delta p_{res} - \Delta p_{tub} = p_{tf}$$

$$\text{Outflow:} \quad p_{sep} + \Delta p_{fl} = p_{tf}$$

The solution procedure is:

1. Assume various values of q_{sc} and determine the corresponding p_{wf} from inflow performance methods.
2. Using the tubing pressure drop equations, determine the wellhead pressure, p_{tf} corresponding to each q_{sc} and p_{wf} determined in Step 1.
3. Plot p_{tf} versus q_{sc}.
4. Using a fixed separator pressure and the pipeline flow equations, calculate p_{tf} for several assumed flow rates.
5. Plot p_{tf} versus q_{sc} on the same graph as used in Step 3. The intersection gives the only values of p_{tf} and q_{sc} that will satisfy both subsystems.

Example 6-2:

A 10,000 ft vertical well is equipped with 1.995 ID tubing and a 6000 ft flowline. Separator pressure is fixed at 1000 psia. Using the following data, determine the flow capacity for the system for both a 1.995 ID flowline and a 2.441 ID flowline.

$\gamma_g = 0.67$, $T_R = 220°F$, $T_{tf} = 100°F$,
$T_{sep} = 60°F$, $\mu_g = 0.012$ cp, $\epsilon = 0.0018$ in.,
$n = 0.83$, $\bar{Z} = 0.95$, $C = 0.0295$ Mscfd/psia$^{1.86}$,
$\bar{p}_R = 1952$ psia

Solution:

1. The reservoir inflow performance equation is

$$q_{sc} = C(\bar{p}_R^2 - p_{wf}^2)^n, \tag{3-31}$$

or

$$p_{wf} = \left[\bar{p}_R^2 - \left(\frac{q_{sc}}{C}\right)^{1/n}\right]^{0.5}$$

$$- \left[1952^2 - \left(\frac{q_{sc}}{0.0295}\right)^{1.205}\right]^{0.5}.$$

Assuming flow rates of 1, 2, 3, and 4 MMscfd, calculate the corresponding values of p_{wf}. These values are given in the table at the end of the solution.

2. The average pressure and temperature method was used to calculate p_{tf} for each q_{sc} and p_{wf}. The equation is:

$$p_{tf} = \left[\frac{p_{wf}^2 - [25\gamma_g\, q_{sc}^2\, \bar{T}\, \bar{Z}\, f\, H\,(\text{EXP}(S) - 1)]/Sd^5}{\text{EXP}(S)}\right]^{0.5}. \tag{4-27}$$

The values of p_{tf} are given in the table.

3. The values for each q_{sc} and p_{tf} for the reservoir-tubing subsystem are plotted on Figure 6-4.
4. The same flow rates as assumed in Step 1 are used to calculate p_{tf} at a fixed p_{sep} of 1000 psia for both the 1.995 and 2.441 in flowlines. The equation used is:

$$p_{tf} = [p_{sep}^2 + (25\, \gamma_g\, q_{sc}^2\, \bar{T}\, \bar{Z}\, f\, L)/d^5]^{0.5}. \tag{4-35}$$

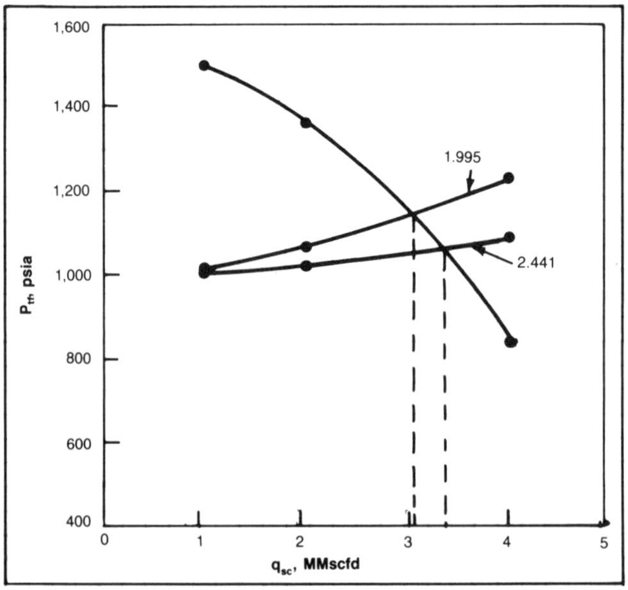

Fig. 6-4. Example 6-2 Solution.

These values are also given in the table.

5. The values for q_{sc} and p_{tf} for each line size for the flowline-separator subsystem are also plotted on Figure 6-4. The intersections with the reservoir-tubing line give flow capacities of 3080 and 3360 Mscfd for the 1.995 and 2.441 in. flowlines, respectively.

Inflow			*Outflow*	
q_{sc}, *Mscfd*	p_{wf}	p_{tf} *(tubing)*	p_{tf} *(1.995)*	p_{tf} *(2.441)*
1000	1877	1500	1016	1006
2000	1774	1362	1062	1022
3000	1653	1158	1134	1049
4000	1512	840	1227	1085

SEPARATOR PRESSURE EFFECT

The effect of separator pressure on the system deliverability can best be determined by dividing the system at the separator. The subsystems then consist of: (1) the separator and (2) the combination of the reservoir, the tubing, and the flowline. The solution is obtained by plotting q_{sc} versus p_{sep}, with p_{sep} being calculated as

$$p_{sep} = \bar{p}_R - \Delta p_{res} - \Delta p_{tub} - \Delta p_{fl}.$$

The solution procedure is illustrated by reworking Example 6-2 using the 1.995 in. flowline.

1. Calculate p_{wf} for various q_{sc} using the reservoir inflow equation.
2. Calculate p_{tf} for each p_{wf} and q_{sc} using Equation 4-27 or Cullender and Smith.
3. Calculate p_{sep} for each p_{tf} and q_{sc} using Equation 4-35 or 4-37.
4. Plot p_{sep} versus q_{sc} and determine q_{sc} for various values of p_{sep}.

Example 6-3:

Determine the flow capacity for the well system of Example 6-2 for the 1.995 in flowline at separator pressures of 1200, 1000, 800, and 500 psia.

Solution:

The separator pressures are tabulated in the following table and plotted in Figure 6-5. The flow capacities for the various separator pressures are listed below.

q_{sc}, *Mscfd*	p_{wf}	p_{tf}	p_{sep}
1000	1877	1500	1490
2000	1774	1362	1320
3000	1653	1158	1042
4000	1512	840	504

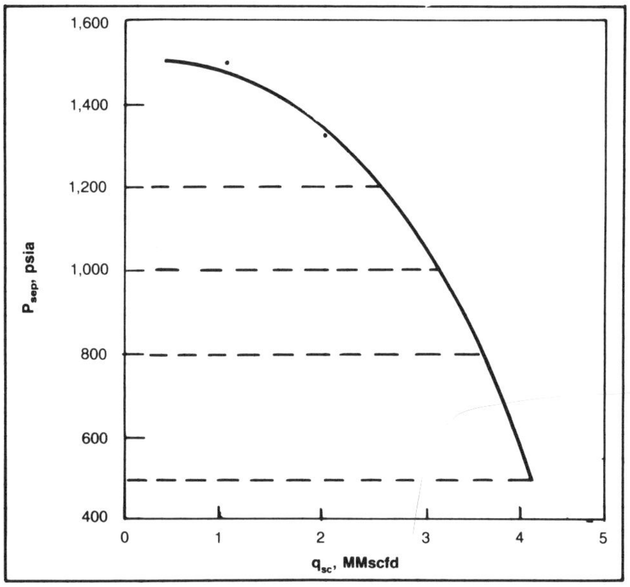

Fig. 6-5. Example 6-3 Solution.

p_{sep}	*Flow capacity, Mscfd*
1200	2560
1000	3080
800	3540
500	4000

COMPRESSOR SELECTION

The selection or sizing of a compressor to increase the deliverability of a well system requires knowing the required suction and discharge pressures and the volume of gas to be pumped. The sales line pressure is usually fixed, but the gas may have to travel a considerable distance from the compressor to the sales line. The compressor discharge pressure would therefore be a function of gas flow rate. Compressor sizing is best accomplished by dividing the system at the compressor, or at the separator if the compressor is to be located near the separator.

The following procedure may be used to determine the necessary design parameters and power required to deliver a quantity of gas to a fixed sales line pressure.

1. Starting at $\bar{p}_R$, determine $\bar{p}_{sep}$ for various values of q_{sc} using the procedure for determining separator pressure effect.
2. Plot p_{sep} versus q_{sc}.
3. Starting at the sales line pressure, determine the discharge pressure required at the compressor, $p_{dis.}$ for various flow rates.
4. Plot $p_{dis.}$ versus q_{sc} on the same graph as used in Step 2. The intersection of these curves (if there is one)

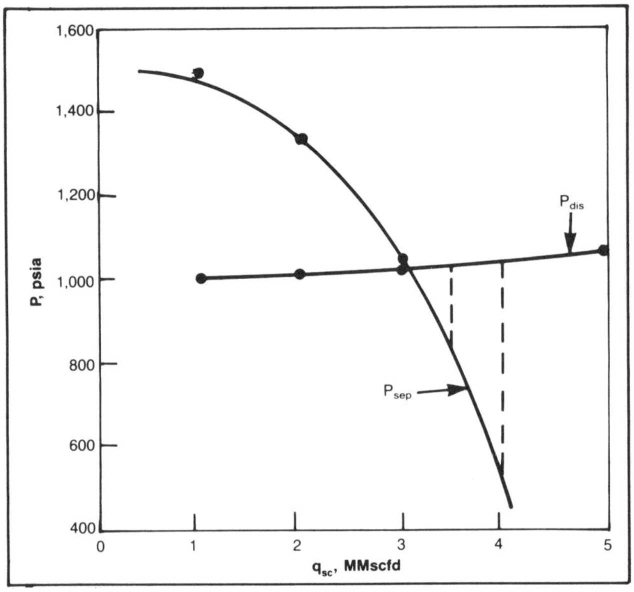

Fig. 6-6. Example 6-4 Solution.

gives the flow capacity or deliverability for no compressor in the system.

5. Select values of q_{sc} and determine the values of $p_{dis.}$, p_{sep} and $\Delta p = p_{dis.} - p_{sep}$ for each q_{sc}.
6. Calculate the required compression ratio, $r = p_{dis.}/p_{sep}$ and the power required using Equation 5-9.

Example 6-4:

The well system described in Examples 6-2 and 6-3 is to deliver gas to a sales line pressure of 1000 psia through a 3.068 in. ID pipeline. The compressor is located at the separator, 10,000 ft from the sales line. Determine the compression ratio and horsepower required for delivery rates of 3.5 and 4.0 MMscfd.

Solution:

The separator or compressor suction pressures for various flow rates were calculated in Example 6-3 and plotted on Figure 6-5. These are replotted on Figure 6-6. Starting at the sales line pressure of 1000 psia, the following required compressor discharge pressures were calculated using Equation 4-35.

q_{sc}, Mscfd	p_{dis}, psia
1000	1002
2000	1010
3000	1021
4000	1037
5000	1057

Plotting these points on Figure 6-6 gives an intersection at q_{sc} = 3040 for no compressor. The following values are read from Figure 6-6 at the flow rates of interest:

q_{sc}, Mscfd	p_{sep}	p_{dis}	$r = p_{dis}/p_{sep}$	Z_1
3500	810	1030	1.27	0.86
4000	500	1040	2.08	0.92

To calculate the horsepower required, use Equation 5-9 with $k = 1.3$, $p_{sc} = 14.7$ psia, $T_{sc} = 520°R$, $T_1 = 540°R$.

For $q_{sc} = 3500$ Mscfd:

$$w = \frac{3.027\, p_{sc}\, T_1\, k}{T_{sc}\,(k-1)}[r^{Z_1(k-1)/k} - 1] \tag{5-9}$$

$$w = \frac{3.027(14.7)(540)(1.3)}{520(0.3)}[(1.27)^{.86(.3)/1.3} - 1]$$

$$w = 200.27(0.049) = 9.8 \text{ Hp/MMscfd}$$

$$Hp = (9.8)(3.5) = 34 \text{ Hp}$$

For q_{sc} = 4000 Mscfd:

$$w = 200.27\,[(2.08)^{.92(.3)/1.3} - 1] = 200.27(0.168)$$

$$= 33.6 \text{ Hp/MMscfd}$$

$$Hp = (33.6)(4.0) = 134 \text{ hp}$$

SUBSURFACE SAFETY VALVE SELECTION

Most offshore wells will require that some type of subsurface safety valve (SSSV) be installed in order to shut the well in if the wellhead pressure becomes too low. These may be velocity actuated or surface actuated type. In any event, unless the valve is tubing retrievable, it will not be full-opening and will therefore create a restriction to the well's flow capacity. In order to select the appropriate SSSV, the well system is divided at the SSSV, Node 4 in Figure 6-2. The two subsystems are then: (1) the reservoir plus the tubing below the SSSV and (2) the separator, flowline, surface choke, and the tubing above the SSSV.

The solution procedure is:

1. Using the inflow performance and tubing equations, calculate the pressure upstream of the SSSV, p_{v1}, for several flow rates.
2. Plot p_{v1} versus q_{sc}.
3. Using the flowline equations, the surface choke equations if applicable, and the tubing equations, calculate the pressure downstream of the SSSV, p_{v2}, for several flow rates.
4. Plot p_{v2} versus q_{sc} on the same graph. The intersection of the two curves gives the flow rate for no SSSV.
5. Determine the Δp across the SSSV for several flow rates, $\Delta p = p_{v1} - p_{v2}$ and plot Δp versus q_{sc}.

6. Calculate the pressure drop across the SSSV for several q_{sc} using Equation 4-62.
7. Plot Δp versus q_{sc} on the same graph as used in Step 5. The intersection of the two curves gives the flow capacity of the system for the particular size valve used in the calculations of Step 6. Repeat Steps 6 and 7 to determine the effect of SSSV size.

The solution could also be obtained by including the SSSV in the inflow expression. This would result in a different inflow curve for each SSSV size considered. The inflow and outflow expressions would be:

Inflow: $\bar{p}_R - \Delta p_{res} - \Delta p_{tub}$ (below SSSV) $- \Delta p_{SSSV} = p_{v2}$

Outflow: $p_{sep} + \Delta p_{fl} + \Delta p_{tub}$ (above SSSV) $= p_{v2}$

Example 6-5:

A 10,000 ft vertical well is to have a SSSV installed at a depth of 3000 ft from the surface. The tubing size is 2.441 in. ID Other data are given in Example 6-2. Determine the flow capacity for SSSV's having inside diameters of 0.4 in. and 0.5 in. The flowline diameter is also 2.441 in. ID.

Solution:

The well performance is tabulated below for the lower subsystem:

q_{sc}, Mscfd	p_{wf}, psia	p_{v1}, psia
1000	1877	1606
2000	1774	1506
3000	1653	1378
4000	1512	1218

The performance for the upper subsystem is calculated by starting at a separator pressure of 1000 psia:

q_{sc}, Mscfd	p_{tf}, psia	p_{v2}, psia
1000	1006	1088
2000	1022	1113
3000	1049	1154
4000	1085	1209

The pressures above and below the SSSV are plotted versus q_{sc} on Figure 6-7. The intersection gives an unrestricted flow rate of 4 MMscfd. The pressure drop across the SSSV ($p_{v1} - p_{v2}$) is determined for several flow rates from Figure 6-7. The pressure drop for each flow rate and SSSV size is also calculated using Equation 4-62 and plotted on Figure 6-8. These are tabulated below. An example calculation of Δp using Equation 4-62 is presented for the 0.5 in. SSSV using $C_d = 0.9$ and $Y = 0.85$ for a flow rate of 2 MMscfd.

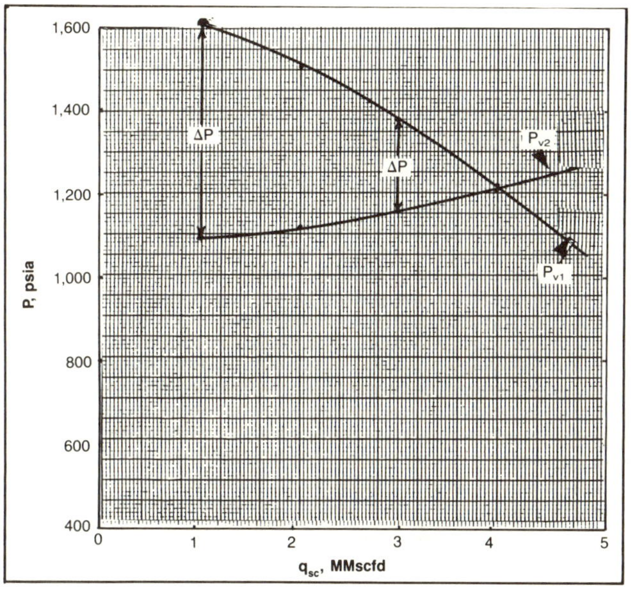

Fig. 6-7. Example 6-5 Solution.

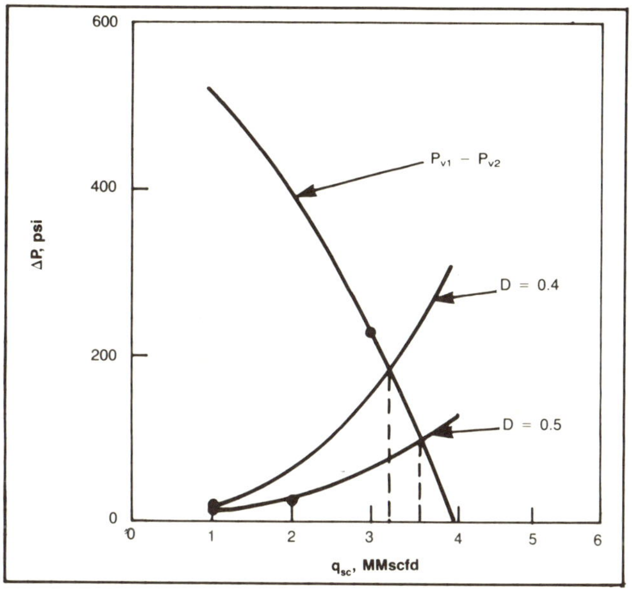

Fig. 6-8. Example 6-5 Solution.

$$\Delta p = \frac{2.7\, \gamma_g\, p_1}{Z_1\, T_1}(1 - \beta^4) \cdot \left[\frac{6.23 \times 10^{-4}\, Z_1\, T_1\, q_{sc}}{p_1\, d^2\, C_d\, Y}\right]^2 \quad (4\text{-}62)$$

$$\beta = \frac{0.5}{2.441} = 0.20$$

$$\Delta p = \frac{2.7(.67)(1506)(1 - .2^4)}{0.86(578)} \cdot \left[\frac{6.23 \times 10^{-4}(.86)(578)(2000)}{1506(.5)^2(0.9)(0.85)}\right]^2$$

$$\Delta p = 5.467\ (4.628) = 25.3 \text{ psi}$$

			Δp, psi	
q_{sc}, Mscfd	p_{v1}, psia	$p_{v1} - p_{v2}$	$D = 0.4$	$D = 0.5$
1000	1606	518	15	6
2000	1506	393	62	25
3000	1378	224	152	62
4000	1218	9	306	125

The results are obtained from Figure 6-8. The intersections of the Δp curves indicate flow capacities of 3230 and 3600 Mscfd for the 0.4 and 0.5 in. SSSV's respectively.

EFFECT OF PERFORATING DENSITY

In many areas of the world, especially the Gulf Coast region of the United States, high capacity gas wells are completed in unconsolidated, high permeability formations. In order to control production of sand, the wells are completed by gravel packing. The flow capacity of the well may be controlled by the number of perforations, since the actual reservoir flow capacity is extremely high.

Another consideration in the well completion design is the pressure drop across the gravel pack. If it is too large, the gravel pack may be destroyed. The consensus is that this pressure drop should be less than about 300 psi, which means that the number of perforations required to meet this limit must be determined.

The method of analysis is the same as that used to analyze subsurface safety valve effect, except that the system is divided at the perforations. The equation for determining the Δp across the gravel pack was given in Chapter 3 (Eq. 3-94).

The two subsystems are: (1) the reservoir, including any actual skin damage and (2) the separator, flowline, surface choke, SSSV, and tubing.

The solution procedure is:

1. Using the reservoir gas flow equation (Eq. 3-11), determine the pressure at the sandface, upstream of the gravel pack, p_{wfs} for various flow rates, q_{sc}.
2. Plot p_{wfs} versus q_{sc}.
3. Starting at the separator pressure, determine the bottom-hole flowing pressure required for the upper subsystem, p_{wf} for various flow rates, q_{sc}.
4. Plot p_{wf} versus q_{sc} on the same graph as used in Step 2.
5. Read values of p_{wfs} and p_{wf} at various q_{sc} and plot $\Delta p_1 = p_{wfs} - p_{wf}$ versus q_{sc}.
6. Assume various q_{sc} and calculate the pressure drop across the gravel pack, Δp_2 using Equation 3-94. This should be done for various perforating densities.
7. Plot Δp_2 versus q_{sc} on the same graph as used in Step 5. The intersection of the curves gives the flow capacity and the gravel pack pressure drop for various perforating densities.

Example 6-6:

The following data apply to the gas well shown in Figure 6-9 (see figs. 6-10 and 6-11 for solution). Determine the flow capacities and gravel pack pressure drops for 4, 12, and 24 perforations per foot for both 1.995 and 2.441 in. tubing.

$p_{tf} = 1200$ psia, $T_s = 100°F$, $\gamma_g = 0.82$,

$\epsilon = 0.0006$ in., $H = 13350$ ft, $T_R = 273°F$,

$\mu_g = 0.012$ cp, $r_e = 1040$ ft, $r_w = 0.50$ ft,

$S = 0$, $\bar{p}_R = 5400$ psia,

$h = 20$ ft, Perforation diameter $= 0.7$ in., $\bar{Z} = .97$

Formation permeability = 130 md

Gravel permeability = 45 darcys

Screen outside diameter = 3.06 in.

Hole diameter = 12.25 in.

$N = (spf)h = (spf)(20)$

Solution:

1. Calculation of p_{wfs}:

$$\beta = \frac{2.33 \times 10^{10}}{k^{1.2}} = \frac{2.33 \times 10^{10}}{(130)^{1.2}}$$

$$= 6.77 \times 10^7$$

$$A = \frac{1422\, T \bar{\mu}\, \bar{Z}\, (\ln(.472\, r_e/r_w) + S)}{k h}$$

$$A = \frac{1422(733)(0.012)(0.97)[\ln(.472(1040)/.50] + 0)}{130(20)}$$

$$A = 32.15$$

$$B = \frac{3.161 \times 10^{-12}\, \beta\, \gamma_g\, \bar{Z}\, T}{h^2\, r_w}$$

$$= \frac{3.161 \times 10^{-12}(6.77 \times 10^7)(.82)(.97)(733)}{(20)^2(.50)}$$

$$B = 6.24 \times 10^{-4}$$

$$p_{wfs} = [\bar{p}_R^{\,2} - A\, q_{sc} - B\, q_{sc}^{\,2}]^{0.5}$$

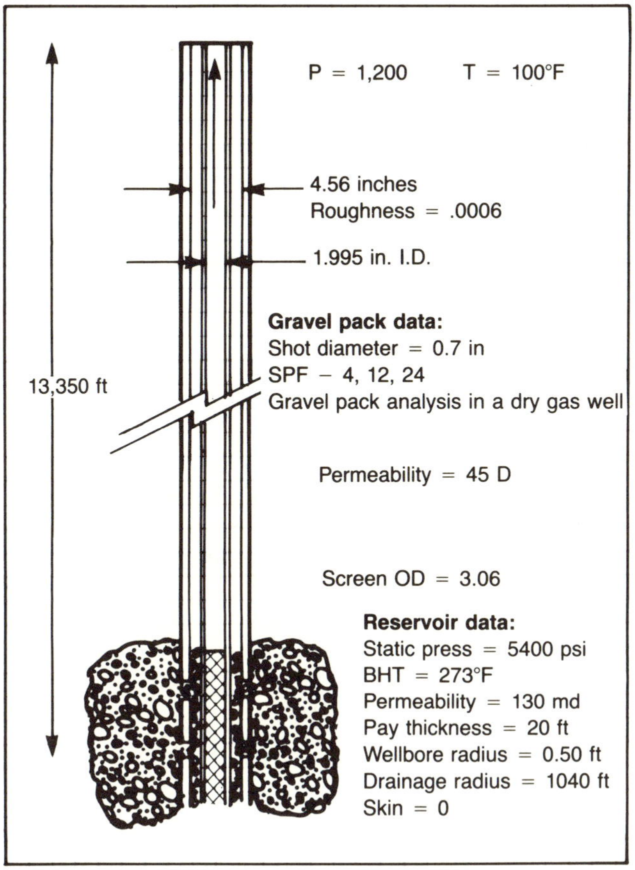

Fig. 6-9. Gravel pack analysis in a dry gas well, Example 6-6.

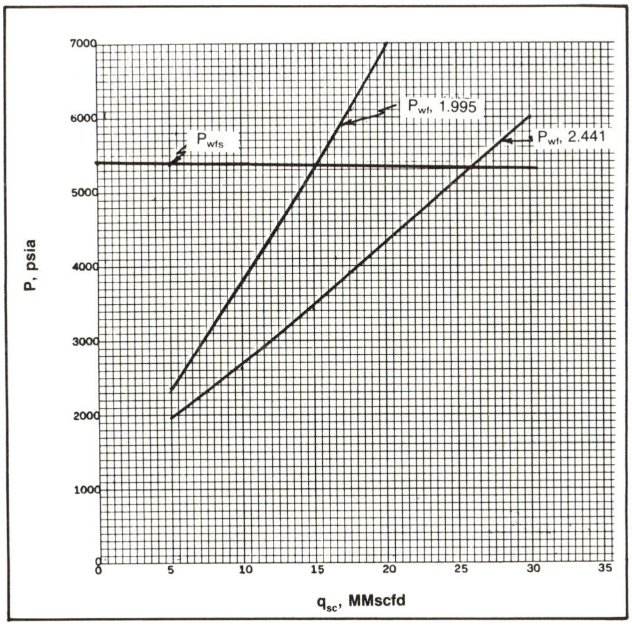

Fig. 6-10. Example 6-6 Solution.

$$p_{wfs} = [(5400)^2 - 32.15\ q_{sc} - 6.24 \times 10^{-4}\ q_{sc}^{\ 2}]^{0.5}$$

2.

q_{sc}, Mscfd	p_{wfs}, psia
5000	5384
10000	5364
20000	5317
30000	5257

3. Equation 4-27 is used to calculate p_{wf} using a fixed p_{tf} of 1200 psia.

4.

	p_{wf}, psia	
q_{sc}, Mscfd	$d = 1.995$	$d = 2.441$
5000	2385	1974
10000	3786	2606
15000	5358	3405
20000	6987	4168
30000	10080	5998

5.

	Δp_1, psi	
q_{sc}, Mscfd	$d = 1.995$	$d = 2.441$
25000	—	170
20000	—	1170
15000	0	1900
10000	1580	2760
5000	3000	3410

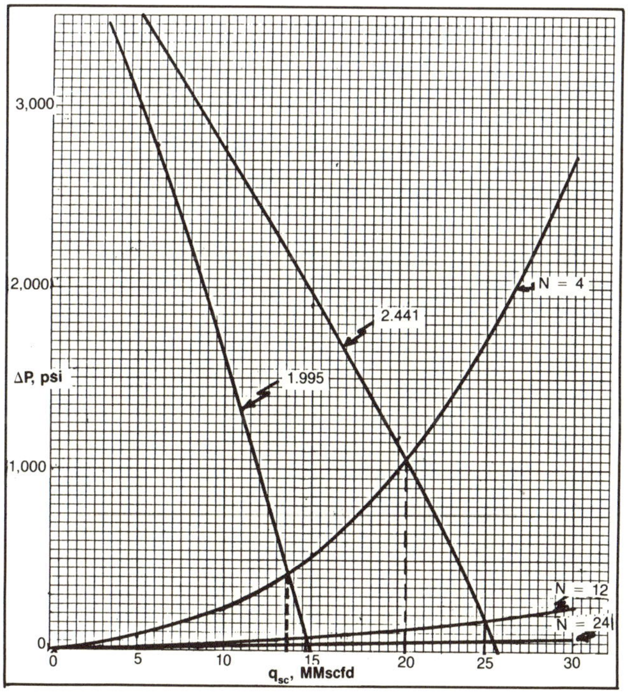

Fig. 6-11. Example 6-6 Solution.

6. Calculation of pressure drop across the gravel pack for 4, 12, and 24 shots per foot:

$$A = \frac{2844\,\bar{Z}\,\mu\,T\,L}{kN\,r_p^2} = \frac{2844(.97)(.012)(733)(.383)}{(45000)N(.029)^2}$$

$$A = 245.57/N$$

$$\beta_G = \frac{1.47 \times 10^7}{(45000)^{0.55}} = 4.05 \times 10^4 \text{ ft}^{-1}$$

$$B = \frac{1.263 \times 10^{-11}\,\beta_G\,\gamma_g\,\bar{Z}\,T\,L}{N^2\,r_p^4}$$

$$= \frac{1.263 \times 10^{-11}(4.05 \times 10^4)(.82)(.97)(733)(.383)}{N^2(.029)^4}$$

$$B = 161.49/N^2$$

$$p_{wf} = [p_{wfs}^2 - A\,q_{sc} - B\,q_{sc}^2]^{0.5}$$

$$\Delta p_2 = p_{wfs} - p_{wf}$$

7.

q_{sc}, Mscfd	p_{wfs}, psia	Δp_2, psi N = 80	N = 240	N = 480
30000	5257	3058	248	62
20000	5317	1061	108	27
15000	5342	566	61	15
10000	5364	244	27	7
5000	5384	60	7	2

The results of the system analysis are tabulated below:

Tubing ID	Flow Capacity, Mscfd SPF = 4	SPF = 12	SPF = 24
1.995	13700	14500	14600
2.441	21000	25000	26000

The results show that the well is severely restricted by the smaller tubing size. The only reason for shooting more than 4 shots per foot if the small tubing is used is to reduce the pressure drop across the gravel pack to less than 300 psi. If the larger tubing is used, 12 shots per foot would be sufficient, giving a rate of 25 MMscfd at a pressure drop across the gravel pack of about 170 psi.

The inflow performance of this well would certainly justify tubing larger than the 2.441 size.

The analysis for determining the effect of perforating density on deliverability can also be performed by selecting other nodes. For example, if p_{wf} were selected as the node pressure the inflow expression would include the pressure drop across the perforations. This would be calculated using Equation 3-88 for gravel packed completions with A_P and B_P equal to zero.

When placing a newly gravel packed well on production it is advisable to be sure that the critical drop across the pack (approximately 300 psi) not be exceeded. This can be accomplished by calculating the expected wellhead pressure for various perforating densities and production rates and plotting p_{tf} versus q_{sc}. The wellhead pressure is obtained from

$$\bar{p}_R - \Delta p_{\text{res}} - \Delta p_{\text{perf}} - \Delta p_{\text{tub}} = p_{tf}.$$

A plot of p_{tf} versus q_{sc} for various perforating densities is illustrated in Figure 6-12. The values of flow rate at which the critical pressure drop is exceeded can be obtained from the plot of Δp versus q_{sc} (Fig. 6-11). The actual wellhead pressure can be measured as the well's rate is increased and plotted on Figure 6-12. This will indicate the number of perforations that are effective and also allow the operator to determine the maximum rate at which he may produce the well without damaging the gravel pack.

EFFECT OF DEPLETION

All of the previous analyses were performed at a constant reservoir pressure, $\bar{p}_R$, that is, at a particular time in the life of the reservoir. As gas is produced and $\bar{p}_R$ declines, the deliverability of the total system will decline. In order to maintain production at a constant rate, the flowing bottom-hole pressure p_{wf} must be reduced as $\bar{p}_R$ declines. This can be accomplished by installing a compressor to lower separator pressure or by installing a larger flowline and tubing to reduce the pressure drop in the piping system.

Inflow performance curves for declining $\bar{p}_R$ are illustrated in Figure 3-19 for the reservoir described in Example 3-7. The decline in flow capacity for a given pip-

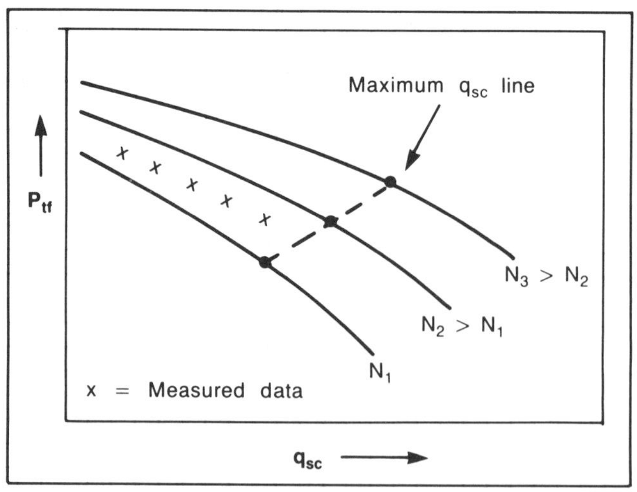

Fig. 6-12. Gravel-packed well performance.

ing system can be determined by plotting the piping system performance on the same graph. The intersections give the flow capacity at various times as the reservoir pressure declines.

The following example illustrates how reducing the separator or wellhead pressure can help to maintain the deliverability as depletion occurs.

Example 6-7:

Determine the deliverability of the well described in Example 3-7 at values of $\bar{p}_R$ = 1952, 1500, and 1000 psia at wellhead pressures of 1000 and 500 psia if the well is equipped with 3.5 in. (2.992 ID) tubing. Other data are:

H = 10000 ft, γ_g = 0.67, $\bar{T}$ = 160°F,
μ_g = 0.012 cp, ϵ = 0.0018 in.

Solution:

The system is divided at the sandface. The inflow performance was calculated in Example 3-7 and is replotted in Figure 6-13. The values of p_{wf} for the upper subsystem are tabulated below and plotted on Fig. 6-13. The flow capacities are also tabulated.

	p_{wf}, psia	
q_{sc}, Mscfd	p_{tf} = 1000	p_{tf} = 500
1000	1256	635
3000	1283	685
5000	1337	780
7000	1415	905
9000	1510	1045

	Flow Capacity, Mscfd	
$\bar{p}_R$	p_{tf} = 1000	p_{tf} = 500
1952	4950	7000
1500	1950	4500
1000	—	1975

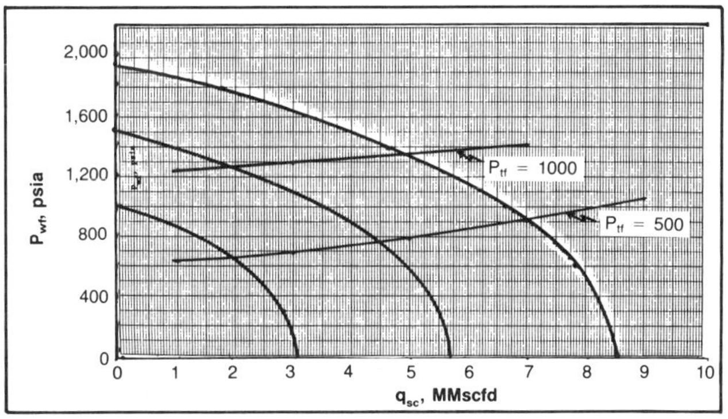

Fig. 6-13. Example 6-7 Solution.

RELATING PERFORMANCE TO TIME

The deliverability or producing capacity of a well or field with respect to time must be known for economic evaluations and for planning equipment purchases. The producing capacity depends on the average reservoir pressure $\bar{p}_R$, which is a function of the cumulative gas produced, G_p. Also, the time required to produce a quantity of gas, ΔG_p, depends on the flow capacity, q_{sc}, during the time period Δt in which the gas was produced. If the production rate is restricted by allowables, the rate of increase in G_p will be constant until the producing

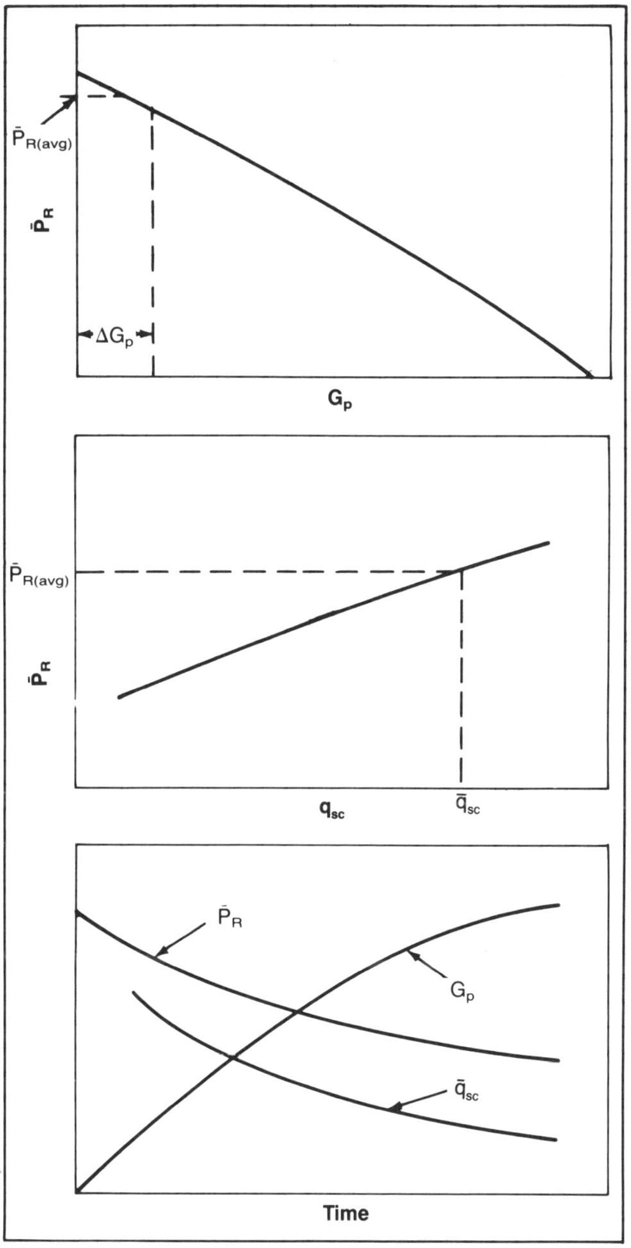

Fig. 6-14. Relating performance to time.

capacity drops below the allowable rate.

Prediction of the rate versus time and cumulative production versus time requires plots of $\bar{p}_R$ or $\bar{p}_R/Z$ versus G_p and $\bar{p}_R$ versus q_{sc}. A plot of $\bar{p}_R$ versus q_{sc} can be obtained from a systems analysis of the effect of depletion, as illustrated in Example 6-7 and Figure 6-13. The total field flow capacity can be obtained by summing the flow capacities of individual wells. A plot of $\bar{p}_R/Z$ or $\bar{p}_R$ versus G_p is obtained from a material balance calculation, as described in Chapter 3.

The procedure involves selecting time periods small enough such that $\bar{p}_R$ can be considered constant during the period. The smaller the time period, or the smaller the value of ΔG_p selected, the more accurate the prediction will be. The procedure is:

1. Construct graphs of $\bar{p}_R$ versus G_p and $\bar{p}_R$ versus q_{sc} (Fig. 6-14).
2. Select a small quantity of gas produced, ΔG_p and determine the average value of $\bar{p}_R$ applicable during this interval from the $\bar{p}_R$ versus G_p plot.
3. From the plot of $\bar{p}_R$ versus q_{sc}, find the average production rate, $\bar{q}_{sc}$ corresponding to the average $\bar{p}_R$ found in Step 2.
4. Calculate the time interval required to produce ΔG_p from

$$\Delta t = \frac{\Delta G_p}{\bar{q}_{sc}}.$$

5. Calculate $t = \Sigma \Delta t$ and $G_p = \Sigma \Delta G_p$. Plot G_p, $\bar{p}_R$ and q_{sc} versus t. Repeat until the economic limit is reached.

SUMMARY

The procedures and examples presented in this chapter demonstrate how the effect of any variable in the total well system may be isolated and analyzed. This was accomplished by dividing the total system into two subsystems, the division being made at various points or nodes, usually as close as possible to the component being considered.

Even though each subsystem is analyzed separately, any change in a component of the subsystem affects the total system performance, therefore, to obtain any meaningful design parameters, this method should be used.

7 Flow Measuring

INTRODUCTION

AN essential part of any gas production operation is the accurate determination of volumetric flow rates. This is evident from all the material presented earlier in which almost all of the design procedures involve the determination of pressure loss for a particular flow rate. Therefore, unless accurate values for flow rate can be obtained, one cannot expect a system to perform according to its design.

This chapter presents the most common methods used in the field to measure gas flow rates. The orifice meter is by far the most widely used because of its simplicity, ruggedness, and accuracy. Application of orifice meters has been well documented in the past, and the standard reference is the American Gas Association Report Number 3.[1] Other types of gas meters are briefly discussed, but their application is limited.

Most of the tables in this chapter are taken directly from the AGA Report 3. Many of the tables in the report are not included because the parameters tabulated involve only a simple calculation, and these properties can be obtained faster with a hand calculator than by referring to tables.

ORIFICE METERING

An orifice metering system consists of means for measuring the pressure drop caused by a change in velocity of the gas as it passes through a restriction placed in the pipe. The energy balance equation may be reduced to the following when all of the terms that do not apply are eliminated:

$$p_1 V_1 - p_2 V_2 = \frac{v_1^2 - v_2^2}{2g}, \tag{7-1}$$

where conditions 1 and 2 refer to conditions in the pipe and conditions in the restriction or orifice, respectively. Assuming that $V_1 = V_2 = V$, expressing the velocities in terms of flow rates, $v = q/A$ and expressing the pressure change in head of flowing fluid results in

$$q = \frac{A_2}{\sqrt{1 - \beta^4}} \sqrt{2\,g\,h}, \tag{7-2}$$

where

q = volumetric flow rate,
A_2 = area of the orifice opening, $\pi d_2^2/4$,
$\beta = d_2/d_1$,
g = acceleration of gravity,
h = pressure drop across the orifice, head,
d_1 = pipe diameter, and
d_2 = orifice or restriction diameter.

Equation 7-2 may be modified by specifying the pressure and temperature conditions at which the flow rate is measured, expressing h in inches of water and including a factor to account for the fact that irreversible losses were ignored.

$$q_{sc} = 218.44\, K_o\, d_2^2 \frac{p_f T_{sc}}{p_{sc} T_f} \left(\frac{h_w T_f}{p_f \gamma_g}\right)^{0.5}, \tag{7-3}$$

where

q_{sc} = gas flow rate at p_{sc}, T_{sc}, scf/hour,
d_2 = orifice size, in.,
p_f = flowing pressure, psia,
T_f = flowing temperature, °R,

h_w = differential pressure across the orifice, inches of water,
γ_g = gas specific gravity (air = 1), and
K_o = an efficiency factor.

Further modification is usually made by combining terms, assuming $\gamma_g = 1$, $p_{sc} = 14.73$, and $T_{sc} = 60°F$ to obtain

$$q_{sc} = C' \sqrt{h_w p_f}, \tag{7-4}$$

where

$$C' = F_b F_{pb} F_{tb} F_g F_{tf} F_r F_{pv} F_m Y. \tag{7-5}$$

The term C' is known as the orifice constant, the value of which depends primarily on the basic orifice factor $F_b = 338.17\ K_o d_2^2$. Many of the other terms are negligible or essentially equal to one. Values for most of these constants are tabulated for various orifice sizes and flowing conditions. The F_b factors are determined empirically and are periodically updated by the AGA.

Orifice Constants

The values of the constants in Equation 7-5 depend on the points between which the differential pressure h_w is measured. Two standards are provided in gas measurement—flange taps and pipe taps. With the former, the flange or orifice holder is so tapped that the center of the upstream and downstream taps is 1 in. from the respective orifice-plate surfaces. For standard pipe taps, the upstream tap is located 2-1/2 pipe diameters upstream and 8 pipe diameters downstream. The location of the taps makes an obvious difference in the values obtained. Tables are provided for both configurations. The relative locations of the taps are shown in Figure 7-1.

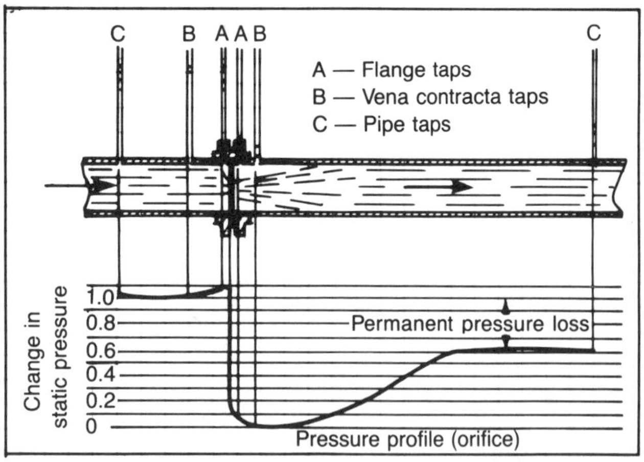

Fig. 7-1. Relative locations of taps. Courtesy John M. Campbell.

Basic Orifice Factor F_b. The F_b factor, as noted above, is based on those assumptions necessary to convert Equation 7-3 to 7-4. These are $T_b = 520°R$, $\gamma_g = 1.00$, and $T_f = 520°R$. It is a function of the experimental constant K_o, which means that it depends on the location of the differential taps and the internal pipe diameter in addition to the orifice diameter. The value of F_b may be found from tables if the meter run is of standard internal diameter. For gas-measurement work involving sale and purchase by a gas-transmission or distribution company, corrections are provided for values of F_b not fitting the standard tables. These are not shown here, for they are not normally used in production operations. The charts show values of F_b for both flange and pipe taps.

Pressure-base Factor F_{pb}. The F_{pb} factor corrects the value of F_b for cases where the pressure base used is not 14.73 psia. It may be determined by the equation $F_{pb} = 14.73/p_b$.

Temperature-base Factor F_{tb}. The F_{tb} factor corrects for any contract wherein the base temperature is not 520°R (60°F). This factor may be computed by the formula $F_{tb} = T_b/520$.

Specific-gravity Factor F_g. The F_g factor is to correct the basic orifice equation for those cases where the specific gravity of the gas is other than 1.00. The equation is $F_g = \gamma_g^{-0.5}$.

Flowing-temperature Factor F_{tf}. The F_{tf} factor corrects for those cases where the flowing temperature of the gas is other than 60°F. The equation is $F_{tf} = (520/T_f)^{0.5}$.

Reynolds-number Factor F_r. The F_r factor takes into account the variation of the discharge coefficient with Reynolds number. In gas measurement the variation is slight and is often ignored in production operations. Values are shown in the charts. It has been assumed in preparation of these charts that gas viscosity is substantially constant. The constant b shown in the charts is then primarily a function of pipe diameter, orifice diameter, and the location of the differential-pressure taps.

Expansion Factor Y. The Y factor accounts for the change in gas density as the pressure changes across the orifice. Inasmuch as the differential involved is usually small, this correction is small and often ignored. The value used depends on which of the differential-pressure taps is used to measure static pressure and the location of the tap. The additional primary variables involved are (1) β, (2) ratio of differential pressure to absolute pressure, and (3) the specific-heat ratio C_p/C_v. In the standard chart the last variable is taken as constant and equal to 1.3. Tables of this factor are shown in the charts.

Supercompressibility Factor F_{pv}. The variation from the ideal-gas laws of an actual gas is corrected by the F_{pv} factor. The factor may be measured experimentally or determined by detailed methods outlined in AGA Report 3. The correction is usually small and is often ignored. It may be estimated from the equation $F_{pv} = (1/Z)^{0.5}$ where Z is equal to the compressibility factor obtained from standard correlations.

Manometer Factor F_m. The F_m factor is used only with mercury-type meters to correct for the slight error in measurement caused by having different heads of gas above the two legs of the manometer. For all practical purposes it is insignificant.

Metering System Design

Inasmuch as standard tables are used, careful attention must be paid to the physical setup of the metering system. Otherwise the results would vary with the installation.

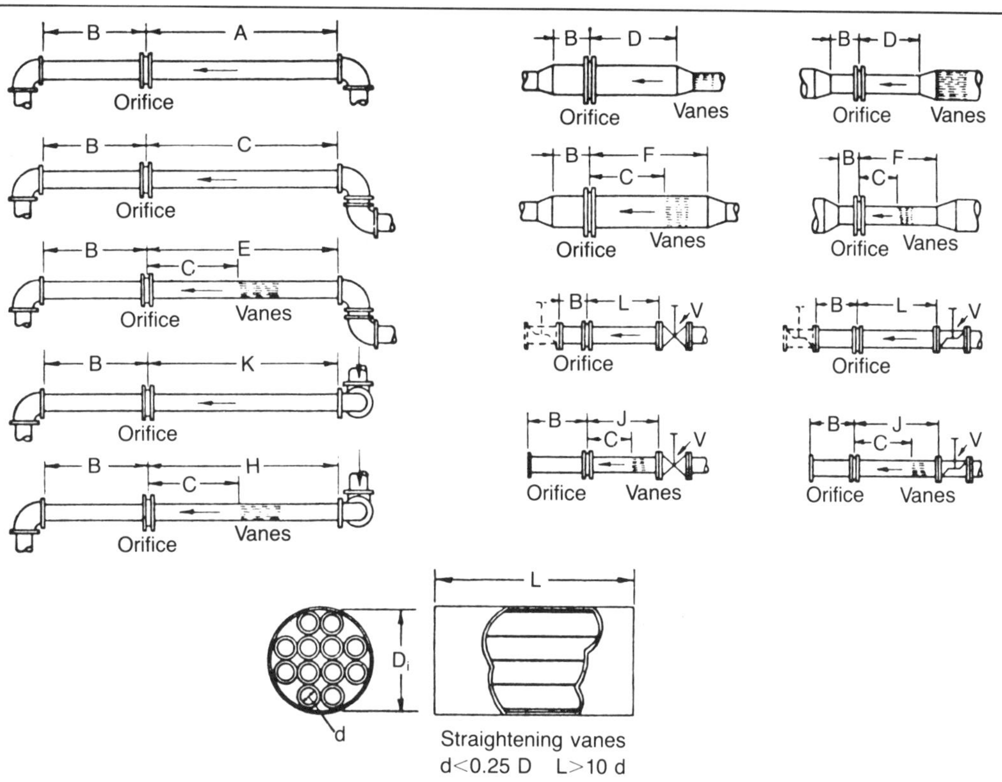

MINIMUM LENGTHS OF STRAIGHT PIPE

β	A	B	C	D	E	F	G	H	J	K	L
0.10	6.0	2.4	5.0	6.0	8.5	8.5	8.5	9.5	9.5	14.0	16.0
0.20	6.0	2.7	5.0	6.0	8.5	8.5	8.5	9.5	9.5	14.5	17.5
0.30	6.0	3.0	5.0	6.0	8.5	8.5	8.5	9.6	9.5	16.0	19.5
0.40	6.0	3.3	5.0	6.0	8.5	8.5	8.5	9.6	9.5	17.8	21.6
0.50	6.8	3.6	5.3	7.4	9.2	9.3	10.0	10.0	10.3	20.2	25.0
0.60	9.0	3.9	5.7	9.4	10.4	10.4	13.4	11.3	12.0	24.7	30.0
0.70	14.0	4.2	6.6	12.0	12.2	12.0	18.5	13.0	15.0	30.5	38.0
0.80	20.0	4.5	8.0	15.0	15.5	15.0	25.0	16.5	19.5	40.0	50.0

β = ratio diameter of orifice to inside diameter of pipe.

Fig. 7-2. Proper installation of a meter run for an orifice meter. Courtesy Gas Processors Suppliers Association.

Straightening Vanes. The purpose of straightening vanes is to minimize the effect of swirls, eddy currents, or irregular velocity distribution on meter accuracy. These vanes are built into a nipple and consist of a bundle of small pipe or tubing as shown in Figure 7-2. The diameter of each tube d should not exceed 1/4 the inside pipe diameter D. The length should be at least $10d$. Unless necessary, vanes should not be used, since they are susceptible to erosion, introduce additional pressure loss, and clog easily.

Orifice Location. Figure 7-2 shows the minimum distance from valves and fittings that the orifice should be placed in order that proper metering might result.

Size of Orifice and Meter Run. The meter run should be sized so that the anticipated maximum and minimum flow rates may be handled within the satisfactory ratio of orifice to pipe diameter. In doing this it should be kept in mind that:

1. The differential h_w should not exceed the static pressure p_f;
2. The meter run ID should be at least one-third larger than the orifice opening;
3. Both the differential and static-pressure pens should preferably operate within the middle 60% of the recording chart range.

As a practical matter, the meter run ID should not be less than 3 in. in nominal diameter regardless of the small quantity of gas flowing. As a first approximation Equation 7-4 should be solved for C' at the flow rate, h_w and p_f desired. Then

$$\text{Orifice ID (in.)} = \left(\frac{C'}{250}\right)^{0.5}. \tag{7-6}$$

If the maximum anticipated flow rate is used to find C', multiplying the orifice size found in Equation 7-6 by 1.5 gives the approximate minimum pipe diameter needed. Once this has been established it is a relatively simple matter to change orifice plates as the flow varies to keep the values of h_w and p_f in the desired range.

Standard sharp-edged orifice plates should be used, the thickness of which is at least 1/16 in. For pipes larger than 4 in. the thickness of the plate should be at least

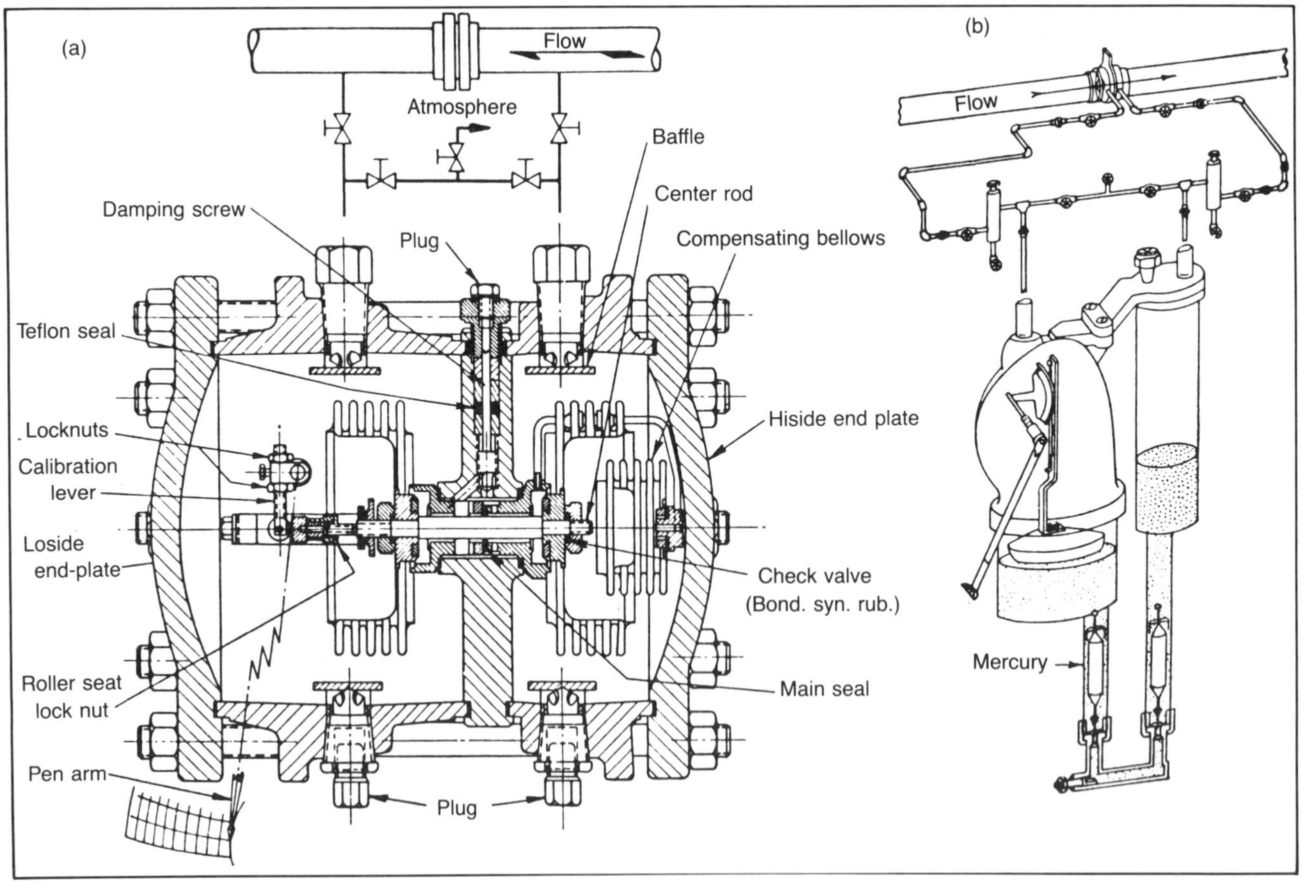

Fig. 7-3. Views showing construction of meters. (a) Printed with permission of the American Meter Division of the Singer Company. (b) Courtesy The Foxboro Company.

1/8 in. The thickness should not exceed one-eighth the orifice opening.

Recorder. The normal orifice meter is equipped with a two-pen recorder for measuring both static and differential pressure. The differential pen is normally actuated through a mechanical system using either a mercury manometer or a bellows. Both types are shown in Figure 7-3.

With the former a slight change in flow rate changes the level in the mercury manometer. A large float resting on the mercury changes level correspondingly and transmits this movement to the chart pen through a system of levers.

The bellows-type meter consists of two bellows filled with some fluid such as glycol. As the pressure differential changes this fluid moves between the bellows through a damping device, causing the bellows to expand or contract. The pen is actuated by a center rod that is connected to the free ends of the bellows. The small liquid-filled bellows located on the high-pressure side serves as an expansion device to correct for changes in ambient temperature.

Both types of instruments are equipped with check valves to protect them from differential pressures that exceed the range of the instrument. This is more of a problem with the mercury meter, since excessive differential pressure can blow mercury out of the meter, which destroys its calibration. In the bellows meter, soft-seated check valves prevent the flow of fluid between the bellows. Then the bellows are not likely to rupture since they are supported internally by the fluid contained within them.

The bellows meter has gained increasing popularity, particularly in field operations. Even though the initial cost is higher, many operators feel that this is compensated for by decreasing maintenance. It also offers the advantage of operating either properly or not at all, with little range in between. The problems of mercury loss and worrying about the change in calibration with the amount of mercury are eliminated. Until it fails, experience has shown that this meter needs little routine calibration.

With both types the static pressure is read from a pen actuated by a bourdon tube. The lead for this may come off either the upstream or downstream tap. In most installations it is good practice to install a bourdon tube having about twice the maximum pressure anticipated. This minimizes the problem of distortion, which destroys the original calibration of the tube. These tubes are available in many materials, the type depending on the service and the pressure rating desired.

Most meters are installed using a standard five-valve manifold of the type shown in Figure 7-4. One valve is placed on each of the leads to the pressure taps, two valves on the bypass, plus a vent valve. When the meter is placed in operation the bypass valves would be open, as would the two main valves, with the vent valve closed. The meter is then placed in service by slowly closing the bypass valves, followed by opening of the vent valve. This procedure prevents momentary pressure surges that might damage the meter or upset the calibration. The reverse procedure would then be used in taking the meter out of service. When the leads are bypassed the vent valve is an excellent place to obtain a gas sample.

Standard calibration procedure on the differential pen involves connecting one side of a water manometer into the high-pressure side of the meter. The low-pressure side of the meter is then opened to the atmosphere. This gives two manometers in parallel. By superimposing pressure on the high-pressure cell with a hand pump or similar device one may compare the meter reading with that of the manometer. If they are different the meter must be adjusted accordingly.

The bourdon tube may be calibrated simply by placing a dead-weight tester or a calibrated pressure gage in the lead line to the bourdon tube. By proper manipulation of the valves, the vent valve may be used to make this connection.

The circular charts used usually cover a time period of either 24 hours or 7 days. Many meters have clocks that may be easily adjusted for either time period. In most production operations 7-day periods are preferred, since this minimizes the cost of changing charts and the number of charts to be handled.

The charts used to record the static and differential pressures may be either the standard type, which record directly the values of h_w and p_f, or the square root or L-10 type, which record the square root of the static pressure and differential pressure. Since the zero reading on the L-10 chart corresponds to zero pounds absolute, the pressure pen must be set at local atmospheric pressure. The position of the pen for zero gage pressure (atmospheric pressure) depends on the atmospheric pressure and the range of the static spring in the recorder. It may be calculated from:

$$p_{\text{zero}} = 10(p_a/R_p)^{0.5}, \tag{7-7}$$

where

p_{zero} = pen setting for zero line pressure, psi,
p_a = local atmospheric pressure, psia, and
R_p = pressure range of meter, psi.

The actual scale of the L-10 charts is from zero to ten; therefore, the correct values of static and differential pressures depend on the ranges of the static and differential springs in the meter. The actual flow rate is calculated from:

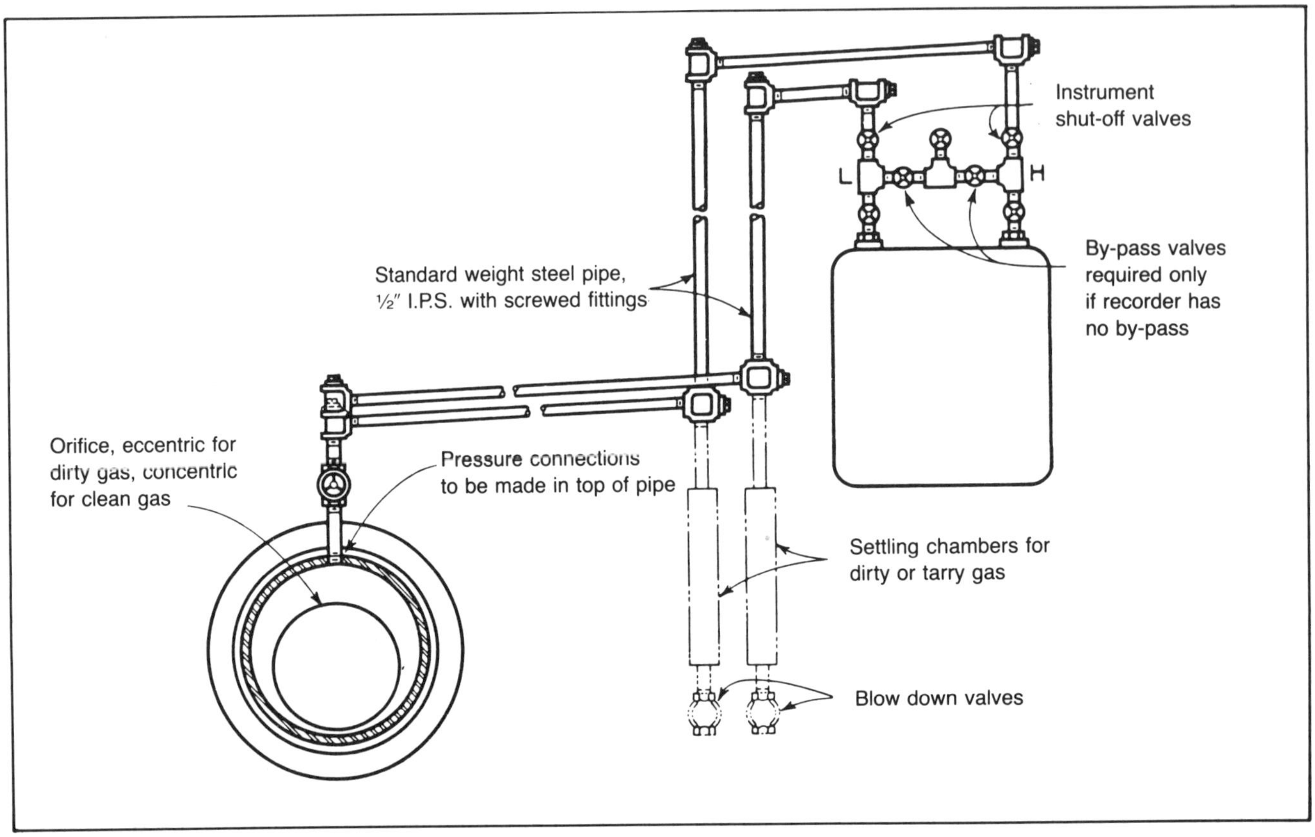

Fig. 7-4. Piping arrangements for orifice meters. Courtesy The American Society of Mechanical Engineers. From Fluid Meters, 6th ed.

$$q_{sc} = 0.01\sqrt{R_h R_p}\, C'\, h_u\, p_u, \tag{7-8}$$

or

$$q_{sc} = C'\, M\, h_u\, p_u, \tag{7-9}$$

where

q_{sc} = flow rate, scf/hr,
R_h = range of differential element, inches,
R_p = range of static element, psi,
C' = orifice constant (Eq. 7-5),
h_u = chart differential reading,
p_u = chart static reading, and
M = $0.01\sqrt{R_h R_p}$, which is constant for a particular meter.

Example 7-1:

A meter run that is equipped with flange taps and a 3.000 inch orifice has an inside diameter of 6.065 inches. The static pressure, obtained from the downstream tap, reads 80 psia and the average differential pressure is 49.5 inches of water. If the pressure and temperature bases are 14.9 psia and 60°F respectively, calculate the flow rate in cubic feet per hour. The gas specific gravity is 0.60 and the flowing temperature is 65°F.

Solution:

1. Calculate $C' = F_b\, F_{pb}\, F_{tb}\, F_g\, F_{tf}\, F_r\, F_{pv}\, F_m\, Y$.
From the tables, for d = 3.000 and D = 6.065, F_b = 1891.9.

$$F_{pb} = \frac{14.73}{p_b} = \frac{14.73}{14.9} = 0.989$$

$$F_{tb} = 1.000$$

$$F_g = (\gamma_g)^{-.5} = (.6)^{-.5} = 1.291$$

$$F_{tf} = (520/T_f)^{0.5} = (520/525)^{0.5} = 0.995$$

From the tables, b = 0.0332,

$$F_r = 1 + \frac{b}{\sqrt{h_w\, p_f}} = 1 + \frac{.0332}{\sqrt{(49.5)(80)}} = 1.001$$

For γ_g = 0.6, T = 65°F, p = 80 psia, the value calculated for Z is 0.990.

$$F_{pv} = (Z)^{-0.5} = (.990)^{-0.5} = 1.005$$

$$F_m = 1.000$$

To determine Y:

$$\beta = \frac{d}{D} = \frac{3.000}{6.065} = 0.495$$

$$\frac{h_w}{p_f} = \frac{49.5}{80} = 0.619$$

From the tables, $Y = 1.0037$ (requires interpolation).

$$C' = 1891.9(0.989)(1)(1.291)(0.995)(1.001)(1.005)(1.0037)$$

$$C' = 2426.9$$

2. Calculate q_{sc}:

$$q_{sc} = C' \sqrt{h_w p_f} = 2426.9 \sqrt{(49.5)(80)}$$

$$q_{sc} = 152{,}721 \text{ cu. ft/hour} = 3.665 \text{ MMscfd}$$

Example 7-2:

A metering system is required to measure approximately 8.5 MMscfd of 0.62 gravity gas at a line pressure of 250 psig. The meter run is to be made of 8 in. pipe (7.981 in. ID). Determine the size of the orifice plate to give a differential of about 50 inches. Flowing temperature averages about 80°F.

Solution:

For $h_w = 50$, $C' = \dfrac{8.5 \times 10^6}{24\sqrt{50(264.7)}} = 3079$

For an approximation, all of the terms in Equation 7-5 except F_g and F_{tf} can be ignored in this case. Therefore,

$$F_b = \frac{C'}{F_g F_{tf}} = \frac{3079}{(1.27)(.981)} = 2471.$$

From the F_b tables for flange taps, for $D = 7.981$ one obtains:

d	F_b
3.375	2352.0
3.500	2537.7

Therefore, a 3.500 inch orifice plate would be selected to obtain an h_w reading of approximately 50 inches at the design flow rate.

Example 7-3:

The metering system described in Example 7-1 is equipped with an *L*-10 recorder that has a differential element range of 100 inches and a static element range of 250 psi. Calculate the flow rate if the following readings are obtained:

$h_u = 7.00, \quad p_u = 5.7$

Solution:

$$q_{sc} = C' \, M \, h_u \, p_u$$

$$M = 0.01 \sqrt{R_h R_p} = 0.01 \sqrt{100(250)} = 1.581$$

From Example 7-1, $\quad C' = 2426.9$

$$q_{sc} = 2426.9(1.581)(7.00)(5.7) = 153{,}000 \text{ scf/hr.}$$

Chart-Reading Accuracy

In most cases when charts are obtained from the field the values of h_w and p_f, or h_u and p_u will not be constant for the entire period covered by the chart. It then becomes necessary to either divide the chart into small enough increments such that the readings can be considered constant or to use integrating devices to obtain the cumulative flow in standard cubic feet for the total chart period.

Various devices are used to mechanically compute the total flow from the chart records, reducing the work required to obtain such a figure from a visual inspection of the chart. Some of these devices are quite complex and expensive but are inherently more accurate.

The simplest device is the linear, or averaging type, planimeter. This instrument merely obtains the area under the curve corresponding to one of the chart records. Consequently, this record could only be used to obtain an average value for the chart record integrated. Since flow varies as the square root of that record in absolute units, the preferable value would be the average square root—instead of being the square root of a direct average. The average square root and the square root of the average value are quite close if the record is fairly constant.

A somewhat more complex instrument is available in the form of a planimeter fitted with a calibrated actuating cam that extracts the square root of the chart reading. This square root planimeter is intended for use with *L*-10 charts. Traversing the circle corresponding to line 7 will result in a reading of 7.00 for a complete chart revolution; a reading of 3.00 will be obtained from tracing a circle corresponding to line 3, etc. This instrument gives a fundamentally more correct analysis of the chart records—being correct, within the limits of measurement, as long as one of the chart records is constant to within 10%. This is the limitation normally put on this instrument. For gas or steam being measured at constant pressures, or in the measurement of liquids, the results obtained with a square root planimeter are theoretically correct.

To obtain a theoretically correct integration, when both chart records vary considerably, the integration must be performed on both records simultaneously. An integrator of this type is inherently more complex and expensive, but its use is justified when the number of charts to be processed is quite large. Such an integrator fulfills the theoretical requirement that the integration be a summation of the square roots of the products of the pen readings, which necessitates the simultaneous integration of the chart records. The integration involves the absolute static pressure, and since most chart records give this variable as a gage pressure, it is necessary to make an adjustment on the arm of the pen that traces the static pressure record to account for the average barometric pressure.

Conditions Affecting Accuracy

Various conditions can exist in the field that adversely affect the measuring accuracy of an orifice metering system. It is common practice to install a metering system and then to accept the results as long as no obvious malfunctions occur.

Condition of the Orifice Edge. The greatest source of error in the primary measuring elements is probably the possible deviation from the specification that the up-stream edge of the orifice plate be square and sharp. A slight rounding of the edge can produce a considerable increase in the discharge coefficient, which results in low measurement. This is especially true with the smaller orifices in the smaller line sizes since the effect of the edge imperfection is relative. A wire-edge burr, or fin, on the orifice edge is also undesirable since it can alter the flow pattern of the stream from that corresponding to proper measurement.

Condition of the Meter Tube. Some error can be introduced as a result of a variation of the finish of the inside of the meter tube. The accepted orifice discharge coefficients were obtained from tests with meter tubes constructed of commercial iron pipe with the corresponding inside surface roughness. Too smooth an inside surface can introduce a slight error in the measurement, just as an error can be produced by too rough a surface, since either condition constitutes a deviation from the conditions under which the accepted discharge coefficients were determined. Measurement would be high with too smooth an inside pipe surface and low with too rough an inside surface.

Pulsation. The effects of pressure and velocity pulsations in the vicinity of the orifice constitute a very indefinite phase in the measurement of gas with an orifice meter. This pulsation can be of a low frequency form such as might result from reciprocating compressors, undamped pressure regulators, chattering valves, or liquid surging back and forth at low points in the line. It might also be a high frequency pulsation caused by resonance of the pipe lines themselves. The pulsations of lower frequency probably have a greater effect on the measurement; however, no conclusive information is at present available by which the pulsation errors can be completely correlated with pulsation frequency or with the wave form and amplitude of those pulsations. This problem has been recognized; considerable research is under way to determine means of either eliminating the effect of pulsations or to determine the degree of error that the pulsation might produce.

In some instances, the effect of pulsation has been largely reduced by installing restrictions in the line that produce pressure drops up to 10% of the line pressure. In many cases, resistances that create much smaller pressure drops have been found satisfactory. In other instances, the effect has been minimized by installing volume tanks, baffles, or dampeners in the line. When any of these devices are used they should be installed between the meter tube and the source of the pulsation.

Other problems can be created by operations that might be employed to obtain a smoother chart record, such as pinching gage lines or over-damping a meter. None of these operations eliminates the basic error corresponding to the pulsation in the line. Pinching gage line valves might even increase the error in the measurement. Furthermore, the pulsation might be of sufficiently high frequency so that, even with an undamped meter, the instrument pens would not follow the fluctuations. Under these conditions a considerable error might exist without any visible indication on the instrument. Probably the best policy would be to install the meter at a point as remote as possible from the source of any pulsation.

In summary, to obtain reliable measurements, it is necessary to suppress the pulsations. The following items, in general, are valuable in diminishing pulsation and/or its effect on orifice flow measurement.

1. Locating the meter tube more favorably with regard to the source of pulsation such as at the inlet side of regulators, or increasing the distance from the source of pulsation.
2. Inserting capacity (volume) or specially designed filters in the line between the source of pulsation and the meter tube in order to reduce the amplitude of the pulsation.
3. Operating at differentials as high as is practical by installing a smaller orifice or by concentrating flow, in a multiple tube installation, through a limited number of the tubes.
4. Using smaller-sized tubes and keeping essentially the same size of orifice, while still maintaining the highest practical limit on the differential.

Effect of Water Vapor. In the measurement of gas containing moisture in a vapor state, the effect of the moisture depends largely on the specific gravity of the gas. Natural gas is quite dry, and its specific gravity is usually quite close to that of the water vapor, about 0.62. For this reason the only appreciable correction would be a direct volume correction based upon the partial pressure of the water vapor at flowing conditions.

In the measurement of certain gases of known composition, the specific gravity is often calculated from the molecular weight or is determined from a moisture-free sample. Under these conditions, especially with very light or very heavy gases, a correction must be made for the erroneous specific gravity obtained by having neglected the moisture content of the gas.

Wet Gas Measurement. The effect of liquid in the gas stream on measurement is a problem that has never been completely solved. Various arrangements of meter tubes, gage line piping, and drip pots have been used in an effort to minimize the errors resulting from liquid accumulating ahead of the orifice plate, at low points in the gage lines, or in the chambers of the meter manometer. An accumulation of liquid ahead of the orifice plate disturbs the normal flow pattern and alters the discharge coefficient for the orifice. Liquid trapped in the gage lines distorts the differential pressure and causes the manometer to give an incorrect indication. Liquid in the chambers of a mercury manometer can cause an unbalance that will alter the zero setting of the instrument by introducing a "wet-leg," or condition of unbalance, in the manometer. Bellows-type manometers, with gage lines attached to the bottom of the manometer chambers, are self-draining and, therefore, eliminate that portion of the error corresponding to the liquid that might find its way into the manometer. With bellows-type meters no error results from liquid being in the manometer chambers, because of a "wet-leg" effect. There would be an error, however, if a difference in liquid levels existed in the manometer chambers, or in the gage lines, under conditions of partial flooding. With either type manometer it is recommended that the manometer be installed above the line, that 1/2 in. (or larger) gage lines be used, and that the gage lines be well graded to permit drainage of liquid back into the meter tube.

The effect of the mechanically entrained liquid that flows through the orifice in the form of a mist is a subject of considerable conjecture. It is difficult to determine anything other than an average value for the amount of liquid entrainment such as might exist over a period of days or weeks. Such an average indication and the corresponding density increase might be obtained from ratios of barrels of liquid per million cubic feet of gas, etc. There is some question, however, as to whether the flow actually varies in strict accordance with the change in average density.

The common practice has been to determine an average density for the mixture and presume that the flow rate corresponds to a fluid of that density.

OTHER METERING METHODS

For well-test purposes other forms of velocity meters are used: orifice well testers, critical-flow provers, and pitot tubes. These methods are not as accurate as the meters discussed above but are convenient and often yield results accurate enough for the purpose.

Orifice Well Tester

The orifice well tester consists of a 2-in. nipple with provision for attaching different sharp-edged orifice plates on the end. The static pressure just upstream from this plate may then be measured. The accuracy of this device is limited but it is suitable for measuring the amount of gas being produced where the pressures are relatively low and the production is to the atmosphere.

Tables for converting the meter reading to flow rates are provided with the meter and may differ among manufacturers.

Critical-flow Prover

The critical-flow prover is a device that also exhausts the gas to the atmosphere. It is also a special pipe nipple with a flange for holding special plates to the end. It is based on the principle that the velocity of sound represents the maximum speed at which a pressure effect may be propagated through a gas; i.e., once this velocity is reached further increase in the pressure differential will not increase the pressure at the throat. This means that the mass rate of flow would not increase as the pressure ratio p_2/p_1 was decreased below this critical value. With ideal gases the critical ratio is 0.49 for monatomic gases and 0.53 for diatomic gases, with slightly higher values for complex gases. Saturated steam, for example, shows a critical pressure ratio of 0.55. The range of these values is the source of the common rule of thumb that once the pressure reduction is twofold critical-flow phenomena limit the mass rate of flow.

When metering gas under such conditions the volume rate of flow will be a function of upstream pressure, gas gravity, and gas temperature since it is a compressible fluid.

The orifice used differs from that in the orifice meter. It is thicker and has a rounded edge. This edge is placed toward the flow, since experience has shown that a sharp-edged orifice does not give reproducible results under critical-flow conditions. In fact sharp-edged orifices do not conform to existing theories and correlations. The general equation for a critical-flow prover is

$$q_{sc} = \frac{Cp}{(\gamma_g T)^{0.5}}, \tag{7-10}$$

where

p = pressure on prover, psia,
C = orifice coefficient for prover,
q_{sc} = rate of flow, (Mcf/D), measured at 14.4 psia and 60°F,
γ_g = specific gravity of gas (air = 1.00), and
T = absolute temperature, °R.

The critical-flow prover is one of the basic devices used for determining the gas flow rate in the open-flow testing of gas wells. Values of the coefficient C may be found in Table 7-1.

Pitot Tube

The pitot tube is another measuring device used extensively during preliminary well tests. It works by measuring the difference between the impact pressure at the tip and the static pressure in the flowing stream. This impact pressure results from conversion of the kinetic energy of the flowing gas to pressure. If the conversion efficiency is relatively constant a conversion between pressure and flow rate is possible. The pitot tube is normally made of 1/8-in. ID pipe and is inserted in the center of a nipple at least 8 diameters long.

TABLE 7-1
Orifice Coefficients for Critical-Flow Provers

Size of Orifice, in.	Orifice Coefficient	
	2-in. Prover	4-in. Prover
1/16	1.524	
3/32	3.355	
1/8	6.301	
5/16	14.47	
7/32	19.97	
1/4	25.86	24.92
5/16	39.77	
3/8	56.68	56.01
7/16	81.09	
1/2	101.8	100.2
5/8	154.0	156.1
3/4	224.9	223.7
7/8	309.3	304.2
1	406.7	396.3
1 1/8	520.8	499.2
1 1/4	657.5	616.4
1 3/8	807.8	742.1
1 1/2	1,002	884.3
1 3/4		1,208
2		1,596
2 1/2		2,566
3		3,904

A pitot tube is used largely for temporary flow measurement since it is small and easy to handle. Very few permanent installations are made, since it produces low-pressure differentials, is difficult to calibrate, and often clogs.

A pitot tube measures a point velocity, i.e., the velocity at only one point across the cross-section of the pipe. In the absence of unknown disturbing elements such as pipe burrs or undue roughness, the velocity at the center is theoretically 20% greater than the mean velocity. For approximate measurement, standard tables use this factor to convert readings taken at the center of the pipe to a volume flow rate. For exact work the mean velocity would be found by taking a series of experimental velocity measurements across the pipe diameter.

A similar method is used for large gas-flow rates and/or where debris produced with the gas makes other methods unfeasible. This method consists simply of measuring the static pressure through a side opening 4 diameters from the end of a nipple at least 8 diameters long. This is the least desirable of the methods discussed because of the potential error involved. Further data on the use of these devices for the measurement of gas flow may be found in U.S. Bureau of Mines Monograph 7.

Turbine Meters

A turbine meter is a velocity responsive meter that is connected in the pipeline such that the entire gas stream passes through the meter. A propeller in the meter turns at a velocity which is proportional to the velocity of the fluid flowing through it. The secondary element may be a revolution counter or other means of sensing and totalizing the revolutions of the propeller or rotor. Turbine meters have been more widely used for measuring liquid flow than gas flow because the driving torque for the rotor depends on the density and the square of the fluid velocity. Therefore it is necessary to have the resisting torque very low for gas measuring; yet it is necessary that this torque be constant throughout the operating range of the meter.

The range of flow rates that can be handled with a turbine meter depends on the size of the meter, the density of the fluid and the viscosity of the fluid. The normal range is about 10 to 1 with an accuracy band of plus or minus 0.5%. Turbine meters used for gas measurement are usually calibrated with a critical flow prover or with an orifice meter of known accuracy. Velocity fluctuations occurring in a turbine meter, caused by turbulence or pressure fluctuations, will cause a turbine meter to register too high.

Although the orifice meter is certainly the most commonly used device for measuring large quantities of gas, various other types of meters have been proposed or are

TABLE 7-2

FLANGE TAPS—BASIC ORIFICE FACTORS—F_b

Orifice Diam., d, Inches	Internal Diameter of Pipe, D, Inches								
	2			3				4	
	1.689	1.939	2.067	2.300	2.626	2.900	3.068	3.152	3.438
.250	12.695	12.708	12.711	12.714	12.712	12.708	12.705	12.703	12.697
.375	28.474	28.440	28.427	28.411	28.393	28.382	28.376	28.373	28.364
.500	50.777	50.587	50.521	50.435	50.356	50.313	50.292	50.283	50.258
.625	80.090	79.508	79.311	79.052	78.817	78.687	78.625	78.599	78.523
.750	117.09	115.62	115.14	114.52	113.99	113.70	113.56	113.50	113.33
.875	162.95	159.56	158.47	157.12	156.00	155.41	155.14	155.03	154.71
1.000	219.77	212.47	210.22	207.44	205.18	204.04	203.54	203.33	202.75
1.125	290.99	276.19	271.70	266.35	262.06	259.95	259.04	258.65	257.63
1.250	385.78	353.58	345.13	335.12	327.39	323.63	322.03	321.37	319.61
1.375	....	448.59	433.50	415.75	402.18	395.80	393.09	391.97	389.03
1.500	....	....	542.27	510.86	487.98	477.36	472.96	471.14	466.39
1.625	....	....	....	623.91	586.82	569.65	562.58	559.72	552.31
1.750	....	....	....	....	701.26	674.44	663.42	658.96	647.54
1.875	....	....	....	....	834.88	793.88	777.18	770.44	753.17
2.000	....	....	....	....	....	930.65	906.01	896.06	870.59
2.125	....	....	....	....	....	1091.2	1052.5	1038.1	1001.4
2.250	....	....	....	....	....	....	1223.2	1199.9	1147.7
2.375	....	....	....	....	....	....	....	....	1311.7
2.500	....	....	....	....	....	....	....	....	1498.4

Orifice Diam., d, Inches	Internal Diameter of Pipe, D, Inches								
	4		6				8		
	3.826	4.026	4.897	5.189	5.761	6.065	7.625	7.981	8.071
.250	12.688	12.683	....	....	....	....	....	....	....
.375	28.353	28.348	....	....	....	....	....	....	....
.500	50.234	50.224	50.197	50.191	50.182	50.178	....	....	....
.625	78.450	78.421	78.338	78.321	78.296	78.287	....	....	....
.750	113.15	113.08	112.87	112.82	112.75	112.72	....	....	....
.875	154.40	154.27	153.88	153.78	153.63	153.56	153.34	153.31	153.31
1.000	202.20	201.99	201.34	201.19	200.96	200.85	200.46	200.40	200.38
1.125	256.69	256.33	255.31	255.08	254.72	254.56	253.99	253.89	253.87
1.250	318.03	317.45	315.83	315.48	314.95	314.72	313.91	313.78	313.74
1.375	386.45	385.51	382.99	382.47	381.70	381.37	380.25	380.06	380.02
1.500	462.27	460.79	456.93	456.16	455.03	454.57	453.02	452.78	452.72
1.625	545.89	543.61	537.77	536.64	535.03	534.38	532.27	531.95	531.87
1.750	637.83	634.39	625.73	624.09	621.79	620.89	618.02	617.60	617.50
1.875	738.75	733.68	721.03	718.69	715.44	714.18	710.32	709.77	709.64
2.000	849.41	842.12	823.99	820.68	816.13	814.41	809.22	808.50	808.34
2.125	970.95	960.48	934.97	930.35	924.07	921.71	914.78	913.85	913.64
2.250	1104.7	1089.9	1054.4	1048.1	1039.5	1036.3	1027.1	1025.9	1025.6
2.375	1252.1	1231.7	1182.9	1174.2	1162.6	1158.3	1146.2	1144.7	1144.3
2.500	1415.0	1387.2	1320.9	1309.3	1293.8	1288.2	1272.3	1270.3	1269.8
2.625	1595.6	1558.2	1469.2	1453.9	1433.5	1426.0	1405.4	1402.9	1402.3
2.750	1797.1	1746.7	1628.9	1608.7	1582.0	1572.3	1545.7	1542.5	1541.8
2.875	....	1955.5	1801.0	1774.5	1740.0	1727.5	1693.4	1689.3	1688.4
3.000	....	2195.0	1986.6	1952.4	1907.8	1891.9	1848.6	1843.5	1842.3
3.125	....	....	2187.2	2143.4	2086.4	2066.1	2011.6	2005.2	2003.8
3.250	....	....	2404.2	2348.8	2276.5	2250.8	2182.6	2174.6	2172.9
3.375	....	....	2639.5	2569.8	2479.1	2446.8	2361.8	2352.0	2349.9
3.500	....	....	2895.5	2808.1	2695.1	2654.9	2549.7	2537.7	2535.0
3.625	....	....	3180.8	3065.3	2925.7	2876.0	2746.5	2731.8	2728.6
3.750	....	....	....	3345.5	3172.1	3111.2	2952.6	2934.8	2930.8
3.875	....	....	....	3657.7	3435.8	3361.5	3168.3	3146.9	3142.1
4.000	....	....	....	....	3718.2	3628.2	3394.3	3368.5	3362.9
4.250	....	....	....	....	4354.8	4216.6	3879.4	3842.3	3834.2
4.500	....	....	....	....	....	4900.9	4412.8	4360.5	4349.0
4.750	....	....	....	....	....	....	5000.7	4928.1	4912.2
5.000	....	....	....	....	....	....	5650.0	5551.1	5529.5
5.250	....	....	....	....	....	....	6369.3	6236.4	6207.3
5.500	....	....	....	....	....	....	7170.9	6992.0	6953.6
5.750	....	....	....	....	....	....	....	7830.0	7777.8
6.000	....	....	....	....	....	....	....	....	8707.0

FLANGE TAPS—BASIC ORIFICE FACTORS—F_b

Orifice Diam., d, Inches	Internal Diameter of Pipe, D, Inches								
	10			12			16		
	9.564	10.020	10.136	11.376	11.938	12.090	14.688	15.000	15.250
1.000	200.20	....	....	....	....	....	....	....	....
1.125	253.56	253.48	253.47	....	....	....	....	....	....
1.250	313.31	313.20	313.18	312.94	312.85	312.83	....	....	....
1.375	379.44	379.29	379.26	378.94	378.82	378.79	....	....	....
1.500	451.95	451.76	451.72	451.30	451.14	451.10	450.53	450.48	....
1.625	530.87	530.63	530.57	530.04	529.84	529.78	529.06	528.99	528.94
1.750	616.20	615.90	615.83	615.16	614.91	614.84	613.94	613.85	613.78
1.875	707.98	707.60	707.51	706.68	706.36	706.28	705.18	705.07	704.99
2.000	806.23	805.76	805.65	804.61	804.23	804.13	802.78	802.65	802.55
2.125	910.97	910.38	910.24	908.98	908.51	908.39	906.77	906.61	906.49
2.250	1022.2	1021.5	1021.3	1019.8	1019.2	1019.1	1017.1	1017.0	1016.8
2.375	1140.1	1139.2	1139.0	1137.1	1136.4	1136.2	1133.9	1133.7	1133.5
2.500	1264.5	1263.4	1263.1	1260.8	1260.0	1259.8	1257.1	1256.8	1256.6
2.625	1395.6	1394.2	1393.9	1391.1	1390.1	1389.9	1386.7	1386.4	1386.1
2.750	1533.4	1531.7	1531.3	1528.0	1526.8	1526.5	1522.7	1522.4	1522.1
2.875	1678.0	1675.9	1675.4	1671.4	1670.0	1669.6	1665.2	1664.8	1664.5
3.000	1829.4	1826.9	1826.3	1821.4	1819.7	1819.3	1814.1	1813.7	1813.3
3.125	1987.8	1984.7	1984.0	1978.1	1976.1	1975.6	1969.6	1969.0	1968.6
3.250	2153.2	2149.5	2148.6	2141.5	2139.2	2138.6	2131.5	2130.9	2130.4
3.375	2325.7	2321.2	2320.2	2311.7	2308.9	2308.2	2299.9	2299.2	2298.7
3.500	2505.6	2500.1	2498.9	2488.7	2485.4	2484.6	2474.9	2474.1	2473.5
3.625	2692.8	2686.2	2684.7	2672.6	2668.7	2667.7	2656.4	2655.5	2654.8
3.750	2887.6	2879.7	2877.9	2863.5	2858.8	2857.7	2844.6	2843.5	2842.7
3.875	3090.1	3080.7	3078.5	3061.4	3055.9	3054.6	3039.4	3038.1	3037.2
4.000	3300.6	3289.3	3286.8	3266.4	3260.0	3258.5	3240.8	3239.4	3238.3
4.250	3746.1	3730.2	3726.7	3698.4	3689.6	3687.5	3663.8	3661.9	3660.5
4.500	4226.0	4204.1	4199.2	4160.4	4148.4	4145.5	4113.9	4111.5	4109.7
4.750	4742.7	4712.8	4706.2	4653.4	4637.2	4633.4	4591.5	4588.4	4586.0
5.000	5298.6	5258.4	5249.6	5179.0	5157.4	5152.3	5097.2	5093.1	5090.1
5.250	5897.4	5843.6	5831.8	5738.5	5710.0	5703.3	5631.4	5626.1	5622.2
5.500	6543.1	6471.9	6456.3	6333.8	6296.6	6287.9	6194.8	6188.1	6183.1
5.750	7240.0	7146.9	7126.5	6966.9	6919.0	6907.8	6788.1	6779.6	6773.3
6.000	7993.3	7872.9	7846.6	7640.4	7579.0	7564.7	7412.3	7401.5	7393.6
6.250	8808.9	8654.8	8621.1	8357.3	8278.9	8260.7	8068.3	8054.8	8044.8
6.500	9693.3	9498.1	9455.3	9121.0	9021.7	8998.7	8757.3	8740.3	8727.9
6.750	10654	10409	10355	9935.2	9810.5	9781.6	9480.4	9459.4	9444.0
7.000	11711	11394	11327	10804	10649	10613	10239	10213	10194
7.250	....	12467	12381	11732	11540	11496	11035	11003	10980
7.500	....	13656	13541	12725	12489	12434	11869	11831	11803
7.750	....	....	....	13787	13500	13433	12745	12698	12664
8.000	....	....	....	14927	14578	14498	13664	13607	13566
8.250	....	....	....	16158	15730	15633	14628	14560	14510
8.500	....	....	....	17505	16963	16845	15642	15560	15500
8.750	....	....	....	....	18297	18148	16706	16609	16538
9.000	....	....	....	....	....	19566	17826	17711	17627
9.250	....	....	....	....	....	....	19004	18868	18769
9.500	....	....	....	....	....	....	20245	20085	19969
9.750	....	....	....	....	....	....	21552	21365	21229
10.000	....	....	....	....	....	....	22930	22712	22554
10.250	....	....	....	....	....	....	24385	24132	23947
10.500	....	....	....	....	....	....	25924	25628	25414
10.750	....	....	....	....	....	....	27567	27210	26960
11.000	....	....	....	....	....	....	29331	28899	28598
11.250	....	....	....	....	....	....	....	30710	30346

FLANGE TAPS—BASIC ORIFICE FACTORS—F_b

Orifice Diam., d, Inches	Internal Diameter of Pipe, D, Inches: 20			24			30		
	18.814	19.000	19.250	22.626	23.000	23.250	28.628	29.000	29.250
2.000	801.40	801.35	801.29	....	....	....	....	....	....
2.125	905.11	905.05	904.98	....	....	....	....	....	....
2.250	1015.2	1015.1	1015.0	....	....	....	....	....	....
2.375	1131.6	1131.5	1131.4	1130.2	1130.1	1130.0	....	....	....
2.500	1254.4	1254.3	1254.2	1252.8	1252.6	1252.6	....	....	....
2.625	1383.6	1383.5	1383.3	1381.7	1381.5	1381.4	....	....	....
2.750	1519.1	1519.0	1518.8	1517.0	1516.8	1516.7	....	....	....
2.875	1661.0	1660.9	1660.7	1658.6	1658.4	1658.3	1656.0	....	....
3.000	1809.4	1809.2	1809.0	1806.6	1806.4	1806.2	1803.7	1803.5	1803.4
3.125	1964.1	1963.9	1963.7	1961.0	1960.7	1960.6	1957.7	1957.5	1957.4
3.250	2125.3	2125.1	2124.8	2121.7	2121.5	2121.3	2118.0	2117.9	2117.7
3.375	2292.9	2292.6	2292.3	2288.9	2288.6	2288.4	2284.7	2284.5	2284.4
3.500	2466.9	2466.6	2466.3	2462.4	2462.1	2461.8	2457.8	2457.6	2457.4
3.625	2647.3	2647.0	2646.6	2642.4	2642.0	2641.7	2637.3	2637.0	2636.8
3.750	2834.2	2833.9	2833.5	2828.7	2828.3	2828.0	2823.1	2822.8	2822.6
3.875	3027.5	3027.3	3026.8	3021.5	3021.0	3020.7	3015.2	3014.9	3014.7
4.000	3227.5	3227.1	3226.5	3220.6	3220.1	3219.8	3213.8	3213.5	3213.3
4.250	3646.7	3646.2	3645.6	3638.3	3637.7	3637.2	3630.1	3629.7	3629.4
4.500	4092.1	4091.5	4090.6	4081.8	4081.0	4080.5	4071.9	4071.4	4071.1
4.750	4563.7	4562.9	4561.9	4551.1	4550.1	4549.5	4539.3	4538.8	4538.4
5.000	5061.8	5060.8	5059.6	5046.4	5045.2	5044.5	5032.4	5031.8	5031.4
5.250	5586.6	5585.4	5583.8	5567.7	5566.4	5565.5	5551.3	5550.5	5550.0
5.500	6138.2	6136.7	6134.8	6115.3	6113.6	6112.6	6095.8	6094.9	6094.4
5.750	6717.1	6715.2	6712.8	6689.1	6687.2	6685.9	6666.2	6665.2	6664.5
6.000	7323.4	7321.1	7318.2	7289.4	7287.1	7285.6	7262.5	7261.3	7260.5
6.250	7957.5	7954.7	7951.2	7916.4	7913.7	7911.9	7884.7	7883.4	7882.5
6.500	8619.9	8616.5	8612.2	8570.2	8566.9	8564.8	8533.0	8531.4	8530.4
6.750	9311.1	9306.9	9301.6	9251.1	9247.2	9244.7	9207.4	9205.6	9204.4
7.000	10031	10026	10020	9959.3	9954.6	9951.7	9908.0	9905.9	9904.6
7.250	10782	10776	10768	10695	10689	10686	10635	10633	10631
7.500	11562	11555	11546	11459	11452	11448	11388	11386	11384
7.750	12374	12365	12354	12250	12243	12238	12168	12165	12163
8.000	13218	13207	13194	13071	13062	13056	12975	12971	12969
8.250	14095	14082	14066	13920	13910	13903	13809	13805	13802
8.500	15005	14990	14971	14799	14787	14779	14669	14665	14661
8.750	15950	15933	15911	15708	15693	15684	15557	15552	15548
9.000	16932	16911	16885	16647	16630	16620	16473	16466	16462
9.250	17950	17926	17895	17618	17598	17585	17416	17409	17404
9.500	19007	18979	18943	18620	18597	18582	18387	18379	18373
9.750	20104	20071	20030	19655	19628	19611	19386	19377	19371
10.000	21243	21205	21157	20723	20692	20672	20414	20403	20396
10.250	22426	22382	23325	21825	21789	21767	21471	21458	21450
10.500	23654	23603	23538	22962	22921	22895	22556	22542	22533
10.750	24931	24872	24797	24134	24088	24058	23672	23656	23646
11.000	26257	26190	26104	25344	25290	25257	24817	24799	24787
11.250	27636	27559	27460	26592	26531	26492	25992	25972	25959
11.500	29070	28982	28870	27878	27809	27766	27199	27176	27161
11.750	30563	30462	30334	29205	29126	29077	28436	28411	28394
12.000	32116	32001	31856	30574	30485	30429	29706	29677	29659
12.500	35417	35270	35084	33444	33330	33259	32343	32306	32283
13.000	39003	38817	38581	36502	36357	36267	35114	35068	35039
13.500	42913	42673	42375	39762	39581	39467	38025	37968	37932
14.000	47244	46921	46523	43241	43015	42874	41082	41012	40968
14.500	....	....	....	46958	46679	46505	44291	44206	44152
15.000	....	....	....	50934	50591	50378	47661	47557	47490
15.500	....	....	....	55192	54774	54513	51202	51075	50994
16.000	....	....	....	59759	59251	58935	54923	54769	54671
16.500	....	....	....	64701	64060	63671	58835	58649	58531
17.000	....	....	....	....	69288	68792	62950	62728	62586
17.500	....	....	....	....	....	....	67282	67017	66848
18.000	....	....	....	....	....	....	71844	71530	71330
18.500	....	....	....	....	....	....	76653	76282	76046
19.000	....	....	....	....	....	....	81725	81289	81011
19.500	....	....	....	....	....	....	87078	86568	86244
20.000	....	....	....	....	....	....	92734	92140	91761
20.500	....	....	....	....	....	....	98728	98025	97584
21.000	....	....	....	....	....	....	105134	104282	103752
21.500	....	....	....	....	....	....	....	110983	110340

"b" VALUES FOR REYNOLDS NUMBER FACTOR, F_r—FLANGE TAPS

$$F_r = 1 + \frac{b}{\sqrt{h_w p_f}}$$

Orifice Diam., d, Inches	Internal Diameter of Pipe, D, Inches: 2			3				4	
	1.689	1.939	2.067	2.300	2.626	2.900	3.068	3.152	3.438
.250	.0879	.0911	.0926	.0950	.0979	.0999	.1010	.1014	.1030
.375	.0677	.0709	.0726	.0755	.0792	.0820	.0836	.0844	.0867
.500	.0562	.0576	.0588	.0612	.0648	.0677	.0695	.0703	.0730
.625	.0520	.0505	.0506	.0516	.0541	.0566	.0583	.0591	.0618
.750	.0536	.0485	.0471	.0462	.0470	.0486	.0498	.0504	.0528
.875	.0595	.0506	.0478	.0445	.0429	.0433	.0438	.0442	.0460
1.000	.0677	.0559	.0515	.0458	.0416	.0403	.0402	.0403	.0411
1.125	.0762	.0630	.0574	.0495	.0427	.0396	.0386	.0383	.0380
1.250	.0824	.0707	.0646	.0550	.0456	.0408	.0388	.0381	.0365
1.375	....	.0772	.0715	.0614	.0501	.0435	.0406	.0394	.0365
1.500	....	....	.0773	.0679	.0554	.0474	.0436	.0420	.0378
1.625	....	....	....	.0735	.0613	.0522	.0477	.0457	.0402
1.750	....	....	....	....	.0669	.0575	.0524	.0500	.0434
1.875	....	....	....	....	.0717	.0628	.0574	.0549	.0473
2.000	....	....	....	....	....	.0676	.0624	.0598	.0517
2.125	....	....	....	....	....	.0715	.0669	.0644	.0563
2.250	....	....	....	....	....	....	.0706	.0685	.0607
2.375	....	....	....	....	....	....	....	....	.0648
2.500	....	....	....	....	....	....	....	....	.0683

Orifice Diam., d, Inches	Internal Diameter of Pipe, D, Inches: 4		6				8		
	3.826	4.026	4.897	5.189	5.761	6.065	7.625	7.981	8.071
.250	.1047	.1054	....	....	....	....			
.375	.0894	.0907	....	....	....	....			
.500	.0763	.0779	.0836	.0852	.0880	.0892	....	....	....
.625	.0653	.0670	.0734	.0753	.0785	.0801	....	....	....
.750	.0561	.0578	.0645	.0665	.0701	.0718	....	....	....
.875	.0487	.0502	.0567	.0587	.0625	.0643	.0723	.0738	.0742
1.000	.0430	.0442	.0500	.0520	.0557	.0576	.0660	.0676	.0680
1.125	.0388	.0396	.0444	.0462	.0498	.0517	.0602	.0619	.0623
1.250	.0361	.0364	.0399	.0414	.0447	.0464	.0549	.0566	.0571
1.375	.0347	.0344	.0363	.0375	.0403	.0419	.0501	.0518	.0523
1.500	.0345	.0336	.0336	.0344	.0367	.0381	.0457	.0474	.0479
1.625	.0354	.0338	.0318	.0322	.0337	.0348	.0418	.0435	.0439
1.750	.0372	.0350	.0307	.0306	.0314	.0322	.0383	.0399	.0403
1.875	.0398	.0370	.0305	.0298	.0298	.0303	.0353	.0366	.0371
2.000	.0430	.0395	.0308	.0296	.0287	.0288	.0327	.0340	.0343
2.125	.0467	.0427	.0318	.0300	.0281	.0278	.0304	.0315	.0318
2.250	.0507	.0462	.0334	.0310	.0281	.0274	.0286	.0295	.0297
2.375	.0548	.0501	.0354	.0324	.0286	.0274	.0271	.0278	.0280
2.500	.0589	.0540	.0378	.0342	.0295	.0279	.0259	.0264	.0265
2.625	.0626	.0579	.0406	.0365	.0308	.0287	.0251	.0253	.0254
2.750	.0659	.0615	.0436	.0391	.0324	.0300	.0246	.0245	.0245
2.875	....	.0647	.0468	.0418	.0343	.0314	.0244	.0240	.0240
3.000	....	.0673	.0500	.0448	.0366	.0332	.0245	.0238	.0237
3.125	....	....	.0533	.0479	.0389	.0353	.0248	.0239	.0237
3.250	....	....	.0564	.0510	.0416	.0375	.0254	.0242	.0240
3.375	....	....	.0594	.0541	.0443	.0400	.0263	.0248	.0244
3.500	....	....	.0620	.0569	.0472	.0426	.0273	.0255	.0251
3.625	....	....	.0643	.0597	.0500	.0452	.0286	.0265	.0260
3.750	....	....	....	.0621	.0527	.0479	.0300	.0276	.0271
3.875	....	....	....	.0640	.0553	.0505	.0316	.0289	.0283
4.000	....	....	....	....	.0578	.0531	.0334	.0304	.0297
4.250	....	....	....	....	.0620	.0579	.0372	.0338	.0330
4.500	....	....	....	....	....	.0618	.0414	.0375	.0366
4.750	....	....	....	....	....	....	.0457	.0416	.0405
5.000	....	....	....	....	....	....	.0500	.0457	.0446
5.250	....	....	....	....	....	....	.0539	.0497	.0487
5.500	....	....	....	....	....	....	.0574	.0535	.0524
5.750	....	....	....	....	....	....	....	.0569	.0559
6.000	....	....	....	....	....	....	....	....	.0588

"*b*" VALUES FOR REYNOLDS NUMBER FACTOR, F_r—FLANGE TAPS

$$F_r = 1 + \frac{b}{\sqrt{h_w p_f}}$$

Orifice Diam., *d*, Inches	Internal Diameter of Pipe, *D*, Inches								
	10			12			16		
	9.564	10.020	10.136	11.376	11.938	12.090	14.688	15.000	15.250
1.000	.0738	....	....	....	....	....	....	....	....
1.125	.0685	.0701	.0705	....	....	....	....	....	....
1.250	.0635	.0652	.0656	.0698	.0714	.0718	....	....	....
1.375	.0588	.0606	.0610	.0654	.0671	.0676	....	....	....
1.500	.0545	.0563	.0568	.0612	.0631	.0635	.0706	.0713	....
1.625	.0504	.0523	.0527	.0573	.0592	.0597	.0670	.0678	.0684
1.750	.0467	.0485	.0490	.0536	.0555	.0560	.0636	.0644	.0650
1.875	.0433	.0451	.0455	.0501	.0521	.0526	.0604	.0612	.0618
2.000	.0401	.0419	.0423	.0469	.0488	.0492	.0572	.0581	.0587
2.125	.0372	.0389	.0393	.0438	.0458	.0463	.0542	.0551	.0558
2.250	.0346	.0362	.0360	.0410	.0429	.0434	.0514	.0523	.0529
2.375	.0322	.0337	.0341	.0383	.0402	.0407	.0487	.0496	.0502
2.500	.0302	.0315	.0319	.0359	.0377	.0382	.0461	.0470	.0476
2.625	.0283	.0296	.0299	.0336	.0354	.0358	.0436	.0445	.0452
2.750	.0267	.0278	.0281	.0316	.0332	.0336	.0413	.0422	.0428
2.875	.0254	.0263	.0265	.0297	.0312	.0317	.0391	.0399	.0406
3.000	.0243	.0250	.0252	.0278	.0294	.0298	.0370	.0378	.0385
3.125	.0234	.0239	.0241	.0264	.0278	.0282	.0350	.0358	.0365
3.250	.0226	.0230	.0231	.0251	.0263	.0266	.0331	.0339	.0346
3.375	.0221	.0223	.0224	.0239	.0250	.0253	.0314	.0321	.0328
3.500	.0219	.0218	.0218	.0229	.0238	.0241	.0298	.0305	.0311
3.625	.0218	.0214	.0214	.0221	.0228	.0230	.0282	.0290	.0295
3.750	.0218	.0213	.0212	.0214	.0219	.0221	.0268	.0275	.0281
3.875	.0221	.0213	.0211	.0208	.0212	.0213	.0255	.0262	.0267
4.000	.0225	.0214	.0212	.0204	.0206	.0207	.0243	.0249	.0254
4.250	.0238	.0222	.0219	.0200	.0198	.0198	.0223	.0228	.0232
4.500	.0256	.0236	.0231	.0201	0195	.0194	.0206	.0210	.0213
4.750	.0279	.0254	.0249	.0207	.0196	.0194	.0193	.0196	.0198
5.000	.0307	.0277	.0270	.0217	.0202	.0199	.0184	.0185	.0187
5.250	.0337	.0303	.0295	.0231	.0212	.0208	.0178	.0178	.0179
5.500	.0370	.0332	.0323	.0249	.0226	.0221	.0176	.0174	.0174
5.750	.0404	.0363	.0354	.0270	.0243	.0237	.0176	.0174	.0172
6.000	.0438	.0396	.0386	.0294	.0263	.0255	.0180	.0176	.0173
6.250	.0473	.0429	.0418	.0320	.0285	.0277	.0186	.0180	.0177
6.500	.0505	.0462	.0451	.0347	.0309	.0300	.0195	.0188	.0183
6.750	.0536	.0493	.0483	.0376	.0335	.0325	.0206	.0198	.0192
7.000	.0562	.0523	.0513	.0406	.0362	.0351	.0220	.0210	.0202
7.250	....	.0550	.0540	.0435	.0390	.0379	.0235	.0224	.0216
7.500	....	.0572	.0564	.0463	.0418	.0407	.0252	.0240	.0230
7.750	....	....	....	.0491	.0446	.0434	.0271	.0257	.0246
8.000	....	....	....	.0517	.0473	.0461	.0291	.0276	.0264
8.250	....	....	....	.0540	.0498	.0487	.0312	.0296	.0283
8.500	....	....	....	.0560	.0522	.0511	.0334	.0317	.0303
8.750	....	....	....	....	.0543	.0534	.0357	.0338	.0324
9.000	....	....	....	....	....	.0553	.0380	.0361	.0346
9.250	....	....	....	....	....	....	.0402	.0383	.0368
9.500	....	....	....	....	....	....	.0425	.0406	.0390
9.750	....	....	....	....	....	....	.0447	.0428	.0412
10.000	....	....	....	....	....	....	.0469	.0449	.0434
10.250	....	....	....	....	....	....	.0489	.0470	.0455
10.500	....	....	....	....	....	....	.0508	.0490	.0475
10.750	....	....	....	....	....	....	.0526	.0509	.0495
11.000	....	....	....	....	....	....	.0541	.0526	.0513
11.250	....	....	....	....	....	....	....	.0541	.0528

"*b*" VALUES FOR REYNOLDS NUMBER FACTOR, F_r—FLANGE TAPS

$$F_r = 1 + \frac{b}{\sqrt{h_w p_f}}$$

Orifice Diam., *d*, Inches	Internal Diameter of Pipe, *D*, Inches								
	20			24			30		
	18.814	19.000	19.250	22.626	23.000	23.250	28.628	29.000	29.250
2.000	.0667	.0671	.0676	....	....	....	....	....	....
2.125	.0640	.0644	.0649	....	....	....	....	....	....
2.250	.0614	.0618	.0622	....	....	....	....	....	....
2.375	.0588	.0592	.0597	.0659	.0665	.0669	....	....	....
2.500	.0563	.0568	.0573	.0636	.0642	.0646	....	....	....
2.625	.0540	.0544	.0549	.0614	.0620	.0624	....	....	....
2.750	.0517	.0521	.0526	.0592	.0599	.0603	....	....	....
2.875	.0494	.0499	.0504	.0571	.0578	.0582	.0662	....	....
3.000	.0473	.0477	.0483	.0551	.0557	.0562	.0644	.0649	.0652
3.125	.0452	.0457	.0462	.0531	.0538	.0542	.0626	.0631	.0634
3.250	.0433	.0437	.0442	.0511	.0518	.0523	.0608	.0613	.0616
3.375	.0414	.0418	.0423	.0493	.0500	.0504	.0590	.0596	.0599
3.500	.0395	.0399	.0405	.0474	.0481	.0486	.0574	.0579	.0582
3.625	.0378	.0382	.0387	.0457	.0464	.0468	.0557	.0562	.0566
3.750	.0361	.0365	.0370	.0440	.0447	.0451	.0541	.0546	.0550
3.875	.0345	.0349	.0354	.0423	.0430	.0435	.0525	.0530	.0534
4.000	.0329	.0333	.0339	.0407	.0414	.0419	.0509	.0515	.0518
4.250	.0301	.0304	.0310	.0376	.0384	.0388	.0479	.0485	.0488
4.500	.0275	.0279	.0283	.0348	.0355	.0360	.0450	.0456	.0460
4.750	.0252	.0256	.0260	.0322	.0328	.0333	.0423	.0429	.0433
5.000	.0232	.0235	.0239	.0297	.0304	.0308	.0397	.0403	.0407
5.250	.0214	.0217	.0220	.0275	.0281	.0285	.0373	.0378	.0382
5.500	.0199	.0201	.0204	.0254	.0260	.0264	.0349	.0355	.0359
5.750	.0186	.0188	.0191	.0236	.0241	.0245	.0327	.0333	.0337
6.000	.0176	.0177	.0179	.0219	.0224	.0228	.0306	.0312	.0316
6.250	.0167	.0168	.0170	.0204	.0208	.0212	.0287	.0292	.0296
6.500	.0161	.0162	.0163	.0191	.0195	.0198	.0269	.0274	.0277
6.750	.0157	.0157	.0157	.0179	.0183	.0185	.0252	.0257	.0260
7.000	.0155	.0155	.0154	.0169	.0172	.0174	.0236	.0240	.0244
7.250	.0155	.0154	.0153	.0161	.0163	.0165	.0221	.0226	.0229
7.500	.0157	.0155	.0154	.0154	.0156	.0157	.0208	.0212	.0215
7.750	.0160	.0158	.0156	.0148	.0150	.0151	.0195	.0199	.0202
8.000	.0166	.0163	.0160	.0144	.0145	.0146	.0184	.0187	.0190
8.250	.0172	.0169	.0165	.0142	.0142	.0142	.0174	.0177	.0179
8.500	.0180	.0177	.0172	.0141	.0140	.0140	.0164	.0168	.0170
8.750	.0190	.0186	.0180	.0141	.0140	.0139	.0156	.0159	.0161
9.000	.0201	.0196	.0190	.0143	.0141	.0140	.0149	.0152	.0153
9.250	.0213	.0208	.0201	.0146	.0143	.0141	.0143	.0145	.0146
9.500	.0226	.0220	.0213	.0150	.0146	.0144	.0138	.0139	.0141
9.750	.0240	.0234	.0226	.0155	.0150	.0147	.0133	.0135	.0136
10.000	.0256	.0249	.0240	.0161	.0155	.0152	.0130	.0131	.0132
10.250	.0271	.0264	.0255	.0168	.0162	.0158	.0128	.0128	.0128
10.500	.0288	.0280	.0270	.0176	.0169	.0164	.0126	.0126	.0126
10.750	.0305	.0297	.0286	.0185	.0176	.0172	.0125	.0125	.0125
11.000	.0322	.0314	.0303	.0194	.0186	.0181	.0125	.0124	.0124
11.250	.0340	.0332	.0320	.0205	.0196	.0190	.0126	.0125	.0124
11.500	.0358	.0349	.0338	.0216	.0207	.0200	.0128	.0126	.0125
11.750	.0376	.0367	.0355	.0228	.0218	.0211	.0130	.0128	.0127
12.000	.0394	.0385	.0373	.0241	.0230	.0223	.0134	.0131	.0129
12.500	.0429	.0420	.0408	.0267	.0255	.0248	.0142	.0138	.0136
13.000	.0463	.0454	.0442	.0296	.0282	.0274	.0153	.0148	.0145
13.500	.0494	.0485	.0474	.0326	.0311	.0302	.0166	.0160	.0157
14.000	.0520	.0512	.0502	.0356	.0341	.0331	.0182	.0175	.0171
14.500	....	....	....	.0386	.0370	.0360	.0199	.0192	.0187
15.000	....	....	....	.0415	.0400	.0390	.0218	.0209	.0204
15.500	....	....	....	.0443	.0428	.0418	.0239	.0230	.0224
16.000	....	....	....	.0470	.0455	.0446	.0260	.0250	.0244
16.500	....	....	....	.0494	.0480	.0471	.0283	.0273	.0266
17.000	....	....	....	....	.0503	.0494	.0307	.0296	.0288
17.500	....	....	....	....	....	....	.0331	.0319	.0312
18.000	....	....	....	....	....	....	.0355	.0343	.0335
18.500	....	....	....	....	....	....	.0379	.0366	.0358
19.000	....	....	....	....	....	....	.0402	.0390	.0382
19.500	....	....	....	....	....	....	.0424	.0412	.0404
20.000	....	....	....	....	....	....	.0446	.0434	.0426
20.500	....	....	....	....	....	....	.0466	.0455	.0448
21.000	....	....	....	....	....	....	.0485	.0475	.0467
21.500	....	....	....	....	....	....	....	.0492	.0485

"b" VALUES FOR REYNOLDS NUMBER FACTOR, F_r—FLANGE TAPS

$$F_r = 1 + \frac{b}{\sqrt{h_w p_f}}$$

Internal Diameter of Pipe, *D*, inches

β	2			3				4			
	1.689	1.939	2.067	2.300	2.626	2.900	3.068	3.152	3.438	3.826	4.026
0.10	0.1062	0.1027	0.1012	0.0987	0.0958	0.0928	0.0927	0.0921	0.0905	0.0886	0.0877
0.11	0.1020	0.0985	0.0970	0.0945	0.0916	0.0896	0.0885	0.0880	0.0863	0.0844	0.0835
0.12	0.0981	0.0945	0.0930	0.0905	0.0876	0.0856	0.0845	0.0840	0.0823	0.0804	0.0795
0.13	0.0943	0.0907	0.0892	0.0867	0.0838	0.0818	0.0807	0.0802	0.0785	0.0766	0.0757
0.14	0.0906	0.0871	0.0856	0.0831	0.0802	0.0782	0.0771	0.0765	0.0749	0.0730	0.0721
0.15	0.0872	0.0837	0.0822	0.0797	0.0768	0.0748	0.0736	0.0731	0.0715	0.0696	0.0687
0.16	0.0840	0.0805	0.0789	0.0764	0.0736	0.0715	0.0704	0.0699	0.0682	0.0663	0.0654
0.17	0.0809	0.0774	0.0759	0.0734	0.0705	0.0685	0.0673	0.0668	0.0652	0.0633	0.0624
0.18	0.0780	0.0745	0.0730	0.0705	0.0676	0.0656	0.0645	0.0639	0.0623	0.0604	0.0595
0.19	0.0753	0.0718	0.0703	0.0678	0.0649	0.0629	0.0617	0.0612	0.0596	0.0577	0.0568
0.20	0.0728	0.0693	0.0677	0.0653	0.0624	0.0603	0.0592	0.0587	0.0570	0.0551	0.0542
0.21	0.0703	0.0668	0.0652	0.0628	0.0599	0.0579	0.0567	0.0562	0.0546	0.0527	0.0518
0.22	0.0681	0.0646	0.0630	0.0606	0.0577	0.0556	0.0545	0.0540	0.0524	0.0505	0.0496
0.23	0.0660	0.0625	0.0610	0.0585	0.0556	0.0536	0.0525	0.0520	0.0503	0.0484	0.0475
0.24	0.0642	0.0606	0.0591	0.0566	0.0538	0.0517	0.0506	0.0501	0.0484	0.0465	0.0457
0.25	0.0624	0.0589	0.0574	0.0549	0.0520	0.0500	0.0489	0.0483	0.0467	0.0448	0.0439
0.26	0.0607	0.0572	0.0557	0.0532	0.0504	0.0483	0.0472	0.0467	0.0451	0.0431	0.0423
0.27	0.0593	0.0558	0.0543	0.0518	0.0489	0.0469	0.0458	0.0452	0.0436	0.0417	0.0408
0.28	0.0580	0.0545	0.0530	0.0505	0.0476	0.0456	0.0445	0.0440	0.0423	0.0404	0.0395
0.29	0.0569	0.0534	0.0518	0.0494	0.0465	0.0444	0.0433	0.0428	0.0412	0.0393	0.0384
0.30	0.0559	0.0524	0.0508	0.0484	0.0455	0.0434	0.0423	0.0418	0.0402	0.0383	0.0374
0.31	0.0549	0.0514	0.0499	0.0474	0.0445	0.0425	0.0414	0.0409	0.0392	0.0373	0.0364
0.32	0.0541	0.0506	0.0491	0.0467	0.0438	0.0417	0.0406	0.0401	0.0385	0.0366	0.0357
0.33	0.0534	0.0499	0.0484	0.0460	0.0431	0.0411	0.0399	0.0394	0.0378	0.0359	0.0350
0.34	0.0529	0.0495	0.0479	0.0455	0.0426	0.0406	0.0395	0.0389	0.0373	0.0354	0.0345
0.35	0.0525	0.0490	0.0475	0.0450	0.0422	0.0401	0.0390	0.0385	0.0369	0.0350	0.0341
0.36	0.0521	0.0487	0.0471	0.0447	0.0418	0.0398	0.0387	0.0382	0.0366	0.0347	0.0338
0.37	0.0520	0.0485	0.0470	0.0446	0.0417	0.0397	0.0386	0.0380	0.0364	0.0345	0.0337
0.38	0.0519	0.0484	0.0469	0.0445	0.0416	0.0396	0.0385	0.0380	0.0363	0.0344	0.0336
0.39	0.0520	0.0485	0.0470	0.0445	0.0417	0.0397	0.0386	0.0380	0.0364	0.0345	0.0337
0.40	0.0521	0.0486	0.0471	0.0446	0.0418	0.0398	0.0387	0.0382	0.0365	0.0346	0.0338
0.41	0.0523	0.0488	0.0473	0.0449	0.0420	0.0400	0.0389	0.0384	0.0368	0.0349	0.0340
0.42	0.0526	0.0491	0.0476	0.0452	0.0423	0.0403	0.0392	0.0387	0.0371	0.0352	0.0343
0.43	0.0530	0.0495	0.0480	0.0456	0.0427	0.0407	0.0396	0.0391	0.0375	0.0356	0.0348
0.44	0.0534	0.0500	0.0485	0.0461	0.0432	0.0412	0.0401	0.0396	0.0380	0.0361	0.0353
0.45	0.0540	0.0506	0.0490	0.0466	0.0438	0.0418	0.0407	0.0402	0.0386	0.0367	0.0359
0.46	0.0546	0.0512	0.0497	0.0473	0.0445	0.0425	0.0414	0.0409	0.0393	0.0374	0.0365
0.47	0.0553	0.0519	0.0504	0.0479	0.0451	0.0431	0.0421	0.0415	0.0400	0.0381	0.0372
0.48	0.0561	0.0527	0.0512	0.0488	0.0459	0.0440	0.0429	0.0424	0.0408	0.0389	0.0380
0.49	0.0569	0.0535	0.0520	0.0496	0.0468	0.0448	0.0437	0.0432	0.0416	0.0397	0.0389
0.50	0.0578	0.0544	0.0529	0.0505	0.0477	0.0457	0.0446	0.0441	0.0426	0.0407	0.0398
0.51	0.0587	0.0553	0.0538	0.0514	0.0486	0.0467	0.0456	0.0451	0.0435	0.0417	0.0408
0.52	0.0598	0.0564	0.0549	0.0525	0.0497	0.0478	0.0467	0.0462	0.0446	0.0427	0.0419
0.53	0.0608	0.0574	0.0559	0.0535	0.0508	0.0488	0.0477	0.0472	0.0457	0.0438	0.0430
0.54	0.0617	0.0584	0.0569	0.0546	0.0518	0.0498	0.0488	0.0483	0.0467	0.0449	0.0440
0.55	0.0629	0.0595	0.0580	0.0557	0.0529	0.0510	0.0499	0.0494	0.0479	0.0460	0.0452
0.56	0.0639	0.0606	0.0591	0.0568	0.0540	0.0521	0.0511	0.0506	0.0490	0.0472	0.0464
0.57	0.0651	0.0618	0.0603	0.0580	0.0553	0.0534	0.0523	0.0518	0.0503	0.0484	0.0476
0.58	0.0663	0.0630	0.0615	0.0592	0.0565	0.0545	0.0535	0.0530	0.0515	0.0496	0.0488
0.59	0.0675	0.0642	0.0628	0.0605	0.0578	0.0558	0.0548	0.0543	0.0528	0.0510	0.0501
0.60	0.0687	0.0654	0.0640	0.0617	0.0590	0.0571	0.0560	0.0556	0.0540	0.0522	0.0514
0.61	0.0698	0.0666	0.0651	0.0628	0.0602	0.0583	0.0572	0.0567	0.0552	0.0534	0.0526
0.62	0.0711	0.0678	0.0664	0.0641	0.0614	0.0595	0.0585	0.0580	0.0565	0.0548	0.0539
0.63	0.0722	0.0690	0.0676	0.0653	0.0626	0.0608	0.0597	0.0593	0.0578	0.0560	0.0552
0.64	0.0733	0.0701	0.0687	0.0665	0.0638	0.0620	0.0610	0.0605	0.0590	0.0572	0.0564
0.65	0.0745	0.0713	0.0699	0.0677	0.0651	0.0632	0.0622	0.0617	0.0602	0.0585	0.0577
0.66	0.0756	0.0724	0.0710	0.0688	0.0662	0.0643	0.0633	0.0629	0.0614	0.0597	0.0589
0.67	0.0766	0.0735	0.0721	0.0699	0.0673	0.0655	0.0645	0.0640	0.0626	0.0608	0.0601
0.68	0.0777	0.0745	0.0732	0.0710	0.0684	0.0666	0.0656	0.0652	0.0637	0.0620	0.0612
0.69	0.0786	0.0755	0.0741	0.0720	0.0694	0.0676	0.0667	0.0662	0.0648	0.0631	0.0623
0.70	0.0795	0.0765	0.0751	0.0730	0.0704	0.0687	0.0677	0.0672	0.0658	0.0641	0.0634
0.71	0.0803	0.0773	0.0759	0.0738	0.0713	0.0695	0.0686	0.0681	0.0667	0.0651	0.0643
0.72	0.0811	0.0781	0.0768	0.0747	0.0722	0.0704	0.0695	0.0690	0.0676	0.0660	0.0652
0.73	0.0818	0.0789	0.0776	0.0755	0.0730	0.0713	0.0704	0.0699	0.0685	0.0669	0.0662
0.74	0.0825	0.0795	0.0782	0.0762	0.0738	0.0721	0.0711	0.0707	0.0693	0.0677	0.0670
0.75	0.0829	0.0800	0.0788	0.0767	0.0743	0.0726	0.0718	0.0713	0.0699	0.0683	0.0677

"b" VALUES FOR REYNOLDS NUMBER FACTOR, F_r—FLANGE TAPS

$$F_r = 1 + \frac{b}{\sqrt{h_w p_f}}$$

Internal Diameter of Pipe, *D*, inches

β	6				8			10			12		
	4.897	5.189	5.761	6.065	7.625	7.981	8.071	9.564	10.020	10.136	11.376	11.938	12.090
0.10	0.0845	0.0836	0.0821	0.0814	0.0784	0.0778	0.0777	0.0757	0.0752	0.0751	0.0739	0.0734	0.0733
0.11	0.0803	0.0795	0.0779	0.0772	0.0742	0.0736	0.0735	0.0716	0.0710	0.0709	0.0697	0.0692	0.0691
0.12	0.0763	0.0755	0.0739	0.0732	0.0702	0.0696	0.0695	0.0676	0.0670	0.0669	0.0657	0.0652	0.0651
0.13	0.0725	0.0717	0.0701	0.0694	0.0664	0.0658	0.0657	0.0638	0.0632	0.0631	0.0619	0.0614	0.0613
0.14	0.0689	0.0680	0.0665	0.0658	0.0628	0.0622	0.0621	0.0601	0.0596	0.0595	0.0583	0.0578	0.0577
0.15	0.0655	0.0646	0.0631	0.0624	0.0594	0.0588	0.0587	0.0567	0.0562	0.0561	0.0549	0.0544	0.0543
0.16	0.0623	0.0614	0.0599	0.0591	0.0561	0.0556	0.0554	0.0535	0.0530	0.0528	0.0516	0.0512	0.0510
0.17	0.0592	0.0583	0.0568	0.0561	0.0531	0.0525	0.0524	0.0504	0.0499	0.0498	0.0486	0.0481	0.0480
0.18	0.0563	0.0554	0.0539	0.0532	0.0502	0.0495	0.0495	0.0475	0.0470	0.0469	0.0457	0.0452	0.0451
0.19	0.0536	0.0527	0.0512	0.0505	0.0475	0.0469	0.0468	0.0448	0.0443	0.0442	0.0430	0.0425	0.0424
0.20	0.0511	0.0502	0.0487	0.0479	0.0449	0.0444	0.0442	0.0423	0.0418	0.0416	0.0404	0.0400	0.0398
0.21	0.0486	0.0477	0.0462	0.0455	0.0425	0.0419	0.0418	0.0398	0.0393	0.0392	0.0380	0.0375	0.0374
0.22	0.0464	0.0455	0.0440	0.0433	0.0403	0.0397	0.0396	0.0376	0.0371	0.0370	0.0358	0.0353	0.0352
0.23	0.0444	0.0435	0.0420	0.0412	0.0382	0.0377	0.0375	0.0356	0.0351	0.0350	0.0338	0.0333	0.0332
0.24	0.0425	0.0416	0.0401	0.0393	0.0364	0.0358	0.0357	0.0337	0.0332	0.0331	0.0319	0.0314	0.0313
0.25	0.0407	0.0399	0.0383	0.0376	0.0346	0.0341	0.0339	0.0320	0.0315	0.0313	0.0301	0.0297	0.0295
0.26	0.0391	0.0382	0.0367	0.0360	0.0330	0.0324	0.0323	0.0303	0.0298	0.0297	0.0285	0.0280	0.0279
0.27	0.0377	0.0368	0.0353	0.0345	0.0315	0.0310	0.0309	0.0289	0.0284	0.0283	0.0271	0.0266	0.0265
0.28	0.0364	0.0355	0.0340	0.0332	0.0303	0.0297	0.0296	0.0276	0.0271	0.0270	0.0258	0.0253	0.0252
0.29	0.0352	0.0343	0.0328	0.0321	0.0291	0.0286	0.0284	0.0265	0.0260	0.0258	0.0246	0.0242	0.0240
0.30	0.0342	0.0333	0.0318	0.0311	0.0281	0.0275	0.0274	0.0255	0.0249	0.0248	0.0236	0.0231	0.0230
0.31	0.0333	0.0324	0.0309	0.0302	0.0272	0.0266	0.0265	0.0245	0.0240	0.0239	0.0227	0.0222	0.0221
0.32	0.0325	0.0317	0.0301	0.0294	0.0264	0.0259	0.0257	0.0238	0.0233	0.0232	0.0220	0.0215	0.0214
0.33	0.0319	0.0310	0.0295	0.0288	0.0258	0.0252	0.0251	0.0231	0.0226	0.0225	0.0213	0.0208	0.0207
0.34	0.0314	0.0305	0.0290	0.0283	0.0253	0.0247	0.0246	0.0226	0.0221	0.0220	0.0208	0.0203	0.0202
0.35	0.0310	0.0301	0.0286	0.0278	0.0249	0.0243	0.0242	0.0222	0.0217	0.0216	0.0204	0.0199	0.0198
0.36	0.0306	0.0298	0.0283	0.0275	0.0246	0.0240	0.0239	0.0219	0.0214	0.0213	0.0201	0.0196	0.0195
0.37	0.0305	0.0296	0.0281	0.0274	0.0244	0.0239	0.0237	0.0218	0.0213	0.0212	0.0200	0.0195	0.0194
0.38	0.0304	0.0285	0.0280	0.0273	0.0244	0.0238	0.0237	0.0217	0.0212	0.0211	0.0199	0.0195	0.0193
0.39	0.0305	0.0296	0.0281	0.0274	0.0244	0.0239	0.0238	0.0218	0.0213	0.0212	0.0200	0.0195	0.0194
0.40	0.0306	0.0296	0.0283	0.0276	0.0246	0.0240	0.0239	0.0220	0.0215	0.0213	0.0202	0.0197	0.0196
0.41	0.0309	0.0300	0.0285	0.0278	0.0248	0.0243	0.0241	0.0222	0.0217	0.0216	0.0204	0.0199	0.0198
0.42	0.0312	0.0303	0.0288	0.0281	0.0252	0.0246	0.0245	0.0226	0.0221	0.0219	0.0208	0.0203	0.0202
0.43	0.0316	0.0308	0.0293	0.0286	0.0256	0.0251	0.0249	0.0230	0.0225	0.0224	0.0212	0.0207	0.0206
0.44	0.0321	0.0313	0.0298	0.0291	0.0261	0.0256	0.0254	0.0235	0.0230	0.0229	0.0217	0.0213	0.0211
0.45	0.0327	0.0319	0.0304	0.0297	0.0267	0.0262	0.0261	0.0241	0.0236	0.0235	0.0223	0.0219	0.0217
0.46	0.0334	0.0326	0.0311	0.0304	0.0274	0.0269	0.0267	0.0248	0.0243	0.0242	0.0230	0.0226	0.0224
0.47	0.0341	0.0333	0.0318	0.0311	0.0282	0.0276	0.0275	0.0256	0.0251	0.0249	0.0238	0.0233	0.0232
0.48	0.0350	0.0341	0.0326	0.0319	0.0290	0.0284	0.0283	0.0264	0.0259	0.0258	0.0246	0.0242	0.0240
0.49	0.0358	0.0349	0.0335	0.0328	0.0299	0.0293	0.0292	0.0273	0.0268	0.0267	0.0255	0.0250	0.0249
0.50	0.0368	0.0359	0.0344	0.0337	0.0308	0.0303	0.0302	0.0283	0.0278	0.0276	0.0265	0.0260	0.0259
0.51	0.0378	0.0369	0.0354	0.0347	0.0318	0.0313	0.0312	0.0293	0.0288	0.0287	0.0275	0.0270	0.0269
0.52	0.0388	0.0380	0.0365	0.0358	0.0329	0.0324	0.0323	0.0304	0.0299	0.0298	0.0286	0.0281	0.0280
0.53	0.0399	0.0391	0.0376	0.0369	0.0340	0.0335	0.0334	0.0315	0.0310	0.0309	0.0297	0.0293	0.0291
0.54	0.0410	0.0402	0.0387	0.0380	0.0351	0.0346	0.0345	0.0326	0.0321	0.0320	0.0308	0.0304	0.0303
0.55	0.0422	0.0413	0.0399	0.0392	0.0363	0.0358	0.0357	0.0338	0.0333	0.0332	0.0321	0.0316	0.0315
0.56	0.0433	0.0425	0.0411	0.0404	0.0375	0.0370	0.0369	0.0350	0.0345	0.0344	0.0333	0.0328	0.0327
0.57	0.0446	0.0438	0.0423	0.0416	0.0388	0.0383	0.0381	0.0363	0.0358	0.0357	0.0346	0.0341	0.0340
0.58	0.0458	0.0450	0.0436	0.0429	0.0401	0.0395	0.0394	0.0376	0.0371	0.0370	0.0358	0.0354	0.0353
0.59	0.0472	0.0463	0.0449	0.0442	0.0414	0.0409	0.0408	0.0389	0.0384	0.0383	0.0372	0.0367	0.0366
0.60	0.0484	0.0476	0.0462	0.0455	0.0427	0.0422	0.0421	0.0402	0.0398	0.0397	0.0385	0.0381	0.0380
0.61	0.0497	0.0489	0.0474	0.0468	0.0440	0.0435	0.0433	0.0415	0.0411	0.0409	0.0398	0.0394	0.0393
0.62	0.0510	0.0502	0.0488	0.0481	0.0402	0.0448	0.0447	0.0429	0.0424	0.0423	0.0412	0.0407	0.0406
0.63	0.0523	0.0515	0.0501	0.0494	0.0466	0.0461	0.0460	0.0442	0.0437	0.0436	0.0425	0.0421	0.0420
0.64	0.0535	0.0527	0.0513	0.0507	0.0479	0.0474	0.0473	0.0455	0.0451	0.0449	0.0438	0.0434	0.0433
0.65	0.0548	0.0540	0.0526	0.0520	0.0492	0.0487	0.0486	0.0468	0.0464	0.0463	0.0452	0.0447	0.0446
0.66	0.0560	0.0552	0.0538	0.0532	0.0505	0.0500	0.0499	0.0481	0.0476	0.0475	0.0465	0.0460	0.0459
0.67	0.0572	0.0564	0.0551	0.0544	0.0517	0.0512	0.0511	0.0494	0.0489	0.0488	0.0477	0.0473	0.0472
0.68	0.0584	0.0576	0.0563	0.0556	0.0530	0.0525	0.0523	0.0506	0.0502	0.0500	0.0490	0.0486	0.0484
0.69	0.0595	0.0587	0.0574	0.0567	0.0541	0.0536	0.0535	0.0518	0.0513	0.0512	0.0502	0.0497	0.0496
0.70	0.0606	0.0598	0.0585	0.0579	0.0552	0.0548	0.0546	0.0529	0.0525	0.0524	0.0513	0.0509	0.0508
0.71	0.0616	0.0608	0.0595	0.0589	0.0563	0.0558	0.0557	0.0540	0.0535	0.0534	0.0524	0.0520	0.0519
0.72	0.0625	0.0618	0.0605	0.0599	0.0573	0.0568	0.0567	0.0550	0.0546	0.0545	0.0535	0.0530	0.0529
0.73	0.0635	0.0627	0.0614	0.0608	0.0583	0.0578	0.0577	0.0560	0.0556	0.0555	0.0545	0.0541	0.0540
0.74	0.0643	0.0636	0.0623	0.0617	0.0592	0.0587	0.0586	0.0570	0.0565	0.0564	0.0554	0.0550	0.0549
0.75	0.0650	0.0643	0.0630	0.0624	0.0599	0.0594	0.0594	0.0577	0.0573	0.0572	0.0562	0.0558	0.0557

"b" VALUES FOR REYNOLDS NUMBER FACTOR, F_r—FLANGE TAPS

$$F_r = 1 + \frac{b}{\sqrt{h_w p_f}}$$

β	Internal Diameter of Pipe, D, inches											
	16			20			24			30		
	14.688	15.000	15.250	18.814	19.000	19.250	22.626	23.000	23.250	28.628	29.000	29.250
0.10	0.0715	0.0713	0.0711	0.0694	0.0693	0.0692	0.0680	0.0679	0.0678	0.0664	0.0663	0.0663
0.11	0.0673	0.0671	0.0670	0.0652	0.0651	0.0650	0.0638	0.0637	0.0636	0.0622	0.0621	0.0621
0.12	0.0633	0.0631	0.0630	0.0612	0.0611	0.0610	0.0598	0.0597	0.0596	0.0582	0.0581	0.0581
0.13	0.0595	0.0593	0.0592	0.0574	0.0573	0.0572	0.0560	0.0559	0.0558	0.0544	0.0543	0.0543
0.14	0.0559	0.0557	0.0555	0.0538	0.0537	0.0536	0.0524	0.0523	0.0522	0.0508	0.0507	0.0507
0.15	0.0525	0.0523	0.0521	0.0504	0.0503	0.0502	0.0490	0.0489	0.0488	0.0474	0.0473	0.0473
0.16	0.0492	0.0490	0.0489	0.0471	0.0471	0.0459	0.0457	0.0456	0.0455	0.0441	0.0441	0.0440
0.17	0.0461	0.0460	0.0458	0.0441	0.0440	0.0439	0.0427	0.0426	0.0425	0.0411	0.0410	0.0409
0.18	0.0433	0.0431	0.0429	0.0412	0.0411	0.0410	0.0398	0.0397	0.0396	0.0382	0.0381	0.0381
0.19	0.0406	0.0404	0.0402	0.0385	0.0384	0.0383	0.0371	0.0370	0.0369	0.0355	0.0354	0.0354
0.20	0.0380	0.0378	0.0377	0.0359	0.0359	0.0358	0.0345	0.0344	0.0344	0.0330	0.0329	0.0328
0.21	0.0356	0.0354	0.0353	0.0335	0.0334	0.0333	0.0321	0.0320	0.0319	0.0305	0.0305	0.0304
0.22	0.0334	0.0332	0.0330	0.0313	0.0312	0.0311	0.0299	0.0298	0.0297	0.0283	0.0282	0.0282
0.23	0.0313	0.0311	0.0310	0.0293	0.0292	0.0291	0.0279	0.0278	0.0277	0.0263	0.0262	0.0261
0.24	0.0294	0.0293	0.0291	0.0274	0.0273	0.0272	0.0260	0.0259	0.0258	0.0244	0.0243	0.0243
0.25	0.0277	0.0275	0.0274	0.0256	0.0256	0.0255	0.0242	0.0241	0.0241	0.0227	0.0226	0.0225
0.26	0.0261	0.0259	0.0258	0.0240	0.0239	0.0238	0.0226	0.0225	0.0224	0.0210	0.0210	0.0209
0.27	0.0247	0.0245	0.0243	0.0226	0.0225	0.0224	0.0212	0.0211	0.0210	0.0196	0.0195	0.0195
0.28	0.0234	0.0232	0.0230	0.0213	0.0212	0.0211	0.0199	0.0198	0.0197	0.0183	0.0182	0.0182
0.29	0.0222	0.0220	0.0219	0.0201	0.0201	0.0200	0.0188	0.0186	0.0186	0.0172	0.0171	0.0170
0.30	0.0212	0.0210	0.0209	0.0191	0.0191	0.0190	0.0177	0.0176	0.0176	0.0162	0.0161	0.0160
0.31	0.0203	0.0201	0.0200	0.0182	0.0182	0.0180	0.0168	0.0167	0.0167	0.0153	0.0152	0.0151
0.32	0.0195	0.0194	0.0192	0.0175	0.0174	0.0173	0.0161	0.0160	0.0159	0.0145	0.0144	0.0144
0.33	0.0189	0.0187	0.0186	0.0168	0.0168	0.0167	0.0155	0.0153	0.0153	0.0139	0.0138	0.0137
0.34	0.0184	0.0182	0.0181	0.0163	0.0163	0.0162	0.0150	0.0148	0.0148	0.0134	0.0133	0.0132
0.35	0.0180	0.0178	0.0177	0.0159	0.0159	0.0158	0.0146	0.0144	0.0144	0.0130	0.0129	0.0128
0.36	0.0177	0.0175	0.0174	0.0157	0.0156	0.0155	0.0143	0.0142	0.0141	0.0127	0.0126	0.0126
0.37	0.0176	0.0174	0.0173	0.0155	0.0155	0.0154	0.0142	0.0140	0.0140	0.0126	0.0125	0.0124
0.38	0.0175	0.0173	0.0172	0.0155	0.0154	0.0153	0.0141	0.0140	0.0139	0.0125	0.0124	0.0124
0.39	0.0176	0.0174	0.0173	0.0156	0.0155	0.0154	0.0142	0.0141	0.0140	0.0126	0.0125	0.0125
0.40	0.0178	0.0176	0.0174	0.0157	0.0156	0.0155	0.0143	0.0142	0.0141	0.0128	0.0127	0.0126
0.41	0.0180	0.0178	0.0177	0.0160	0.0159	0.0158	0.0146	0.0145	0.0144	0.0130	0.0129	0.0129
0.42	0.0184	0.0182	0.0180	0.0163	0.0162	0.0161	0.0150	0.0148	0.0148	0.0134	0.0133	0.0133
0.43	0.0188	0.0186	0.0185	0.0168	0.0167	0.0166	0.0154	0.0153	0.0152	0.0138	0.0138	0.0137
0.44	0.0193	0.0192	0.0190	0.0173	0.0172	0.0171	0.0159	0.0158	0.0158	0.0144	0.0143	0.0142
0.45	0.0200	0.0198	0.0196	0.0179	0.0179	0.0178	0.0166	0.0165	0.0164	0.0150	0.0149	0.0149
0.46	0.0207	0.0205	0.0203	0.0186	0.0186	0.0185	0.0173	0.0172	0.0171	0.0157	0.0156	0.0156
0.47	0.0214	0.0212	0.0211	0.0194	0.0193	0.0192	0.0180	0.0179	0.0178	0.0165	0.0164	0.0163
0.48	0.0223	0.0221	0.0219	0.0202	0.0202	0.0201	0.0189	0.0188	0.0187	0.0173	0.0173	0.0172
0.49	0.0231	0.0230	0.0228	0.0211	0.0211	0.0210	0.0198	0.0197	0.0196	0.0182	0.0182	0.0181
0.50	0.0241	0.0240	0.0238	0.0221	0.0220	0.0220	0.0208	0.0207	0.0206	0.0192	0.0192	0.0191
0.51	0.0252	0.0250	0.0248	0.0231	0.0231	0.0230	0.0218	0.0217	0.0216	0.0203	0.0202	0.0201
0.52	0.0263	0.0261	0.0259	0.0243	0.0242	0.0241	0.0229	0.0228	0.0227	0.0214	0.0213	0.0213
0.53	0.0274	0.0272	0.0271	0.0254	0.0253	0.0252	0.0241	0.0240	0.0239	0.0225	0.0225	0.0224
0.54	0.0285	0.0283	0.0282	0.0265	0.0265	0.0264	0.0252	0.0251	0.0250	0.0237	0.0236	0.0236
0.55	0.0297	0.0296	0.0294	0.0278	0.0277	0.0276	0.0264	0.0263	0.0263	0.0249	0.0248	0.0248
0.56	0.0310	0.0308	0.0306	0.0290	0.0289	0.0288	0.0277	0.0276	0.0275	0.0262	0.0261	0.0260
0.57	0.0323	0.0321	0.0319	0.0303	0.0302	0.0301	0.0290	0.0289	0.0288	0.0275	0.0274	0.0274
0.58	0.0335	0.0334	0.0332	0.0316	0.0315	0.0314	0.0303	0.0302	0.0301	0.0288	0.0287	0.0287
0.59	0.0349	0.0347	0.0346	0.0330	0.0329	0.0328	0.0317	0.0316	0.0315	0.0302	0.0301	0.0300
0.60	0.0363	0.0361	0.0359	0.0343	0.0342	0.0342	0.0330	0.0329	0.0328	0.0315	0.0315	0.0314
0.61	0.0376	0.0374	0.0373	0.0356	0.0356	0.0355	0.0343	0.0342	0.0342	0.0329	0.0328	0.0327
0.62	0.0389	0.0388	0.0386	0.0370	0.0370	0.0369	0.0357	0.0356	0.0356	0.0343	0.0342	0.0341
0.63	0.0403	0.0401	0.0400	0.0384	0.0383	0.0382	0.0371	0.0370	0.0369	0.0356	0.0356	0.0355
0.64	0.0416	0.0415	0.0413	0.0397	0.0397	0.0396	0.0384	0.0384	0.0383	0.0370	0.0369	0.0369
0.65	0.0430	0.0428	0.0427	0.0411	0.0410	0.0409	0.0398	0.0397	0.0397	0.0384	0.0383	0.0383
0.66	0.0443	0.0441	0.0440	0.0424	0.0423	0.0422	0.0411	0.0410	0.0410	0.0397	0.0396	0.0396
0.67	0.0456	0.0454	0.0452	0.0437	0.0436	0.0435	0.0425	0.0424	0.0423	0.0410	0.0410	0.0409
0.68	0.0468	0.0467	0.0465	0.0450	0.0449	0.0448	0.0438	0.0436	0.0436	0.0423	0.0423	0.0422
0.69	0.0480	0.0479	0.0477	0.0462	0.0461	0.0460	0.0450	0.0449	0.0448	0.0436	0.0435	0.0435
0.70	0.0492	0.0491	0.0489	0.0474	0.0473	0.0472	0.0462	0.0461	0.0460	0.0448	0.0447	0.0447
0.71	0.0503	0.0501	0.0500	0.0485	0.0484	0.0484	0.0473	0.0472	0.0471	0.0459	0.0459	0.0458
0.72	0.0514	0.0512	0.0511	0.0496	0.0495	0.0495	0.0484	0.0483	0.0483	0.0471	0.0470	0.0469
0.73	0.0524	0.0523	0.0522	0.0507	0.0506	0.0505	0.0495	0.0494	0.0493	0.0482	0.0481	0.0480
0.74	0.0534	0.0532	0.0531	0.0517	0.0516	0.0515	0.0505	0.0504	0.0503	0.0492	0.0491	0.0490
0.75	0.0542	0.0540	0.0540	0.0525	0.0522	0.0523	0.0513	0.0512	0.0512	0.0500	0.0499	0.0499

EXPANSION FACTORS—FLANGE TAPS—Y_1

Static Pressure Taken from Upstream Taps

$\frac{h_w}{p_{f1}}$ Ratio	$\beta = \frac{d}{D}$ Ratio												
	.1	.2	.3	.4	.45	.50	.52	.54	.56	.58	.60	.61	.62
0.0	1.0000	1.0000	1.0000	1.0000	1.0000	1.0000	1.0000	1.0000	1.0000	1.0000	1.0000	1.0000	1.0000
0.1	.9989	.9989	.9989	.9988	.9988	.9988	.9988	.9988	.9988	.9988	.9987	.9987	.9987
0.2	.9977	.9977	.9977	.9977	.9976	.9976	.9976	.9976	.9975	.9975	.9975	.9975	.9974
0.3	.9966	.9966	.9966	.9955	.9965	.9964	.9964	.9963	.9963	.9963	.9962	.9962	.9962
0.4	.9954	.9954	.9954	.9953	.9953	.9952	.9952	.9951	.9951	.9950	.9949	.9949	.9949
0.5	.9943	.9943	.9943	.9942	.9941	.9940	.9940	.9939	.9938	.9938	.9937	.9936	.9936
0.6	.9932	.9932	.9931	.9930	.9929	.9928	.9927	.9927	.9926	.9925	.9924	.9924	.9923
0.7	.9920	.9920	.9920	.9919	.9918	.9916	.9915	.9915	.9914	.9913	.9912	.9911	.9910
0.8	.9909	.9909	.9908	.9907	.9906	.9904	.9903	.9902	.9901	.9900	.9899	.9898	.9897
0.9	.9898	.9897	.9897	.9895	.9894	.9892	.9891	.9890	.9889	.9888	.9886	.9885	.9885
1.0	.9886	.9886	.9885	.9884	.9882	.9880	.9879	.9878	.9877	.9875	.9874	.9873	.9872
1.1	.9875	.9875	.9874	.9872	.9870	.9868	.9867	.9866	.9864	.9863	.9861	.9860	.9859
1.2	.9863	9863	.9862	.9860	.9859	.9856	.9855	.9853	.9852	.9850	.9848	.9847	.9846
1.3	.9852	.9852	.9851	.9849	.9847	.9844	.9843	.9841	.9840	.9838	.9836	.9835	.9833
1.4	.9841	.9840	.9840	.9837	.9835	.9832	.9831	.9829	.9827	.9825	.9823	.9822	.9821
1.5	.9829	.9829	.9828	.9826	.9823	.9820	.9819	.9817	.9815	.9813	.9810	.9809	.9808
1.6	.9818	.9818	.9817	.9814	.9811	.9808	.9806	.9805	.9803	.9800	.9798	.9796	.9795
1.7	.9806	.9806	.9805	.9802	.9800	.9796	.9794	.9792	.9790	.9788	.9785	.9784	.9782
1.8	.9795	.9795	.9794	.9791	.9788	.9784	.9782	.9780	.9778	.9775	.9772	.9771	.9769
1.9	.9784	.9783	.9782	.9779	.9776	.9772	.9770	.9768	.9766	.9763	.9760	.9758	.9756
2.0	.9772	.9772	.9771	.9767	.9764	.9760	.9758	.9756	.9753	.9750	.9747	.9745	.9744
2.1	.9761	.9761	.9759	.9756	.9753	.9748	.9746	.9744	.9741	.9738	.9734	.9733	.9731
2.2	.9750	.9749	.9748	.9744	.9741	.9736	.9734	.9731	.9729	.9725	.9722	.9720	.9718
2.3	.9738	.9738	.9736	.9732	.9729	.9724	.9722	.9719	.9716	.9713	.9709	.9707	.9705
2.4	.9727	.9726	.9725	.9721	.9717	.9712	.9710	.9707	.9704	.9700	.9697	.9694	.9692
2.5	.9715	.9715	.9713	.9709	.9705	.9700	.9698	.9695	.9692	.9688	.9684	.9582	.9680
2.6	.9704	.9704	.9702	.9698	.9594	.9588	.9686	.9683	.9679	.9675	.9671	.9669	.9667
2.7	.9593	.9592	.9691	.9686	.9582	.9676	.9573	.9670	.9667	.9663	.9659	.9656	.9654
2.8	.9681	.9681	.9679	.9674	.9670	.9564	.9661	.9658	.9654	.9650	.9646	.9644	.9641
2.9	.9670	.9669	.9568	.9663	.9658	.9652	.9649	.9546	.9642	.9638	.9633	.9631	.9628
3.0	.9658	.9558	.9656	.9651	.9647	.9540	.9637	.9534	.9630	.9626	.9621	.9618	.9615
3.1	.9647	.9647	.9645	.9639	.9635	.9528	.9625	.9522	.9617	.9613	.9608	.9605	.9603
3.2	.9636	.9635	.9633	.9628	.9623	.9616	.9613	.9609	.9605	.9601	.9595	.9593	.9590
3.3	.9624	.9624	.9622	.9616	.9611	.9504	.9501	.9597	.9593	.9588	.9583	.9580	.9577
3.4	.9613	.9612	.9610	.9604	.9599	.9592	.9589	.9585	.9580	.9576	.9570	.9567	.9564
3.5	.9602	.9601	.9599	.9593	.9588	.9580	.9577	.9573	.9568	.9563	.9558	.9554	.9551
3.6	.9590	.9590	.9587	.9581	.9576	.9558	.9565	.9550	.9556	.9551	.9545	.9542	.9538
3.7	.9579	.9578	.9576	.9570	.9564	.9555	.9553	.9548	.9543	.9538	.9532	.9529	.9526
3.8	.9557	.9567	.9564	.9558	.9552	.9544	.9540	.9536	.9531	.9526	.9520	.9516	.9513
3.9	.9556	.9555	.9553	.9546	.9540	.9532	.9528	.9524	.9519	.9513	.9507	.9504	.9500
4.0	.9545	.9544	.9542	.9535	.9529	.9520	.9516	.9512	.9506	.9501	.9494	.9491	.9487

EXPANSION FACTORS—FLANGE TAPS—Y_1
Static Pressure Taken from Upstream Taps

$\frac{h_w}{p_{f1}}$ Ratio	$\beta = \frac{d}{D}$ Ratio .63	.64	.65	.66	.67	.68	.69	.70	.71	.72	.73	.74	.75
0.0	1.0000	1.0000	1.0000	1.0000	1.0000	1.0000	1.0000	1.0000	1.0000	1.0000	1.0000	1.0000	.10000
0.1	.9987	.9987	.9987	.9987	.9987	.9987	.9986	.9986	.9986	.9986	.9986	.9986	.9986
0.2	.9974	.9974	.9974	.9974	.9973	.9973	.9973	.9973	.9972	.9972	.9972	.9971	.9971
0.3	.9961	.9961	.9961	.9960	.9960	.9960	.9959	.9959	.9958	.9958	.9958	.9957	.9957
0.4	.9948	.9948	.9948	.9947	.9947	.9946	.9946	.9945	.9945	.9944	.9943	.9943	.9942
0.5	.9935	.9935	.9934	.9934	.9933	.9933	.9932	.9931	.9931	.9930	.9929	.9929	.9928
0.6	.9923	.9922	.9921	.9921	.9920	.9919	.9918	.9918	.9917	.9916	.9915	.9914	.9913
0.7	.9910	.9909	.9908	.9907	.9907	.9906	.9905	.9904	.9903	.9902	.9901	.9900	.9899
0.8	.9897	.9896	.9895	.9894	.9893	.9892	.9891	.9890	.9889	.9888	.9887	.9886	.9884
0.9	.9884	.9883	.9882	.9881	.9880	.9879	.9878	.9877	.9875	.9874	.9873	.9871	.9870
1.0	.9871	.9870	.9869	.9868	.9867	.9865	.9864	.9863	.9861	.9860	.9859	.9857	.9855
1.1	.9858	.9857	.9856	.9854	.9853	.9852	.9851	.9849	.9848	.9846	.9844	.9843	.9841
1.2	.9845	.9844	.9843	.9841	.9840	.9838	.9837	.9835	.9834	.9832	.9830	.9828	.9826
1.3	.9832	.9831	.9829	.9828	.9827	.9825	.9823	.9822	.9820	.9818	.9816	.9814	.9812
1.4	.9819	.9818	.9816	.9815	.9813	.9812	.9810	.9808	.9806	.9804	.9802	.9800	.9798
1.5	.9806	.9805	.9803	.9802	.9800	.9798	.9796	.9794	.9792	.9790	.9788	.9786	.9783
1.6	.9793	.9792	.9790	.9788	.9787	.9785	.9783	.9781	.9778	.9776	.9774	.9771	.9769
1.7	.9780	.9779	.9777	.9775	.9773	.9771	.9769	.9767	.9764	.9762	.9760	.9757	.9754
1.8	.9768	.9766	.9764	.9762	.9760	.9758	.9755	.9753	.9751	.9748	.9745	.9743	.9740
1.9	.9755	.9753	.9751	.9749	.9747	.9744	.9742	.9739	.9737	.9734	.9731	.9728	.9725
2.0	.9742	.9740	.9738	.9735	.9733	.9731	.9728	.9726	.9723	.9720	.9717	.9714	.9711
2.1	.9729	.9727	.9725	.9722	.9720	.9717	.9715	.9712	.9709	.9706	.9703	.9700	.9696
2.2	.9716	.9714	.9711	.9709	.9706	.9704	.9701	.9698	.9695	.9692	.9689	.9685	.9682
2.3	.9703	.9701	.9698	.9696	.9693	.9690	.9688	.9685	.9681	.9678	.9675	.9671	.9667
2.4	.9690	.9688	.9685	.9683	.9680	.9677	.9674	9671	.9668	.9664	.9661	.9657	.9653
2.5	.9677	.9675	.9672	.9669	.9666	.9663	.9660	.9657	.9654	.9650	.9646	.9643	.9639
2.6	.9664	.9662	.9659	.9656	.9653	.9650	.9647	.9643	.9640	.9636	.9632	.9628	.9624
2.7	.9651	.9649	.9646	.9643	.9640	.9637	.9633	.9630	.9626	.9622	.9618	.9614	.9610
2.8	.9638	.9636	.9633	.9630	.9626	.9623	.9620	.9616	.9612	.9608	.9604	.9600	.9595
2.9	.9625	.9623	.9620	.9616	.9613	.9610	.9606	.9602	.9598	.9594	.9590	.9585	.9581
3.0	.9613	.9610	.9606	.9603	.9600	.9596	.9592	.9588	.9584	.9580	.9576	.9571	.9566
3.1	.9600	.9597	.9593	.9590	.9586	.9583	.9579	.9575	.9571	.9566	.9562	.9557	.9552
3.2	.9587	.9584	.9580	.9577	.9573	.9569	.9565	.9561	.9557	.9552	.9547	.9542	.9537
3.3	.9574	.9571	.9567	.9564	.9560	.9555	.9552	.9547	.9543	.9538	.9533	.9528	.9523
3.4	.9561	.9558	.9554	.9550	.9546	.9542	.9538	.9534	.9529	.9524	.9519	.9514	.9508
3.5	.9548	.9545	.9541	.9537	.9533	.9529	.9524	.9520	.9515	.9510	.9505	.9500	.9494
3.6	.9535	.9532	.9528	.9524	.9520	.9515	.9511	.9506	.9501	.9496	.9491	.9485	.9480
3.7	.9522	.9518	.9515	.9511	.9506	.9502	.9497	.9492	.9487	.9482	.9477	.9471	.9465
3.8	.9509	.9505	.9502	.9497	.9493	.9488	.9484	.9479	.9474	.9468	.9463	.9457	.9451
3.9	.9496	.9492	.9488	.9484	.9480	.9475	.9470	.9465	.9460	.9454	.9448	.9442	.9436
4.0	.9483	.9479	.9475	.9471	.9466	.9462	.9457	.9451	.9446	.9440	.9434	.9428	.9422

EXPANSION FACTORS—FLANGE TAPS—Y_2
Static Pressure Taken from Downstream Taps

$\frac{h_w}{p_{f2}}$ Ratio	$\beta = \frac{d}{D}$ Ratio .1	.2	.3	.4	.45	.50	.52	.54	.56	.58	.60	.61	.62
0.0	1.0000	1.0000	1.0000	1.0000	1.0000	1.0000	1.0000	1.0000	1.0000	1.0000	1.0000	1.0000	1.0000
0.1	1.0007	1.0007	1.0006	1.0006	1.0006	1.0006	1.0006	1.0006	1.0006	1.0006	1.0005	1.0005	1.0005
0.2	1.0013	1.0013	1.0013	1.0013	1.0012	1.0012	1.0012	1.0012	1.0011	1.0011	1.0011	1.0011	1.0010
0.3	1.0020	1.0020	1.0020	1.0019	1.0019	1.0018	1.0018	1.0018	1.0017	1.0017	1.0016	1.0016	1.0016
0.4	1.0027	1.0027	1.0026	1.0026	1.0025	1.0024	1.0024	1.0023	1.0023	1.0022	1.0022	1.0021	1.0021
0.5	1.0033	1.0033	1.0033	1.0032	1.0031	1.0030	1.0030	1.0029	1.0029	1.0028	1.0027	1.0027	1.0026
0.6	1.0040	1.0040	1.0040	1.0039	1.0038	1.0036	1.0036	1.0035	1.0034	1.0034	1.0033	1.0032	1.0032
0.7	1.0047	1.0047	1.0046	1.0045	1.0044	1.0043	1.0042	1.0041	1.0040	1.0039	1.0038	1.0038	1.0037
0.8	1.0054	1.0053	1.0053	1.0052	1.0050	1.0049	1.0048	1.0047	1.0046	1.0045	1.0044	1.0043	1.0042
0.9	1.0060	1.0060	1.0060	1.0058	1.0057	1.0055	1.0054	1.0053	1.0052	1.0050	1.0049	1.0048	1.0048
1.0	1.0067	1.0067	1.0066	1.0065	1.0063	1.0061	1.0060	1.0059	1.0058	1.0056	1.0055	1.0054	1.0053
1.1	1.0074	1.0074	1.0073	1.0071	1.0069	1.0067	1.0066	1.0065	1.0063	1.0062	1.0060	1.0059	1.0058
1.2	1.0080	1.0080	1.0080	1.0078	1.0076	1.0073	1.0072	1.0071	1.0069	1.0068	1.0066	1.0065	1.0064
1.3	1.0087	1.0087	1.0086	1.0084	1.0082	1.0080	1.0078	1.0077	1.0075	1.0073	1.0071	1.0070	1.0069
1.4	1.0094	1.0094	1.0093	1.0091	1.0089	1.0086	1.0084	1.0083	1.0081	1.0079	1.0077	1.0076	1.0074
1.5	1.0101	1.0101	1.0100	1.0097	1.0095	1.0092	1.0090	1.0089	1.0087	1.0085	1.0082	1.0081	1.0080
1.6	1.0108	1.0107	1.0106	1.0104	1.0101	1.0098	1.0096	1.0095	1.0093	1.0090	1.0088	1.0087	1.0085
1.7	1.0114	1.0114	1.0113	1.0110	1.0108	1.0104	1.0103	1.0101	1.0099	1.0096	1.0094	1.0092	1.0091
1.8	1.0121	1.0121	1.0120	1.0117	1.0114	1.0111	1.0109	1.0107	1.0104	1.0102	1.0099	1.0098	1.0096
1.9	1.0128	1.0128	1.0126	1.0123	1.0121	1.0117	1.0115	1.0113	1.0110	1.0108	1.0105	1.0103	1.0102
2.0	1.0135	1.0134	1.0133	1.0130	1.0127	1.0123	1.0121	1.0119	1.0116	1.0114	1.0110	1.0109	1.0107
2.1	1.0142	1.0141	1.0140	1.0136	1.0134	1.0129	1.0127	1.0125	1.0122	1.0119	1.0116	1.0114	1.0112
2.2	1.0148	1.0148	1.0147	1.0143	1.0140	1.0136	1.0133	1.0131	1.0128	1.0125	1.0122	1.0120	1.0118
2.3	1.0155	1.0155	1.0154	1.0150	1.0146	1.0142	1.0140	1.0137	1.0134	1.0131	1.0127	1.0126	1.0124
2.4	1.0162	1.0162	1.0160	1.0156	1.0153	1.0148	1.0146	1.0143	1.0140	1.0137	1.0133	1.0131	1.0129
2.5	1.0169	1.0168	1.0167	1.0163	1.0159	1.0154	1.0152	1.0149	1.0146	1.0142	1.0139	1.0137	1.0134
2.6	1.0176	1.0175	1.0174	1.0170	1.0166	1.0161	1.0158	1.0155	1.0152	1.0148	1.0144	1.0142	1.0140
2.7	1.0182	1.0182	1.0180	1.0176	1.0172	1.0167	1.0164	1.0161	1.0158	1.0154	1.0150	1.0148	1.0146
2.8	1.0189	1.0189	1.0187	1.0183	1.0179	1.0173	1.0170	1.0167	1.0164	1.0160	1.0156	1.0154	1.0151
2.9	1.0196	1.0196	1.0194	1.0189	1.0185	1.0180	1.0177	1.0173	1.0170	1.0166	1.0162	1.0159	1.0157
3.0	1.0203	1.0203	1.0201	1.0196	1.0192	1.0186	1.0183	1.0180	1.0176	1.0172	1.0167	1.0165	1.0162
3.1	1.0210	1.0210	1.0208	1.0203	1.0198	1.0192	1.0189	1.0186	1.0182	1.0178	1.0173	1.0170	1.0168
3.2	1.0217	1.0216	1.0214	1.0209	1.0205	1.0198	1.0195	1.0192	1.0188	1.0184	1.0179	1.0176	1.0173
3.3	1.0224	1.0223	1.0221	1.0216	1.0211	1.0205	1.0202	1.0198	1.0194	1.0189	1.0184	1.0182	1.0179
3.4	1.0230	1.0230	1.0228	1.0223	1.0218	1.0211	1.0208	1.0204	1.0200	1.0195	1.0190	1.0187	1.0184
3.5	1.0237	1.0237	1.0235	1.0229	1.0224	1.0217	1.0214	1.0210	1.0206	1.0201	1.0196	1.0193	1.0190
3.6	1.0244	1.0244	1.0242	1.0236	1.0231	1.0224	1.0220	1.0216	1.0212	1.0207	1.0202	1.0199	1.0196
3.7	1.0251	1.0251	1.0248	1.0243	1.0237	1.0230	1.0226	1.0222	1.0218	1.0213	1.0207	1.0204	1.0201
3.8	1.0258	1.0258	1.0255	1.0249	1.0244	1.0236	1.0233	1.0229	1.0224	1.0219	1.0213	1.0210	1.0207
3.9	1.0265	1.0264	1.0262	1.0256	1.0250	1.0243	1.0239	1.0235	1.0230	1.0225	1.0219	1.0216	1.0213
4.0	1.0272	1.0271	1.0269	1.0263	1.0257	1.0249	1.0245	1.0241	1.0236	1.0231	1.0225	1.0222	1.0218

EXPANSION FACTORS—FLANGE TAPS—Y_2
Static Pressure Taken from Downstream Taps

$\frac{h_w}{p_{f2}}$ Ratio	$\beta = \frac{d}{D}$ Ratio												
	.63	.64	.65	.66	.67	.68	.69	.70	.71	.72	.73	.74	.75
0.0	1.0000	1.0000	1.0000	1.0000	1.0000	1.0000	1.0000	1.0000	1.0000	1.0000	1.0000	1.0000	1.0000
0.1	1.0005	1.0005	1.0005	1.0005	1.0005	1.0004	1.0004	1.0004	1.0004	1.0004	1.0004	1.0004	1.0004
0.2	1.0010	1.0010	1.0010	1.0010	1.0009	1.0009	1.0009	1.0009	1.0008	1.0008	1.0008	1.0008	1.0007
0.3	1.0015	1.0015	1.0015	1.0014	1.0014	1.0014	1.0013	1.0013	1.0013	1.0012	1.0012	1.0011	1.0011
0.4	1.0021	1.0020	1.0020	1.0019	1.0019	1.0018	1.0018	1.0017	1.0017	1.0016	1.0016	1.0015	1.0014
0.5	1.0026	1.0025	1.0025	1.0024	1.0024	1.0023	1.0022	1.0022	1.0021	1.0020	1.0020	1.0019	1.0018
0.6	1.0031	1.0030	1.0030	1.0029	1.0028	1.0028	1.0027	1.0026	1.0025	1.0025	1.0024	1.0023	1.0022
0.7	1.0036	1.0036	1.0035	1.0034	1.0033	1.0032	1.0032	1.0031	1.0030	1.0029	1.0028	1.0027	1.0026
0.8	1.0042	1.0041	1.0040	1.0039	1.0038	1.0037	1.0036	1.0035	1.0034	1.0033	1.0032	1.0030	1.0029
0.9	1.0047	1.0046	1.0045	1.0044	1.0043	1.0042	1.0041	1.0040	1.0038	1.0037	1.0036	1.0034	1.0033
1.0	1.0052	1.0051	1.0050	1.0049	1.0048	1.0047	1.0045	1.0044	1.0043	1.0041	1.0040	1.0038	1.0037
1.1	1.0057	1.0056	1.0055	1.0054	1.0053	1.0051	1.0050	1.0049	1.0047	1.0046	1.0044	1.0042	1.0041
1.2	1.0062	1.0061	1.0060	1.0059	1.0058	1.0056	1.0055	1.0053	1.0052	1.0050	1.0048	1.0046	1.0044
1.3	1.0068	1.0066	1.0065	1.0064	1.0062	1.0061	1.0059	1.0058	1.0056	1.0054	1.0052	1.0050	1.0048
1.4	1.0073	1.0072	1.0070	1.0069	1.0067	1.0066	1.0064	1.0062	1.0060	1.0058	1.0056	1.0054	1.0052
1.5	1.0078	1.0077	1.0076	1.0074	1.0072	1.0070	1.0069	1.0067	1.0065	1.0063	1.0060	1.0058	1.0056
1.6	1.0084	1.0082	1.0081	1.0079	1.0077	1.0075	1.0073	1.0071	1.0069	1.0067	1.0065	1.0062	1.0060
1.7	1.0089	1.0088	1.0086	1.0084	1.0082	1.0080	1.0078	1.0076	1.0074	1.0071	1.0069	1.0066	1.0064
1.8	1.0094	1.0093	1.0091	1.0089	1.0087	1.0085	1.0083	1.0080	1.0078	1.0076	1.0073	1.0070	1.0068
1.9	1.0100	1.0098	1.0096	1.0094	1.0092	1.0090	1.0088	1.0085	1.0083	1.0080	1.0077	1.0074	1.0071
2.0	1.0105	1.0103	1.0101	1.0099	1.0097	1.0095	1.0092	1.0090	1.0087	1.0084	1.0081	1.0078	1.0075
2.1	1.0111	1.0109	1.0106	1.0104	1.0102	1.0100	1.0097	1.0094	1.0092	1.0089	1.0086	1.0083	1.0079
2.2	1.0116	1.0114	1.0112	1.0109	1.0107	1.0104	1.0102	1.0099	1.0096	1.0093	1.0090	1.0087	1.0083
2.3	1.0121	1.0119	1.0117	1.0114	1.0112	1.0109	1.0106	1.0104	1.0101	1.0098	1.0094	1.0091	1.0087
2.4	1.0127	1.0124	1.0122	1.0120	1.0117	1.0114	1.0111	1.0108	1.0105	1.0102	1.0098	1.0095	1.0091
2.5	1.0132	1.0130	1.0127	1.0125	1.0122	1.0119	1.0116	1.0113	1.0110	1.0106	1.0103	1.0099	1.0095
2.6	1.0138	1.0135	1.0133	1.0130	1.0127	1.0124	1.0121	1.0118	1.0114	1.0111	1.0107	1.0103	1.0099
2.7	1.0143	1.0140	1.0138	1.0135	1.0132	1.0129	1.0126	1.0122	1.0119	1.0115	1.0111	1.0107	1.0103
2.8	1.0148	1.0146	1.0143	1.0140	1.0137	1.0134	1.0131	1.0127	1.0124	1.0120	1.0116	1.0112	1.0107
2.9	1.0154	1.0151	1.0148	1.0145	1.0142	1.0139	1.0136	1.0132	1.0128	1.0124	1.0120	1.0116	1.0111
3.0	1.0160	1.0157	1.0154	1.0150	1.0147	1.0144	1.0140	1.0137	1.0133	1.0129	1.0124	1.0120	1.0116
3.1	1.0165	1.0162	1.0159	1.0156	1.0152	1.0149	1.0145	1.0141	1.0137	1.0133	1.0129	1.0124	1.0120
3.2	1.0170	1.0167	1.0164	1.0161	1.0158	1.0154	1.0150	1.0146	1.0142	1.0138	1.0133	1.0128	1.0124
3.3	1.0176	1.0173	1.0170	1.0166	1.0163	1.0159	1.0155	1.0151	1.0147	1.0142	1.0138	1.0133	1.0128
3.4	1.0181	1.0178	1.0175	1.0171	1.0168	1.0164	1.0160	1.0156	1.0151	1.0147	1.0142	1.0137	1.0132
3.5	1.0187	1.0184	1.0180	1.0177	1.0173	1.0169	1.0165	1.0160	1.0156	1.0151	1.0146	1.0141	1.0136
3.6	1.0192	1.0189	1.0186	1.0182	1.0178	1.0174	1.0170	1.0165	1.0161	1.0156	1.0151	1.0146	1.0140
3.7	1.0198	1.0195	1.0191	1.0187	1.0183	1.0179	1.0175	1.0170	1.0165	1.0160	1.0155	1.0150	1.0144
3.8	1.0204	1.0200	1.0196	1.0192	1.0188	1.0184	1.0180	1.0175	1.0170	1.0165	1.0160	1.0154	1.0148
3.9	1.0209	1.0206	1.0202	1.0198	1.0194	1.0189	1.0185	1.0180	1.0175	1.0170	1.0164	1.0159	1.0153
4.0	1.0215	1.0211	1.0207	1.0203	1.0199	1.0194	1.0190	1.0185	1.0180	1.0174	1.0169	1.0163	1.0157

PIPE TAPS—BASIC ORIFICE FACTORS—F_b

Orifice Diam., d, Inches	Internal Diameter of Pipe, D, Inches								
	2			3				4	
	1.689	1.939	2.067	2.300	2.626	2.900	3.068	3.152	3.438
.250	12.850	12.813	12.800	12.782	12.764	12.753	12.748	12.745	12.737
.375	29.359	29.098	29.005	28.882	28.771	28.710	28.681	28.669	28.634
.500	53.703	52.816	52.481	52.019	51.591	51.353	51.243	51.196	51.064
.625	87.212	84.918	84.083	82.922	81.795	81.142	80.835	80.702	80.332
.750	132.23	126.86	124.99	122.45	120.06	118.67	118.00	117.70	116.86
.875	192.74	181.02	177.09	171.92	167.23	164.58	163.31	162.76	161.17
1.000	276.45	251.10	243.27	233.30	224.56	219.76	217.52	216.55	213.79
1.125	391.93	342.98	327.98	309.43	293.79	285.48	281.66	280.02	275.42
1.250	....	465.99	437.99	404.52	377.36	363.41	357.12	354.44	347.03
1.375	....	....	583.96	524.68	478.68	455.82	445.74	441.48	429.83
1.500	....	....	....	679.10	602.45	565.79	549.94	543.31	525.40
1.625	....	....	....	....	755.34	697.43	672.95	662.81	635.76
1.750	....	....	....	....	946.99	856.37	819.05	803.77	763.51
1.875	....	....	....	....	....	1050.4	993.98	971.19	911.98
2.000	....	....	....	....	....	1290.7	1205.6	1171.8	1085.5
2.125	....	....	....	....	....	....	1465.1	1415.0	1289.7
2.250	....	....	....	....	....	....	....	....	1532.0
2.375	....	....	....	....	....	....	....	....	1822.8

Orifice Diam., d, Inches	Internal Diameter of Pipe, D, Inches								
	4		6				8		
	3.826	4.026	4.897	5.189	5.761	6.065	7.625	7.981	8.071
.250	12.727	12.722	....	....	....	....	....	....	....
.375	28.599	28.584	....	....	....	....	....	....	....
.500	50.936	50.886	50.739	50.705	50.652	50.628	....	....	....
.625	79.974	79.835	79.436	79.349	79.217	79.162	....	....	....
.750	116.05	115.73	114.81	114.61	114.32	114.20	....	....	....
.875	159.57	158.94	157.11	156.71	156.13	155.89	155.11	154.99	154.96
1.000	211.03	209.91	206.62	205.91	204.84	204.41	203.00	202.80	202.75
1.125	270.90	269.09	263.71	262.51	260.71	259.98	257.61	257.28	257.20
1.250	339.87	337.05	328.73	326.85	324.02	322.86	319.10	318.56	318.44
1.375	418.80	414.51	402.06	399.30	395.08	393.33	387.62	386.81	386.62
1.500	508.76	502.38	484.20	480.23	474.20	471.69	463.39	462.19	461.92
1.625	611.11	601.80	575.73	570.14	561.73	558.24	546.61	544.92	544.53
1.750	727.54	714.16	677.38	669.63	658.08	653.33	637.51	635.19	634.65
1.875	860.17	841.19	789.99	779.40	763.77	757.39	736.34	733.23	732.52
2.000	1011.7	985.04	914.57	900.28	879.38	870.93	843.34	839.29	838.35
2.125	1185.3	1148.4	1052.3	1033.2	1005.6	994.52	958.78	953.58	952.38
2.250	1385.4	1334.4	1204.7	1179.4	1143.2	1128.8	1083.0	1076.4	1074.9
2.375	1617.2	1547.3	1373.4	1340.2	1293.1	1274.6	1216.3	1208.0	1206.1
2.500	1887.6	1792.3	1560.5	1517.2	1456.4	1432.7	1359.2	1348.8	1346.5
2.625	2206.0	2075.9	1768.3	1712.3	1634.3	1604.3	1512.0	1499.2	1496.3
2.750	....	2407.0	1999.8	1927.6	1828.3	1790.3	1675.4	1659.7	1656.1
2.875	....	....	2258.5	2165.9	2039.9	1992.2	1849.8	1830.6	1826.3
3.000	....	....	2548.6	2430.2	2271.2	2211.6	2036.0	2012.7	2007.3
3.125	....	....	2875.2	2724.4	2524.3	2450.1	2234.7	2206.4	2199.9
3.250	....	....	3244.8	3052.8	2801.8	2709.9	2446.5	2412.4	2404.7
3.375	....	....	3665.6	3420.9	3106.9	2993.3	2672.5	2631.6	2622.3
3.500	....	....	....	3835.7	3443.0	3303.0	2913.7	2864.7	2853.7
3.625	....	....	....	4305.7	3814.4	3642.3	3171.1	3112.7	3099.6
3.750	....	....	....	....	4226.3	4014.8	3446.0	3376.6	3361.0
3.875	....	....	....	....	4684.9	4425.1	3739.9	3657.6	3639.2
4.000	....	....	....	....	5197.7	4878.4	4054.2	3957.0	3935.2
4.250	....	....	....	....	....	....	4751.4	4616.6	4586.6
4.500	....	....	....	....	....	....	5554.7	5369.0	5327.9
4.750	....	....	....	....	....	....	6485.3	6231.1	6175.2
5.000	....	....	....	....	....	....	7571.4	7224.3	7148.7
5.250	....	....	....	....	....	....	8850.3	8376.4	8274.0
5.500	....	....	....	....	....	....	....	9723.8	9585.1

PIPE TAPS—BASIC ORIFICE FACTORS—F_b

Orifice Diam., d, Inches	Internal Diameter of Pipe, D, Inches								
	10			12			16		
	9.564	10.020	10.136	11.376	11.938	12.090	14.688	15.000	15.250
1.000	202.16	....	....	....	....	....	....	....	....
1.125	256.22	256.01	255.96	....	....	....	....	....	....
1.250	316.90	316.56	316.49	315.81	315.57	315.51	....	....	....
1.375	384.29	383.79	383.68	382.66	382.30	382.22	....	....	....
1.500	458.52	457.79	457.63	456.16	455.64	455.52	453.92	453.78	....
1.625	539.72	538.69	538.45	536.38	535.66	535.48	533.27	533.07	532.93
1.750	628.03	626.61	626.29	623.44	622.45	622.20	619.18	618.92	618.73
1.875	723.61	721.70	721.26	717.43	716.10	715.78	711.73	711.39	711.13
2.000	826.63	824.12	823.54	818.48	816.73	816.30	810.99	810.53	810.19
2.125	937.27	934.02	933.27	926.71	924.44	923.88	917.01	916.43	915.99
2.250	1055.7	1051.6	1050.6	1042.3	1039.4	1038.7	1029.9	1029.1	1028.6
2.375	1182.2	1177.0	1175.8	1165.3	1161.6	1160.7	1149.7	1148.8	1148.1
2.500	1316.9	1310.5	1309.0	1295.9	1291.4	1290.2	1276.5	1275.4	1274.5
2.625	1460.0	1452.1	1450.3	1434.3	1428.7	1427.4	1410.5	1409.1	1408.0
2.750	1611.7	1602.3	1600.1	1580.7	1573.9	1572.2	1551.7	1549.9	1548.6
2.875	1772.5	1761.0	1758.4	1735.1	1726.9	1724.9	1700.1	1698.0	1696.5
3.000	1942.5	1928.8	1925.6	1897.8	1888.1	1885.7	1856.1	1853.6	1851.7
3.125	2122.1	2105.7	2101.9	2069.0	2057.5	2054.7	2019.5	2016.6	2014.3
3.250	2311.6	2292.2	2287.8	2248.9	2235.4	2232.0	2190.7	2187.2	2184.5
3.375	2511.5	2488.6	2483.4	2437.7	2421.8	2418.0	2369.6	2365.5	2362.4
3.500	2722.3	2695.3	2689.1	2635.6	2617.2	2612.6	2556.5	2551.7	2548.1
3.625	2944.3	2912.7	2905.5	2843.0	2821.6	2816.3	2751.4	2745.9	2741.7
3.750	3178.1	3141.2	3132.7	3060.2	3035.3	3029.3	2954.5	2948.1	2943.3
3.875	3424.3	3381.3	3371.5	3287.4	3258.7	3251.7	3165.9	3158.6	3153.1
4.000	3683.5	3633.5	3622.1	3524.9	3492.0	3484.0	3385.8	3377.5	3371.2
4.250	4243.8	4176.8	4161.6	4032.8	3989.5	3979.0	3851.6	3840.9	3832.8
4.500	4865.1	4776.2	4756.1	4587.0	4530.8	4517.2	4353.4	4339.8	4329.6
4.750	5554.9	5437.9	5411.5	5191.5	5119.0	5101.5	4892.9	4875.8	4862.9
5.000	6322.2	6169.2	6134.9	5850.6	5757.8	5735.5	5471.9	5450.5	5434.3
5.250	7177.7	6978.9	6934.4	6569.5	6451.5	6423.2	6092.5	6065.9	6045.9
5.500	8134.1	7877.2	7820.0	7354.1	7205.1	7169.5	6757.0	6724.1	6699.5
5.750	9207.0	8876.4	8803.1	8211.4	8024.2	7979.6	7468.0	7427.7	7397.4
6.000	10415	9991.2	9897.7	9149.5	8915.4	8859.8	8228.5	8179.2	8142.3
6.250	11783	11240	11121	10178	9886.2	9817.2	9041.6	8981.7	8937.0
6.500	13340	12644	12492	11307	10945	10860	9911.2	9838.7	9784.7
6.750	....	14230	14038	12550	12103	11998	10841	10754	10689
7.000	....	16035	15790	13923	13371	13242	11837	11732	11654
7.250	....	....	....	15442	14762	14604	12902	12777	12684
7.500	....	....	....	17131	16294	16101	14044	13894	13783
7.750	....	....	....	19017	17986	17750	15268	15090	14959
8.000	....	....	....	....	19861	19572	16583	16371	16216
8.250	....	....	....	....	21947	21593	17996	17746	17561
8.500	....	....	....	....	....	....	19517	19221	19003
8.750	....	....	....	....	....	....	21156	20807	20551
9.000	....	....	....	....	....	....	22926	22515	22214
9.250	....	....	....	....	....	....	24841	24356	24003
9.500	....	....	....	....	....	....	26917	26346	25932
9.750	....	....	....	....	....	....	29172	28501	28014
10.000	....	....	....	....	....	....	31629	30839	30268
10.250	....	....	....	....	....	....	34315	33383	32713
10.500	....	....	....	....	....	....	....	36160	35372

PIPE TAPS—BASIC ORIFICE FACTORS—F_b

Orifice Diam., d, Inches	Internal Diameter of Pipe, D, Inches								
	20			24			30		
	18.814	19.000	19.250	22.626	23.000	23.250	28.628	29.000	29.250
2.000	806.71	806.57	806.40	....	....	....	....	....	....
2.125	911.52	911.35	911.13	....	....	....	....	....	....
2.250	1022.9	1022.7	1022.4	....	....	....	....	....	....
2.375	1141.0	1140.7	1140.4	1136.8	1136.5	1136.3	....	....	....
2.500	1265.7	1265.4	1265.0	1260.6	1260.2	1259.9	....	....	....
2.625	1397.2	1396.8	1396.3	1390.9	1390.5	1390.2	....	....	....
2.750	1535.5	1535.0	1534.4	1527.9	1527.3	1527.0	....	....	....
2.875	1680.7	1680.1	1679.3	1671.5	1670.9	1670.4	1663.8	....	....
3.000	1832.7	1832.1	1831.2	1821.9	1821.1	1820.6	1812.7	1812.3	1812.0
3.125	1991.8	1991.0	1990.0	1978.9	1978.0	1977.4	1968.1	1967.7	1967.4
3.250	2158.0	2157.0	2155.8	2142.8	2141.7	2141.0	2130.2	2129.6	2129.3
3.375	2331.3	2330.2	2328.7	2313.5	2312.3	2311.5	2298.8	2298.2	2297.7
3.500	2511.9	2510.6	2508.8	2491.2	2489.7	2488.8	2474.1	2473.3	2472.8
3.625	2699.7	2698.2	2696.2	2675.8	2674.1	2673.0	2656.0	2655.1	2654.6
3.750	2895.0	2893.2	2890.9	2867.4	2865.4	2864.1	2844.6	2843.7	2843.0
3.875	3097.7	3095.7	3093.0	3066.0	3063.8	3062.3	3040.0	3038.9	3038.2
4.000	3308.0	3305.7	3302.7	3271.8	3269.2	3267.6	3242.1	3240.9	3240.1
4.250	3751.6	3748.7	3744.8	3705.0	3701.7	3699.6	3666.9	3665.3	3664.3
4.500	4226.8	4223.0	4218.1	4167.6	4163.4	4160.7	4119.3	4117.3	4116.0
4.750	4734.1	4729.4	4723.3	4660.0	4654.8	4651.4	4599.6	4597.1	4595.4
5.000	5274.6	5268.7	5261.2	5183.0	5176.4	5172.3	5108.2	5105.0	5103.0
5.250	5849.0	5841.9	5832.6	5737.0	5729.1	5723.9	5645.4	5641.6	5639.1
5.500	6458.6	6449.9	6438.7	6322.9	6313.2	6307.0	6211.8	6207.2	6204.2
5.750	7104.5	7094.0	7080.4	6941.3	6929.7	6922.2	6807.7	6802.1	6798.5
6.000	7787.9	7775.4	7759.1	7592.8	7579.0	7570.1	7433.6	7426.9	7422.6
6.250	8510.4	8495.4	8476.0	8278.4	8262.0	8251.5	8089.9	8082.0	8076.9
6.500	9273.4	9255.6	9232.5	8998.3	8979.5	8967.1	8777.2	8767.9	8761.9
6.750	10079	10058	10030	9755.0	9732.4	9717.9	9496.0	9485.2	9478.1
7.000	10928	10903	10871	10548	10522	10505	10247	10234	10226
7.250	11824	11794	11756	11379	11348	11329	11030	11016	11006
7.500	12767	12733	12689	12249	12214	12191	11847	11830	11819
7.750	13762	13722	13670	13160	13119	13093	12697	12678	12665
8.000	14810	14763	14703	14113	14065	14035	13582	13560	13546
8.250	15914	15860	15791	15109	15054	15020	14501	14477	14461
8.500	17078	17015	16935	16150	16087	16048	15457	15429	15411
8.750	18305	18232	18139	17237	17166	17121	16450	16418	16397
9.000	19598	19515	19408	18373	18292	18241	17480	17444	17421
9.250	20963	20867	20743	19560	19468	19409	18548	18508	18482
9.500	22402	22292	22151	20800	20695	20628	19656	19611	19582
9.750	23923	23796	23634	22094	21976	21900	20805	20754	20721
10.000	25529	25384	25198	23447	23312	23227	21995	21938	21902
10.250	27227	27061	26849	24859	24708	24612	23228	23165	23124
10.500	29023	28834	28592	26335	26164	26056	24505	24434	24389
10.750	30925	30709	30434	27878	27685	27563	25827	25749	25698
11.000	32940	32695	32382	29490	29273	29136	27196	27109	27052
11.250	35078	34798	34443	31175	30932	30779	28613	28516	28453
11.500	37348	37030	36626	32938	32666	32494	30080	29972	29903
11.750	39762	39400	38941	34783	34478	34285	31598	31479	31402
12.000	42330	41920	41399	36714	36373	36158	33169	33038	32953
12.500	47991	47460	46790	40855	40429	40161	36479	36318	36215
13.000	54463	53778	52914	45406	44877	44544	40024	39829	39704
13.500	....	....	....	50420	49763	49352	43824	43589	43437
14.000	....	....	....	55960	55147	54638	47898	47615	47434
14.500	....	....	....	62099	61094	60468	52271	51932	51714
15.000	....	....	....	68929	67687	66915	56967	56562	56301
15.500	....	....	....	76562	75025	74074	62017	61533	61223
16.000	....	....	....	....	83231	82055	67454	66878	66509
16.500	....	....	....	....	....	....	73314	72630	72193
17.000	....	....	....	....	....	....	79641	78831	78313
17.500	....	....	....	....	....	....	86485	85525	84913
18.000	....	....	....	....	....	....	93900	92765	92042
18.500	....	....	....	....	....	....	101954	100612	99759
19.000	....	....	....	....	....	....	110722	109135	108128
19.500	....	....	....	....	....	....	120295	118417	117228
20.000	....	....	....	....	....	....	130781	128556	127150

"b" VALUES FOR REYNOLDS NUMBER FACTOR, F_r—PIPE TAPS

$$F_r = 1 + \frac{b}{\sqrt{h_w p_f}}$$

Orifice Diam., d, Inches	Internal Diameter of Pipe, D, Inches								
	2			3				4	
	1.689	1.939	2.067	2.300	2.626	2.900	3.068	3.152	3.438
.250	.1105	.1091	.1087	.1081	.1078	.1078	.1080	.1081	.1084
.375	.0890	.0878	.0877	.0879	.0888	.0898	.0905	.0908	.0918
.500	.0758	.0734	.0729	.0728	.0737	.0750	.0758	.0763	.0778
.625	.0693	.0647	.0635	.0624	.0624	.0634	.0642	.0646	.0662
.750	.0675	.0608	.0586	.0559	.0546	.0548	.0552	.0555	.0568
.875	.0684	.0602	.0570	.0528	.0497	.0488	.0488	.0489	.0496
1.000	.0702	.0614	.0576	.0522	.0473	.0452	.0445	.0443	.0443
1.125	.0708	.0635	.0595	.0532	.0469	.0435	.0422	.0417	.0407
1.250	....	.0650	.0616	.0552	.0478	.0434	.0414	.0406	.0387
1.375	....	....	.0629	.0574	.0496	.0443	.0418	.0408	.0379
1.500	....	....	....	.0590	.0518	.0460	.0431	.0418	.0382
1.625	....	....	....	....	.0539	.0482	.0450	.0435	.0392
1.750	....	....	....	....	.0553	.0504	.0471	.0456	.0408
1.875	....	....	....	....	....	.0521	.0492	.0477	.0427
2.000	....	....	....	....	....	.0532	.0508	.0495	.0448
2.125	....	....	....	....	....	....	.0519	.0509	.0467
2.250	....	....	....	....	....	....	....	....	.0483
2.375	....	....	....	....	....	....	....	....	.0494

Orifice Diam., d, Inches	Internal Diameter of Pipe, D, Inches								
	4		6				8		
	3.826	4.026	4.897	5.189	5.761	6.065	7.625	7.981	8.071
.250	.1087	.1091	....	....	....	....	....	....	....
.375	.0932	.0939	....	....	....	....	....	....	....
.500	.0799	.0810	.0850	.0862	.0883	.0895	....	....	....
.625	.0685	.0697	.0747	.0762	.0789	.0802	....	....	....
.750	.0590	.0602	.0655	.0672	.0703	.0718	....	....	....
.875	.0513	.0524	.0575	.0592	.0625	.0642	.0716	.0730	.0733
1.000	.0453	.0461	.0506	.0523	.0556	.0573	.0652	.0668	.0662
1.125	.0408	.0412	.0448	.0464	.0495	.0512	.0592	.0609	.0613
1.250	.0376	.0377	.0401	.0413	.0442	.0458	.0538	.0555	.0560
1.375	.0358	.0353	.0363	.0373	.0397	.0412	.0489	.0506	.0510
1.500	.0350	.0340	.0334	.0340	.0360	.0372	.0445	.0462	.0466
1.625	.0351	.0336	.0313	.0315	.0329	.0339	.0404	.0421	.0425
1.750	.0358	.0340	.0300	.0298	.0304	.0311	.0369	.0384	.0388
1.875	.0371	.0349	.0293	.0287	.0285	.0290	.0338	.0352	.0355
2.000	.0388	.0363	.0292	.0281	.0273	.0273	.0311	.0323	.0327
2.125	.0407	.0380	.0297	.0281	.0265	.0262	.0288	.0298	.0301
2.250	.0427	.0398	.0305	.0285	.0261	.0258	.0268	.0277	.0280
2.375	.0445	.0417	.0316	.0293	.0262	.0253	.0252	.0259	.0261
2.500	.0460	.0435	.0330	.0304	.0267	.0254	.0239	.0244	.0246
2.625	.0472	.0450	.0345	.0317	.0274	.0258	.0230	.0232	.0233
2.750	....	.0462	.0362	.0331	.0284	.0265	.0224	.0224	.0224
2.875	....	....	.0379	.0347	.0295	.0274	.0220	.0218	.0218
3.000	....	....	.0395	.0364	.0308	.0285	.0219	.0214	.0213
3.125	....	....	.0410	.0380	.0323	.0297	.0220	.0213	.0211
3.250	....	....	.0422	.0394	.0338	.0311	.0223	.0214	.0212
3.375	....	....	.0432	.0408	.0353	.0325	.0228	.0216	.0214
3.500	....	....	....	.0419	.0367	.0339	.0235	.0221	.0218
3.625	....	....	....	.0428	.0381	.0354	.0243	.0227	.0224
3.750	....	....	....	....	.0393	.0367	.0252	.0234	.0230
3.875	....	....	....	....	.0404	.0380	.0262	.0243	.0238
4.000	....	....	....	....	.0413	.0391	.0273	.0252	.0246
4.250	....	....	....	....	....	....	.0296	.0273	.0268
4.500	....	....	....	....	....	....	.0321	.0296	.0290
4.750	....	....	....	....	....	....	.0344	.0320	.0314
5.000	....	....	....	....	....	....	.0364	.0342	.0336
5.250	....	....	....	....	....	....	.0381	.0361	.0356
5.500	....	....	....	....	....	....	....	.0377	.0372

"b" VALUES FOR REYNOLDS NUMBER FACTOR, F_r—PIPE TAPS

$$F_r = 1 + \frac{b}{\sqrt{h_w p_f}}$$

Orifice Diam., d, Inches	Internal Diameter of Pipe, D, Inches								
	10			12			16		
	9.564	10.020	10.136	11.376	11.938	12.090	14.688	15.000	15.250
1.000	.0728	....	....	....	....	....	....	....	....
1.125	.0674	.0690	.0694	....	....	....	....	....	....
1.250	.0624	.0641	.0646	.0687	.0704	.0708	....	....	....
1.375	.0576	.0594	.0599	.0643	.0661	.0666	....	....	....
1.500	.0532	.0550	.0555	.0601	.0620	.0625	.0697	.0705	....
1.625	.0490	.0509	.0514	.0561	.0580	.0585	.0662	.0670	.0676
1.750	.0452	.0471	.0476	.0523	.0543	.0548	.0628	.0636	.0642
1.875	.0417	.0436	.0440	.0488	.0508	.0513	.0594	.0603	.0610
2.000	.0385	.0403	.0407	.0454	.0475	.0480	.0563	.0572	.0578
2.125	.0355	.0372	.0377	.0423	.0443	.0449	.0532	.0541	.0548
2.250	.0329	.0345	.0349	.0394	.0414	.0419	.0503	.0512	.0519
2.375	.0305	.0320	.0324	.0367	.0387	.0392	.0475	.0484	.0492
2.500	.0283	.0298	.0301	.0342	.0361	.0366	.0449	.0458	.0466
2.625	.0265	.0277	.0281	.0319	.0337	.0342	.0424	.0433	.0440
2.750	.0248	.0260	.0262	.0298	.0316	.0320	.0400	.0409	.0417
2.875	.0234	.0244	.0246	.0279	.0295	.0300	.0378	.0387	.0394
3.000	.0222	.0230	.0232	.0262	.0277	.0281	.0356	.0365	.0372
3.125	.0212	.0218	.0220	.0244	.0260	.0264	.0336	.0345	.0352
3.250	.0204	.0209	.0210	.0232	.0245	.0249	.0317	.0326	.0332
3.375	.0199	.0201	.0202	.0220	.0232	.0235	.0300	.0308	.0314
3.500	.0195	.0195	.0196	.0210	.0220	.0222	.0283	.0291	.0297
3.625	.0193	.0191	.0191	.0200	.0209	.0212	.0268	.0275	.0281
3.750	.0192	.0188	.0188	.0193	.0200	.0202	.0254	.0261	.0267
3.875	.0193	.0187	.0186	.0187	.0192	.0194	.0240	.0247	.0253
4.000	.0195	.0187	.0186	.0182	.0185	.0187	.0228	.0235	.0240
4.250	.0203	.0192	.0189	.0176	.0176	.0177	.0207	.0213	.0217
4.500	.0215	.0200	.0197	.0175	.0172	.0171	.0190	.0194	.0198
4.750	.0230	.0212	.0208	.0178	.0171	.0170	.0176	.0180	.0182
5.000	.0248	.0228	.0223	.0185	.0174	.0173	.C.66	.0168	.0170
5.250	.0267	.0244	.0239	.0194	.0181	.0178	.0160	.0161	.0162
5.500	.0287	.0263	.0257	.0207	.0190	.0186	.0156	.0156	.0156
5.750	.0307	.0282	.0276	.0221	.0202	.0197	.0155	.0154	.0153
6.000	.0326	.0302	.0295	.0231	.0215	.0210	.0157	.0154	.0153
6.250	.0343	.0320	.0316	.0253	.0230	.0224	.0161	.0157	.0154
6.500	.0358	.0336	.0331	.0270	.0246	.0239	.0167	.0162	.0159
6.750	....	.0351	.0346	.0288	.0262	.0256	.0174	.0169	.0164
7.000	....	.0363	.0359	.0304	.0279	.0272	.0184	.0177	.0172
7.250	....	....	....	.0320	.0295	.0288	.0195	.0187	.0181
7.500	....	....	....	.0334	.0310	.0304	.0206	.0198	.0191
7.750	....	....	....	.0347	.0325	.0318	.0219	.0209	.0202
8.000	....	....	....	....	.0338	.0332	.0232	.0222	.0214
8.250	....	....	....	....	.0349	.0344	.0246	.0235	.0227
8.500	....	....	....	....	....	....	.0259	.0248	.0240
8.750	....	....	....	....	....	....	.0273	.0262	.0253
9.000	....	....	....	....	....	....	.0286	.0276	.0267
9.250	....	....	....	....	....	....	.0299	.0288	.0280
9.500	....	....	....	....	....	....	.0311	.0300	.0292
9.750	....	....	....	....	....	....	.0322	.0312	.0304
10.000	....	....	....	....	....	....	.0332	.0323	.0315
10.250	....	....	....	....	....	....	.0341	.0333	0326
10.500	....	....	....	..	....	...	....	.0341	.0335

"b" VALUES FOR REYNOLDS NUMBER FACTOR, F_r—PIPE TAPS

$$F_r = 1 + \frac{b}{\sqrt{h_w p_f}}$$

Orifice Diam., d, Inches	Internal Diameter of Pipe, D, Inches								
	20			24			30		
	18.814	19.000	19.250	22.626	23.000	23.250	28.628	29.000	29.250
2.000	.0663	.0667	.0672	....	....	....	....	....	....
2.125	.0635	.0639	.0644	....	....	....	....	....	....
2.250	.0609	.0613	.0618	....	....	....	....	....	....
2.375	.0583	.0588	.0593	.0658	.0665	.0669	....	....	....
2.500	.0558	.0562	.0568	.0635	.0642	.0646	....	....	....
2.625	.0534	.0539	.0544	.0613	.0620	.0624	....	....	....
2.750	.0510	.0515	.0520	.0591	.0598	.0603	....	....	....
2.875	.0488	.0492	.0498	.0570	.0577	.0582	.0667	....	....
3.000	.0466	.0470	.0476	.0549	.0556	.0561	.0649	.0654	.0657
3.125	.0445	.0449	.0455	.0529	.0536	.0541	.0630	.0636	.0639
3.250	.0425	.0429	.0435	.0509	.0516	.0521	.0613	.0616	.0622
3.375	.0406	.0410	.0416	.0490	.0497	.0502	.0595	.0601	.0604
3.500	.0387	.0391	.0397	.0471	.0479	.0484	.0578	.0584	.0587
3.625	.0369	.0373	.0379	.0454	.0461	.0466	.0561	.0567	.0571
3.750	.0352	.0356	.0362	.0436	.0444	.0449	.0545	.0550	.0554
3.875	.0336	.0340	.0346	.0419	.0427	.0432	.0528	.0534	.0538
4.000	.0320	.0324	.0330	.0403	.0411	.0416	.0513	.0518	.0522
4.250	.0291	.0295	.0301	.0372	.0380	.0385	.0482	.0488	.0492
4.500	.0265	.0269	.0274	.0343	.0351	.0356	.0453	.0459	.0463
4.750	.0242	.0246	.0250	.0316	.0324	.0328	.0425	.0431	.0435
5.000	.0221	.0225	.0229	.0292	.0299	.0303	.0399	.0405	.0409
5.250	.0203	.0206	.0210	.0269	.0276	.0280	.0374	.0380	.0384
5.500	.0188	.0190	.0194	.0248	.0255	.0259	.0350	.0356	.0360
5.750	.0175	.0177	.0180	.0230	.0236	.0240	.0328	.0334	.0338
6.000	.0164	.0165	.0168	.0212	.0218	.0222	.0307	.0313	.0317
6.250	.0155	.0156	.0158	.0197	.0202	.0206	.0287	.0293	.0297
6.500	.0148	.0149	.0150	.0184	.0189	.0192	.0269	.0274	.0278
6.750	.0143	.0144	.0145	.0172	.0176	.0179	.0252	.0257	.0260
7.000	.0141	.0141	.0141	.0162	.0166	.0168	.0236	.0241	.0244
7.250	.0140	.0140	.0139	.0153	.0156	.0158	.0221	.0226	.0229
7.500	.0140	.0140	.0139	.0146	.0148	.0150	.0207	.0212	.0215
7.750	.0142	.0141	.0140	.0140	.0142	.0144	.0195	.0199	.0202
8.000	.0146	.0144	.0142	.0136	.0138	.0138	.0183	.0187	.0190
8.250	.0151	.0148	.0146	.0133	.0134	.0132	.0173	.0177	.0179
8.500	.0156	.0154	.0151	.0132	.0132	.0130	.0164	.0167	.0169
8.750	.0163	.0160	.0157	.0131	.0130	.0130	.0155	.0158	.0161
9.000	.0171	.0168	.0163	.0131	.0130	.0130	.0148	.0151	.0153
9.250	.0180	.0176	.0171	.0133	.0131	.0130	.0142	.0144	.0146
9.500	.0189	.0185	.0180	.0136	.0133	.0132	.0136	.0138	.0140
9.750	.0198	.0194	.0189	.0139	.0136	.0134	.0132	.0133	.0134
10.000	.0209	.0204	.0198	.0143	.0140	.0138	.0128	.0129	.0130
10.250	.0219	.0214	.0208	.0148	.0144	.0142	.0125	.0126	.0127
10.500	.0230	.0225	.0219	.0154	.0150	.0147	.0123	.0124	.0124
10.750	.0241	.0236	.0229	.0160	.0155	.0152	.0122	.0122	.0122
11.000	.0252	.0247	.0240	.0168	.0162	.0158	.0121	.0121	.0121
11.250	.0263	.0261	.0251	.0175	.0169	.0165	.0122	.0121	.0121
11.500	.0273	.0268	.0262	.0183	.0176	.0172	.0122	.0121	.0122
11.750	.0284	.0278	.0272	.0191	.0184	.0180	.0124	.0123	.0122
12.000	.0293	.0288	.0282	.0200	.0192	.0190	.0126	.0124	.0123
12.500	.0312	.0307	.0301	.0218	.0210	.0204	.0132	.0130	.0128
13.000	.0327	.0323	.0318	.0236	.0228	.0222	.0140	.0137	.0135
13.500	....	....	....	.0254	.0246	.0240	.0150	.0146	.0143
14.000	....	....	....	.0272	.0264	.0258	.0161	.0156	.0153
14.500	....	....	....	.0289	.0280	.0275	.0173	.0168	.0165
15.000	....	....	....	.0304	.0296	.0291	.0186	.0181	.0177
15.500	....	....	....	.0318	.0311	.0306	.0200	.0194	.0190
16.000	....	....	....	....	.0323	.0318	.0215	.0209	.0204
16.500	....	....	....	....	....	....	.0230	.0223	.0219
17.000	....	....	....	....	....	....	.0244	.0238	.0233
17.500	....	....	....	....	....	....	.0259	.0252	.0248
18.000	....	....	....	....	....	....	.0272	.0266	.0261
18.500	....	....	....	....	....	....	.0286	.0279	.0275
19.000	....	....	....	....	....	....	.0298	.0292	.0288
19.500	....	....	....	....	....	....	.0309	.0303	.0299
20.000	....	....	....	....	....	....	.0318	.0313	.0310

"b" VALUES FOR REYNOLDS NUMBER FACTOR, F_r—PIPE TAPS

$$F_r = 1 + \frac{b}{\sqrt{h_w p_f}}$$

β	Internal Diameter of Pipe, D, inches										
	2			3				4			
	1.689	1.939	2.067	2.300	2.626	2.900	3.068	3.152	3.438	3.826	4.026
0.10	0.1295	0.1209	0.1173	0.1118	0.1058	0.1017	0.0996	0.0986	0.0957	0.0924	0.0909
0.11	0.1253	0.1167	0.1132	0.1077	0.1016	0.0976	0.0954	0.0945	0.0915	0.0882	0.0867
0.12	0.1212	0.1126	0.1090	0.1035	0.0975	0.0934	0.0913	0.0903	0.0874	0.0841	0.0826
0.13	0.1172	0.1086	0.1051	0.0996	0.0935	0.0895	0.0874	0.0864	0.0835	0.0802	0.0787
0.14	0.1134	0.1049	0.1013	0.0958	0.0898	0.0858	0.0837	0.0827	0.0797	0.0765	0.0750
0.15	0.1098	0.1013	0.0978	0.0923	0.0863	0.0823	0.0801	0.0792	0.0762	0.0729	0.0715
0.16	0.1065	0.0980	0.0944	0.0889	0.0829	0.0789	0.0768	0.0758	0.0729	0.0696	0.0682
0.17	0.1033	0.0948	0.0912	0.0856	0.0798	0.0758	0.0737	0.0727	0.0698	0.0665	0.0650
0.18	0.1001	0.0916	0.0881	0.0827	0.0767	0.0727	0.0706	0.0696	0.0667	0.0634	0.0620
0.19	0.0973	0.0888	0.0853	0.0799	0.0739	0.0699	0.0678	0.0669	0.0639	0.0607	0.0592
0.20	0.0945	0.0861	0.0825	0.0771	0.0712	0.0672	0.0651	0.0642	0.0613	0.0580	0.0566
0.21	0.0919	0.0835	0.0800	0.0746	0.0687	0.0647	0.0626	0.0617	0.0588	0.0555	0.0541
0.22	0.0894	0.0811	0.0776	0.0722	0.0663	0.0623	0.0603	0.0593	0.0564	0.0532	0.0518
0.23	0.0872	0.0788	0.0753	0.0700	0.0641	0.0602	0.0581	0.0571	0.0543	0.0510	0.0496
0.24	0.0851	0.0767	0.0733	0.0679	0.0621	0.0581	0.0561	0.0551	0.0523	0.0490	0.0476
0.25	0.0831	0.0748	0.0714	0.0660	0.0602	0.0563	0.0542	0.0533	0.0504	0.0472	0.0458
0.26	0.0812	0.0730	0.0695	0.0642	0.0584	0.0545	0.0524	0.0515	0.0487	0.0455	0.0441
0.27	0.0796	0.0713	0.0679	0.0626	0.0568	0.0530	0.0509	0.0500	0.0471	0.0440	0.0426
0.28	0.0781	0.0699	0.0665	0.0612	0.0554	0.0516	0.0495	0.0486	0.0458	0.0426	0.0412
0.29	0.0766	0.0685	0.0651	0.0598	0.0541	0.0502	0.0482	0.0473	0.0445	0.0413	0.0399
0.30	0.0753	0.0672	0.0638	0.0586	0.0529	0.0490	0.0470	0.0461	0.0433	0.0401	0.0388
0.31	0.0741	0.0661	0.0627	0.0575	0.0518	0.0480	0.0460	0.0451	0.0423	0.0391	0.0378
0.32	0.0730	0.0650	0.0616	0.0564	0.0508	0.0470	0.0450	0.0441	0.0413	0.0382	0.0368
0.33	0.0720	0.0640	0.0606	0.0555	0.0499	0.0461	0.0441	0.0432	0.0405	0.0374	0.0360
0.34	0.0712	0.0632	0.0599	0.0548	0.0492	0.0455	0.0435	0.0426	0.0398	0.0368	0.0354
0.35	0.0705	0.0626	0.0593	0.0542	0.0486	0.0449	0.0429	0.0420	0.0393	0.0362	0.0349
0.36	0.0699	0.0620	0.0587	0.0537	0.0481	0.0444	0.0424	0.0416	0.0388	0.0358	0.0345
0.37	0.0593	0.0615	0.0582	0.0532	0.0477	0.0440	0.0421	0.0412	0.0385	0.0355	0.0341
0.38	0.0689	0.0611	0.0578	0.0529	0.0474	0.0437	0.0418	0.0409	0.0382	0.0352	0.0339
0.39	0.0684	0.0607	0.0575	0.0525	0.0471	0.0435	0.0415	0.0407	0.0380	0.0350	0.0337
0.40	0.0681	0.0604	0.0572	0.0523	0.0469	0.0433	0.0414	0.0405	0.0379	0.0349	0.0336
0.41	0.0678	0.0603	0.0571	0.0522	0.0468	0.0432	0.0414	0.0405	0.0379	0.0349	0.0336
0.42	0.0677	0.0602	0.0570	0.0521	0.0468	0.0433	0.0414	0.0405	0.0379	0.0350	0.0337
0.43	0.0676	0.0601	0.0570	0.0522	0.0469	0.0434	0.0415	0.0402	0.0381	0.0352	0.0339
0.44	0.0675	0.0601	0.0570	0.0522	0.0470	0.0435	0.0417	0.0408	0.0382	0.0354	0.0341
0.45	0.0675	0.0602	0.0571	0.0524	0.0472	0.0437	0.0419	0.0410	0.0385	0.0357	0.0344
0.46	0.0676	0.0603	0.0572	0.0526	0.0474	0.0440	0.0422	0.0413	0.0388	0.0360	0.0347
0.47	0.0676	0.0604	0.0574	0.0528	0.0476	0.0442	0.0424	0.0416	0.0391	0.0363	0.0351
0.48	0.0677	0.0606	0.0576	0.0530	0.0479	0.0446	0.0428	0.0420	0.0395	0.0367	0.0355
0.49	0.0678	0.0608	0.0578	0.0532	0.0482	0.0449	0.0431	0.0423	0.0399	0.0372	0.0359
0.50	0.0680	0.0511	0.0581	0.0536	0.0486	0.0453	0.0436	0.0428	0.0404	0.0377	0.0365
0.51	0.0682	0.0613	0.0584	0.0539	0.0490	0.0458	0.0440	0.0433	0.0409	0.0382	0.0370
0.52	0.0684	0.0615	0.0587	0.0543	0.0494	0.0462	0.0445	0.0437	0.0413	0.0387	0.0375
0.53	0.0687	0.0619	0.0590	0.0547	0.0499	0.0467	0.0450	0.0442	0.0419	0.0393	0.0381
0.54	0.0689	0.0622	0.0594	0.0551	0.0503	0.0472	0.0455	0.0448	0.0424	0.0398	0.0387
0.55	0.0592	0.0625	0.0598	0.0555	0.0508	0.0477	0.0460	0.0453	0.0430	0.0404	0.0393
0.56	0.0694	0.0628	0.0601	0.0559	0.0513	0.0482	0.0466	0.0458	0.0435	0.0410	0.0399
0.57	0.0696	0.0632	0.0605	0.0563	0.0517	0.0487	0.0471	0.0464	0.0441	0.0416	0.0405
0.58	0.0699	0.0635	0.0608	0.0567	0.0522	0.0492	0.0476	0.0469	0.0447	0.0422	0.0412
0.59	0.0701	0.0638	0.0612	0.0571	0.0527	0.0497	0.0482	0.0474	0.0453	0.0428	0.0418
0.60	0.0704	0.0641	0.0615	0.0575	0.0532	0.0502	0.0487	0.0480	0.0458	0.0434	0.0424
0.61	0.0705	0.0643	0.0618	0.0578	0.0535	0.0506	0.0491	0.0484	0.0463	0.0439	0.0429
0.62	0.0706	0.0645	0.0620	0.0581	0.0539	0.0510	0.0495	0.0489	0.0468	0.0444	0.0434
0.63	0.0707	0.0648	0.0623	0.0585	0.0543	0.0515	0.0500	0.0493	0.0473	0.0450	0.0440
0.64	0.0708	0.0649	0.0625	0.0587	0.0546	0.0518	0.0504	0.0497	0.0477	0.0454	0.0444
0.65	0.0709	0.0651	0.0627	0.0590	0.0549	0.0522	0.0508	0.0501	0.0481	0.0459	0.0449
0.66	0.0708	0.0651	0.0628	0.0591	0.0551	0.0525	0.0511	0.0504	0.0485	0.0463	0.0453
0.67	0.0708	0.0653	0.0629	0.0594	0.0554	0.0528	0.0514	0.0508	0.0489	0.0467	0.0458
0.68	0.0708	0.0653	0.0630	0.0595	0.0556	0.0531	0.0517	0.0511	0.0492	0.0471	0.0462
0.69	0.0705	0.0652	0.0629	0.0595	0.0557	0.0532	0.0518	0.0512	0.0494	0.0473	0.0464
0.70	0.0704	0.0651	0.0629	0.0595	0.0558	0.0533	0.0520	0.0514	0.0496	0.0476	0.0467

"b" VALUES FOR REYNOLDS NUMBER FACTOR, F_r—PIPE TAPS

$$F_r = 1 + \frac{b}{\sqrt{h_w p_f}}$$

β	Internal Diameter of Pipe, D, inches: 6				8			10			12		
	4.897	5.189	5.761	6.065	7.625	7.981	8.071	9.564	10.020	10.136	11.376	11.938	12.090
0.10	0.0859	0.0846	0.0825	0.0815	0.0777	0.0771	0.0769	0.0747	0.0742	0.0741	0.0729	0.0724	0.0725
0.11	0.0818	0.0805	0.0783	0.0773	0.0735	0.0729	0.0727	0.0706	0.0700	0.0699	0.0687	0.0682	0.0681
0.12	0.0777	0.0764	0.0742	0.0732	0.0695	0.0688	0.0687	0.0665	0.0660	0.0658	0.0646	0.0642	0.0640
0.13	0.0738	0.0725	0.0703	0.0694	0.0656	0.0649	0.0648	0.0626	0.0621	0.0619	0.0607	0.0603	0.0602
0.14	0.0701	0.0688	0.0666	0.0657	0.0619	0.0612	0.0611	0.0589	0.0584	0.0583	0.0571	0.0566	0.0565
0.15	0.0666	0.0653	0.0631	0.0622	0.0584	0.0578	0.0576	0.0554	0.0549	0.0548	0.0536	0.0531	0.0530
0.16	0.0633	0.0620	0.0598	0.0589	0.0551	0.0545	0.0543	0.0522	0.0516	0.0515	0.0503	0.0498	0.0497
0.17	0.0601	0.0589	0.0567	0.0558	0.0520	0.0514	0.0512	0.0491	0.0485	0.0484	0.0472	0.0467	0.0466
0.18	0.0571	0.0558	0.0537	0.0527	0.0490	0.0484	0.0482	0.0461	0.0455	0.0454	0.0442	0.0438	0.0436
0.19	0.0544	0.0531	0.0510	0.0500	0.0463	0.0456	0.0455	0.0433	0.0428	0.0427	0.0415	0.0410	0.0409
0.20	0.0517	0.0504	0.0483	0.0474	0.0436	0.0430	0.0428	0.0407	0.0402	0.0401	0.0389	0.0384	0.0383
0.21	0.0492	0.0480	0.0459	0.0449	0.0412	0.0405	0.0404	0.0383	0.0377	0.0376	0.0364	0.0360	0.0359
0.22	0.0469	0.0457	0.0436	0.0426	0.0389	0.0383	0.0381	0.0360	0.0355	0.0353	0.0342	0.0337	0.0336
0.23	0.0448	0.0435	0.0414	0.0405	0.0368	0.0362	0.0360	0.0339	0.0334	0.0333	0.0321	0.0316	0.0315
0.24	0.0428	0.0416	0.0395	0.0385	0.0349	0.0342	0.0341	0.0320	0.0314	0.0313	0.0302	0.0297	0.0296
0.25	0.0410	0.0398	0.0377	0.0367	0.0331	0.0324	0.0323	0.0302	0.0297	0.0296	0.0284	0.0279	0.0278
0.26	0.0393	0.0381	0.0360	0.0350	0.0314	0.0308	0.0306	0.0285	0.0280	0.0279	0.0267	0.0263	0.0262
0.27	0.0378	0.0366	0.0345	0.0336	0.0299	0.0293	0.0292	0.0271	0.0266	0.0264	0.0253	0.0248	0.0247
0.28	0.0365	0.0352	0.0332	0.0322	0.0286	0.0280	0.0278	0.0258	0.0252	0.0251	0.0240	0.0235	0.0234
0.29	0.0352	0.0340	0.0319	0.0310	0.0274	0.0268	0.0266	0.0245	0.0240	0.0239	0.0228	0.0223	0.0222
0.30	0.0341	0.0328	0.0308	0.0299	0.0263	0.0257	0.0255	0.0235	0.0230	0.0228	0.0217	0.0213	0.0211
0.31	0.0331	0.0319	0.0298	0.0289	0.0254	0.0247	0.0246	0.0225	0.0220	0.0219	0.0208	0.0203	0.0202
0.32	0.0322	0.0310	0.0290	0.0280	0.0245	0.0239	0.0237	0.0217	0.0212	0.0211	0.0199	0.0195	0.0194
0.33	0.0314	0.0302	0.0282	0.0273	0.0238	0.0232	0.0230	0.0210	0.0205	0.0204	0.0192	0.0188	0.0187
0.34	0.0308	0.0296	0.0276	0.0267	0.0232	0.0226	0.0225	0.0204	0.0199	0.0198	0.0187	0.0183	0.0182
0.35	0.0303	0.0291	0.0272	0.0263	0.0228	0.0221	0.0220	0.0200	0.0195	0.0194	0.0183	0.0178	0.0177
0.36	0.0299	0.0287	0.0268	0.0259	0.0224	0.0218	0.0216	0.0196	0.0192	0.0190	0.0179	0.0175	0.0174
0.37	0.0296	0.0284	0.0265	0.0256	0.0221	0.0215	0.0214	0.0194	0.0189	0.0188	0.0177	0.0173	0.0172
0.38	0.0294	0.0282	0.0263	0.0254	0.0220	0.0214	0.0212	0.0193	0.0188	0.0187	0.0176	0.0171	0.0170
0.39	0.0293	0.0281	0.0262	0.0253	0.0219	0.0213	0.0211	0.0192	0.0187	0.0186	0.0175	0.0171	0.0170
0.40	0.0292	0.0281	0.0261	0.0253	0.0219	0.0213	0.0212	0.0192	0.0187	0.0186	0.0175	0.0171	0.0170
0.41	0.0292	0.0281	0.0262	0.0253	0.0220	0.0214	0.0213	0.0193	0.0189	0.0187	0.0177	0.0173	0.0172
0.42	0.0294	0.0282	0.0263	0.0255	0.0222	0.0216	0.0214	0.0195	0.0191	0.0189	0.0179	0.0175	0.0174
0.43	0.0296	0.0285	0.0266	0.0257	0.0224	0.0218	0.0217	0.0198	0.0193	0.0192	0.0182	0.0178	0.0177
0.44	0.0298	0.0287	0.0268	0.0260	0.0227	0.0221	0.0220	0.0201	0.0197	0.0196	0.0185	0.0181	0.0180
0.45	0.0301	0.0290	0.0272	0.0263	0.0231	0.0225	0.0224	0.0205	0.0201	0.0200	0.0189	0.0185	0.0184
0.46	0.0305	0.0294	0.0276	0.0268	0.0235	0.0230	0.0228	0.0210	0.0205	0.0204	0.0194	0.0190	0.0189
0.47	0.0309	0.0298	0.0280	0.0272	0.0240	0.0235	0.0233	0.0215	0.0210	0.0209	0.0199	0.0195	0.0194
0.48	0.0314	0.0303	0.0285	0.0277	0.0245	0.0240	0.0238	0.0220	0.0216	0.0215	0.0205	0.0201	0.0200
0.49	0.0319	0.0308	0.0290	0.0282	0.0251	0.0245	0.0244	0.0226	0.0222	0.0221	0.0211	0.0207	0.0206
0.50	0.0324	0.0314	0.0296	0.0288	0.0257	0.0252	0.0250	0.0233	0.0228	0.0227	0.0217	0.0213	0.0212
0.51	0.0330	0.0319	0.0302	0.0294	0.0263	0.0258	0.0257	0.0239	0.0235	0.0234	0.0224	0.0220	0.0219
0.52	0.0335	0.0325	0.0308	0.0300	0.0270	0.0265	0.0263	0.0246	0.0242	0.0241	0.0231	0.0227	0.0226
0.53	0.0342	0.0332	0.0315	0.0307	0.0277	0.0272	0.0270	0.0253	0.0249	0.0248	0.0238	0.0235	0.0234
0.54	0.0348	0.0338	0.0321	0.0314	0.0284	0.0279	0.0278	0.0261	0.0256	0.0255	0.0246	0.0242	0.0241
0.55	0.0355	0.0345	0.0328	0.0321	0.0291	0.0286	0.0285	0.0268	0.0264	0.0263	0.0254	0.0250	0.0249
0.56	0.0361	0.0351	0.0335	0.0327	0.0298	0.0293	0.0292	0.0276	0.0272	0.0271	0.0261	0.0258	0.0257
0.57	0.0368	0.0358	0.0342	0.0335	0.0306	0.0301	0.0300	0.0283	0.0279	0.0278	0.0269	0.0266	0.0265
0.58	0.0375	0.0365	0.0349	0.0342	0.0313	0.0309	0.0307	0.0291	0.0287	0.0286	0.0277	0.0274	0.0273
0.59	0.0381	0.0372	0.0356	0.0349	0.0321	0.0316	0.0315	0.0299	0.0295	0.0294	0.0285	0.0282	0.0281
0.60	0.0388	0.0378	0.0363	0.0356	0.0328	0.0323	0.0322	0.0307	0.0303	0.0302	0.0293	0.0290	0.0289
0.61	0.0393	0.0384	0.0369	0.0362	0.0335	0.0330	0.0329	0.0314	0.0310	0.0309	0.0300	0.0297	0.0296
0.62	0.0399	0.0390	0.0375	0.0368	0.0342	0.0337	0.0336	0.0321	0.0317	0.0316	0.0307	0.0304	0.0303
0.63	0.0405	0.0396	0.0381	0.0375	0.0348	0.0344	0.0343	0.0328	0.0324	0.0323	0.0315	0.0311	0.0311
0.64	0.0411	0.0402	0.0387	0.0381	0.0355	0.0350	0.0349	0.0334	0.0331	0.0330	0.0322	0.0318	0.0318
0.65	0.0416	0.0407	0.0393	0.0386	0.0361	0.0357	0.0355	0.0341	0.0337	0.0336	0.0328	0.0325	0.0324
0.66	0.0421	0.0412	0.0398	0.0391	0.0366	0.0362	0.0361	0.0347	0.0343	0.0342	0.0334	0.0331	0.0331
0.67	0.0425	0.0417	0.0403	0.0397	0.0372	0.0368	0.0367	0.0352	0.0349	0.0349	0.0341	0.0338	0.0337
0.68	0.0430	0.0422	0.0408	0.0402	0.0377	0.0373	0.0372	0.0358	0.0355	0.0354	0.0346	0.0344	0.0343
0.69	0.0433	0.0425	0.0411	0.0405	0.0381	0.0377	0.0376	0.0363	0.0359	0.0359	0.0351	0.0348	0.0347
0.70	0.0436	0.0428	0.0415	0.0409	0.0386	0.0382	0.0381	0.0367	0.0364	0.0363	0.0356	0.0353	0.0352

"b" VALUES FOR REYNOLDS NUMBER FACTOR, F_r—PIPE TAPS

$$F_r = 1 + \frac{b}{\sqrt{h_w p_f}}$$

β	Internal Diameter of Pipe, D, inches: 16			20			24			30		
	14.688	15.000	15.250	18.814	19.000	19.250	22.626	23.000	23.250	28.628	29.000	29.250
0.10	0.0706	0.0705	0.0704	0.0690	0.0689	0.0688	0.0680	0.0679	0.0678	0.0669	0.0669	0.0668
0.11	0.0665	0.0663	0.0662	0.0648	0.0647	0.0647	0.0638	0.0637	0.0637	0.0628	0.0627	0.0627
0.12	0.0624	0.0622	0.0621	0.0607	0.0607	0.0606	0.0597	0.0596	0.0596	0.0587	0.0586	0.0586
0.13	0.0585	0.0584	0.0582	0.0569	0.0568	0.0567	0.0558	0.0558	0.0557	0.0548	0.0548	0.0547
0.14	0.0549	0.0548	0.0546	0.0532	0.0531	0.0530	0.0522	0.0521	0.0520	0.0511	0.0511	0.0511
0.15	0.0514	0.0512	0.0511	0.0497	0.0497	0.0496	0.0487	0.0486	0.0486	0.0477	0.0476	0.0476
0.16	0.0481	0.0479	0.0478	0.0464	0.0464	0.0463	0.0454	0.0453	0.0453	0.0444	0.0443	0.0443
0.17	0.0450	0.0448	0.0447	0.0433	0.0433	0.0432	0.0423	0.0423	0.0422	0.0413	0.0413	0.0412
0.18	0.0420	0.0419	0.0417	0.0404	0.0403	0.0402	0.0394	0.0393	0.0392	0.0383	0.0383	0.0383
0.19	0.0393	0.0391	0.0390	0.0376	0.0376	0.0375	0.0366	0.0366	0.0365	0.0356	0.0356	0.0355
0.20	0.0367	0.0365	0.0364	0.0350	0.0350	0.0349	0.0340	0.0340	0.0339	0.0330	0.0330	0.0329
0.21	0.0343	0.0341	0.0340	0.0326	0.0326	0.0325	0.0316	0.0316	0.0315	0.0306	0.0306	0.0305
0.22	0.0320	0.0318	0.0317	0.0304	0.0303	0.0302	0.0294	0.0293	0.0293	0.0284	0.0283	0.0283
0.23	0.0299	0.0298	0.0296	0.0283	0.0282	0.0281	0.0273	0.0272	0.0272	0.0263	0.0262	0.0262
0.24	0.0280	0.0278	0.0277	0.0264	0.0263	0.0262	0.0254	0.0253	0.0253	0.0244	0.0243	0.0243
0.25	0.0262	0.0261	0.0260	0.0246	0.0246	0.0245	0.0236	0.0236	0.0235	0.0226	0.0226	0.0226
0.26	0.0246	0.0244	0.0243	0.0230	0.0229	0.0228	0.0220	0.0219	0.0219	0.0210	0.0210	0.0209
0.27	0.0231	0.0230	0.0229	0.0215	0.0215	0.0214	0.0206	0.0205	0.0204	0.0196	0.0195	0.0195
0.28	0.0218	0.0217	0.0216	0.0202	0.0202	0.0201	0.0193	0.0192	0.0191	0.0183	0.0182	0.0182
0.29	0.0206	0.0205	0.0204	0.0191	0.0190	0.0189	0.0181	0.0180	0.0180	0.0171	0.0171	0.0170
0.30	0.0196	0.0194	0.0193	0.0180	0.0179	0.0179	0.0170	0.0170	0.0169	0.0161	0.0160	0.0160
0.31	0.0187	0.0185	0.0184	0.0171	0.0170	0.0170	0.0161	0.0161	0.0160	0.0152	0.0151	0.0151
0.32	0.0179	0.0177	0.0176	0.0163	0.0162	0.0162	0.0153	0.0153	0.0152	0.0144	0.0143	0.0143
0.33	0.0172	0.0170	0.0169	0.0156	0.0156	0.0155	0.0147	0.0146	0.0146	0.0137	0.0137	0.0136
0.34	0.0166	0.0165	0.0164	0.0151	0.0150	0.0150	0.0142	0.0141	0.0140	0.0132	0.0131	0.0131
0.35	0.0162	0.0161	0.0160	0.0147	0.0146	0.0145	0.0137	0.0137	0.0136	0.0128	0.0127	0.0127
0.36	0.0159	0.0157	0.0156	0.0144	0.0143	0.0142	0.0134	0.0134	0.0133	0.0125	0.0124	0.0124
0.37	0.0157	0.0155	0.0154	0.0141	0.0141	0.0140	0.0132	0.0132	0.0131	0.0123	0.0122	0.0122
0.38	0.0155	0.0154	0.0153	0.0140	0.0140	0.0139	0.0131	0.0130	0.0130	0.0122	0.0121	0.0121
0.39	0.0155	0.0154	0.0153	0.0140	0.0139	0.0139	0.0131	0.0130	0.0130	0.0122	0.0121	0.0121
0.40	0.0156	0.0154	0.0153	0.0141	0.0140	0.0139	0.0132	0.0131	0.0130	0.0122	0.0122	0.0122
0.41	0.0157	0.0155	0.0154	0.0142	0.0142	0.0141	0.0133	0.0132	0.0132	0.0124	0.0124	0.0123
0.42	0.0159	0.0158	0.0157	0.0144	0.0144	0.0143	0.0136	0.0135	0.0134	0.0126	0.0126	0.0126
0.43	0.0162	0.0161	0.0160	0.0148	0.0147	0.0146	0.0139	0.0138	0.0138	0.0130	0.0129	0.0129
0.44	0.0166	0.0164	0.0163	0.0151	0.0151	0.0150	0.0143	0.0142	0.0141	0.0134	0.0133	0.0133
0.45	0.0170	0.0169	0.0168	0.0156	0.0155	0.0155	0.0147	0.0146	0.0146	0.0138	0.0138	0.0137
0.46	0.0175	0.0174	0.0173	0.0161	0.0160	0.0160	0.0152	0.0151	0.0151	0.0143	0.0143	0.0143
0.47	0.0180	0.0179	0.0178	0.0166	0.0166	0.0165	0.0158	0.0157	0.0157	0.0149	0.0149	0.0148
0.48	0.0186	0.0185	0.0184	0.0172	0.0172	0.0171	0.0164	0.0163	0.0163	0.0155	0.0155	0.0154
0.49	0.0192	0.0191	0.0190	0.0178	0.0178	0.0177	0.0170	0.0169	0.0169	0.0162	0.0161	0.0161
0.50	0.0199	0.0198	0.0197	0.0185	0.0185	0.0184	0.0177	0.0176	0.0176	0.0169	0.0168	0.0168
0.51	0.0206	0.0205	0.0204	0.0193	0.0192	0.0191	0.0184	0.0184	0.0183	0.0176	0.0176	0.0175
0.52	0.0213	0.0212	0.0211	0.0200	0.0199	0.0199	0.0192	0.0191	0.0191	0.0183	0.0183	0.0183
0.53	0.0221	0.0220	0.0219	0.0208	0.0207	0.0207	0.0200	0.0199	0.0199	0.0191	0.0191	0.0191
0.54	0.0229	0.0227	0.0226	0.0215	0.0215	0.0214	0.0208	0.0207	0.0207	0.0199	0.0199	0.0199
0.55	0.0237	0.0235	0.0234	0.0224	0.0223	0.0223	0.0216	0.0215	0.0215	0.0208	0.0207	0.0207
0.56	0.0244	0.0243	0.0242	0.0232	0.0231	0.0231	0.0224	0.0223	0.0223	0.0216	0.0216	0.0215
0.57	0.0253	0.0251	0.0250	0.0240	0.0239	0.0239	0.0232	0.0232	0.0231	0.0224	0.0224	0.0224
0.58	0.0261	0.0260	0.0259	0.0248	0.0248	0.0247	0.0241	0.0240	0.0240	0.0233	0.0233	0.0232
0.59	0.0269	0.0268	0.0267	0.0256	0.0256	0.0255	0.0249	0.0248	0.0248	0.0241	0.0241	0.0241
0.60	0.0277	0.0276	0.0275	0.0265	0.0264	0.0264	0.0257	0.0257	0.0256	0.0250	0.0249	0.0249
0.61	0.0284	0.0283	0.0282	0.0272	0.0272	0.0271	0.0265	0.0265	0.0264	0.0258	0.0257	0.0257
0.62	0.0292	0.0291	0.0290	0.0280	0.0280	0.0279	0.0273	0.0272	0.0272	0.0266	0.0265	0.0265
0.63	0.0299	0.0298	0.0297	0.0288	0.0287	0.0287	0.0281	0.0280	0.0280	0.0273	0.0273	0.0273
0.64	0.0306	0.0305	0.0304	0.0295	0.0295	0.0294	0.0288	0.0288	0.0287	0.0281	0.0281	0.0280
0.65	0.0313	0.0312	0.0312	0.0302	0.0302	0.0301	0.0295	0.0295	0.0294	0.0288	0.0288	0.0288
0.66	0.0320	0.0319	0.0318	0.0309	0.0308	0.0308	0.0302	0.0301	0.0301	0.0295	0.0295	0.0295
0.67	0.0326	0.0325	0.0324	0.0315	0.0315	0.0314	0.0309	0.0308	0.0308	0.0302	0.0302	0.0302
0.68	0.0332	0.0331	0.0330	0.0322	0.0321	0.0321	0.0315	0.0315	0.0314	0.0308	0.0308	0.0308
0.69	0.0337	0.0336	0.0335	0.0327	0.0326	0.0326	0.0320	0.0320	0.0320	0.0314	0.0314	0.0313
0.70	0.0342	0.0341	0.0340	0.0332	0.0332	0.0331	0.0326	0.0325	0.0325	0.0319	0.0319	0.0319

EXPANSION FACTORS—PIPE TAPS—Y_1
Static Pressure Taken from Upstream Taps

$\frac{h_w}{p_{f1}}$ Ratio	$\beta = \frac{d}{D}$ Ratio										
	.1	.2	.3	.4	.45	.50	.52	.54	.56	.58	.60
0.0	1.0000	1.0000	1.0000	1.0000	1.0000	1.0000	1.0000	1.0000	1.0000	1.0000	1.0000
0.1	.9990	.9989	.9988	.9985	.9984	.9982	.9981	.9980	.9979	.9978	.9977
0.2	.9981	.9979	.9976	.9971	.9968	.9964	.9962	.9961	.9959	.9957	.9954
0.3	.9971	.9968	.9964	.9956	.9952	.9946	.9944	.9941	.9938	.9935	.9931
0.4	.9962	.9958	.9951	.9942	.9936	.9928	.9925	.9921	.9917	.9913	.9908
0.5	.9952	.9947	.9939	.9927	.9919	.9910	.9906	.9902	.9897	.9891	.9885
0.6	.9943	.9937	.9927	.9913	.9903	.9892	.9887	.9882	.9876	.9870	.9862
0.7	.9933	.9926	.9915	.9898	.9887	.9874	.9869	.9862	.9856	.9848	.9840
0.8	.9923	.9916	.9903	.9883	.9871	.9857	.9850	.9843	.9835	.9826	.9817
0.9	.9914	.9905	.9891	.9869	.9855	.9839	.9831	.9823	.9814	.9805	.9794
1.0	.9904	.9895	.9878	.9854	.9839	.9821	.9812	.9803	.9794	.9783	.9771
1.1	.9895	.9884	.9866	.9840	.9823	.9803	.9794	.9784	.9773	.9761	.9748
1.2	.9885	.9874	.9854	.9825	.9807	.9785	.9775	.9764	.9752	.9739	.9725
1.3	.9876	.9863	.9842	.9811	.9791	.9767	.9756	.9744	.9732	.9718	.9702
1.4	.9866	.9853	.9830	.9796	.9775	.9749	.9737	.9725	.9711	.9696	.9679
1.5	.9857	.9842	.9818	.9782	.9758	.9731	.9719	.9705	.9690	.9674	.9656
1.6	.9847	.9832	.9805	.9767	.9742	.9713	.9700	.9685	.9670	.9652	.9633
1.7	.9837	.9821	.9793	.9752	.9726	.9695	.9681	.9666	.9649	.9631	.9610
1.8	.9828	.9811	.9781	.9738	.9710	.9677	.9662	.9646	.9628	.9609	.9587
1.9	.9818	.9800	.9769	.9723	.9694	.9659	.9643	.9626	.9608	.9587	.9565
2.0	.9809	.9790	.9757	.9709	.9678	.9641	.9625	.9607	.9587	.9566	.9542
2.1	.9799	.9779	.9745	.9694	.9662	.9623	.9606	.9587	.9566	.9544	.9519
2.2	.9790	.9768	.9732	.9680	.9645	.9605	.9587	.9567	.9546	.9522	.9496
2.3	.9780	.9758	.9720	.9665	.9630	.9587	.9568	.9548	.9525	.9500	.9473
2.4	.9770	.9747	.9708	.9650	.9613	.9570	.9550	.9528	.9505	.9479	.9450
2.5	.9761	.9737	.9696	.9636	.9597	.9552	.9531	.9508	.9484	.9457	.9427
2.6	.9751	.9726	.9684	.9621	.9581	.9534	.9512	.9489	.9463	.9435	.9404
2.7	.9742	.9716	.9672	.9607	.9565	.9516	.9493	.9469	.9443	.9414	.9381
2.8	.9732	.9705	.9659	.9592	.9549	.9498	.9475	.9449	.9422	.9392	.9358
2.9	.9723	.9695	.9647	.9578	.9533	.9480	.9456	.9430	.9401	.9370	.9335
3.0	.9713	.9684	.9635	.9563	.9517	.9462	.9437	.9410	.9381	.9348	.9312
3.1	.9704	.9674	.9623	.9549	.9501	.9444	.9418	.9390	.9360	.9327	.9290
3.2	.9694	.9663	.9611	.9534	.9485	.9426	.9400	.9371	.9339	.9305	.9267
3.3	.9684	.9653	.9599	.9519	.9469	.9408	.9381	.9351	.9319	.9283	.9244
3.4	.9575	.9642	.9587	.9505	.9452	.9390	.9362	.9331	.9298	.9261	.9221
3.5	.9665	.9632	.9574	.9490	.9436	.9372	.9343	.9312	.9277	.9240	.9198
3.6	.9656	.9621	.9562	.9476	.9420	.9354	.9324	.9292	.9257	.9218	.9175
3.7	.9646	.9611	.9550	.9461	.9404	.9336	.9306	.9272	.9236	.9196	.9152
3.8	.9637	.9600	.9538	.9447	.9388	.9318	.9287	.9253	.9216	.9175	.9129
3.9	.9627	.9590	.9526	.9432	.9372	.9301	.9268	.9233	.9195	.9153	.9106
4.0	.9617	.9579	.9514	.9417	.9356	.9283	.9249	.9213	.9174	.9131	.9083

EXPANSION FACTORS—PIPE TAPS—Y_1
Static Pressure Taken from Upstream Taps

$\frac{h_w}{p_{f1}}$ Ratio	$\beta = \frac{d}{D}$ Ratio									
	.61	.62	.63	.64	.65	.66	.67	.68	.69	.70
0.0	1.0000	1.0000	1.0000	1.0000	1.0000	1.0000	1.0000	1.0000	1.0000	1.0000
0.1	.9976	.9976	.9975	.9974	.9973	.9972	.9971	.9970	.9969	.9968
0.2	.9953	.9951	.9950	.9948	.9947	.9945	.9943	.9941	.9938	.9935
0.3	.9929	.9927	.9925	.9923	.9920	.9917	.9914	.9911	.9907	.9903
0.4	.9906	.9903	.9900	.9897	.9893	.9890	.9886	.9881	.9876	.9871
0.5	.9882	.9879	.9875	.9871	.9867	.9862	.9857	.9851	.9845	.9839
0.6	.9859	.9854	.9850	.9845	.9840	.9834	.9828	.9822	.9814	.9806
0.7	.9835	.9830	.9825	.9819	.9813	.9807	.9800	.9792	.9784	.9774
0.8	.9811	.9606	.9800	.9794	.9787	.9779	.9771	.9762	.9753	.9742
0.9	.9788	.9782	.9775	.9768	.9760	.9752	.9742	.9733	.9722	.9710
1.0	.9764	.9757	.9750	.9742	.9733	.9724	.9714	.9703	.9691	.9677
1.1	.9741	.9733	.9725	.9716	.9707	.9696	.9685	.9673	.9660	.9645
1.2	.9717	.9709	.9700	.9690	.9680	.9669	.9657	.9643	.9629	.9613
1.3	.9694	.9685	.9675	.9664	.9653	.9641	.9628	.9614	.9598	.9581
1.4	.9670	.9660	.9650	.9639	.9627	.9614	.9599	.9584	.9567	.9548
1.5	.9646	.9636	.9625	.9613	.9600	.9586	.9571	.9554	.9536	.9516
1.6	.9623	.9612	.9600	.9587	.9573	.9558	.9542	.9525	.9505	.9484
1.7	.9599	.9587	.9575	.9561	.9547	.9531	.9514	.9495	.9474	.9452
1.8	.9576	.9563	.9550	.9535	.9520	.9503	.9485	.9465	.9443	.9419
1.9	.9552	.9539	.9525	.9510	.9493	.9476	.9456	.9435	.9412	.9387
2.0	.9529	.9515	.9500	.9484	.9467	.9448	.9428	.9406	.9381	.9355
2.1	.9505	.9490	.9475	.9458	.9440	.9420	.9399	.9376	.9351	.9323
2.2	.9481	.9466	.9450	.9432	.9413	.9393	.9371	.9346	.9320	.9290
2.3	.9458	.9442	.9425	.9406	.9387	.9365	.9342	.9317	.9289	.9258
2.4	.9434	.9418	.9400	.9381	.9360	.9338	.9313	.9287	.9258	.9226
2.5	.9411	.9393	.9375	.9355	.9333	.9310	.9285	.9257	.9227	.9194
2.6	.9387	.9369	.9350	.9329	.9307	.9282	.9256	.9227	.9196	.9161
2.7	.9364	.9345	.9325	.9303	.9280	.9255	.9227	.9198	.9165	.9129
2.8	.9340	.9321	.9300	.9277	.9253	.9227	.9199	.9168	.9134	.9097
2.9	.9316	.9296	.9275	.9252	.9227	.9200	.9170	.9138	.9103	.9064
3.0	.9293	.9272	.9250	.9226	.9200	.9172	.9142	.9108	.9072	.9032
3.1	.9269	.9248	.9225	.9200	.9173	.9144	.9113	.9079	.9041	.9000
3.2	.9246	.9223	.9200	.9174	.9147	.9117	.9084	.9049	.9010	.8968
3.3	.9222	.9199	.9175	.9148	.9120	.9089	.9056	.9019	.8979	.8935
3.4	.9199	.9175	.9150	.9122	.9093	.9062	.9027	.8990	.8948	.8903
3.5	.9175	.9151	.9125	.9097	.9067	.9034	.8999	.8960	.8918	.8871
3.6	.9151	.9126	.9100	.9071	.9040	.9006	.8970	.8930	.8887	.8839
3.7	.9128	.9102	.9075	.9045	.9013	.8979	.8941	.8900	.8856	.8806
3.8	.9104	.9078	.9050	.9019	.8987	.8951	.8913	.8871	.8825	.8774
3.9	.9081	.9054	.9025	.8993	.8960	.8924	.8884	.8841	.8794	.8742
4.0	.9057	.9029	.9000	.8968	.8933	.8896	.8856	.8811	.8763	.8710

EXPANSION FACTORS—PIPE TAPS—Y_2
Static Pressure Taken from Downstream Taps

$\frac{h_w}{p_{f2}}$ Ratio	$\beta = \frac{d}{D}$ Ratio										
	.1	.2	.3	.4	.45	.50	.52	.54	.56	.58	.60
0.0	1.0000	1.0000	1.0000	1.0000	1.0000	1.0000	1.0000	1.0000	1.0000	1.0000	1.0000
0.1	1.0008	1.0008	1.0006	1.0003	1.0002	1.0000	.9999	.9998	.9997	.9996	.9995
0.2	1.0017	1.0015	1.0012	1.0007	1.0004	1.0000	.9999	.9997	.9995	.9993	.9990
0.3	1.0025	1.0023	1.0018	1.0010	1.0006	1.0000	.9998	.9995	.9992	.9989	.9986
0.4	1.0034	1.0030	1.0024	1.0014	1.0008	1.0001	.9997	.9994	.9990	.9986	.9981
0.5	1.0042	1.0038	1.0030	1.0018	1.0010	1.0001	.9997	.9992	.9988	.9982	.9976
0.6	1.0051	1.0045	1.0036	1.0021	1.0012	1.0001	.9996	.9991	.9985	.9979	.9972
0.7	1.0059	1.0053	1.0041	1.0025	1.0014	1.0002	.9996	.9990	.9983	.9975	.9967
0.8	1.0068	1.0060	1.0047	1.0028	1.0016	1.0002	.9995	.9988	.9980	.9972	.9962
0.9	1.0076	1.0068	1.0053	1.0032	1.0018	1.0002	.9995	.9987	.9978	.9969	.9958
1.0	1.0085	1.0075	1.0059	1.0036	1.0021	1.0003	.9994	.9986	.9976	.9965	.9954
1.1	1.0093	1.0083	1.0065	1.0039	1.0023	1.0003	.9994	.9984	.9974	.9962	.9949
1.2	1.0102	1.0091	1.0071	1.0043	1.0025	1.0004	.9994	.9983	.9972	.9959	.9945
1.3	1.0110	1.0098	1.0077	1.0047	1.0027	1.0004	.9994	.9982	.9970	.9956	.9941
1.4	1.0119	1.0106	1.0083	1.0051	1.0030	1.0004	.9993	.9981	.9968	.9953	.9936
1.5	1.0127	1.0113	1.0089	1.0054	1.0032	1.0005	.9993	.9980	.9966	.9950	.9932
1.6	1.0136	1.0121	1.0096	1.0058	1.0034	1.0006	.9993	.9979	.9964	.9947	.9928
1.7	1.0144	1.0128	1.0102	1.0062	1.0036	1.0006	.9992	.9978	.9962	.9944	.9924
1.8	1.0153	1.0136	1.0108	1.0066	1.0039	1.0007	.9992	.9977	.9960	.9941	.9920
1.9	1.0161	1.0144	1.0114	1.0070	1.0041	1.0008	.9992	.9976	.9958	.9938	.9916
2.0	1.0170	1.0151	1.0120	1.0073	1.0044	1.0008	.9992	.9975	.9956	.9935	.9912
2.1	1.0178	1.0159	1.0126	1.0077	1.0046	1.0009	.9992	.9974	.9954	.9932	.9908
2.2	1.0187	1.0167	1.0132	1.0081	1.0048	1.0010	.9992	.9973	.9952	.9929	.9904
2.3	1.0195	1.0174	1.0138	1.0085	1.0051	1.0010	.9992	.9972	.9950	.9927	.9900
2.4	1.0204	1.0182	1.0144	1.0089	1.0053	1.0011	.9992	.9971	.9949	.9924	.9896
2.5	1.0212	1.0189	1.0150	1.0093	1.0056	1.0012	.9992	.9971	.9947	.9921	.9893
2.6	1.0221	1.0197	1.0156	1.0097	1.0058	1.0013	.9992	.9970	.9945	.9919	.9889
2.7	1.0229	1.0205	1.0162	1.0101	1.0061	1.0014	.9992	.9969	.9944	.9916	.9885
2.8	1.0238	1.0212	1.0169	1.0104	1.0063	1.0014	.9992	.9968	.9942	.9914	.9882
2.9	1.0246	1.0220	1.0175	1.0108	1.0066	1.0015	.9992	.9968	.9941	.9911	.9878
3.0	1.0255	1.0228	1.0181	1.0112	1.0068	1.0016	.9993	.9967	.9939	.9908	.9874
3.1	1.0264	1.0235	1.0187	1.0116	1.0071	1.0017	.9993	.9966	.9938	.9906	.9871
3.2	1.0272	1.0243	1.0193	1.0120	1.0074	1.0018	.9993	.9966	.9936	.9904	.9867
3.3	1.0280	1.0250	1.0199	1.0124	1.0076	1.0019	.9993	.9965	.9935	.9901	.9864
3.4	1.0289	1.0258	1.0206	1.0128	1.0079	1.0020	.9994	.9965	.9933	.9899	.9860
3.5	1.0298	1.0266	1.0212	1.0133	1.0082	1.0021	.9994	.9964	.9932	.9895	.9857
3.6	1.0306	1.0273	1.0218	1.0137	1.0084	1.0022	.9994	.9964	.9931	.9894	.9854
3.7	1.0314	1.0281	1.0224	1.0141	1.0087	1.0024	.9994	.9963	.9929	.9892	.9850
3.8	1.0323	1.0289	1.0230	1.0145	1.0090	1.0025	.9995	.9963	.9928	.9890	.9847
3.9	1.0332	1.0296	1.0237	1.0149	1.0093	1.0026	.9995	.9963	.9927	.9888	.9844
4.0	1.0340	1.0304	1.0243	1.0153	1.0095	1.0027	.9996	.9962	.9926	.9885	.9840

EXPANSION FACTORS—PIPE TAPS—Y_2
Static Pressure Taken from Downstream Taps

$\frac{h_w}{p_{f2}}$ Ratio	$\beta = \frac{d}{D}$ Ratio									
	.61	.62	.63	.64	.65	.66	.67	.68	.69	.70
0.0	1.0000	1.0000	1.0000	1.0000	1.0000	1.0000	1.0000	1.0000	1.0000	1.0000
0.1	.9994	.9994	.9993	.9992	.9991	.9990	.9989	.9988	.9987	.9986
0.2	.9989	.9988	.9986	.9985	.9983	.9981	.9979	.9977	.9974	.9972
0.3	.9984	.9982	.9979	.9977	.9974	.9972	.9969	.9965	.9962	.9958
0.4	.9978	.9976	.9972	.9969	.9966	.9962	.9958	.9954	.9949	.9944
0.5	.9973	.9970	.9966	.9962	.9958	.9953	.9948	.9942	.9936	.9930
0.6	.9968	.9964	.9959	.9954	.9949	.9944	.9938	.9931	.9924	.9916
0.7	.9962	.9958	.9953	.9947	.9941	.9935	.9928	.9920	.9912	.9902
0.8	.9957	.9952	.9946	.9940	.9933	.9926	.9918	.9909	.9899	.9889
0.9	.9952	.9946	.9940	.9932	.9925	.9917	.9908	.9898	.9887	.9875
1.0	.9947	.9940	.9933	.9925	.9917	.9908	.9898	.9887	.9875	.9862
1.1	.9942	.9935	.9927	.9918	.9909	.9899	.9888	.9876	.9863	.9848
1.2	.9937	.9929	.9920	.9911	.9901	.9890	.9878	.9865	.9851	.9835
1.3	.9932	.9924	.9914	.9904	.9893	.9881	.9868	.9854	.9839	.9822
1.4	.9928	.9918	.9908	.9897	.9885	.9872	.9859	.9844	.9827	.9809
1.5	.9923	.9912	.9902	.9890	.9877	.9864	.9849	.9833	.9815	.9796
1.6	.9918	.9907	.9896	.9883	.9870	.9855	.9840	.9822	.9804	.9783
1.7	.9913	.9902	.9889	.9876	.9862	.9847	.9830	.9812	.9792	.9770
1.8	.9908	.9896	.9883	.9870	.9854	.9838	.9821	.9801	.9780	.9757
1.9	.9904	.9891	.9877	.9863	.9847	.9830	.9811	.9791	.9769	.9744
2.0	.9899	.9886	.9872	.9856	.9840	.9822	.9802	.9781	.9757	.9732
2.1	.9895	.9881	.9866	.9849	.9832	.9813	.9793	.9770	.9746	.9719
2.2	.9890	.9876	.9860	.9843	.9825	.9805	.9784	.9760	.9734	.9706
2.3	.9886	.9870	.9854	.9836	.9817	.9797	.9774	.9750	.9723	.9694
2.4	.9881	.9865	.9848	.9830	.9810	.9789	.9765	.9740	.9712	.9681
2.5	.9877	.9860	.9842	.9823	.9803	.9780	.9756	.9730	.9701	.9669
2.6	.9873	.9855	.9837	.9817	.9796	.9772	.9747	.9720	.9690	.9657
2.7	.9868	.9850	.9831	.9811	.9788	.9764	.9738	.9710	.9679	.9644
2.8	.9864	.9846	.9826	.9804	.9781	.9757	.9730	.9700	.9668	.9632
2.9	.9860	.9841	.9820	.9798	.9774	.9749	.9721	.9690	.9657	.9620
3.0	.9856	.9836	.9815	.9792	.9767	.9741	.9712	.9681	.9646	.9608
3.1	.9852	.9831	.9809	.9786	.9760	.9733	.9703	.9671	.9635	.9596
3.2	.9848	.9826	.9804	.9780	.9754	.9725	.9695	.9661	.9625	.9584
3.3	.9843	.9822	.9798	.9774	.9747	.9718	.9686	.9652	.9614	.9572
3.4	.9839	.9817	.9793	.9768	.9740	.9710	.9678	.9642	.9603	.9561
3.5	.9835	.9812	.9788	.9762	.9733	.9702	.9669	.9633	.9593	.9549
3.6	.9832	.9808	.9783	.9756	.9727	.9695	.9661	.9623	.9582	.9537
3.7	.9828	.9803	.9778	.9750	.9720	.9688	.9652	.9614	.9572	.9526
3.8	.9824	.9799	.9772	.9744	.9713	.9680	.9644	.9605	.9562	.9514
3.9	.9820	.9794	.9767	.9738	.9707	.9673	.9636	.9596	.9551	.9503
4.0	.9816	.9790	.9762	.9732	.9700	.9665	.9628	.9586	.9541	.9491

in the development stage presently. Other than those already discussed are the magnetic flowmeters, thermal meters, anemometers, current meters, vortex shedding meters, and sonic flowmeters. A thorough discussion of all types of fluid meters may be found in Reference 2.

A limited selection of tables for determining the constants for use in orifice metering are included in this chapter. A more complete set of tables and charts may be found in References 1 and 3. A programmable calculator program for use in a Hewlett-Packard model 41-C calculator may be found in Reference 4. A similar program for the Texas Instruments model 59 was published in Reference 5.

REFERENCES

1. American Gas Association: "Orifice Metering of Natural Gas," Gas Measurement Committee Report No. 3 (1969) Revision.
2. *Fluid Meters—Their Theory and Application:* Report of ASME Research Committee on Fluid Meters, 6th Edition, The American Society of Mechanical Engineers, New York (1971).
3. *Orifice Meter Constants:* Handbook E-2, Singer American Meter Division (1973).
4. Belcher, Phil: "Calculator Program Written for Accurate AGA Orifice Metering of Natural Gas," *Oil & Gas J.* (Aug. 8, 1983).
5. Martin, J. R.: "Program Computes Orifice Meter Flow Rate," *Oil & Gas J.* (Oct. 12, 1981).

8 Gas-Condensate Reservoirs

THE importance of gas-condensate reservoirs has increased considerably in the last few years because of the increased demand for all types of hydrocarbon energy sources and because, as deeper drilling occurs, more and more condensate reservoirs are being discovered.

Gas-condensate reservoirs may occur at pressures below 2000 psi and temperatures below 100°F and probably can occur at any higher fluid pressures and temperatures. Most known gas-condensate reservoirs are in the range of 3000 to 6000 psi and 200 to 400°F. These pressure and temperature ranges, together with wide composition ranges, provide a great variety of conditions for the physical behavior of gas-condensate deposits. This emphasizes the need for engineering studies of each gas-condensate reservoir in order to arrive at the best mode of development and operation.

Condensate reservoirs include both wet gas and retrograde condensate types, as described in Chapter 2. The phase diagrams for the two types are shown in Figures 2-28 and 2-29. In both types the fluid exists as a single phase gas at initial reservoir conditions, the distinction being that liquid is formed in the reservoir only for the retrograde type.

A rough classification can be made based on the performance of the reservoir at initial conditions. If initial producing gas-oil ratio (GOR) exceeds about 15,000 scf/STB, and pressure and temperature do not exceed 8000 psia and 225°F, respectively, it is likely that no liquid will form in the reservoir.

Where producing gas-oil ratios exceed 6000 but not 15,000 scf/STB, the reservoir may experience retrograde behavior. Under these conditions it is important that a representative reservoir fluid sample be obtained and a pressure-volume-temperature (pVT) analysis be made. This will permit planning for the most efficient and profitable development of the reservoir. Where the initial gas-oil ratio is between 3000 and 6000 scf/STB, the reservoir may contain either a volatile oil or a retrograde gas condensate. A representative fluid sample can be obtained and pVT studies made to make the distinction between the two types. The distinction is important, since optimum exploitation for the two different types of reservoirs may be substantially different. Where spacing regulations exist, spacing would normally be wider for a "gas" reservoir as opposed to an "oil" reservoir. These regulations recognize the increased mobility of gas, as contrasted with oil and the different migration capabilities of each during producing operations.

Below an initial producing gas-oil ratio of 3000 scf/STB, the reservoir will contain oil, volatile or otherwise. It is theoretically possible to have a retrograde gas condensate reservoir with an initial producing gas-oil ratio as low as 2000 scf/STB, but this would take a unique combination of very high discovery pressure (exceeding 8000 psia), modest temperatures, and high concentrations of intermediates, C_2 through C_6.

Table 8-1 shows the mole compositions of a typical dry gas, single-phase wet gas, retrograde gas condensate, and a volatile oil reservoir.

Figure 8-1 illustrates a pressure-temperature phase diagram for a reservoir fluid. The development of this diagram assumes that volume and the total composition of the fluid are constants. The figure is useful as a guide to relationships among the several reservoir fluid types but is not truly representative of all types since both single phase gas reservoirs and oil reservoirs are represented, with only pressure and temperature allowed to

TABLE 8-1
Mole Composition of Typical Hydrocarbon Reservoir Fluids

Component	Dry Gas	Single Phase Wet Gas	Retrograde Gas Condensate	Volatile Oil
C_1	96.0	90.0	75.0	60.0
C_2	2.0	3.0	7.0	8.0
C_3	1.0	2.0	4.5	4.0
C_4	0.5	2.0	3.0	4.0
C_5	0.5	1.0	2.0	3.0
C_6	—	0.5	2.5	4.0
C_{7+}	—	1.5	6.0	17.0
	100.0	100.0	100.0	100.0
Mol. Wt. C_{7+}	-	115	125	180
GOR, scf/STB	High	26,000	7,000	2,000
Tank Gravity, °API	—	60	55	50
Liquid Color	—	Water White to Light Yellow	Light Yellow to Yellow	Amber to Darker Colors

change. However, it is useful for descriptive purposes.

Single-phase gas reservoirs are shown by point A to the right of the cricondentherm. Pressure depletion with production at constant reservoir temperature results in the gas remaining as gas. However, cooling and pressure drop in the wellbore and surface facilities allow the condensing of hydrocarbons along line A to A_2.

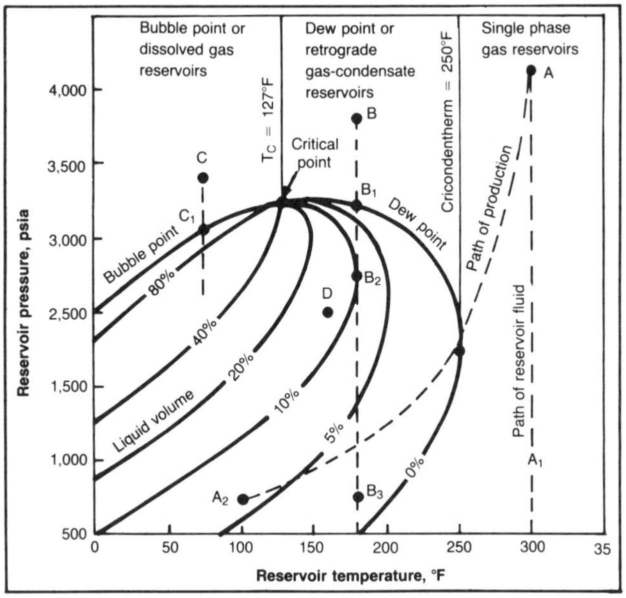

Fig. 8-1. Pressure-temperature diagram of a reservoir fluid. B. C. Craft/M. F. Hawkins, Applied Petroleum Reservoir Engineering, copyright 1959, pp. 64, 66, 74, 76, 78. Reprinted by permission of Prentice-Hall, Inc., Englewood Cliffs, N.J.

Retrograde gas condensate reservoirs exist at pressures sufficient to be at or above the upper boundary of the two-phase envelope and at a temperature between the critical and cricondentherm values. Frequently, retrograde gas condensate reservoirs are found to be on, or very close to, the dew-point line at the time of discovery. Apparently this is due to the presence of a large percentage of intermediates (C_2 to C_6). It is also quite common to find a small volatile oil rim in the reservoir. By definition, in this latter instance the gas cap would be exactly at the dew point. Should the observed dew point in the laboratory duplicate the discovery pressure within a few percent, an oil rim would probably be present.

Figure 8-2 presents descriptive phase diagrams for a gas cap gas and oil zone fluid for the two cases of a retrograde gas cap, and a nonretrograde gas cap. Notice the superimposing of the diagrams at the time of discovery, since phase equilibrium exists.

WELL TESTING AND SAMPLING

Proper testing of gas-condensate wells is essential for ascertaining the state of the hydrocarbon system at reservoir conditions and for planning the best production and recovery program for the reservoir. Without proper well tests and samples, it would be impossible to determine accurately the phase conditions of the reservoir contents at reservoir temperature and pressure and to estimate accurately the amount of hydrocarbon materials in place.

Tests are made on gas condensate wells for a number of specific purposes: to obtain representative samples for laboratory analysis, for identifying the composition and properties of the reservoir fluids; to make field determinations of gas and liquid properties; and to determine

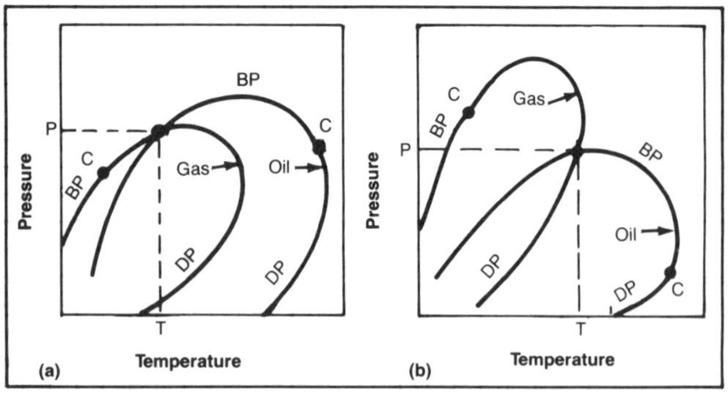

Fig. 8-2. Phase diagrams of a gas cap gas and oil zone liquid for (a) retrograde gas cap gas and (b) nonretrograde gas cap gas. B. C. Craft/M. F. Hawkins, Applied Petroleum Reservoir Engineering, copyright 1959, pp. 64, 66, 74, 76, 78. Reprinted by permission of Prentice-Hall, Inc., Englewood Cliffs, N.J.

formation and well characteristics, including productivity, producibility, and injectivity. The first consideration for selecting wells for gas-condensate fluid samples is that the well be far enough removed from the "black-oil ring" (if present) to minimize any chance that the liquid oil phase will enter the well during the test period. A second and highly important consideration is the selection of wells with productivities as high as possible, so that minimum pressure drawdown will occur in actually acquiring the reservoir fluid samples.

Well Conditioning and Sampling Procedures

Proper well conditioning is essential to obtaining representative samples from the reservoir. The best production rates prior to and during the sampling procedure have to be considered individually for each reservoir and for each well. Usually the best procedure is to employ the lowest rate that results in smooth well operation and the most dependable measurements of surface products. Minimum drawdown of bottom-hole pressure during the conditioning period is desirable, and the produced gas-liquid ratio should remain constant (within about 2%) for several days; the less permeable reservoirs require longer periods. The further the well deviates from a constant producing gas-liquid ratio, the greater the likelihood that the samples will not be representative.

Recombined separator samples from gas condensate wells are usually considered more representative of the original reservoir fluid than are bottom-hole samples. The tubing strings of gas-condensate wells almost invariably contain both liquid and gas, usually with a liquid layer adhering to the tubing walls making it very difficult to obtain a representative sample from such a stream by employing bottom-hole samplers. In addition, the relatively large quantities of material required for adequate laboratory studies favor surface sampling.

The original reservoir fluid cannot be simulated in the laboratory unless accurate field measurements of all the separator streams are taken. If the produced gas-condensate ratio from field measurements is in error by as little as 5%, the dew point pressure determined in the laboratory may be in error by as much as 100 psi.

Separator pressures and temperatures should remain as constant as possible during the well conditioning period; this will help maintain constancy of the stream rates and the observed hydrocarbon gas-liquid ratio.

Laboratory Testing

With laboratory equipment, pressure-depletion tests can be made on a gas-condensate reservoir fluid in such a manner as to simulate pressure depletion of the actual reservoir, usually assuming that retrograde liquid appearing during pressure reduction would remain immobile in the reservoir. This assumption applies in those cases for which volume of condensation is low enough that liquid saturation of the rock pores remains too low to provide significant permeability to the retrograde liquid. However, there are cases wherein reduction of pressure isothermally would condense relatively large volumes of liquid in the pores, raising the liquid saturation high enough for some liquid to migrate to producing wells. If preliminary tests indicate this possibility, it is then necessary to obtain relative-permeability curves for the particular rock-liquid system to adjust the predicted behavior of the reservoir.

The actual test procedure for simulating pressure depletion (with immobile reservoir liquid) in a laboratory visual cell is to place the properly recombined reservoir fluid sample of known total composition in the cell at original reservoir pressure and temperature. Maintaining the cell volume and temperature constant (to simulate a constant-volume reservoir), pressure is reduced stepwise at about 500 psi increments by removing gas. The volumes of gas and liquid (below the dew point) remaining in the cell are recorded. The gas removed at each step is measured and analyzed. The fractional analyses yield composition information for calculating the condensable liquids content of the gases produced at all stages of pressure depletion. Condensible liquids content can also be determined experimentally by subjecting the cell output gas stream to field separation conditions, but this usually is not considered necessary. The computation of natural gas content is useful in designing gasoline plants for recovery of liquid products.

The stepwise pressure-depletion tests are usually carried to some level representing abandonment pressure of the reservoir, after which the quantity and composition of the fluids remaining in the cell are determined. This provides a means of checking the quantity of each component produced in the gases during pressure-depletion, since original system composition is known. An example of a pressure-depletion test starting at the dew point pressure of 4265 psia is shown in Figure 8-3. The gasoline content of the heavier components in the produced gas is illustrated in Figure 8-4.

CALCULATION OF INITIAL IN-PLACE GAS AND CONDENSATE

The initial amounts of both the gas and the condensate in a reservoir can be calculated on a unit volume basis by two methods. The more accurate method requires a compositional analysis of the reservoir fluid, but an estimate can be made if limited data are available. A procedure and example calculation for each case are presented. The example was adapted from Reference 1.

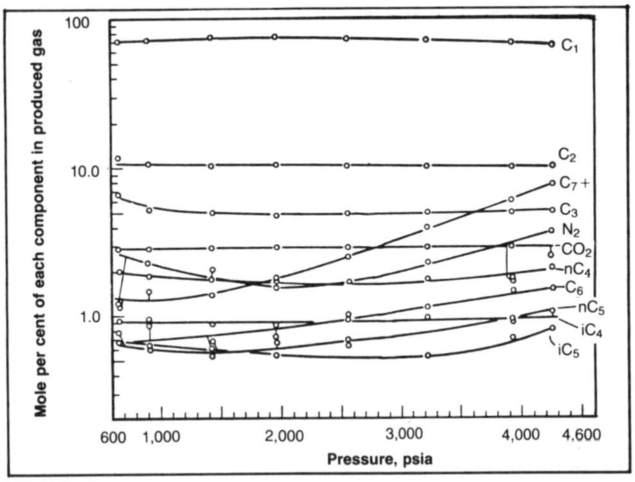

Fig. 8-3. Change in composition of produced gas during laboratory pressure-depletion test of a gas condensate system at 250°F. Permission to publish by the Society of Petroleum Engineers of AIME. Copyright SPE-AIME.[2]

Compositional Analysis Not Available

The initial in-place gas and oil (condensate) for gas-condensate reservoirs, both retrograde and nonretrograde, may be calculated from generally available field data by recombining the produced gas and oil in the correct ratio to find the average specific gravity (air = 1.00) of the total well fluid, which is presumably being produced initially from a one-phase reservoir. The method may also be used to calculate the initial oil and gas in gas caps.

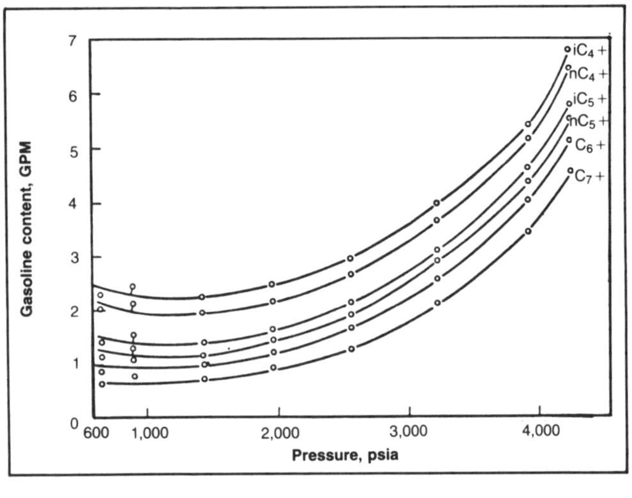

Fig. 8-4. Effect of pressure on gasoline content of produced gas during laboratory pressure-depletion test of a gas-condensate system at 250°F. Permission to publish by the Society of Petroleum Engineers of AIME. Copyright SPE-AIME.[2]

Let R = initial surface gas-oil ratio of the production, scf of dry or residue gas per barrel of oil,

γ_o = specific gravity of the tank oil (water = 1.00),

M_o = molecular weight of the tank oil, and

γ_g = average specific gravity of the gas produced from the surface separator(s), air = 1.00.

Standard conditions are 14.7 psia and 60°F, at which conditions the molar volume is 379.4 scf/mole (other base conditions yield different molar volumes).

Then, on the basis of one barrel of tank oil and R standard cubic feet of separator or residue gas, the mass of total well fluid m_w is

$$m_w = \frac{28.97\, R\, \gamma_g}{379.4} + 350\, \gamma_o$$

$$= 0.07636\, R\gamma_g + 350\, \gamma_o\,. \tag{8-1}$$

The total moles of fluid in one barrel of oil and R cubic feet of gas is

$$n_t = \frac{R}{379.4} + \frac{350\gamma_o}{M_o}$$

$$= 0.002636\, R + 350\, \gamma_o/M_o\,. \tag{8-2}$$

Hence the molecular weight of the total well fluid M_w is

$$M_w = \frac{m_w}{n_t} = \frac{0.07636\, R\gamma_g + 350\gamma_o}{0.002636 + R + 350\gamma_o/M_o}$$

and the specific gravity of the well fluid is $M_w/28.97$, or

$$\gamma_m = \frac{R\gamma_g + 4584\gamma_o}{R + 132{,}800\gamma_o/M_o}. \tag{8-3}$$

If the molecular weight of the tank oil is not known it may be estimated using the following formula developed by Cragoe[3]:

$$M_o = \frac{44.29\gamma_o}{1.03 - \gamma_o} = \frac{6084}{\text{API} - 5.9}. \tag{8-4}$$

Example 8-1:

Calculate the initial oil and gas in place per acre-foot for a gas-condensate reservoir, given:

Initial pressure	2740 psia
Reservoir temperature	215°F
Average porosity	25%
Average connate water	30%
Daily tank oil	242 bbl
Oil gravity, 60°F	48.0° API

Daily separator gas .... 3100 MCF
Separator gas gravity 0.650
Daily tank gas .... 120 MCF
Tank gas gravity 1.20

Solution:

Average gas gravity =

$$\frac{(3100)(0.650) + (120)(1.20)}{3100 + 120} = 0.670$$

$$\gamma_o = \frac{141.5}{48.0 + 131.5} = 0.7883$$

$$M_o = \frac{6084}{48.0 - 5.9} = 144.5$$

$$R = \frac{3100 + 120}{242} \times 1000 = 13{,}300 \text{ scf/STB}$$

$$\gamma_m = \frac{(13{,}300)(0.670) + (4584)(0.7883)}{13{,}300 + (132{,}800)(0.7883/144.5)} = 0.893$$

From Figure 2-9, T_c = 425°R and p_c = 652 psia using the condensate curves. Then T_r = 1.59 and p_r = 4.20, from which the gas deviation factor is 0.82 at 2740 psia and 215°F. Then the total initial gas in place per acre-foot of bulk reservoir rock is

$$G = \frac{379.4 \text{ pV}}{zRT} = \frac{(379.4)(2740)(43{,}560)(0.25)(1 - 0.30)}{(0.82)(10.73)(675)}$$

$$= 1334 \text{ MCF/acre-ft.}$$

Since the volume fraction equals the mole fraction in the gas state, the fraction of the total that is produced on the surface as gas is

$$f_g = \frac{n_g}{n_g + n_o} = \frac{R/379.4}{R/379.4 + 350\gamma_o/M_o}$$

$$= \frac{13{,}300/379.4}{13{,}300/379.4 + (350)(0.7883/144.5)}$$

$$= 0.9483.$$

Then,

Initial gas in place = 0.9483(1334)

$$= 1265 \text{ MCF/acre-ft, and}$$

$$\text{Initial oil in place} = \frac{1265 \times 10^3}{13{,}300} = 95.1 \text{ bbl/acre-ft.}$$

Since the gas production is 94.83% of the total moles produced, the total daily gas-condensate production in standard cubic feet is

$$\Delta G_p = \frac{\text{Daily gas}}{0.9483} = \frac{3100 + 120}{0.9483} = 3396 \text{ MCF/day.}$$

The total daily reservoir voidage by the gas law is

$$\Delta V = 3{,}396{,}000\left(\frac{675}{520}\right)\left(\frac{14.7}{2740}\right)(0.820)$$

$$= 19{,}400 \text{ cu ft/day.}$$

This includes the stock tank vapors rather than just the gas from the main, high-pressure separator; therefore, the gas gravity should be the average of all gases produced. The gas deviation factor at initial reservoir temperature and pressure is estimated from the gas gravity of the recombined oil and gas.

Compositional Analysis Available

If a compositional analysis of the separator gas and liquid is available, a compositional analysis of the total well fluid, from which a more accurate Z-factor can be calculated, can be obtained by recombining the gas and liquid phases in the correct proportions. The following example, adapted from Craft and Hawkins[1], illustrates the procedure.

Example 8-2:

Use the following data and the compositions given in Columns 2 and 3 of Table 8-1 to calculate the initial gas and condensate in the reservoir.

Reservoir pressure .... 4350 psia
Reservoir temperature .... 217°F
Hydrocarbon porosity 17.4%
Std. cond. .... 15.025 psia, 60°F
Separator gas .... 842,600 scf/day
Stock tank oil .... 31.1 STB/day
Mol. wt. C_{7+} in separator liquid 185.0
Sp. gr. C_{7+} in separator liquid 0.8343
Sp. gr. separator liquid at 880 psig
and 60°F 0.7675
Separator liquid volume factor ..... 1.235 bbl at 880 psia/STB, both at 60°F
Compositions of high-pressure
gas and liquid .... Cols. (2) and (3), Table 8-2
Molar volume at 15.025 psia
and 60°F .... 371.2 cu ft/mole.

Solution:

1. Calculate the mole proportions in which to recombine the separator gas and liquid. Multiply the mole fraction of each component in the liquid, Column 3, by its molecular weight, Column 4, and enter the products in Column 5. The sum of Column 5 is the molecular weight of the separator liquid, 127.48. Since the specific gravity of the separator liquid is 0.7675 at 880 psig and 60°F, then the moles per barrel is

$$\frac{(0.7675)(350) \text{ lb/bbl}}{127.48 \text{ lb/mole}} = 2.107 \text{ moles/bbl for the separator liquid.}$$

The separator liquid rate is 31.1 STB/day × 1.235 sep. bbl/STB so that the separator gas-oil ratio is

$$\frac{842{,}600}{(31.1)(1.235)} = 21{,}940 \text{ scf sep. gas/bbl sep. liquid.}$$

Since the 21,940 scf is 21,940/371.2, or 59.11 moles, the separator gas and liquid must be recombined in the ratio of 59.11 moles of gas to 2.107 moles of liquid.

2. Recombine the 59.11 moles of gas and 2.107 moles of liquid. Multiply the mole fraction of each component in the gas, Column 2, by 59.11 moles, and enter in Column 8. Multiply the mole fraction of each component in the liquid, Column 3, by 2.107 moles, and enter in Column 9. Enter the sum of the moles of each component in the gas and liquid, Column 8 plus Column 9, in Column 10. Divide each figure in Column 10 by the sum of Column 10, 61.217, and enter the quotients in Column 11, which is the mole composition of the total well fluid. Calculate the pseudocritical temperature, 379.23°R and pressure 668.23 psia from the composition. From the pseudocriticals find the pseudoreduced values, and then the deviation factor at 4350 psia and 217°F, which is 0.963.
3. Find the gas and oil in place per acre-foot of net reservoir rock. From the gas law, the initial moles per acre-foot at 17.4% hydrocarbon porosity is

$$\frac{pV}{zRT} = \frac{(4350)(43{,}560)(0.174)}{(0.963)(10.73)(677)} = 4713 \text{ moles/acre-ft.}$$

$$\text{Gas mole fraction} = \frac{59.11}{59.11 + 2.107} = 0.966$$

$$\text{Initial gas in place} = \frac{(0.966)(4713)(371.2)}{1000}$$

$$= 1690 \text{ MCF/acre-ft}$$

$$\text{Initial oil in place} = \frac{(1 - 0.966)(4713)}{(2.107)(1.235)}$$

$$= 61.6 \text{ STB/acre-ft}$$

The pseudocritical properties of the C_{7+} fractions used in Table 8-2 were obtained as a function of molecular weight and specific gravity from Figure 8-5.

RECOVERY ESTIMATES

Once the initial hydrocarbons in place are determined it is then necessary to calculate the amounts of gas and condensate that can be recovered, assuming a certain abandonment pressure. As pressure drops below the dew point, liquids will condense in the reservoir and will, in most cases, be unrecoverable. This causes the gas produced to contain fewer heavy components as the reservoir is depleted. An estimate of recoveries by pressure depletion must be made before the feasibility of maintaining pressure by cycling can be established.

Three different methods for predicting depletion performance for a volumetric reservoir are discussed. These are: (1) laboratory simulation, (2) flash calculations, and (3) empirical correlations. The most accurate is the laboratory simulation method, and this method should be used if possible. The added cost is certainly justified when the value of the reserves is considered.

Laboratory Simulation

The laboratory simulation procedure assumes that the separation process occurring in the reservoir is a differ-

TABLE 8-2
Calculations for Example 8-2 on Gas Condensate Fluid

(1)	(2) Mole Composition of Separator Fluids: Gas	(3) Liquid	(4) Mol wt.	(5) (3) × (4) lb/mole	(6) Liquid bbl/mole for Each Component	(7) (3) × (6) bbl Each Component per Mole of sep. Liq.	(8) Moles of Each Component in 59.11 Moles of Gas (2) × 59.11	(9) Moles of Each Component in 2.107 Moles of Liquid (3) × 2.107	(10) Moles of Each Component in 61.217 Moles of Gas and Liquid (8) + (9)	(11) Mole Composition of Total Well Fluid (10) ÷ 61.217	(12) Critical Pressure, psia	(13) Partial Critical Pressure, psia (11) × (12)	(14) Critical Temp. °R	(15) Partial Critical Temp. °R (11) × (14)
CO_2	0.0120	0.0000					0.709	0.0000	0.7090	0.0116	1070	12.41	548	6.36
C_1	0.9404	0.2024	16.04	3.247	0.1317	0.02666	55.587	0.4265	56.0135	0.9150	673	615.80	343	313.85
C_2	0.0305	0.0484	30.07	1.455	0.1771	0.00857	1.803	0.1020	1.9050	0.0311	708	22.02	550	17.11
C_3	0.0095	0.0312	44.09	1.376	0.2480	0.00774	0.562	0.0657	0.6277	0.0102	617	6.29	666	6.79
i-C_4	0.0024	0.0113	58.12	0.657	0.2948	0.00333	0.142	0.0238	0.1658	0.0027	529	1.43	735	1.98
n-C_4	0.0023	0.0196	58.12	1.139	0.2840	0.00557	0.136	0.0413	0.1773	0.0029	550	1.60	766	2.22
i-C_5	0.0006	0.0159	72.15	1.147	0.3298	0.00524	0.035	0.0335	0.0685	0.0011	484	0.53	830	0.91
n-C_5	0.0003	0.0170	72.15	1.227	0.3264	0.00555	0.018	0.0358	0.0538	0.0009	490	0.44	846	0.76
C_6	0.0013	0.0384	86.17	3.309	0.3706	0.01423	0.077	0.0809	0.1579	0.0026	440	1.14	914	2.38
C_{7+}	0.0007	0.6158	185.0	113.923	0.6336*	0.39017	0.041	1.2975	1.3385	0.0219	300	6.57	1227	26.87
	1.0000	1.0000		127.480		0.46706	59.110	2.1070	61.2170	1.0000		668.23		379.23

*185 lb/mole ÷ (0.8343 × 350 lb/bbl) = 0.6336 bbl/mole.

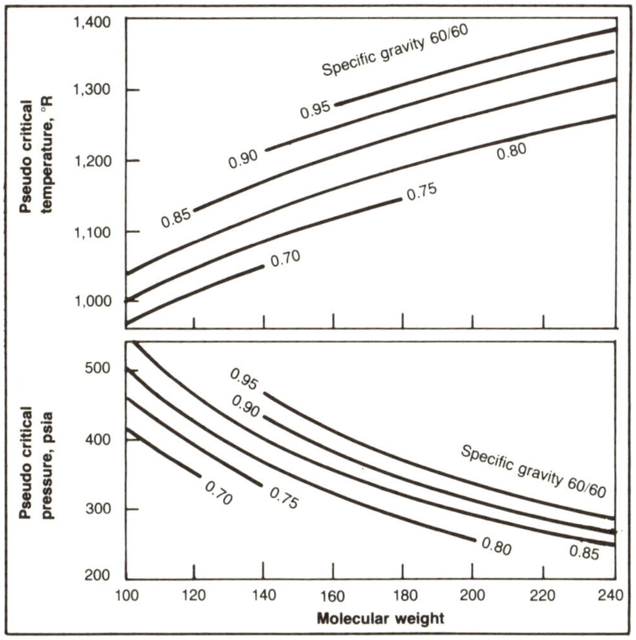

Fig. 8-5. Correlation charts for estimation of the pseudo-critical temperature and pressure of heptanes plus fractions from molecular weight and specific gravity. Courtesy Gas Processors Suppliers Association.

ential process; that is, no liquid phase is produced from the reservoir.

The process consists of starting with a reservoir fluid sample at or above its dew point pressure and at reservoir temperature. Gas is removed to decrease the pressure in reasonable increments, down to the assumed abandonment pressure. Each increment of gas removed is analyzed to determine its composition, and the volume removed is measured. The changing composition with pressure decrease is illustrated in Figure 8-6.

The liquid recovery from the gas increments produced from the cell may be measured by passing the gas through small-scale separators, or it may be calculated from the composition for usual field separation methods or for gasoline plant methods. Liquid recovery of the pentanes plus will be somewhat greater in gasoline plants than by field separation, but much greater for the propanes and butanes, commonly called liquefied petroleum gas, LPG.

The following example, taken from Craft and Hawkins[1], illustrates the calculation procedure. It was assumed that the recoveries of liquid from the produced gas would be 25% of the butanes, 50% of the pentane, 75% of the hexane, and 100% of the heptanes and heavier components.

Example 8-3:

Calculate the performance of the retrograde condensate reservoir based on the results of a differential laboratory separation given in Table 8-3 and the following data:

Initial pressure (dew point)	2960 psia
Abandonment pressure	500 psia
Reservoir temperature	195°F
Connate water	30%
Porosity	25%
Standard conditions	14.7 psia & 60°F
Initial cell volume	947.5 cc
Mol. Wt. of C_{7+} in Initial Fluid (Assumed constant)	114 lb/lb-mole
Sp. Gr. of C_{7+} in Initial Fluid (Assumed constant)	0.775 at 60°F

Solution:

The solution is summarized in Table 8-4 and Figure 8-7. Details of some of the calculations are given below.

(1) Calculate the increments of gross production in Mscf per acre-foot of net bulk reservoir rock, and enter in Column 2 of Table 8-4.

$$V_{HC} = 43{,}560\ Ah\phi(1 - S_w)$$

$$= 43{,}560(1)(1)(0.25)(1 - 0.30)$$

$$V_{HC} = 7623\ \text{ft}^3/\text{acre-ft}$$

For example, in the pressure increment between 2960 and 2500 psia,

$$\Delta V = (7623)\ \frac{175.3\ \text{cu cm}}{947.5\ \text{cu cm}} = 1410\ \text{cu ft/acre-ft at}$$

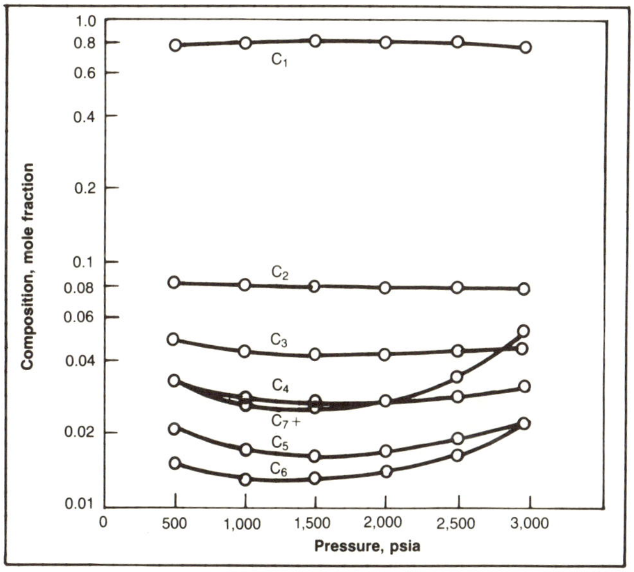

Fig. 8-6. Variations in the composition of the produced gas phase material of a retrograde gas-condensate fluid with pressure decline. B. C. Craft/M. F. Hawkins, Applied Petroleum Reservoir Engineering, copyright 1959, pp. 64, 66, 74, 76, 78. Reprinted by permission of Prentice-Hall, Inc., Englewood Cliffs, N.J.

2500 psia and 195°F

$$\Delta G_P = \frac{379.4\ p\ V}{1000\ zRT}$$

$$= \frac{(379.4)(2500)(1410)}{(1000)(0.794)(10.73)(655)} = 240.1 \text{ Mscf.}$$

Column 3 is the summation of gross gas production in Column 2.

(2) Calculate the Mscf of residue gas and the barrels of liquid obtained from each increment of gross gas production and enter in Columns 4 and 6. Then, the mole fraction recovery as liquid is

$$\Delta n_L = (0.25)(0.028) + (0.50)(0.019) + (0.75)(0.016) + (0.034)(1) = 0.0625 \text{ mole fraction.}$$

Since mole fraction is the same as volume fraction in gas, the Mscf recovered as liquid from 240.1 Mscf is

$$G_L = (0.25)(0.028)(240.1) + (0.50)(0.019)(240.1) + (0.75)(0.016)(240.1) + (0.034)(240.1) = 1.681 + 2.281 + 2.881 + 8.163 = 15.006 \text{ Mscf.}$$

This volume of gas can be converted to gallons of liquid using Table 2-2 of the Gas Properties chapter for C_4, C_5, and C_6. The average of the iso- and normal compounds is used for C_4 and C_5.

For C_{7+}:

$$\frac{(114 \text{ lb/lb-mole})(1000 \text{ scf/Mscf})}{(379.4 \text{ scf/lb-mole})(8.337 \text{ lb/gal})(0.755)} = 47.74 \text{ gal/Mscf.}$$

TABLE 8-3
Volume, Composition, and Gas Deviation Factors for a Retrograde Condensate Fluid

(1) Pressure psia	(2) C_1	(3) C_2	(4) C_3	(5) C_4	(6) C_5	(7) C_6	(8) C_{7+}	(9) Produced Gas, cu cm at 195°F and Cell Pressure	(10) Retrograde Liquid Volume, cu cm Cell Volume, 947.5 cu cm	(11) Retrograde Volume, Per Cent of Hydrocarbon Volume	(12) Gas Deviation Factor at 195°F and Cell Pressure
	Composition of Produced Gas Increments, Mole Fraction										
2960	0.752	0.077	0.044	0.031	0.022	0.022	0.052	0.0	0.0	0.0	0.771
2500	0.783	0.077	0.043	0.028	0.019	0.016	0.034	175.3	62.5	6.6	0.794
2000	0.795	0.078	0.042	0.027	0.017	0.014	0.027	227.0	77.7	8.2	0.805
1500	0.798	0.079	0.042	0.027	0.016	0.013	0.025	340.4	75.0	7.9	0.835
1000	0.793	0.080	0.043	0.028	0.017	0.013	0.026	544.7	67.2	7.1	0.875
500	0.768	0.082	0.048	0.033	0.021	0.015	0.033	1080.7	56.9	6.0	0.945

B. C. Craft/M. F. Hawkins, *Applied Petroleum Reservoir Engineering,* © 1959, pp. 64, 66, 74, 76, 78. Reprinted by permission of Prentice-Hall, Inc., Englewood Cliffs, N.J.

TABLE 8-4
Gas and Liquid Recoveries in Per Cent and Per Acre-Foot for Example 8-3

(1) Pressure, psia	(2) Increments of Gross Gas Production, M SCF	(3) Cumulative Gross Gas Production, M SCF, Σ(2)	(4) Residue Gas in Each Increment, M SCF	(5) Cumulative Residue Gas Production, M SCF Σ(4)	(6) Liquid in Each Increment, bbl	(7) Cumulative Liquid Production, bbl Σ(6)	(8) Average Gas-Oil Ratio of Each Increment, SCF Residue Gas per bbl, (4) ÷ (6)	(9) Cumulative Gross Gas Recovery, per cent, (3) × 100/1580	(10) Cumulative Residue Gas Recovery, per cent, (5) × 100/1441	(11) Cumulative Liquid Recovery, per cent, (7) × 100/143.2
2960	0	0	0	0	0	0	10,600	0	0	0
2500	240.1	240.1	225.1	225.1	15.3	15.3	14,700	15.2	15.6	10.7
2000	245.2	485.3	232.3	457.4	13.1	28.4	17,730	30.7	31.7	19.8
1500	266.0	751.3	252.8	710.2	13.3	41.7	19,010	47.6	49.3	29.1
1000	270.8	1022.1	256.9	967.1	14.0	55.7	18,350	64.7	67.1	38.9
500	248.7	1270.8	233.0	1200.1	15.9	71.6	14,650	80.4	83.3	50.0

B. C. Craft/M. F. Hawkins, *Applied Petroleum Reservoir Engineering,* © 1959, pp 64, 66, 74, 76, 78. Reprinted by permission of Prentice-Hall, Inc., Englewood Cliffs, N.J.

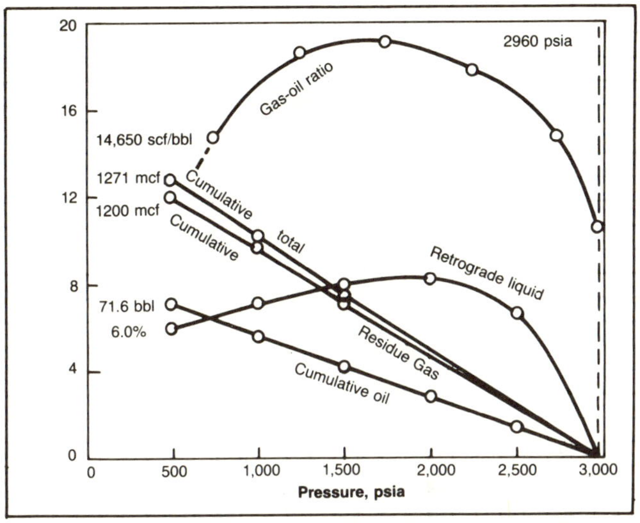

Fig. 8-7. Gas-oil ratios, retrograde liquid volumes, and recoveries for the depletion performance of a retrograde gas-condensate reservoir. B. C. Craft/M. F. Hawkins, Applied Petroleum Reservoir Engineering, copyright 1959, pp. 64, 66, 74, 76, 78. Reprinted by permission of Prentice-Hall, Inc., Englewood Cliffs, N.J.

Then the total liquid recovered from the 240.1 Mscf determined in part (1) is

$$(1.681)(32.04) + (2.281)(36.32) + (2.881)(41.03) + (8.163)(47.74) = 644.6 \text{ gal} = 15.3 \text{ bbl}.$$

The residue gas recovered from the 240.1 Mscf determined in part (1) is

$$(240.1)(1 - 0.0625) = 225.1 \text{ Mscf}.$$

Column 5 is the summation of Column 4, and Column 7 is the summation of Column 6.

(3) Calculate the gas-oil ratio for each increment of gross production in units of residue gas per barrel of liquid. Enter in Column 8. For instance,

$$\frac{(225.1)(1000)}{15.3} = 14{,}700 \text{ scf/bbl}.$$

(4) Calculate the cumulative percent recoveries of gross gas, residue gas, and liquid. Enter in Columns 9, 10, 11. The initial gross gas-in-place is

$$\frac{379.4\,pV}{1000\,zRT} = \frac{(379.4)(2960)(7623)}{(1000)(0.771)(10.73)(655)}$$

$$= 1580 \text{ Mscf/acre-ft}.$$

The liquid mole fraction is 0.088, and the total liquid recovery is 3.808 gal/Mscf of gross gas, which are calculated from the initial composition in the same manner as shown in part (2). Then,

$$G = (1 - 0.088)(1580)$$

$$= 1441 \text{ Mscf residue gas/acre-ft}$$

$$N = \frac{(3.808)(1580)}{42} = 143.2 \text{ bbl/acre-ft}.$$

At 2500 psia, then

$$\text{Gross gas recovery} = \frac{(100)(240.1)}{1580} = 15.2\%.$$

$$\text{Residue gas recovery} = \frac{(100)(225.1)}{1441} = 15.6\%.$$

$$\text{Liquid recovery} = \frac{(100)(15.3)}{143.2} = 10.7\%.$$

Reference to Figure 8-7 will show that the producing gas-oil ratio increases and then declines below 1500 psia due to revaporization of liquids. However, as a practical matter, field experience suggests that essentially all liquids are held to rock surfaces by capillary forces and virtually none is revaporized. Thus, the field producing gas-oil ratio does not normally decline.

Reference to Columns 10 and 11 of Table 8-4 will show that 83.3% of the residue gas is calculated as recovered to an abandonment pressure of 500 psia. However, only 50% of the liquids are recovered due to retrograde condensation. Gas cycling projects are justified on the recovery of the liquids that would be otherwise lost if production operations only permitted pressure depletion of the reservoir involved.

Flash Calculations

In the absence of direct laboratory data on a specific gas-condensate system, estimates of pressure-depletion behavior can be obtained by using vapor-liquid equilibrium ratios to compute the phase behavior when composition of the total gas-condensate system is known. Correlations for estimating phase volumes must also be available.

When multicomponent hydrocarbon gases and liquids exist together under pressure, part of the lighter hydrocarbons are dissolved in the liquid phase, and part of the heavier hydrocarbons are vaporized in the gas phase. A convenient concept for describing quantitatively the behavior of specific components is the equilibrium ratio. The ratios vary considerably with the pressure, temperature, and composition of the system involved.

The use of equilibrium ratios for flash calculations is illustrated in Chapter 2. Determining the proper convergence pressure for selecting a set of equilibrium ratios is a problem because of the changing composition of the fluids as pressure decreases. This problem is discussed extensively by Standing.[4] A step-by-step procedure for calculating performance using equilibrium ratios follows.

1. Assume flashing the original (known) composition from original pressure (and volume) to a lower pressure, at which the compositions and amounts (in moles) of the liquid and gas phases are computed, using the best k values available.
2. Estimate the volume of each phase using methods discussed below.
3. Assume removal of enough vapor phase volume at constant pressure to cause the remaining gas plus all the liquid to conform to the reservoir's original constant volume.
4. Subtract from the original system composition the number of moles of each component in the vapor represented by this gas removal.
5. Using the new total composition from step 4, consider the system flashed to the next lower pressure in the stepwise process and repeat the above procedure. Removal of vapor phase alone is required by the assumption that fluid flowing into the wells will not be accompanied by any liquid phase at any step of the process.

As indicated, the stepwise calculations require knowledge of the volume occupied by each phase at each step. Means of estimating these volumes are described in Chapter 2 and also by Standing.[4] In estimating phase volumes, smoothed values should be used from curves drawn through the points computed from properties of the phase at each known composition.

The above calculations are intended to approximate the experimental procedure used in the pVT cell during a laboratory pressure-depletion study. The number of pressure steps to be used in making such calculations is arbitrary but probably should conform to about 500 psi intervals, with points usually closer together at the start and at the end of the calculations. After the pressure-depletion calculations have been made on a stepwise basis to some assumed abandonment pressure, the resulting information can be used for plotting smoothed curves similar to those obtained from laboratory experiments. Calculated ultimate recoveries of gas and natural gas liquids can then be made by the methods discussed previously.

The prediction of condensate reservoir performance from equilibrium ratios alone is likely to be in considerable error. Nothing takes the place of carefully collected and analyzed reservoir fluid samples obtained early in field life (or early in reservoir life if multiple reservoirs are involved). Standing[4] has discussed this problem. Unfortunately, equilibrium ratios change with pressure, temperature, and composition. Condensation of liquids in the reservoir changes the composition of the flowing hydrocarbons; therefore, the equilibrium values change.

Empirical Correlations for Estimating Performance

Where laboratory studies of the reservoir fluid are not available, it is sometimes possible to use the published results of Jacoby, Koeller, and Berry.[5] This is often the case for small reservoirs. It also is a useful method where estimates of future condensate and gas production are needed in advance of a laboratory analysis, even if such work is ultimately planned. The authors studied several rich natural gas-condensate systems (gas-oil ratios of about 3600 to 60,000 scf/STB), one natural volatile oil system (gas-oil ratio of 2363 scf/STB), and a series of related synthetic reservoir fluid mixtures (gas-oil ratios of 2000 to 25,000 scf/STB). The correlations made were by regression analysis. It was found that total tank oil-in-place could be correlated with initial GOR and that the cumulative oil and separator gas-in-place recovered from the saturation pressure to an abandonment pressure of 500 psi could be correlated with initial GOR, initial tank oil gravity, and reservoir temperature. It was noted that depletion from saturation pressure to 500 psi would recover an average of 92% of the total separator gas-in-place.

The calculated ultimate oil recovery by depletion from the saturation pressure to 500 psi is correlated by the equation

$$N_P = -0.061743 + \frac{143.55}{R_i} + 0.00012184T + 0.0010114\,(°API), \qquad (8\text{-}5)$$

where

N_p = cumulative stock tank oil production from dew point pressure to 500 psia, bbl stock tank oil/bbl HCPS,
R_i = initial separator gas-oil ratio, scf/STB oil,
T = reservoir temperature, °F, and
$°API$ = initial stock tank oil gravity.

In Equation 8-5, it is worth noting that pressure was found not to be an important correlating factor. This might not be the case for lean gases. Equation 8-5 fits the available data with a standard deviation of the percentage errors of 11.3%.

The separator gas-in-place at the saturation pressure was correlated by the equation

$$G = -2229.4 + 148.43\,(R_i/100)^{0.2} + 124{,}130/T + 21.831\,(°API) + 0.26356\,p_{d,b}, \qquad (8\text{-}6)$$

where

G = total initial primary separator gas-in-place, scf, and
$p_{d,b}$ = saturation pressure (dew point or bubble point), psia.

Equation 8-6 fits the available data points with a standard deviation of the percentage errors of 3.8%.

Effects of Water Drive

Very few cases of gas-condensate reservoirs operated under natural water drive have been reported in the literature. To be economically attractive a water drive would have to be sufficiently strong to maintain pressure high enough to minimize condensed hydrocarbon losses in the formation.

Gas-condensate reservoirs (whether exhibiting retrograde behavior or not) may perform volumetrically or may produce under a partial or total water drive. If pressure maintenance occurs, the recovery will depend upon the stabilization pressure and the displacement efficiency of the invading water (i.e., a frontal displacement mechanism). The liquid recovery for retrograde reservoirs will be less since the liquid will usually be immobile and will be trapped with some gas behind the invading water front. Unfortunately, recoveries are usually lower with water influx than with volumetric depletion.

If no oil zone is associated with the gas condensate reservoir, the material balance may be presented as follows:

$$\frac{p_{sc}G_p}{T_{sc}} = \frac{p_iV_i}{z_iT} + \frac{p(V_i - W_e + B_wW_p)}{zT}, \quad (8\text{-}7)$$

or as

$$G(B_g - B_{gi}) + W_e = G_pB_g + B_wW_p. \quad (8\text{-}8)$$

Equations 8-7 and 8-8 can be used to determine water influx, W_e, the original gas-in-place, G, or the initial reservoir gas volume, V_i. Care should be taken in the determination of z at the reservoir pressure of interest, since it must include the equivalent condensate or oil; i.e., it is a two-phase gas deviation factor.

The injection of water into a gas-condensate reservoir to maintain pressure is sometimes considered. A number of factors must be carefully weighed before a decision is reached. The mobility ratio in this case is favorably low because of the very high mobility of the gas, thus tending to provide high areal sweep and pattern efficiencies. However, there is strong evidence that displacement efficiency by the water is not high, and it may be as low as 50%. This is offset to some extent by the improved areal sweep efficiency at low mobility ratio. All things considered, the recovery of gas condensate in the vapor phase by water injection is likely to be appreciably lower than by cycling, and any consideration of water injection for gas-condensate recovery should be accompanied by detailed experimental work on cores from the specific reservoir involved. This will help determine whether the water can in fact accomplish a high enough displacement efficiency to justify its use.

GAS CYCLING

Incentive exists to cycle gas-condensate reservoirs in those instances in which natural depletion of the resource will result in substantial loss of liquid hydrocarbons. This occurs in water drive fields where "wet" gas is trapped or in volumetric-type reservoirs where retrograde behavior exists. Liquid hydrocarbons formed during pressure depletion are not normally revaporized at lower reservoir pressures and thus are lost. Where the reservoir rock has favorable characteristics, cycling with "dry" gas should permit recovery of part of the liquids that would otherwise be lost.

Liquids recovery is determined in the reservoir by displacement efficiencies, which are a product of the vertical, areal, and pore-to-pore fractional recoveries. Most cycling projects recover in the range of 50% of the liquid hydrocarbons that would otherwise be lost.

Recoveries from cycling projects depend on the volume of reservoir contacted by the dry gas and on the fraction of wet gas displaced in the contacted volume. The definitions of and methods for determining the various efficiencies are described.

Areal Sweep Efficiency (E_A)

Areal sweep efficiency is the area enclosed by the leading edge of the dry gas front divided by the total area of the reservoir. E_A can be estimated using analog models or by observing well performance in an actual project.

Vertical Sweep Efficiency (E_V)

Vertical sweep efficiency is the pore space invaded by the injected gas, divided by the hydrocarbon pore space enclosed by the projection (through full reservoir thickness) of the leading edge of the dry-gas front.

Vertical sweep efficiencies can be as high as 90% under favorable conditions. However, invasion is significantly affected by large variations in reservoir flow properties. These might be (1) strictly lateral variations in horizontal permeability of a single bed comprising a reservoir that does not have any variations vertically at any location; (2) strictly layering effects, by which several strata may comprise the reservoir, each being uniform as to properties but differing appreciably in permeability from all the others; or (3) combinations of these extreme cases. Performance of cycling operations can vary appreciably according to what combination of the two extremes may exist for a given reservoir. Analog models can handle permeability variations to a limited extent in representing gas-condensate reservoirs; such techniques are handicapped considerably by lack of detailed data on lateral and vertical permeability distributions in the reservoir since wells are widely spaced and unfortunately are not all cored to provide the necessary subsurface information.

Displacement Efficiency (E_D)

Displacement efficiency E_D is the volume of wet hydrocarbons swept out of individual pores or small groups of pores, divided by the volume of hydrocarbons in the same pores at the start of cycling; note that both volumes must be calculated at the same conditions of pressure and temperature. Displacement efficiency is controlled mainly by the miscibility of the driving and driven fluids and their mobilities. For a cycling operation in which the pressure is being maintained at or above the dew point, the displacement efficiency resulting from the action of the dry gas against the wet gas phase in the individual pores will be virtually 100%.

Reservoir Cycling Efficiency (E_R)

Reservoir cycling efficiency is the overall recovery factor defined as the wet hydrocarbons recovered divided by the wet hydrocarbons initially in place in the reservoir. It is the product of the three previously defined efficiencies. That is,

$$E_R = E_A E_V E_D.$$

Feasibility of Gas Cycling

In choosing between pressure depletion and pressure maintenance as operating methods for a gas-condensate reservoir, detailed analyses must be made for predicting optimum economics. Cycling and gas-processing procedures require sizable plant expenditures. Possible processing methods, whether reservoir fluids are cycled or not, include stabilization, compression, absorption, and fractionation. The latter two recover appreciably more condensibles from wet gas than do the former. If the removal of ethane from a gas stream is desirable, for economic or other reasons, fractionation can accomplish this.

When the reservoir characteristics appear favorable for recovery of condensible hydrocarbons, it must then be considered whether cycling would be economic. The primary comparison is between value of the estimated additional recovery of liquid products by cycling and the actual cycling costs, taking into account deferment of gas income and other factors. Economic analyses of cycling and noncycling are required; these must be carried out in detail for maximum dependability, using information factors and assumptions pertinent to each particular case.

The primary considerations in making a decision on whether or not to cycle are summarized by Frick.[2]

1. Reservoir formation and fluid characteristics: occurrence or absence of black oil; size of reserves of products; properties and composition of reservoir hydrocarbons; productivities and injectivities of wells; permeability variation—controls the degree of bypassing of injected gas; degree of natural water drive existing
2. Reservoir development and operating costs
3. Plant installation and operating costs
4. Market demand for gas and liquid petroleum products
5. Future relative value of the products
6. Existence or absence of competitive producing conditions between operators (in the same reservoir)
7. Taxes: severance, ad valorem, and income
8. Special hazards or risks (limited concession or lease life, political climate, others)
9. Overall economic analysis

REFERENCES

1. Craft, B. C., and Hawkins, M. F.: *Applied Petroleum Reservoir Engineering,* Prentice-Hall, Inc., Englewood Cliffs, N.J. (1959).
2. Frick, T. C.: *Petroleum Production Handbook,* McGraw-Hill, New York (1962).
3. Cragoe, C. S.: "Thermodynamic Properties of Petroleum Products," Bureau of Standards, U.S. Dept. of Commerce (1929) Misc. Publ. No. 97, p. 22.
4. Standing, M. B.: *Volumetric and Phase Behavior of Oil Field Hydrocarbon Systems,* SPE of AIME (1977).
5. Jacoby, R. H., Koeller, R. C., and Berry, V. J., Jr.: "Effect of Composition and Temperature on Phase Behavior and Depletion Performance of Rich Gas-Condensate Systems," *Trans.,* AIME, (1958) **216** 406–411.

9 Field Operation Problems

THE efficient operation of a gas well or gas field requires constant monitoring of the total field and individual well performance in order to detect problems that may seriously reduce gas recovery or producing capacity. Parameters to be monitored include pressure, flow rate, cumulative production, water production, and condensate production.

The best way to detect abnormal behavior is to plot these parameters, either as a function of time or cumulative production, and observe any sudden changes in behavior from the past or any deviation from expected behavior. Several of the most common production problems and the methods to detect them are described in this chapter.

PRESSURE-CUMULATIVE PRODUCTION PLOTS

By plotting total field or reservoir gas recovered versus p/Z on various scales, production problems such as abnormal pressure, water influx, leakage, or loss of gas, or bad data can be detected.

p/Z Versus G_p Plots

As described in Chapter 3, for a volumetric gas reservoir, a plot of p/Z versus G_p on cartesian coordinates should yield a straight line of constant slope.

Once sufficient production history has been obtained under reasonably stabilized operating conditions, it is possible to extrapolate the historical plot to the anticipated abandonment pressure and thus arrive at an estimate of ultimate reserves.

Unfortunately, several factors will affect the validity of this method of estimation. If a full or partial water drive is present, the rate of pressure decline will be less than would have been observed had the reservoir been on a straight pressure depletion, as shown in Figure 9-1. Such a decline would be erroneously interpreted as indicating a much larger reservoir than actually exists. The performance observed must thus be closely tied to the known geological configuration of the reservoir so that the reasons for the particular performance observed can be properly evaluated.

In certain other cases, it may be found that the gas reservoir is actually a gas cap on top of an oil reservoir. In this case, the pressure decline will be influenced by the rate of oil production and the evolution of solution gas from the oil into the gas cap as the pressure declines. In such a case, the entire reservoir must be analyzed, including the effects of the oil production rate and the effect of pressure reduction on solution gas. Transfer between the oil and gas cap can go either way depending on reservoir conditions of temperature, pressure, and fluid composition.

Variation in the reservoir permeability can also affect the observed pressure performance. A typical pattern is a sharp drop in the reservoir pressure during the early depletion, which will yield a very low ultimate recovery if extrapolated in a straight-line manner, but which breaks to a lower slope on a balance between the flow capacity of the unfractured matrix and the fracture system, as shown in the lowest diagram of Figure 9-1. In such cases, it is necessary to pass this breakpoint before any kind of reliable pressure decline estimate can be made.

In order to determine reserves by the pressure decline versus cumulative method, ideally one should have accurate cumulative production histories and accurate bottom-hole pressure data based on a shut-in period that is

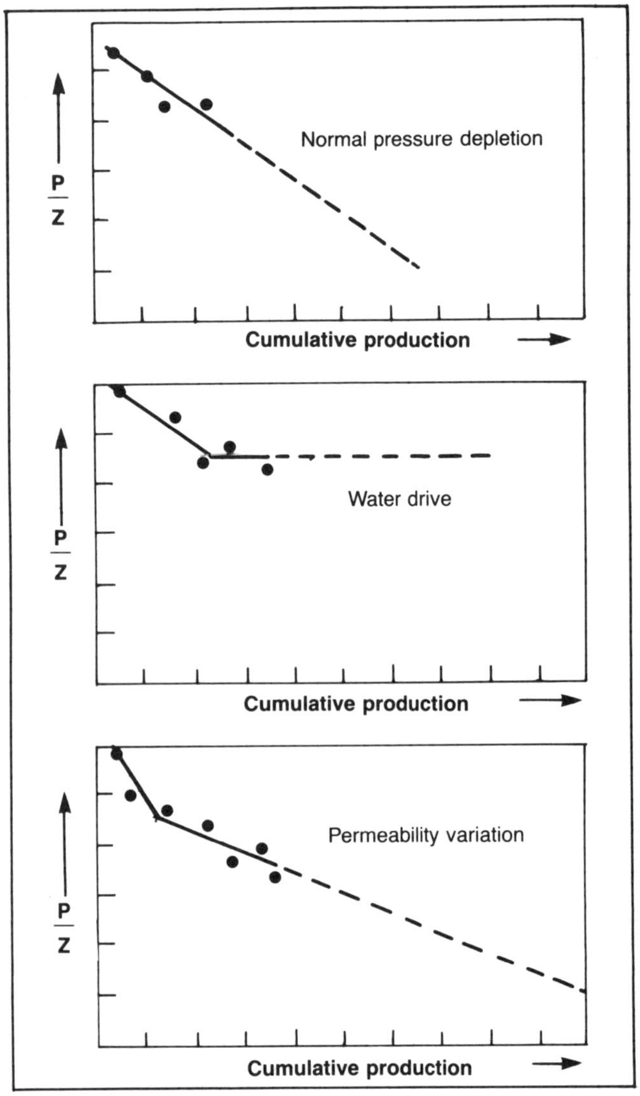

Fig. 9-1. Effect of reservoir characteristics on production performance.

sufficient for the well to achieve static conditions. Although current practices normally yield adequate production data, bottom-hole pressure data is often sparce and, even when available, is based on short shut-in periods. The most usual test provided is a shut-in wellhead pressure test that was obtained to meet regulatory requirements or as a result of the well having been shut-in for some other purpose.

In the event that the well produces a dry gas without condensate or water present, it is a very simple matter to arrive at a reasonable estimate of bottom-hole pressure from a shut-in wellhead pressure. If the well is producing condensate and/or water with the gas stream when it is shut-in, there will usually be an accumulation of liquids in the bottom portion of the wellbore that will account for a substantial part of the bottom-hole pressure. Unless the depth to the liquid is reasonably known, it is impossible to estimate the bottom-hole pressure from the shut-in surface pressure. In such situations, bottom-hole pressure recordings are imperative.

In some cases, the slope of the p/Z versus G_p plot will decrease with increasing G_p. This can result from bad pressure or production data, loss of gas through leakage to some other formation, or abnormal reservoir pressure. A plot of p/Z versus G_p for an abnormally pressured gas reservoir is shown in Figure 9-2. Once the reservoir pressure reaches a normal value, the slope remains fairly constant but is steeper than the initial slope. In this case, if the early data had been extrapolated to the abandonment conditions, a recovery twice the actual would have been predicted.

Energy Plots

In Chapter 3, it was shown that a plot of $(1 - pZ_i/p_iZ)$ versus G_p on log-scales would yield a straight line with a slope of one for volumetric gas reservoirs.

These energy plots are very useful in detecting water influx or abnormal pressure very early in the life of the reservoir. An early estimate of gas-in-place can also be made more accurately than by using p/Z versus G_p plots. The principal reason for the advantage of using energy plots is that the slope of a p/Z plot depends on the initial gas-in-place, which may not be known early in the reservoir's life. Therefore, the magnitude of the slope is of no value in detecting abnormalities. Enough data must be obtained to determine if a straight line plot occurs before problems can be detected.

Conversely, the slope of an energy plot must be one for normal behavior. Therefore, only enough data to detect deviation of the slope from 45° are required to reflect abnormal behavior. This could conceivably be only two data points. An example energy plot is shown in Figure 9-3.

Rate Versus Time Plots

Various types of production problems can be detected by monitoring the performance of a gas well by plotting production rates of gas, condensate and water, and pressure versus time. A typical performance plot is shown in Figure 9-4.

Rate-time performance plots are especially useful when used in conjunction with well status maps. A well status map, such as shown in Figure 9-5, allows the engineer to compare the performance of wells in the same area and to determine if a well is performing abnormally compared to the other wells.

An example of abnormal production rate decline is shown in Figure 9-4. The well's pressure remains high

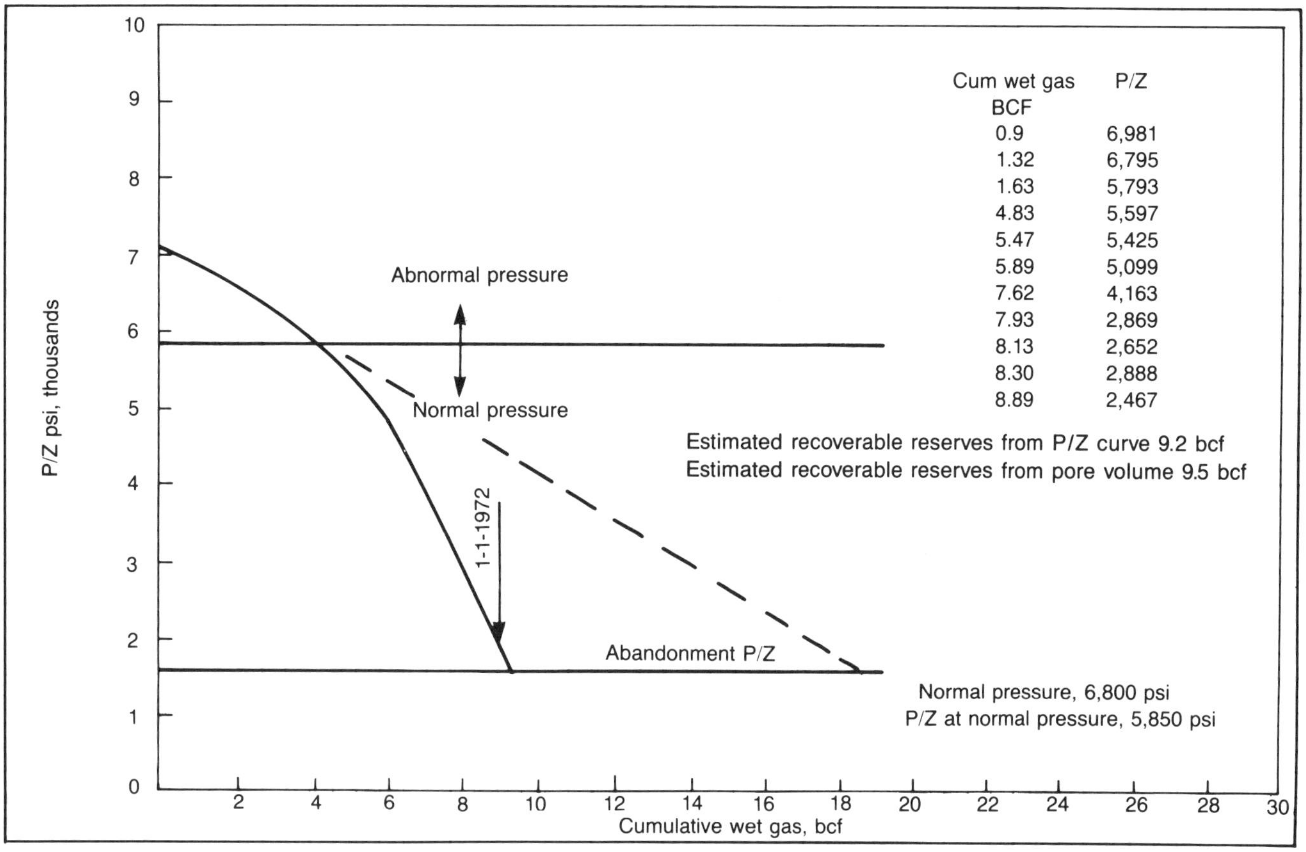

Fig. 9-2. p/z versus cumulative wet gas.

but the gas production rate is declining. Also, by comparing Well No. 42's performance with the surrounding wells on the well status map, it can be seen to be producing at a much lower rate than the other wells. This could be caused by such things as formation damage or

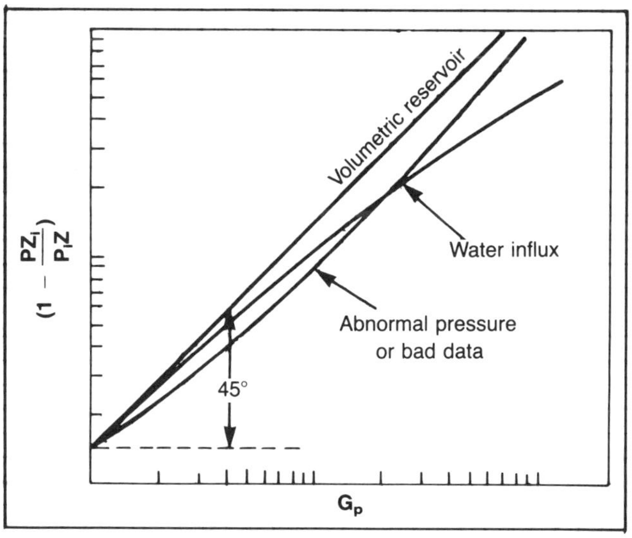

Fig. 9-3. Energy plot.

perforation plugging by scale. After this well was re-perforated, performance improved considerably, as shown in Figure 9-6.

Water production problems can also be analyzed by observing rate-time plots. An increase in water rate without an increase in drawdown can result from normal advance in the gas-water contact in a water drive reservoir.

If the water rate increases after the gas production rate has been increased, this may indicate water coning. The well rate should then be reduced to see if the water rate returns to normal.

Water coning is the term used to describe the entry of water that underlies the gas in a reservoir under excessive drawdown pressures. The coning phenomenon is illustrated in Figure 9-7. If the vertical and horizontal permeabilities are known, a maximum allowable production rate to prevent water coning can be calculated.

Once the coning has occurred, a reduction in producing rate will not necessarily eliminate the problem. This results from the fact that once a high water saturation has built up around the wellbore, the permeability to water will have increased, and therefore a lower drawdown will be required to support the coning process. Even if the well is shut-in, capillary forces will usually prevent the

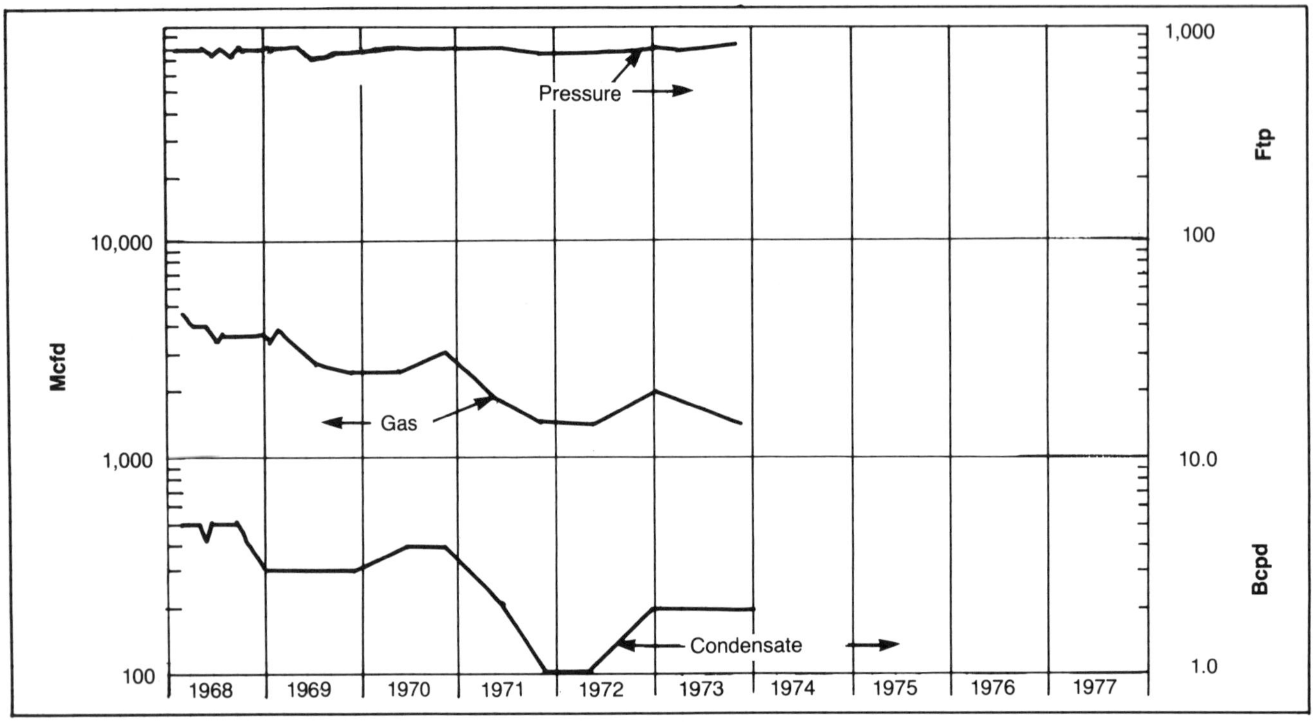

Fig. 9-4. Gas well performance curve.

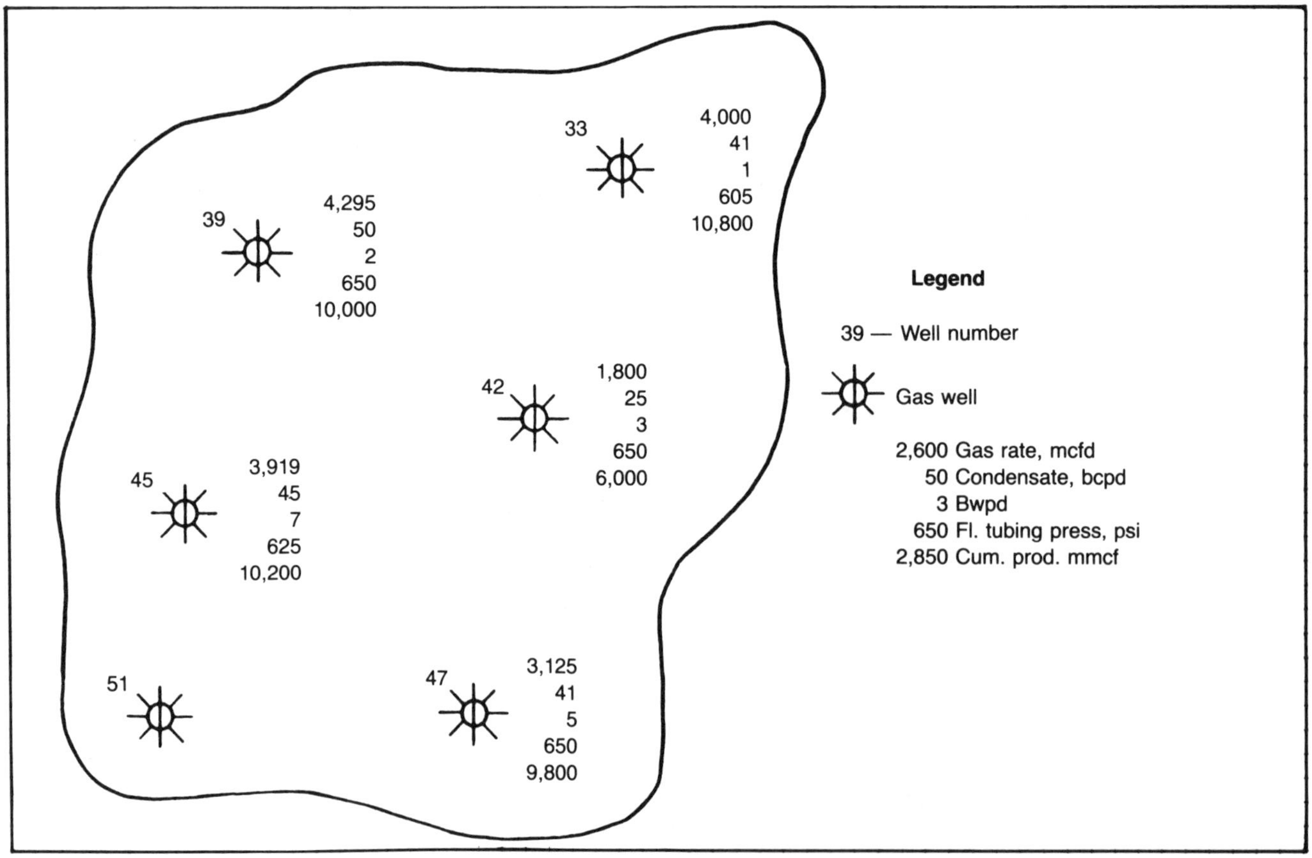

Fig. 9-5. Well status map.

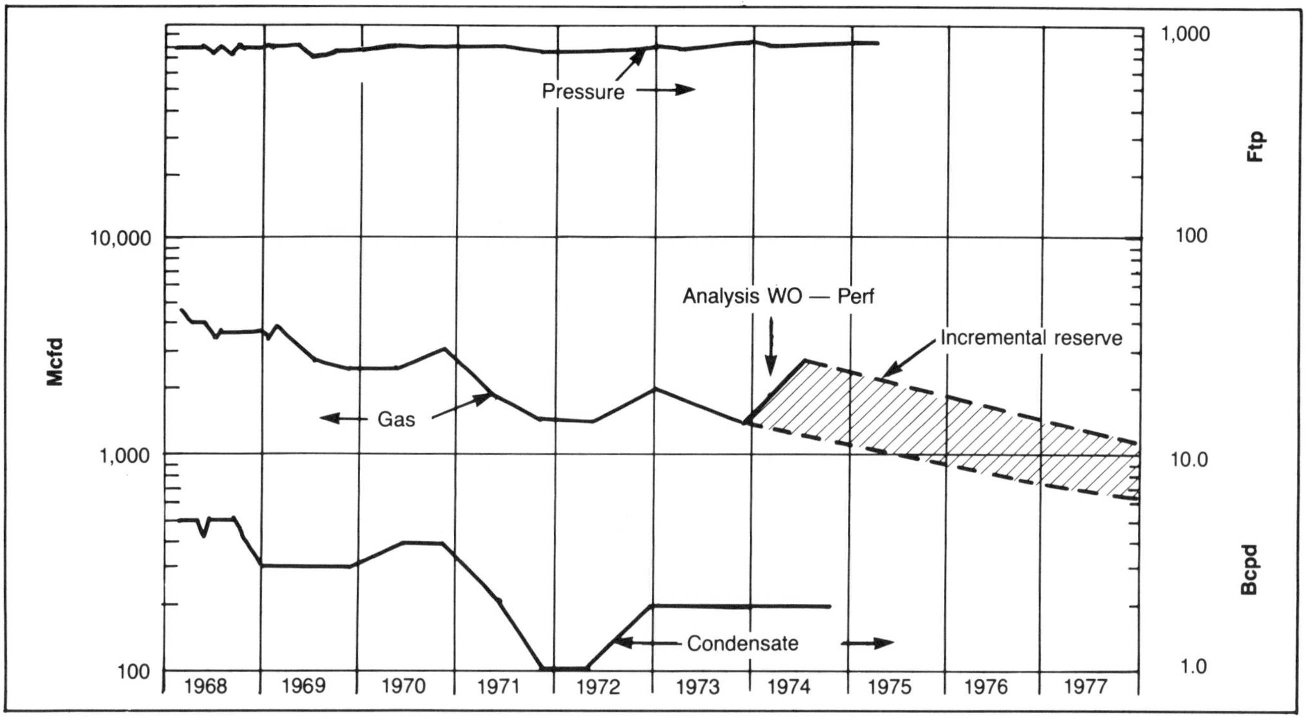

Fig. 9-6. Well 42 performance curve.

water from subsiding to its original level. Therefore, if it is known that a gas reservoir is underlain by water, calculations should be made to estimate the maximum safe production rates for each well.

HYDRATE FORMATION

Most natural gas contains substantial amounts of water vapor at the time it is produced from a well or separated from an associated crude-oil stream. Water vapor must be removed from the gas stream because it will condense into liquid and may cause hydrate formation as the gas is cooled from the high reservoir temperature to the cooler surface temperature. Liquid water almost always accelerates corrosion, and the solid hydrates may pack solidly in gas-gathering systems, resulting in partial or complete blocking of flow lines.

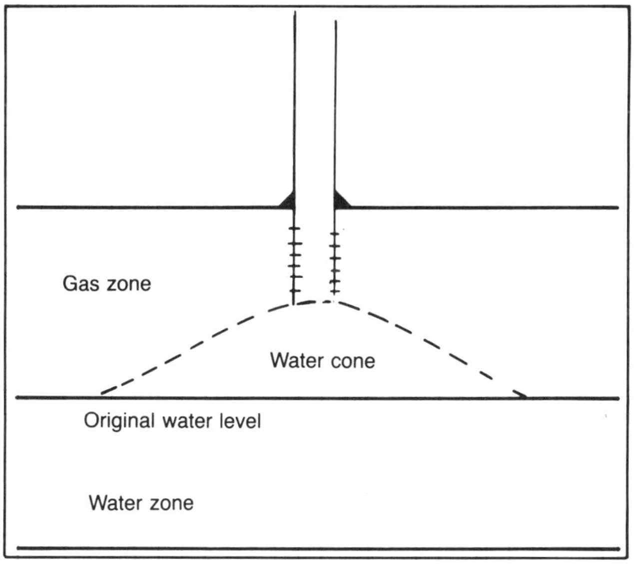

Fig. 9-7. Water coning.

Hydrates are solid compounds that form as crystals and resemble snow in appearance. They are created by a reaction of natural gas with water, and when formed, they are about 10% hydrocarbon and 90% water. Hydrates have a specific gravity of about 0.98 and will usually float in water and sink in hydrocarbon liquids. Water is always necessary for hydrate formation as well as some turbulence in the flowing gas stream.

Causes, Occurrence, and Prediction

The temperature at which hydrates will form depends upon the actual composition of the gas and the pressure of the gas stream. Therefore, the chart shown in Figure 9-8 cannot be completely accurate for all gases, but it is typical for many gases. The chart shows the water content in pounds of water per MMscf of saturated gas at any pressure or temperature. The dotted line crossing the family of curves shows the temperature at which hydrates will probably form at any given pressure. Note that hydrates form more easily at higher pressures. At

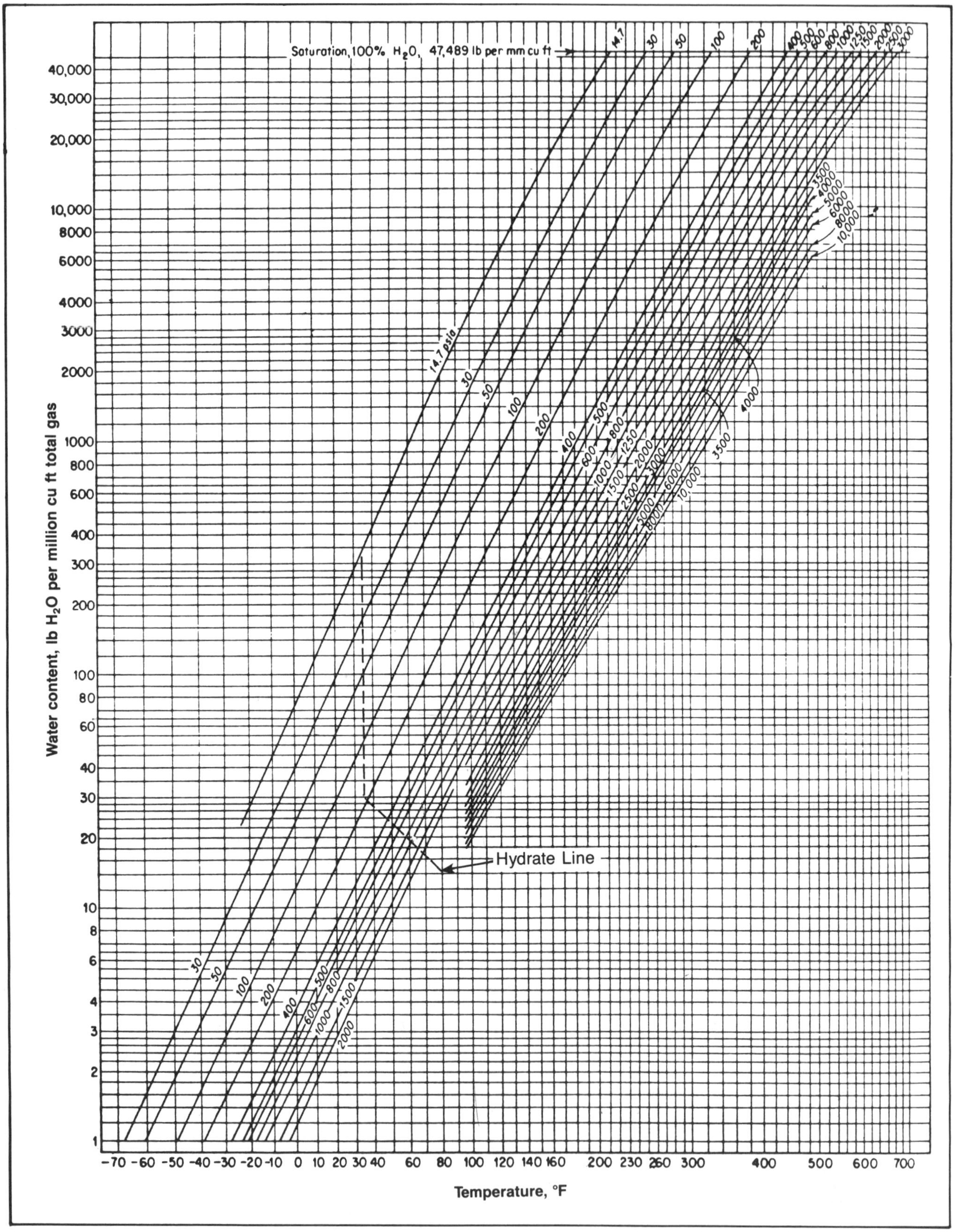

Fig. 9-8. Water-vapor content of natural gas at saturation. Courtesy Gas Processors Suppliers Association.

1500 psig, for example, hydrates may form at 70°F, whereas at 200 psig, hydrates will not form unless the gas is cooled to about 39°F. Each curve on the chart shows the water content of a saturated gas at that pressure when the temperature is at any of the various points shown along the bottom of the chart. For example, at 100 psia and 60°F, each 1 million cu ft of gas would contain about 130 lb of water vapor. The same gas at 100 psia and 20°F (instead of 60°F) could only hold 30 lb per MMscf. At 0°F, this gas could only hold 13 lb per MMscf.

It can thus be seen that as a gas is cooled, it can hold less water in the vapor form. Therefore, cooling a gas will cause some of the water vapor to condense with the balance remaining in the gas as water vapor. Pipeline specifications usually require that the water vapor content of natural gas be 7 lb/MMscf or less in order to minimize the problem of hydrate formation in the transmission lines from the field to ultimate user. In some fields, hydrates form in the tubing and the wellhead valves necessitating the application of heat down the hole to keep the well from freezing up. In most fields, fortunately, the temperature of the gas at the wellhead is 100°F or more; therefore, the hydrate problem does not usually begin until the gas passes through the Christmas Tree.

An all-important factor in the movement of gas saturated with water vapor is the retention of the heat, which is in the gas when it is produced. The temperature is lowered at the wellhead when the gas is expanded through a choke to reduce the pressure and control the rate of flow. After passing the choke, the gas enters the gathering lines, which are cooled by the ground. The effects of the ground temperature are shown in Figure 9-9. This curve was prepared from data taken in the Carthage gas field of east Texas. Gas leaving a heater at 160°F had dropped to 75°F by the time the gas had traveled 1-1/2 miles due to a ground temperature of 55°F. At a higher rate of flow, the gas would travel farther before the formation of hydrates would become a problem. With a line pressure of 900 psi, it could be predicted that hydrates would form at a temperature of about 60°F. Some of the flow lines in this field are as long as 15 miles between the well and the gasoline plant, so it would appear that this gas would need to be heated again before it reached the plant.

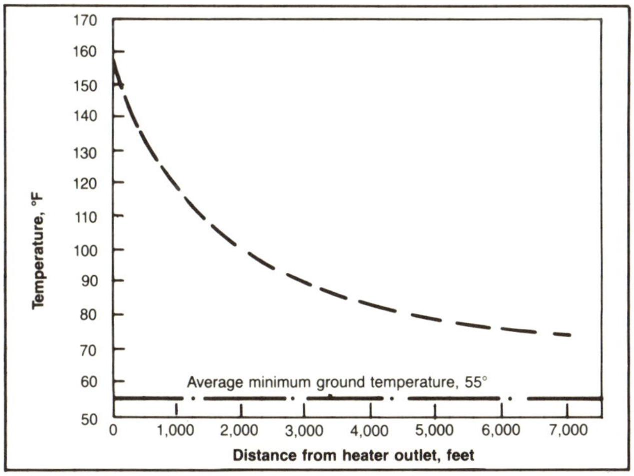

Fig. 9-9. Chart of temperature drop in natural gas flow lines.

Minimum ground temperatures at a depth of 18 in. in the various gas-producing regions of the southwest are as follows:

1. Upper Gulf Coast of Texas and Louisiana, 50°F to 56°F
2. Permian Basin, east Texas, and north Louisiana fields, 45°F
3. Hugoton-Panhandle gas areas, 25°F to 30°F

The figures are approximate and represent minimum values. The ground temperatures may be expected to reach higher levels than this during most of the year, especially in the southern part of Texas and Louisiana. On the other hand, it is probable that in the gas-producing areas of the Rocky Mountain states and in western Canada, the frost line is below the 18-in. level. Gas lines in those areas should be buried at correspondingly greater depths.

It is convenient to divide hydrate formation into two categories: (1) hydrate formation due to a decrease in temperature, with no sudden pressure drop, such as in the flow string or surface lines and (2) hydrate formation where a sudden expansion occurs such as in flow provers, orifices, back-pressure regulators, or chokes.

Hydrate Formation in the Flow String and Surface Lines

As mentioned above, free water is essential to hydrate formation. Free water is almost certain to be present during well testing since gas reservoirs are essentially water saturated, and a decrease in temperature results in a lower solubility of water in gas. The hydrate temperature depends on the pressure and composition of the gas. Figure 9-10 gives approximate values of the hydrate temperature as a function of pressure and specific gravity. Hydrates will form whenever temperature and pressure plot to the left of the hydrate formation line and the water dew point line for the gas in question. The water dew point may be obtained from Figure 9-8 as follows:

1. Find water content W_s at $\bar{p}_R$, T_R.
2. Draw a horizontal line on Figure 9-8 at W_s.
3. Read the dew point temperature for various pressures at the intersection of the horizontal line and each pressure line.

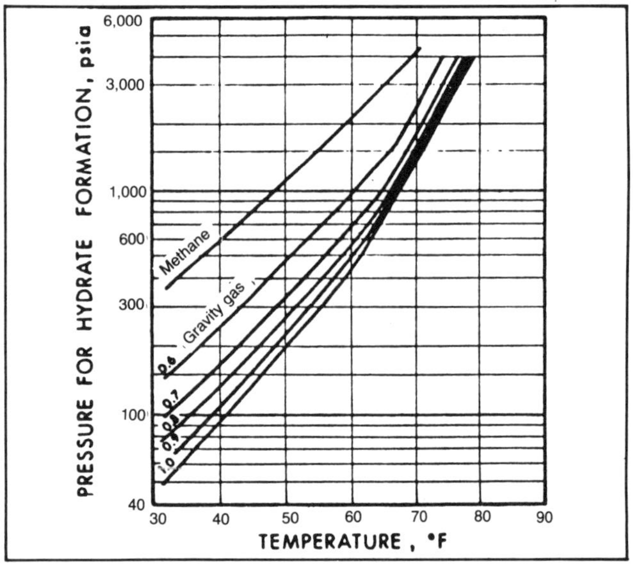

Fig. 9-10. Pressure-temperature drop in natural gas flow lines. Courtesy Gas Processors Suppliers Association.

Example 9-1:

A gas of specific gravity 0.8 is at a pressure of 1000 psia. To what extent can the temperature be lowered without hydrate formation, assuming presence of free water?

Solution:

From Figure 9-10, at a specific gravity of 0.8 and a pressure of 1000 psia, hydrate temperature = 66°F. Hydrates may form at or below 66°F.

Figure 9-10 is applicable only to sweet natural gases. For sour gases it may be used, keeping in mind that the presence of H_2S and CO_2 will increase the hydrate temperature and reduce the pressure above which hydrates will form. In other words, the presence of H_2S or CO_2 enhances the possibility of hydrate formation.

Hydrate Formation in Flow Provers, Orifices, and Back-Pressure Regulators

Sudden expansion in one of these devices is accompanied by a temperature drop, which may cause hydrate formation. Figures 9-11 through 9-15 may be used to approximate the conditions for hydrate formation. The limitations of Figure 9-10, discussed above, also apply to Figures 9-11 through 9-15. Figures 9-11 through 9-15 may be used for gases of other specific gravity by linear interpolation.

Example 9-2:

A 0.8 specific gravity gas is to be expanded from a pressure of 1000 psia at 100°F. Determine the minimum final pressure for no hydrate formation.

Solution:

From Figure 9-13, the intersection of the 1000 psia initial pressure line with the 100°F initial temperature line gives a final pressure of 440 psia. Therefore, lowering the pressure below 440 psia will possibly result in hydrate formation.

Example 9-3:

A 0.6 gravity gas is to be expanded from 1000 psia to 600 psia. What is the minimum initial gas temperature permitted without the danger of hydrate formation?

Solution:

From Figure 9-11, the 1000 psia initial pressure line intersects the 600 psia final pressure line at an initial temperature of approximately 74°F. Therefore, if the gas is cooler than 74°F at 1000 psia, hydrates may form on expansion.

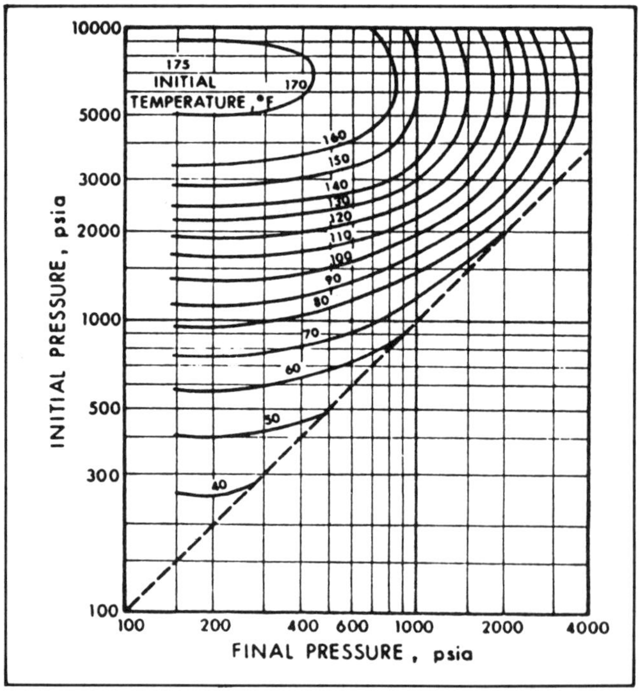

Fig. 9-11. Permissible expansion of a 0.6 gravity natural gas without hydrate formation. Courtesy Gas Processors Suppliers Association.

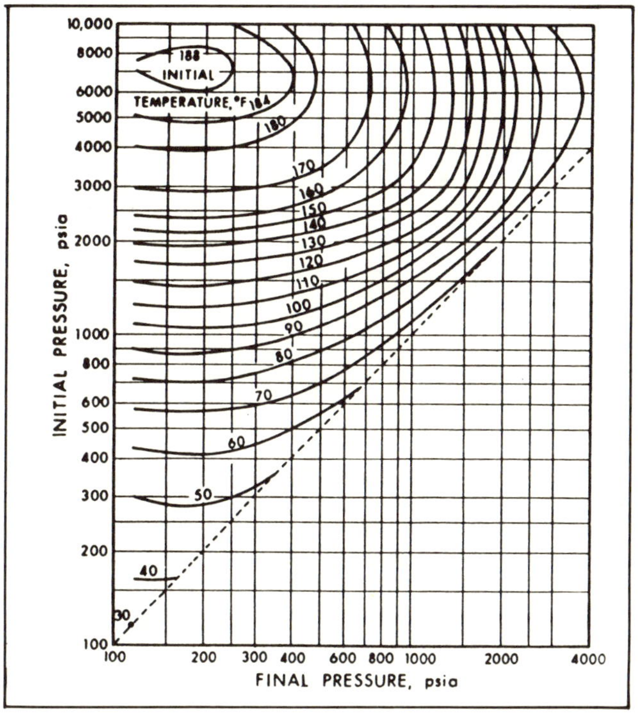

Fig. 9-12. Permissible expansion of a 0.7 gravity natural gas without hydrate formation. Courtesy Gas Processors Suppliers Association.

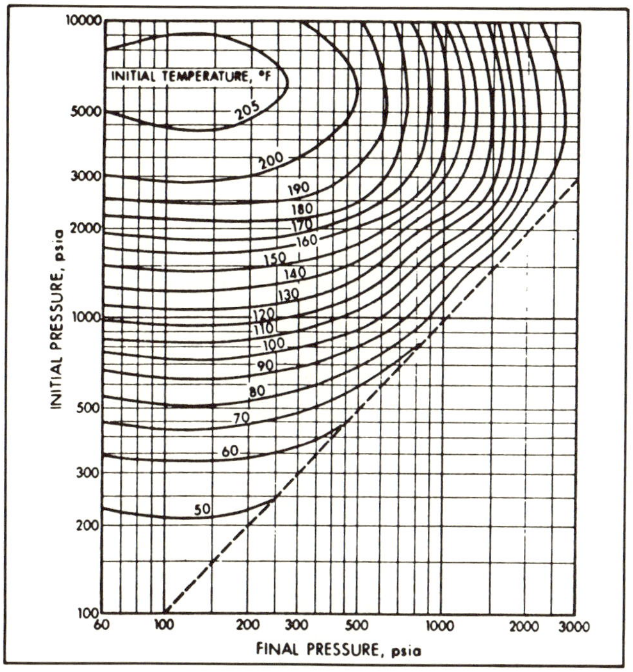

Fig. 9-14. Permissible expansion of a 0.9 gravity natural gas without hydrate formation. Courtesy Gas Processors Suppliers Association.

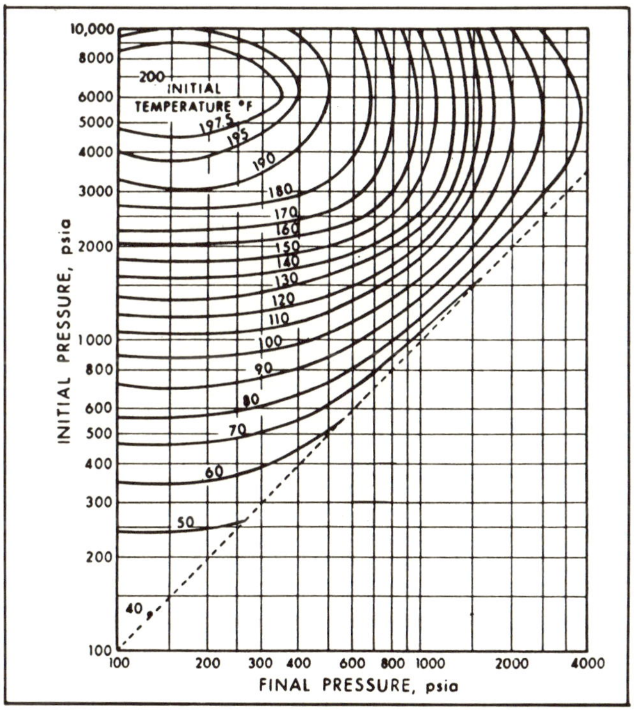

Fig. 9-13. Permissible expansion of a 0.8 gravity natural gas without hydrate formation. Courtesy Gas Processors Suppliers Association.

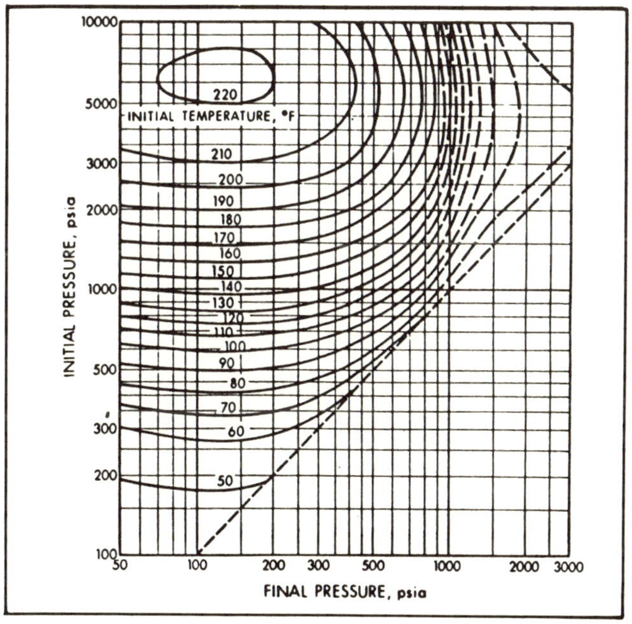

Fig. 9-15. Permissible expansion of a 1.0 gravity natural gas without hydrate formation. Courtesy Gas Processors Suppliers Association.

Hydrate Control

Ammonia, brines, glycol, and methanol have all been used to lower the freezing point of water vapor and thus prevent hydrate formation in flow lines. Low-volume injectors or pumps are used to feed the inhibitor fluid into the gas flow system. Methanol and glycol are the inhibitors most widely used; however, they are expensive. When hydrate problems arise so rarely that the installation of a heater or dehydration equipment is not economically feasible, methanol or glycol are usually used. For example, in some fields gas-producing operations may be carried out without hydrate trouble except for two or three days each year. Injection of inhibitors might conceivably be preferred in such cases over investment in additional equipment. More information about hydrate control by methanol injection can be found in Chapter 10.

In a majority of the situations where dehydration of the gas is not economical and where some measures must be adopted to control hydrate formation, application of heat is the method used. The reasons for this are: initial investment is not excessive, fuel is conveniently available, and the heaters operate with a minimum of attention. Where it is desirable to handle the gas by means of temporarily holding the temperature above the ground temperature, heaters seem to offer the best solution. For long-distance transmission, the gas will eventually reach the ground temperature and, for such gas movement, water will ultimately have to be removed.

Most natural gas is produced at relatively high flowline pressures, ranging from 1500 to 10,000 psig, and then transferred to pipelines operating at 1200 psig and lower. Heat must almost always be used to compensate for the natural refrigeration of the gas, which is caused by this pressure reduction. An indirect heater is the most widely used heater for natural-gas well streams because it is simple, economical, and if properly sized, is usually trouble-free. Field treating of natural gas is described in more detail in Chapter 10.

SOUR GAS PRODUCTION

Many natural gas reservoirs contain hydrogen sulfide (H_2S) and carbon dioxide (CO_2) along with the hydrocarbon gases. These two gases are called acid gases because they form acids or acidic solutions in the presence of water. A gas is called sour if it contains H_2S in amounts above the acceptable industry limits, usually about four parts per million.

Since the early 1970's, many deep, high-pressure sour gas reservoirs have been discovered in France, Germany, Canada, and the United States. Producing these gases involves several problems not encountered in normal gas production operations. Some of these problems are:

Toxicity—H_2S is an extremely poisonous gas, and therefore special safety measures are required.

Sulfur deposition—Sour gases with high H_2S contents often contain desolved elemental sulfur. Under certain conditions this leads to sulfur precipitation in the formation, tubing, and gathering lines.

Corrosion—The extreme corrosivity of sour gases, especially in the presence of brines, causes problems such as tubing and surface equipment failure.

CORROSION

Corrosion can be defined as the deterioration of a metal or its properties because of a reaction with its environment. Corrosion due to H_2S occurs in three ways: (1) in the presence of water, where H_2S forms an acidic solution that causes pitting and/or a continuous metal loss, (2) blisters and pitting resulting from hydrogen diffusion into the metal, and (3) stress corrosion cracking.

The first corrosion type is extremely difficult to define and predict. Without CO_2 and/or oxygen also being present, H_2S will not necessarily be corrosive. H_2S reacts with steel to form a film that tends to stop corrosion. If CO_2 is present, it combines with water to form carbonic acid. This reacts with steel and causes a continuous washing away of the metal. It also washes away the protective H_2S film, and the H_2S corrosion rate is accelerated considerably.

Hydrogen diffusion occurs when the atomic hydrogen (H) from the reacting surface of steel diffuses into the crystal structure, concentrating in voids, inclusions, and minor laminations. The atomic hydrogen changes to nondiffusable molecular hydrogen, building up high localized pressures that lead to blistering and corrosion pitting.

Stress corrosion cracking, sometimes called sulfide stress cracking, also results from embrittlement of the steel by the atomic hydrogen diffusion into the crystal structure. The exact failure mechanism is uncertain, but the failure is normally spontaneous. Three factors must be present before this phenomenon will occur: (1) materials must be susceptible to stress corrosion cracking; this would include any carbon, alloy, or stainless steels with a hardness over 22 on the Rockwell C scale; (2) H_2S and water (in the liquid phase) must be present; and (3) the metal must be under dynamic or static stresses.

The general H_2S corrosion reaction can be stated as:

$$H_2S + F_e + H_2O \rightarrow F_eS + H_2.$$

When CO_2 dissolves in water, it forms carbonic acid,

decreases the pH of the water, and increases its corrosivity.

$$CO_2 + H_2O \rightarrow H_2CO_3$$

$$F_e + H_2CO_3 \rightarrow F_eCO_2 + H_2$$

In a gas-condensate well with few dissolved minerals and at relatively high temperatures, pressure is the controlling factor influencing CO_2 corrosivity. The partial pressure of CO_2 can be used as a criterion for predicting corrosivity of gas wells producing CO_2. The partial pressure of CO_2 can be calculated from:

$$p_{CO_2} = p_{total} \text{ (mole fraction } CO_2\text{)}.$$

Using the partial pressure, the following relationships have been found:

1. p_{CO_2} above 30 psig usually indicates corrosion;
2. p_{CO_2} between 3 psig and 30 psia may indicate corrosion; and
3. p_{CO_2} less than 3 psig generally is not considered corrosive.

The combination of H_2S and CO_2 is even more aggressive than H_2S alone. Also, even minute quantities of oxygen can accelerate corrosivity tremendously.

CORROSION CONTROL WITH INHIBITORS

The types of applications that have been used for gas wells cover a wide range including batch and continuous methods such as: (1) low volume, pump, and fall; (2) tubing displacement; (3) inhibitor carried or pushed with nitrogen; (4) continuous injection; and (5) formation squeeze. In the types of wells encountered in the past, there was a continuously available liquid phase in the tubing bore that carried the inhibitor and distributed it to all of the surfaces that needed protection. In deep, hot wells, the tubing bore, and especially the formation, experiences an essentially dry condition with very little liquid to carry and distribute the inhibitor. Thus, special phase behavior requirements are placed on the inhibitor and on the solvent/carrier used in the application. Because of these latter considerations, two of the more popular modern treatment media are water and nitrogen used either as pusher or as carrier for the inhibitor mixture.

The Short Batch Method of Application

Traditionally, the short batch method has used 5 to 50 barrels of inhibitor solution containing one-half to five drums of inhibitor carried in water (as a dispersion of oil soluble inhibitor or perhaps as a solution of a water soluble compound), or diesel, or sometimes condensate. The inhibitor mix is pumped into the shut-in well and allowed to travel to bottom. Radiotracer investigations have shown that the liquid mix bridges over at the top of the hole and fills the tubing bore. Then a film of liquid falls down the tubing walls proceeding from the bottom of the filled column as a bubble of gas rises through the center of the liquid column. To get proper distribution of inhibitor from top to bottom of the string requires about one drum of inhibitor mixed in 7 to 10 barrels of diluent for each 10,000 feet of depth. If significant evaporation of diluent occurs in the hole, then even larger treatment volumes are required to get even distribution of inhibitor from top to bottom. Of course, if corrosion occurs only high in the hole, proper distribution all the way to bottom and consequent large volumes are not required. Even so, in especially hot wells, there is a tendency for flash evaporation of diluent to form tiny bubbles and a stabilized foam at the bottom edge of the liquid column of the inhibitor mix. Many times these foams prevent proper fall of the mix because they are too viscous to run down the tubing walls. Sometimes the addition of a surfactant helps.

The combination of possible evaporation and of not falling to bottom makes it difficult to achieve good corrosion control by this method when applied to deep, hot wells where corrosion occurs deep in the hole. Possibly with appropriate use of water or of a special solvent with a very high boiling point, the method would be more feasible. Since gases always contain water vapor, when water is used as solvent/carrier, fewer moles are required to achieve complete saturation than if diesel or some other solvent is used that is not contained in the gas. Furthermore, one barrel of water contains 19.6 moles of water; whereas one barrel of diesel contains 1.5 moles of hydrocarbon. Thus, it is ten times easier to increase the mole fraction of carrier in the hole by injection of a few barrels of water than by injection of diesel. Water is also cheaper than diesel or other solvents and less hazardous to pump. However, truly water soluble corrosion inhibitors are sometimes less effective than oil soluble compounds. Thus, frequently oil soluble/water dispersible compounds are chosen for use with water as a carrier.

For wells in the mid to upper range of the deep, hot category, it is desirable to put in enough diluent to displace from 1/4 to 1/2 of the tubing volume (for a 20,000 foot well, the range is from 20 barrels for 1/4-full 2-3/8-in. tubing to 120 barrels for 1/2-full 4-in. tubing). After the mix is pumped, the well is shut-in for four to eight hours for each 10,000 feet of tubing. This recommended volume provides for guaranteed treatment of the top 1/4 to 1/2 of the tubing surface as well as allowing for approximately half the diluent to evaporate as the solution falls, leaving enough liquid to distribute evenly. Of course, the gas that is displaced from the tubing back into the formation does not have to be saturated, and this lowers the evaporation volume requirement. When full coverage is achieved, the indicated corrosion rate will be reduced to at least as low as 25% of the uninhibited value.

The life of a treatment will depend on how much total inhibitor is left in the hole. Many times, analyses of inhibitor returns immediately after the treatment show that 80% or more of what went into the hole is produced back out almost immediately. This leaves only 20% to provide protection in the hole. This condition is especially true when not much of the inhibitor mix falls below the injection depth. As the inhibitor mix falls to bottom, a thin liquid film is formed—approximately 0.030 in. thick. When the well returns to production, some of this liquid is swept off by the gas flow, but much of it is evaporated into the undersaturated formation gas, leaving solvent-free inhibitor behind so that with a successful application, only 20% or so of the inhibitor is brought back out of the hole immediately.

The Tubing Displacement Method

By this method of treatment, a tubing full of fluids is pumped into the well with the objective of displacing inhibitor completely top to bottom to ensure contact with the tubing surface. The most common mode of treatment is to use inhibitor-free solvent to push to bottom 2 to 20 barrels of a concentrated inhibitor solution (5 to 20%; sometimes 100% inhibitor is used). Diesel or condensate are the most frequently chosen solvents for the inhibitor mix, although sometimes special aromatic or higher boiling point solvents are chosen. Water, nitrogen, or natural gas are frequently used to push the mix to bottom. Diesel and condensate are also used.

Although it is fairly certain that all of the surface is contacted with inhibitor when this method is used, it is not certain how much film is left behind. If the volume of inhibitor mix is too small and is pushed with a non-miscible solvent, then there may be a lack of adequate distribution of inhibitor to all the necessary areas of the tubing surface because of spiralling flow down the hole or other factors. These factors are especially likely to occur when neat, oil soluble inhibitor is pushed with water or gas.

The inhibitor mix chosen for these treatments should be relatively nonvolatile in the hole so that as little as possible evaporates and as much of the liquid volume as possible is available to form a film on the tubing surface. When a liquid displacement is made, the gas volume in the hole decreases to a small volume so that there is no problem achieving saturation on the down trip. However, as gas flows out, displacing the push fluid, redistribution of inhibitor is aided by a nonvolatile mix. In order for the treatment to be effective for a long time, the same volatility principles apply to the active ingredient (that is, nonsolvent components) of tubing displacement compounds as for limited batch compounds. The immiscible push mode produces the best retention of inhibitor in the hole but runs the risk of poor distribution so that not all the corrodible areas are covered.

Methods of Inhibitor Application Using Nitrogen Gas

Nitrogen gas is frequently used to push a slug of inhibitor mix to bottom. The principles of this mode are the same as for the other tubing displacement methods. The other mode of nitrogen application is to displace the gas in the hole with nigrogen gas containing a mist of inhibitor solution. One to four drums of inhibitor contained in a 2% to 50% solution are used. Nitrogen is taken to the well site in liquid form and expanded into gaseous form at a pressure slightly greater than that in the well. Displacement is calculated from the gas law modified to account for deviations of nitrogen from an ideal gas. The inhibitor mix is aspirated into the flowing gas as droplets and carried with it to bottom. Rapid displacement helps ensure that the mist droplets penetrate all the way to bottom. After the well is shut-in for a few hours, the inhibitor mist settles onto the tubing walls. It is desirable to resume flow in the well at a relatively slow rate and then build up to full flow rate. In this way, solvent evaporation fixes the active inhibitor ingredient on the tubing walls with only a minimum amount swept back out of the well. Thus, this method offers the best possible utilization of inhibitor. Evaporation and other removal processes operating after a nitrogen treatment are identical to those operating after a short batch or tubing displacement treatment.

Method of Continuous Treatment with Inhibitors

The most effective and reliable method of corrosion control is to continuously add inhibitor to the bottom of the producing string in an adequate amount to cover the surface and with an amount of liquid carrier adequate to produce a liquid phase condition in the hole. When these conditions are met, the periodic corrosion rate highs and lows found in the batch methods are avoided. Thereby overall metal loss is significantly less over the life of the downhole equipment, and thus the chance for premature failure is greatly reduced.

The two methods most widely used for conducting the inhibitor to the bottom of the well are the use of a tubing kill string and injection down the annulus through a bottom-hole chemical injector valve. In some cases, a parallel string of tubing is run and used for chemical injection. Figure 9-16A shows the installation of a kill string.

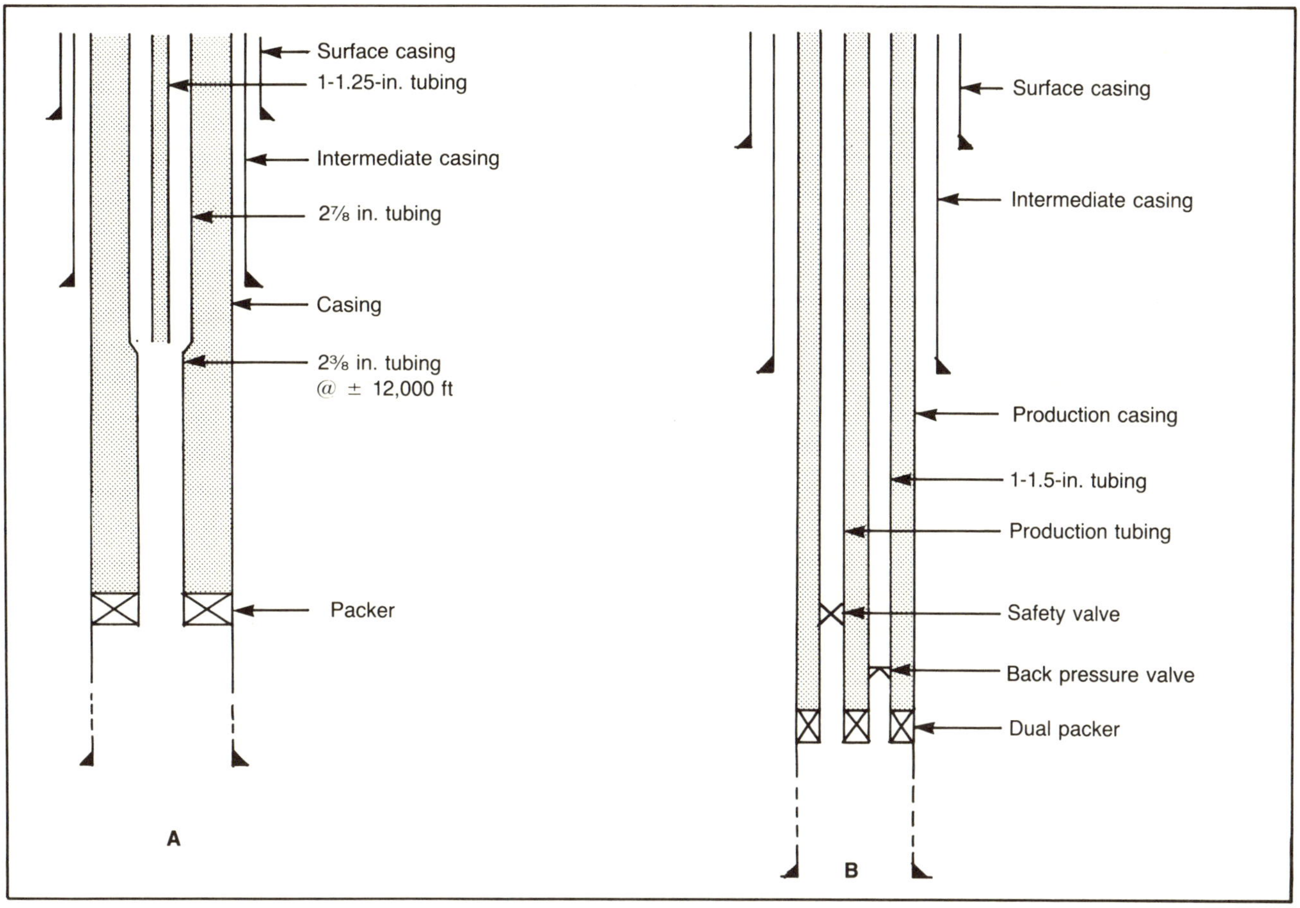

Fig. 9-16. (a) Typical high-pressure gas well concentric string completion; (b) typical parallel string high-pressure gas well completion.

The primary purpose of the kill string is to provide a means to fill the well with a high density fluid that kills the well so that it may be entered safely for workover operations. It also provides a means for continuous or batch application of inhibitor. If the kill string does not extend to the bottom of the tubing, a baked-on plastic coating is usually used for tubing protection below the kill string.

Figure 9-16B also illustrates a parallel kill string that allows inhibitor to be pumped to the bottom of the hole. While this type of completion costs more initially, it eliminates the need to interrupt production to inhibit the well. This would be necessary if both the kill string and the tubing-kill string annulus were normally used to conduct fluids to the surface. Positive injection below the packer can be accomplished by placing a check valve in the kill string and loading the string with inhibitor.

The completion shown in Figure 9-17 utilizes a packer and a bottom-hole chemical injection valve. The tubing-casing annulus is filled with inhibitor. Pressure applied at the surface forces the injection valve to open.

From a technical standpoint, the parallel tubing string concept is preferable to the use of downhole injection valves. Solids can plug injection valves, and the bottom-hole pressure may exceed the desired working pressure of the casing annulus. Another advantage of the parallel tubing string concept is that the inhibitor mixture or the neat inhibitor is not exposed for such long periods to elevated temperatures and pressures. The inhibitor reaches the bottom of the well in a much shorter period, and thermal degradation of the inhibitor is minimized.

High alloy, high strength capillary tubing is available for use in gas wells by coupling it to the tubing at bottom-hole and running it into the hole with the tubing string. A high pressure liquid pump is used at the surface to inject neat chemical at bottom-hole by pumping through the capillary tubing. Because of friction losses, as much as 20,000 psi injection pressure is required to overcome downhole pressure and to put up to 20 gallons per day of chemical into the wellbore. To maintain adequate flow rate, the viscosity of the chemical must remain low while contained in the capillary tubing, and it must be free of solids. For this latter reason, very fine pore size filters are incorporated into the system to prevent plug-ups.

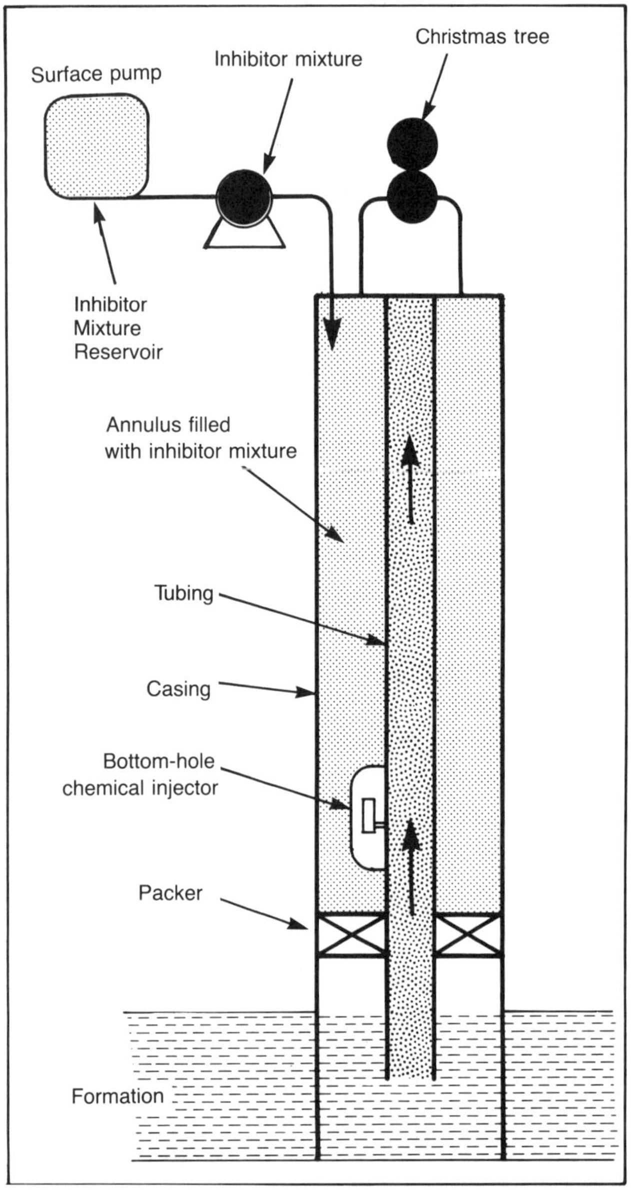

Fig. 9-17. Schematic of a well completion with a bottom-hole chemical injector.

Up to now, these applications have been used for cooler, low pressure wells. However, because of the evidence reported above, where some compounds appear to have a low enough vapor pressure, it now appears possible to use the method for some deep, hot wells in the lower end of the range. It must be recognized that low volatility, good thermal stability, and low solids requirements are extremely stringent for this application.

This system of inhibitor injection has been used successfully in the Gulf Coast area of the U.S. Some of the advantages claimed for the small tubing system are:

1. Assures delivery of clean, debris-free inhibitor to the downhole injection chamber.
2. Capillary volume is small, minimizing time and well temperature effects on the inhibitor.
3. Inhibitor types and injection rates can be quickly changed.
4. Design of the capillary system minimizes possibility of communication between tubing and casing annulus.
5. Capillary can be used for batching of combination treatments, i.e., corrosion and scale inhibitors, foaming agents, cleaning agents, and methanol.
6. Capillary can be used for continuous recording of bottom-hole pressures.

Formation Squeeze

The corrosion inhibitor squeeze treatment involves placement of an inhibitor solution into the producing formation. Its principal advantages are that (1) it can be used in tubingless or multiple completions and (2) treating frequency is reduced. It may be effective for 3 to 18 months, depending on the inhibitor, the formation, the placement technique, and the fluids being produced. In the squeeze treatment, a relatively large volume of inhibitor (1 to 4 drums) is mixed in a compatible diluent (10 to 20% inhibitor). The mixture is displaced down the tubing and into the formation. The amount of displacing fluid is normally 25 to 75 drums plus the tubing volume. When a well is returned to production after a squeeze, the initial concentration of chemical in the returned fluids is high, with a relatively fast decrease in concentration with production. A protective film is thus formed on the metal surfaces during the squeeze operation and during the short time that the chemical concentration in the returned fluid is high. The continuous return of chemical at a lower concentration maintains the initially formed protective film. When the returning inhibitor can no longer repair the protective film as rapidly as it is removed, the effectiveness of the treatment declines.

Practically all types of formation materials will adsorb the various squeeze inhibitors to some degree. The adsorptive capacities of clays are several hundred-fold greater than the capacity of sand or limestone. Adsorption in sand or limestone is reversible. Much of the inhibitor adsorbed by clays is bound so tightly that it is irreversibly adsorbed and unavailable for desorption. However, clays adsorb such a large amount of inhibitor that the fraction that desorbs can still provide substantial corrosion protection. During desorption, the inhibitor concentration in the produced fluid falls rapidly to a few hundred ppm, and then the inhibitor continues to be desorbed at low concentrations for several hundred pore volumes of hydrocarbon production. Because of irreversible adsorption, only a portion of the originally adsorbed inhibitor is desorbed. Obviously, more inhibitor should be used on the initial squeeze (2 to 4 drums).

Significant permeability reduction seldom occurs from the action of corrosion inhibitors on formations; however, clays present in many sandstones are often sensitive to inhibitors and may swell or disperse upon contact with the inhibitor. The resulting permeability changes vary widely, although most are negligible. In areas where formation damage has been known to occur or in which sensitive clays are known or suspected to exist, core samples from the formation should be tested and an inhibitor selected that does not cause damage.

SULFUR DEPOSITION

Sulfur deposition problems can arise in the production of deep sour natural gas reservoirs if a source of elemental sulfur exists and fluid composition and pressure/temperature changes have appropriate characteristics. There is a greater potential for sulfur deposition if:

1. H_2S content is high,
2. condensate (C_{5+}) content is low,
3. aromatic components are not present in the fluid,
4. bottom-hole pressure and temperature are high, and
5. pressure and temperature changes from formation to wellhead are high.

The sulfur-carrying capacity of the produced fluid is principally due to hydrogen polysulfide formation and/or physical dissolution of the sulfur in the dense fluid. The relative importance of these two mechanisms at very high pressures may well determine whether sulfur deposition will be a serious problem in ultra-deep sour gas wells.

Some control of sulfur deposition is possible by limiting pressure/temperature changes through production rate, and a variety of physical and chemical solvents are available for well injection to remove sulfur as it is deposited. Physical solvents are adequate for low or moderate sulfur deposition. Solvents that function by chemical incorporation of the sulfur into the molecular structure of the solvent have much higher sulfur carrying capacity and may be required in cases where sulfur deposition is heavy.

It should be recognized that if elemental sulfur is deposited in formations, tubing, gathering lines, or inlet separators, such depositions may be minimized by heating and maintaining high pressures that discourage sulfur deposition in production equipment and gathering lines, or the sulfur may be removed by continuously or periodically injecting appropriate solvents into the tubing or gathering lines. Solid sulfur that is contained in the inlet well stream normally is carried away, and disposed of, with liquid oil or condensate products.

In the event that a high hydrogen sulfide content lean gas stream is processed, it may be necessary to squeeze the formation or to wash wells periodically with solvent. Such squeezing or washing also removes some of the deposited sulfur from the gathering lines.

SAFETY

High pressure sour gas can be produced safely, but because of the serious consequences that can arise if this gas is released near a populated area, special attention must be given to safety and contingency planning.

Contingency planning is complex and must be thoughtfully detailed for the locale of each sour gas operation. Among other things, the plans should be based on population density, local terrain (including roads), housing and medical facilities, and local weather patterns.

Facilities should be provided to shut-in a well automatically in case of leaks. A typical facility for accomplishing this is shown schematically in Figure 9-18.

Some of the safety procedures to be observed when working around sour gas facilities are listed below.

1. In low concentrations, H_2S has a characteristic odor of rotten eggs and a sweet taste. In higher concentrations, the sense of smell is quickly paralyzed. The sense of smell, therefore, can never be relied upon to indicate the amount of H_2S present.

2. Hydrogen sulfide may be present in natural gas produced alone or in combination with crude oil. Concentrations may be from the faintest odor to a percentage that will result in sudden death. Accordingly, whenever H_2S is present, respiratory protection is of extreme importance.

3. Hydrogen sulfide is heavier than air, having a specific gravity of 1.19 with respect to air, so heavier concentrations will be found at ground or lower levels, such as well cellars, open ditches, and natural topographical low spots.

4. Hydrogen sulfide is highly flammable and has an explosive range of 4.3% to 45.5% concentration by volume in air.

5. When H_2S is known to be present in natural gas or crude oil, all products are handled in systems designed to confine and, when necessary, dispose of the gas in a safe manner.

6. When, through accident, leakage or necessary

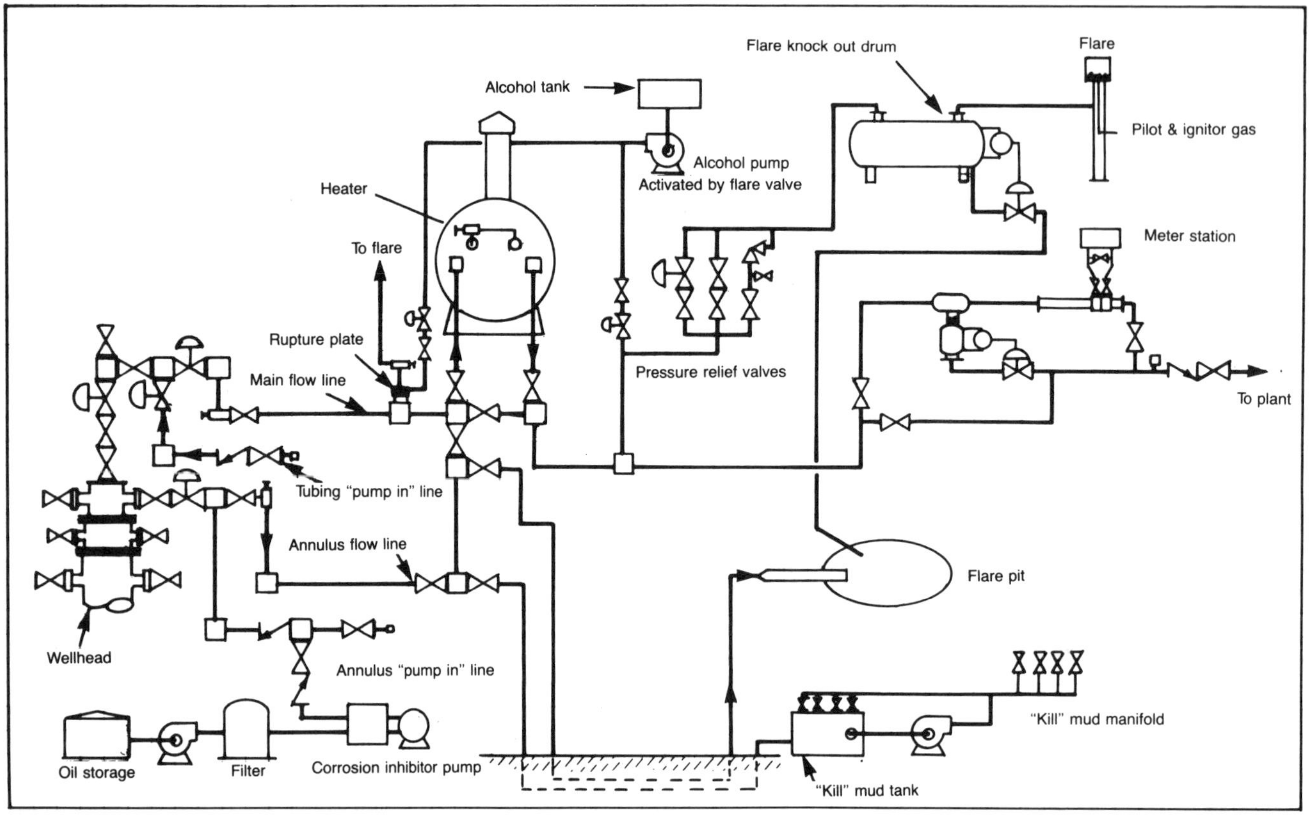

Fig. 9-18. Typical high-pressure sour gas surface facilities.

opening of a closed system, H_2S becomes present in the atmosphere, prescribed respiratory protection should be worn.

7. Whenever hydrogen sulfide is present or thought to be present, the H_2S content must be determined, or an air supplied breathing apparatus must be used.

8. If eyes become irritated or a halo is noticed around an electric light while working in a plant or area that has been determined to be "safe" from H_2S, take the following precautions:

   a. Leave the location at once.
   b. Wash eyes thoroughly with water.
   c. Wear air supplied respiratory equipment with full face protection, if necessary, to return to the location.

9. Permissible limit of concentrations of H_2S for continuous exposure during an 8-hour period is 0.001%, or 10 parts per million by volume in air. Exposure to higher concentrations of H_2S will have the following effect:

—0.001 to 0.010% (10 to 100 ppm) for 30 to 60 minutes results in eye and respiratory tract irritation.

—0.010 to 0.020% (100 to 200 ppm) for 2 to 15 minutes results in loss of sense of smell; exposure for 8 to 48 hours may be fatal.

—0.020 to 0.050% (200 to 500 ppm) for 1 to 4 hours will be fatal.

—0.050 to 0.060 (500 to 600 ppm) for 30 to 60 minutes is fatal.

—0.060 to 0.150% (600 to 1500 ppm) for 2 to 15 minutes is fatal.

10. Since the result of exposure to H_2S is paralysis of the nerves controlling respiration, persons stop breathing and lose consciousness quickly. If the victim is promptly removed to a clear area and artificial respiration is started immediately, the chances of complete recovery are good.

Any delay in the start of artificial respiration appreciably reduces chances of recovery. Even though chances of recovery may seem slim, artificial respiration should be continued until normal breathing is resumed or, if a physician is not available, it should be continued for a minimum of two hours.

Any person overcome by hydrogen sulfide should be treated for shock—that is, kept warm while artificial respiration is being applied and kept quiet until he can be checked and released by a physician.

11. Hydrogen sulfide reacts with iron and steel to form iron sulfide. Iron sulfide reacts with air to form iron oxide. The conversion of iron sulfide to iron oxide creates heat sufficient to ignite flammable vapors.

WELL TESTING

Due to the extreme safety hazards involved in testing sour gas wells, care must be taken to not allow any unflared gas to escape to the atmosphere, especially in populated areas.

If possible, extensive testing should be delayed until the gas can be sent to a pipeline or gas plant. All gas produced during preliminary tests should be flared.

Testing equipment such as separators, pressure gages, and meters should be thoroughly tested before beginning a sour gas well test, especially if the equipment has previously been used in sour gas service.

10 Gas Processing

PETROLEUM as produced from a reservoir is a complex mixture of hundreds of different compounds of hydrogen and carbon, all with different densities, vapor pressures, and other physical characteristics. A typical well stream is a high-velocity, turbulent, constantly expanding mixture of gases and hydrocarbon liquids, intimately mixed with water vapor, free water, solids, and other contaminants. As it flows from the hot, high-pressure petroleum reservoir, the well stream undergoes continuous pressure and temperature reduction. Gases evolve from the liquids, water vapor condenses, and some of the well stream changes in character from liquid to bubbles, mist, and free gas. The high-velocity gas is carrying liquid droplets, and the liquid is carrying gas bubbles.

Stated simply, field processing is required to remove undesirable components and to separate the well stream into salable gas and petroleum liquids, recovering the maximum amounts of each at the lowest possible overall cost. Field processing of natural gas actually consists of four basic processes:

1. Separation of the gas from free liquids such as crude oil, hydrocarbon condensate, water, and entrained solids.
2. Processing the gas to remove condensable and recoverable hydrocarbon vapors.
3. Processing the gas to remove condensable water vapor, which under certain conditions might cause hydrate formation.
4. Processing the gas to remove other undesirable components, such as hydrogen sulfide or carbon dioxide.

Some of these processes are accomplished in the field, but in some cases, the gas goes to a plant facility for further processing. The discussion in this section is divided into two categories, field treatment and plant operations.

FIELD TREATMENT OF NATURAL GAS

Separation of well-stream gas from free liquids is by far the most common of all field-processing operations and, at the same time, one of the most critical. A properly designed separator will provide a clean separation of free gases from the free hydrocarbon liquids. A well-stream separator must perform the following:

1. Cause a primary-phase separation of the mostly liquid hydrocarbons from those that are mostly gas;
2. Refine the primary separation by removing most of the entrained liquid mist from the gas;
3. Further refine the separation by removing the entrained gas from the liquid; and
4. Discharge the separated gas and liquid from the vessel and ensure that no re-entrainment of one into the other occurs.

If these functions are to be accomplished, the basic separator design must:

1. Control and dissipate the energy of the well stream as it enters the separator;
2. Ensure that the gas and liquid velocities are low enough so that gravity segregation and vapor-liquid equilibrium can occur;
3. Minimize turbulence in the gas section of the separator and reduce velocity;

4. Control the accumulation of froths and foams in the vessel;
5. Eliminate re-entrainment of the separated gas and liquid;
6. Provide an outlet for gases, with suitable controls to maintain preset operating pressure;
7. Provide outlets for liquids, with suitable liquid-level controls;
8. If necessary, provide cleanout ports at points where solids may accumulate;
9. Provide relief for excessive pressures in case the gas or liquid outlets should be plugged; and
10. Provide equipment (pressure gages, thermometers, and liquid-level gage-glass assemblies) to check visually for proper operation.

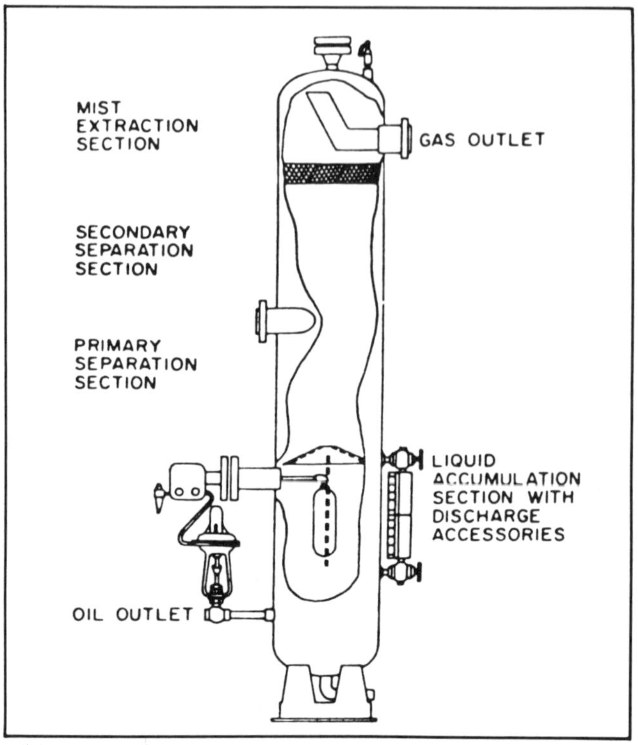

Fig. 10-1. Typical vertical oil and gas separator with mist extractor. Permission to publish by the Society of Petroleum Engineers of AIME. Copyright SPE-AIME.

The process equipment and conditions downstream of a separator will usually dictate the necessary degree of separation and the actual vessel design. Ideally, gases and liquids should come to full equilibrium in the separator, but a compromise must usually be made between the degree of separation achieved and cost of the installation.

The following factors must be considered in sizing and selecting a separator.

1. Liquid flow rate (oil and water), barrels per day and minimum and peak instantaneous.
2. Gas flow rate, million standard cubic feet (MMscf) per day.
3. Specific gravities of oil, water, and gas.
4. Required retention time of fluids within the separator; retention time is a function of physical properties of the fluids.
5. Temperature and pressure at which the separator will operate and design pressure of the vessel.
6. Whether the separator is to be two phase, such as liquid and gas, or three phase—that is, oil, water, and gas.
7. Whether or not there are solid impurities, such as sand or paraffin.
8. Whether or not there are foaming tendencies.

Types of Separators

Three basic types of separators are widely used for gas-liquid separation: (1) vertical (Fig. 10-1), (2) horizontal (Fig. 10-2), and (3) horizontal double-barrel (Fig. 10-3). Each has specific advantages, and selection is usually based on which one will accomplish the desired results at the lowest cost.

A vertical separator is often used on low to intermediate gas-oil ratio well streams and where relatively large slugs of liquid are expected. It can be fitted with a false cone bottom to handle sand production. A vertical separator occupies less floor space, an important consideration where this might be expensive as on an offshore platform. However, because the natural upward flow of gas in a vertical vessel opposes the falling droplets of liquid, a vertical separator for the same capacity may be larger and more expensive than a horizontal unit. In operation, an inlet diverter spreads the inlet fluids against the vertical separator shell in a thin film and at the same

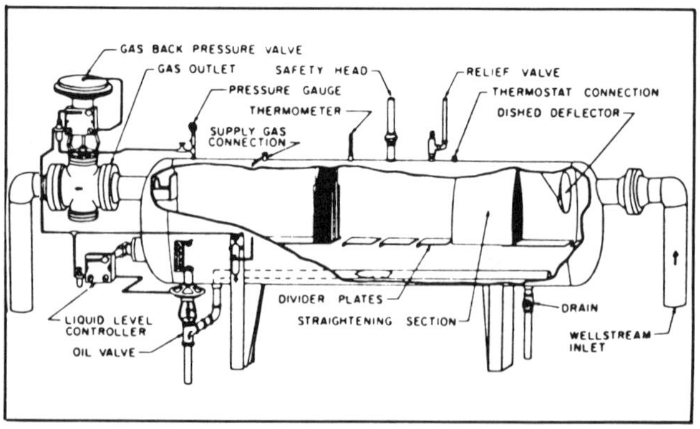

Fig. 10-2. High pressure horizontal oil and gas separator. Courtesy Black, Sivalls, and Bryson, Inc.

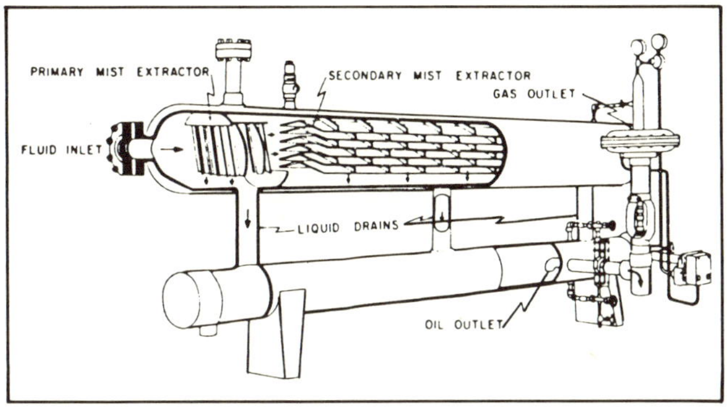

Fig. 10-3. Double-tube horizontal oil and gas separator. Courtesy the American Petroleum Institute.

time imparts a centrifugal motion to the fluids. This provides the desired momentum reduction and allows the gas to escape from the thin oil film. The gas rises to the top of the vessel, and the liquids fall to the bottom. Some small liquid particles will be swept upward with the rising gas stream, and these particles are separated by a centrifugal baffle arrangement below the gas-outlet connection.

The horizontal separator may be the most economical. It may be less expensive than the vertical separator of equal capacity. The horizontal separator has a much greater gas-liquid interface area consisting of a large, long, baffled gas-separation section. Horizontal separators are almost always used for high gas-oil ratio well streams, for foaming well streams, or for liquid-from-liquid separators. A horizontal separator is easier to install, service, and transport. Several separators can be stacked easily into stage-separation assemblies minimizing space requirements. In a horizontal separator, gas flows horizontally and, at the same time, falls toward the liquid surface. Some separators have closely spaced horizontal baffle plates that extend lengthwise down the vessel upon which baffle plates are evenly spaced at a 45° angle to the horizontal. The gas flows in the baffle surfaces and forms a liquid film that is drained away to the liquid section of the separator. The baffles need only be longer than the distance of liquid trajectory travel at the design gas velocity.

Some separators use knitted wire-mesh pads, 4 to 8 in. thick, across the gas section of the separator or across the gas-outlet nozzle. Wire mesh can provide good gas-liquid separation but may be plugged by paraffin or solids in the gas stream. Wire-mesh pads should not be used unless the well stream is so clean that no danger of plugging exists.

A double-barrel horizontal separator has all the advantages of a normal horizontal separator plus a much higher liquid capacity. Incoming free liquid is immediately drained away from the upper section into the lower section. The upper section is filled with baffles, and gas flow is straight through and at higher velocities.

A horizontal three-phase separator (Fig. 10-4) is designed to separate oil, water, and gas and has two liquid outlets. Three-phase separators are used commonly for well testing and in instances where free water readily separates from the oil or condensate. They are identical to two-phase vessels except for the water compartment, an extra level control, and dump valve.

Filter separators (Fig. 10-5) are designed to remove small liquid and/or solid particles from gas streams. These units were designed specifically to handle those applications where, due to the extremely small particle size, conventional separation techniques employing gravitational or centrifugal force are ineffective. Contaminants of this small size can be removed most effectively by passing the gas through a fine, high-quality filtering medium. Several configurations of filter separators are used, depending upon the required efficiency and on whether liquids or solids, or both, are to be removed. Some filter elements have collection efficiencies of 98% of the 1-micron particles and 100% of the 5-micron particles when operated at rated capacity and recommended filter-change intervals. A typical filter separator for removing both liquid and solid contaminants is shown in Figure 10-5.

Separator Controls

Liquid level within the separator must be maintained within reasonable limits to prevent discharging liquid out of the gas line or gas out of the liquid line and to assure proper functioning and flow through the separator internals. Pressure within the separator is usually maintained within a specific pressure range by a gas back-pressure regulating valve. Temperature within most separators is usually not controlled although there are exceptions, such as low-temperature separation systems. Safety and protection against overpressure is provided by a pressure-relief valve set at the design pressure of the separator.

Stage Separation

Stage separation is a process in which gaseous and liquid hydrocarbons are separated into vapor and liquid phases by two or more equilibrium flashes at consecutively lower pressures. As shown in Figure 10-6, two-stage separation involves one separator and a storage tank. Three-stage separation requires two separators and a storage tank. Four-stage separation would require three separators and a storage tank. The tank is always counted as the final stage of vapor-liquid separation because the final equilibrium flash occurs in the tank.

The purpose of stage separation is to reduce the pressure on the reservoir liquids gradually, in steps or stages,

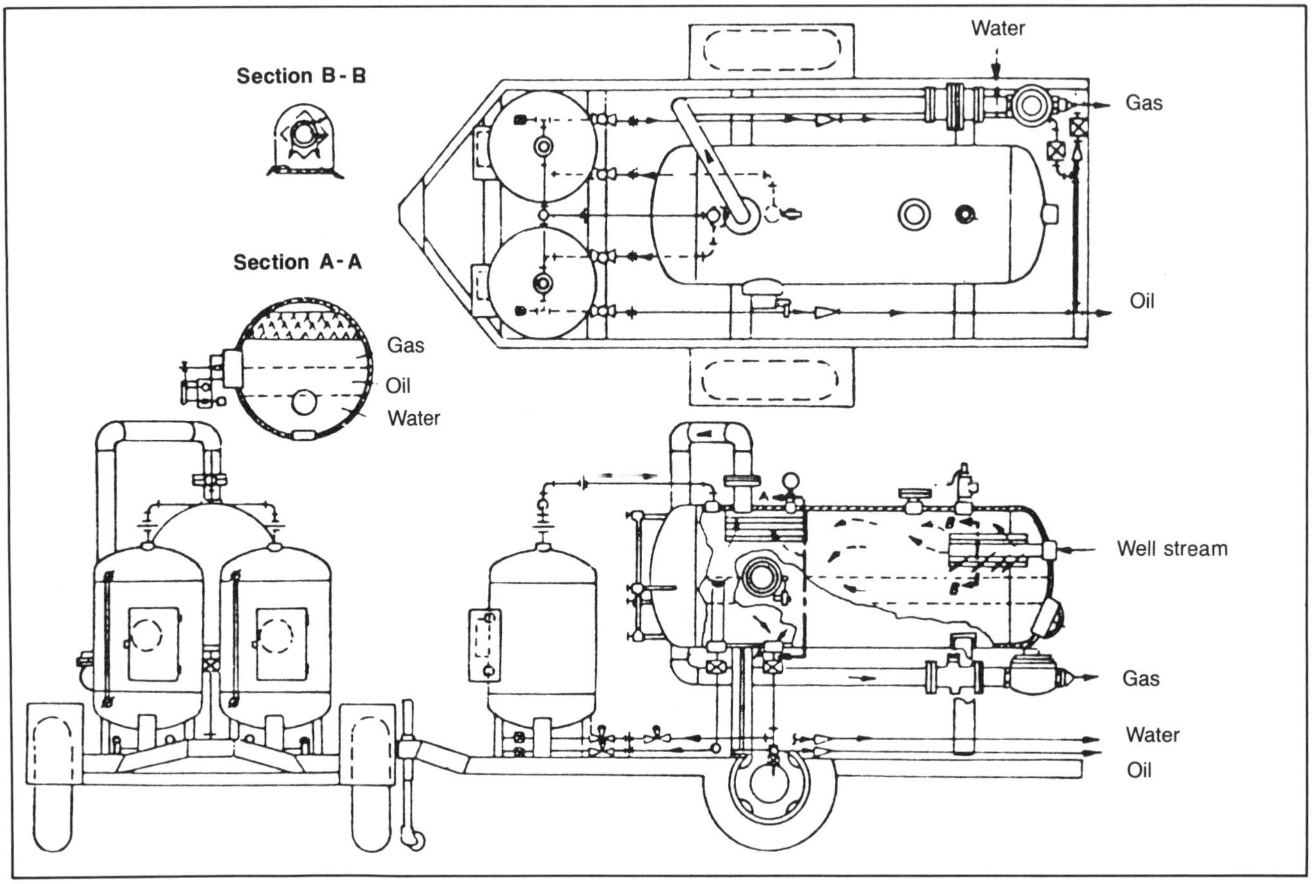

Fig. 10-4. Three-phase portable separator. Permission to publish by McGraw-Hill Book Company.

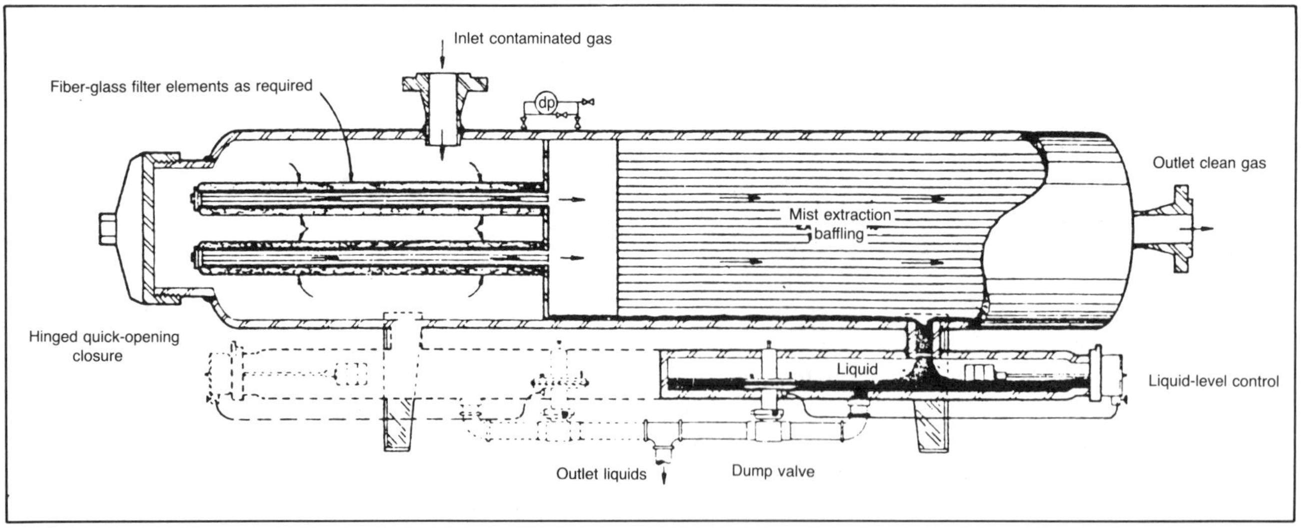

Fig. 10-5. Filter separator for removing liquid and solid contaminates. From Field Handling of Natural Gas, 3rd ed. 1972, Copyright, Petroleum Extension Service, The University of Texas at Austin (PETEX).

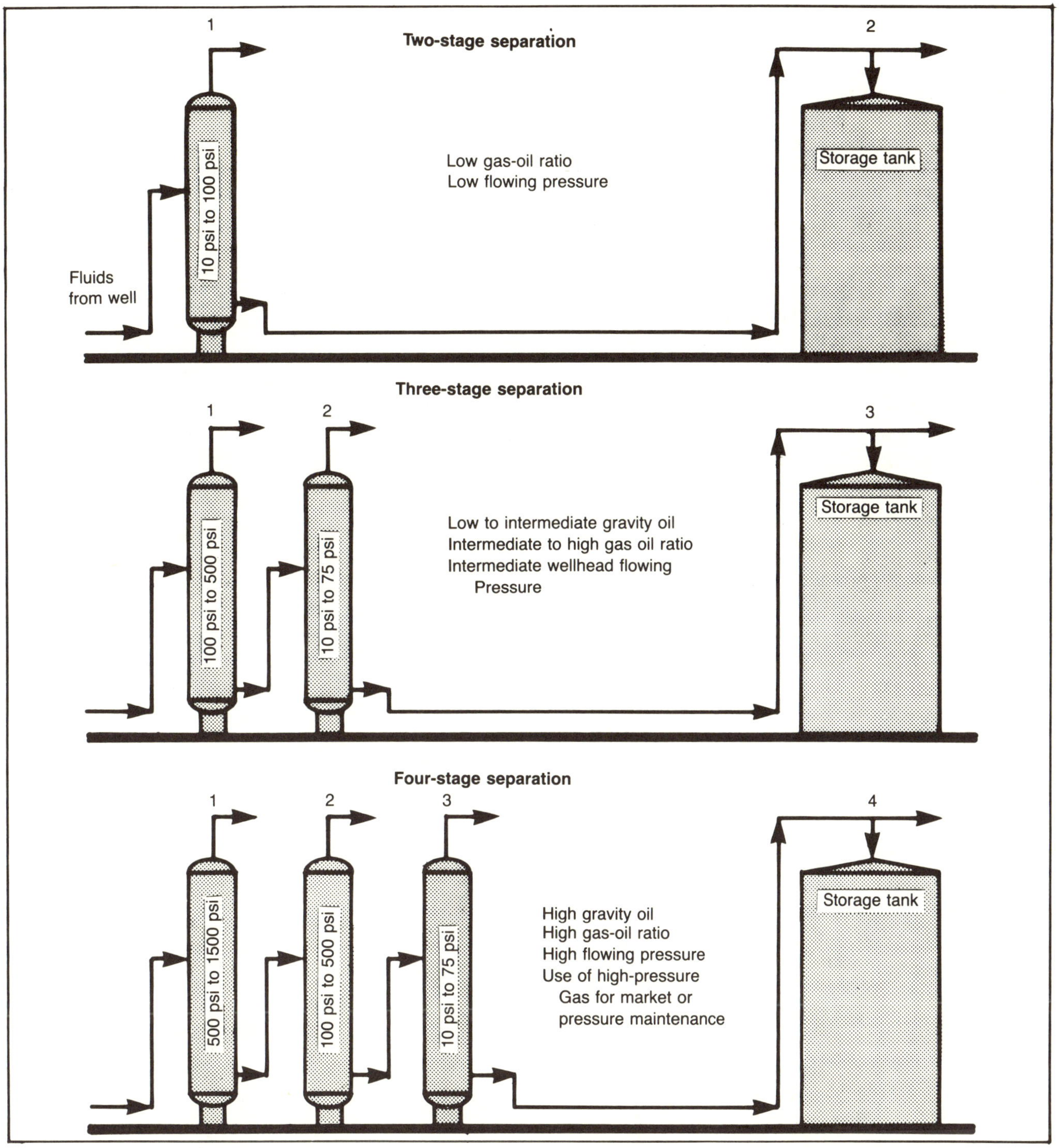

Fig. 10-6. Stage separation.

so that a more stable stock-tank liquid will result. Petroleum liquids at high pressures usually contain large quantities of liquefied propanes, butanes, and pentanes, which will vaporize or flash as the pressure is reduced. This flashing can cause a substantial reduction in stock-tank liquid recovery, depending upon well-stream composition, pressure, temperature, and other factors. For example, if a volatile condensate at 1500 psig was discharged directly into an atmospheric storage tank, most of it would immediately vaporize, leaving very little liquid in the tank.

The ideal method of separation, to attain maximum

liquid recovery, would be that of differential liberation of gas by means of a steady decrease in pressure from that existing in the reservoir to that existing in the storage tanks. With each tiny decrease in pressure, the gas evolved would immediately be removed from the liquid. However, carrying out this differential process would require an infinite number of separation stages, obviously an impractical solution.

Low Temperature Separation

Low temperature separation, probably the most efficient means yet devised for handling high pressure gas and condensate at the wellhead, performs the following functions:

1. Separation of water and hydrocarbon liquid from the inlet well stream;
2. Recovery of more liquids from the gas than can be recovered with normal temperature separators; and
3. Dehydration of gas, usually to pipeline specifications.

The first low temperature units were developed and placed in operation in 1948 to dehydrate gas at the wellhead in remote locations so that high-pressure gas and condensate could be gathered at central locations without the problems of hydrate plugging.

Essentially, the low temperature separation process is one of intentional hydrate formation and controlled melting. The inlet gas is cooled by expansion, due to pressure reduction, causing water and liquid hydrocarbons to condense; if hydrates are formed, they are quickly melted. Dry gas, condensate, and free water are then discharged from the vessel under controlled conditions. Since this separation system permits operating conditions well below hydrate-formation temperatures, the recovery of hydrocarbon liquids is much higher than that possible for conventional separation. Condensation of a higher percentage of the water vapor also is accomplished, resulting in dehydration of the gas.

Since low temperature separation is achieved by the cooling effect of gas expansion—that is, pressure reduction across the choke—how much pressure drop is needed to obtain adequate cooling? Generally speaking, satisfactory operation and dehydration to pipeline specifications can be accomplished with a pressure differential as low as 1000 psi, provided the temperature upstream of the choke can be controlled to near the hydrate point.

The actual minimum differential between flowing and line pressures depends primarily upon the pressure range in which the hydrate temperature falls. Noting the pressure-temperature drop chart (Fig. 10-7), it can be seen that a 1500 psi drop from 3000 psi at 120°F will provide

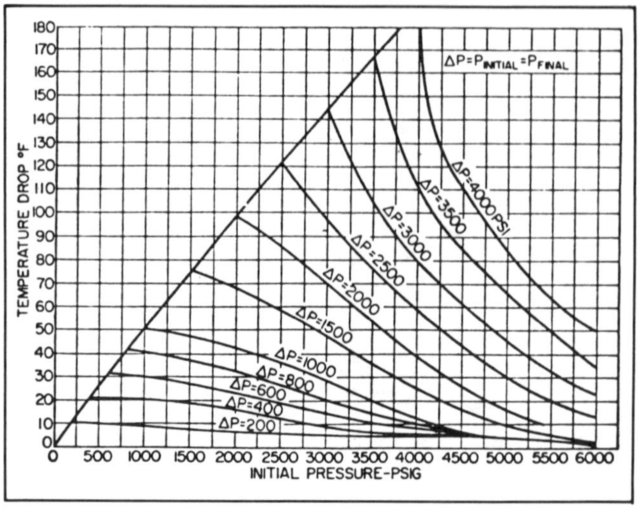

Fig. 10-7. Temperature drop associated with a given pressure. Permission to publish by the McGraw-Hill Book Company.

a final temperature of only 78°F, while a 1500 psi drop from 2000 psi at 120°F will give a final temperature of 49°F. The actual temperature drop will not necessarily correspond to that shown on the chart; composition of the gas stream, flow rate, liquid rates, bath temperature, and ambient temperature will affect the actual temperature drop. Water dew points will average 10°F to 15°F below the indicated temperature due to the adsorptive effects of the hydrates on the vapor-phase water.

Some means must be provided to prevent formation of hydrates in the low temperature separator. This is accomplished by either piping the hot well stream through the separator, as shown in Figure 10-8, or by injecting

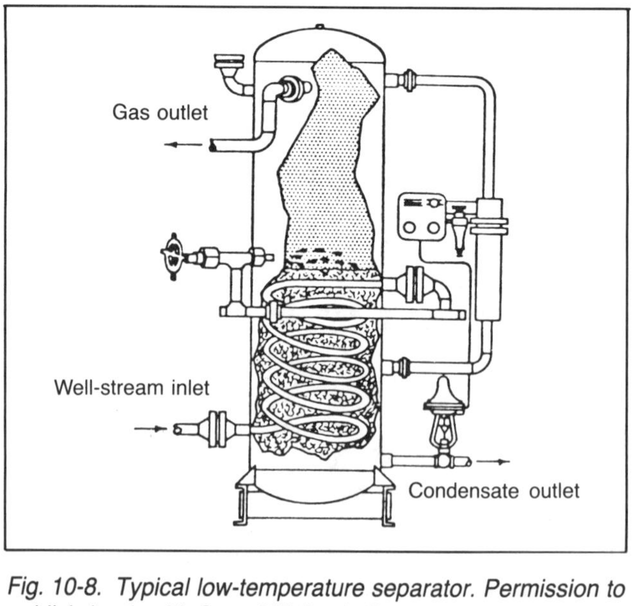

Fig. 10-8. Typical low-temperature separator. Permission to publish by the McGraw-Hill Book Company.

hydrate inhibitors upstream of the separator. In some cases, a line heater is required upstream of the separator for the first method (Fig. 10-9).

A schematic diagram of the second method, using glycol injection, is shown in Figure 10-10.

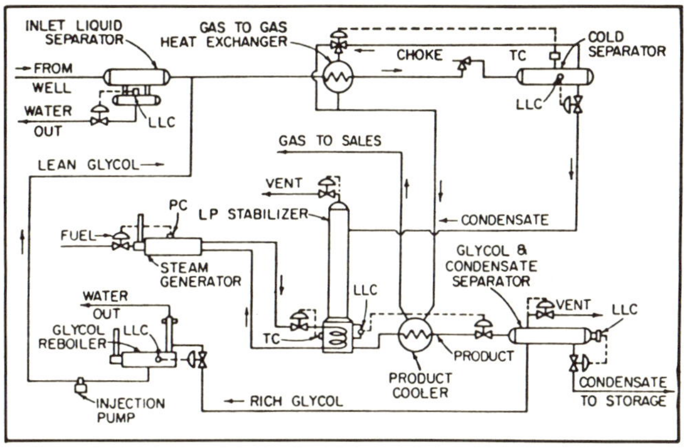

Fig. 10-10. Schematic flow diagram of a glycol injection LTS system with low-temperature stabilization. Permission to publish by the McGraw-Hill Book Company.

Condensate Stabilization

One of the problems in using low temperature separation units of both the mechanical and glycol-injection types is high stock-tank vapor loss. These losses are the result of vaporizing appreciable quantities of liquid propane and butane with dissolved methane and ethane, which are liberated when the pressure on the liquid is reduced from the low temperature separator to storage pressure. When these light ends vaporize in a stock tank, they carry some of the heavier hydrocarbons with them to be burned or lost in the atmosphere. Stabilization is a means of removing these lighter hydrocarbons from the liquid present in the bottom of the low temperature separator with a minimum loss of heavier hydrocarbons. Stabilization results in a larger volume of stock-tank liquids available for sale.

The stabilization system consists of a vertical vessel, which may be packed with ceramic rings or fitted with trays spaced from 12 to 24 in. apart inside the vessel. The liquid in the lower section of the tower is heated by an indirect heater or steam coils. The cold condensate from the bottom of the low-temperature separator flows directly to the top of the stabilizer or may flow to a flash tank prior to entering the stabilizer. From the flash tank, the gas is routed to become fuel or to be disposed of by other means, and the liquid is taken to the top of the stabilizer. Some of the light hydrocarbons are vaporized and pass from the top of this vessel to the gas-sales, fuel, or vent line. The liquid hydrocarbons flow through the packing or down the trays, absorbing some of the heavier gaseous hydrocarbons, which have been vaporized at the bottom of the vessel. At the bottom of the vessel, the heat added from the heater or reboiler vaporizes most of the lighter hydrocarbons. After being cooled, the stabilized liquid flows to storage and the lighter ends flow upward to be reabsorbed or to leave the top of the vessel (Fig. 10-11). The amount of additional condensate that can be recovered by stabilization is dependent upon the pressure and temperature at which the low temperature separator is operated and the composition of the gas being processed.

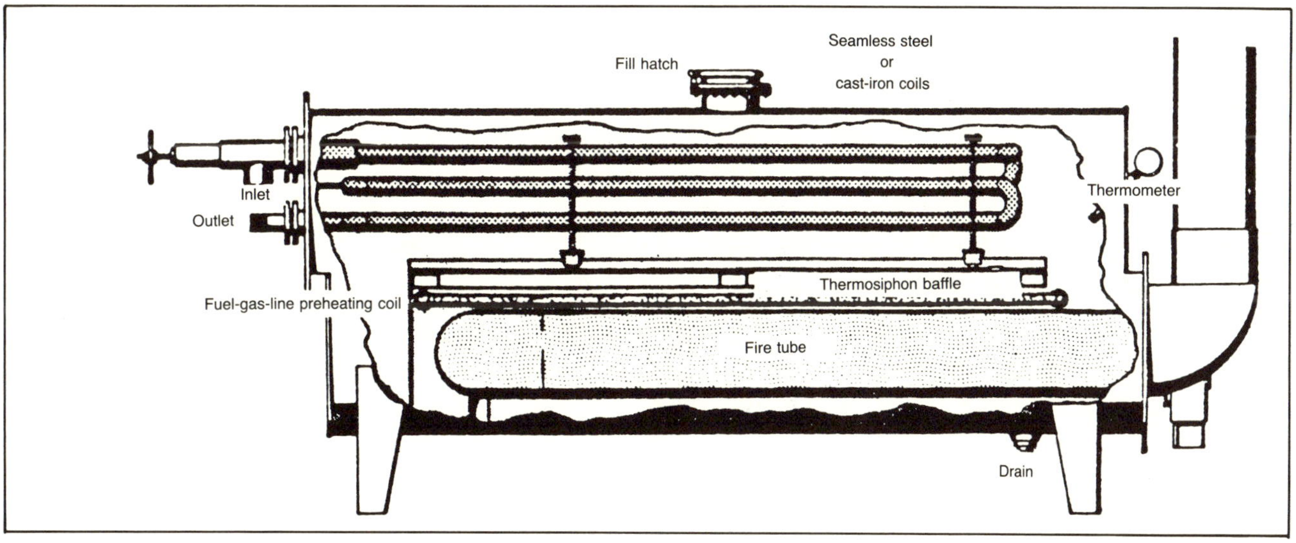

Fig. 10-9. Cutaway view of an indirect heater. From Field Handling of Natural Gas, 3rd ed. 1972, Copyright, Petroleum Extension Service, The University of Texas at Austin (PETEX).

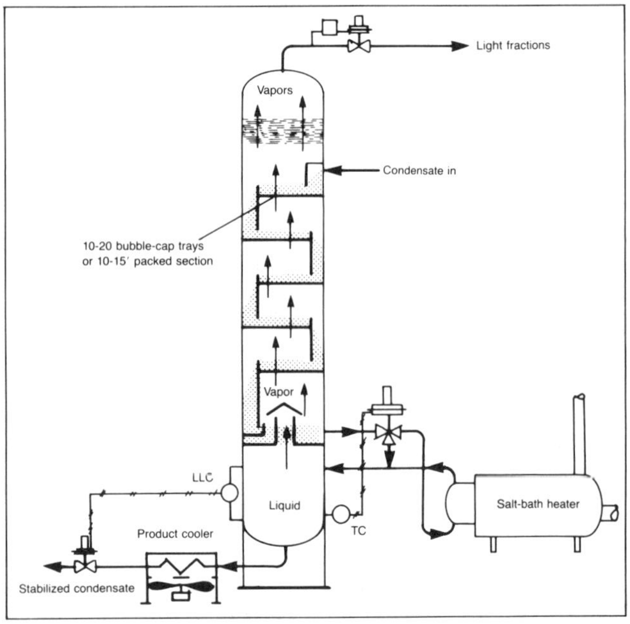

Fig. 10-11. Gas stabilizer.

Additional hydrocarbon liquid recovery that can be gained by use of low temperature separation as compared to conventional separation depends on the exact composition of the well stream and upon the operating temperature of the low temperature separator. In the absence of an analysis, the increased recovery may be estimated by assuming that stock-tank liquid recovery will be increased about 0.5 bbl per MMscf for each 10°F decrease in separator temperature. For example, if separation at 80°F produces 10 bbl/MMscf, then separation at 20°F would produce an additional 3 bbl/MMscf. An additional increase in recovery of about 10% might be expected if the low temperature liquid was then stabilized. For richer gas streams, on the order of 50 bbl/MMscf, the increased liquid recovery would be about 0.75 bbl/MMscf for each 10°F decrease in separator temperature. These figures are merely rules of thumb and are only an approximation of the actual increased recovery. Figure 10-12 illustrates the effect of separation temperature on liquid recovery.

GAS PLANT OPERATIONS

Gas plants and the treatment of natural gas are very important segments of the gas production operation. Untreated or unprocessed natural gas contains many hydrocarbon compounds and a few nonhydrocarbon compounds. Treating natural gas in plant facilities involves removing certain of the compounds that are of considerable value by themselves or because they are contaminants that render the gas unsuitable for sale purposes. The predominate constituent of natural gas is methane with smaller amounts of other hydrocarbons. Table 10-1 gives examples of a few of the common components of natural gas and their phase after processing.

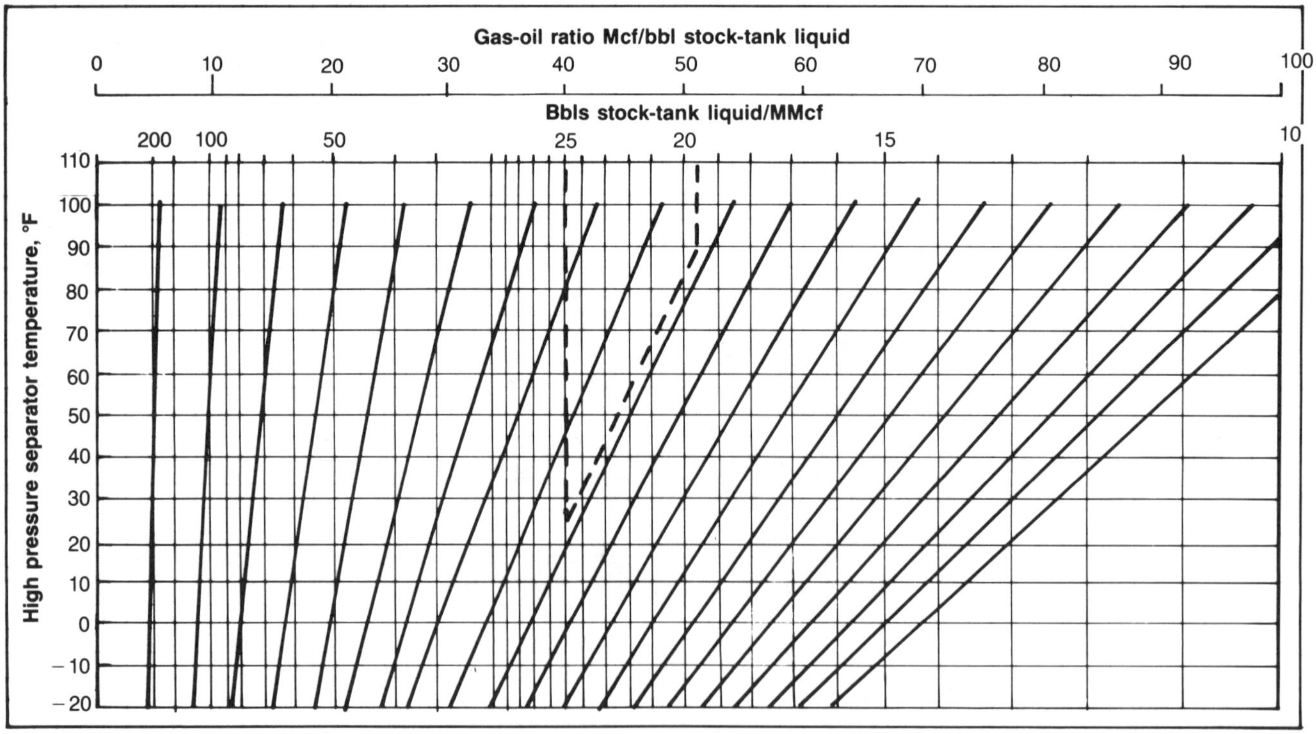

Fig. 10-12. Separator temperature effect.

TABLE 10-1

Component Hydrocarbons	Chemical Formula	How Utilized
Methane	CH_4	Fuel, gaseous state
Ethane	C_2H_6	Mixed with methane as gaseous fuel and alone as chemical feed stock. Exists as gas.
Propane	C_3H_8	Liquid fuel or chemical feed stock. Pressure storage required.
Pentanes	C_5H_{12}	Constituents of natural gasoline. Pressure storage not required.
Hexanes	C_6H_{14}	Constituents of natural gasoline. Pressure storage not required.
Non-Hydrocarbons		
Hydrogen Sulfide (acid gas)	H_2S	Poisonous gas constituent of natural gas. Removed and converted to elemental sulfur.
Carbon Dioxide (acid gas)	CO_2	Removed from natural gas if in excess of sales specification.
Nitrogen	N_2	Ordinarily not removed. Inert gas with no heating value in natural gas.
Water	H_2O	Removed to meet sales gas water dewpoint specification.

Pentanes and hexanes are shown in the table as constituents of natural gasoline. Natural gasoline also includes liquids heavier than hexane. Butanes may also be present in natural gasoline. Natural gas containing hydrogen sulfide in concentrations sufficient to prevent it from meeting sales gas specifications is termed "sour" gas.

The gas sold to gas transmission companies usually must meet certain requirements with respect to water content, hydrocarbon dewpoint, heating value, and hydrogen sulfide content. Some sort of processing plant is required to treat the gas so that the specifications in the sales contract are met. A dehydration plant is provided to control the water content; a gas processing plant of one type or another is provided to remove certain hydrocarbon components to meet hydrocarbon dewpoint specifications; and a gas sweetening plant is required to remove acid gas (when present). Because of environmental pollution, hydrogen sulfide cannot be flared, and is therefore converted to elemental sulfur, a marketable product.

Any given plant installation can have overlapping purposes. Oil production may be made possible or be increased; the associated gas can be conditioned for sale or other use, such as injection; the removal of liquid products can be accomplished; and the environment can be protected by converting hydrogen sulfide removed from the gas into elemental sulfur, all within the same plant.

The removal of the heavier hydrocarbon components as liquid products is accomplished in gas processing facilities by the use of simple physical principles. Gas and liquid or two liquids of different densities are separated by passing them through a vessel large enough for them to decelerate and have time to separate. Hydrocarbons can be condensed and separated as liquids by increasing the pressure and reducing the temperature. They can also be separated by absorbing them in oil, cryogenically separating them, or adsorbing them on a desiccant. Water can be removed by adsorbing it on a desiccant, absorbing it in glycol, or by chilling the gas. Hydrogen sulfide and carbon dioxide can be removed by chemically reacting them with a chemical solution, adsorbing them on a desiccant, or absorbing them in a physical solvent.

Gas processing is simply various combinations of these basic processes. With the necessary piping and equipment to provide heat, cooling, heat exchange, and to contact and separate the streams as they are processed, these processes become the gas plant. A typical combination of processes to form a gas plant is shown in Figure 10-13. The following discussion treats in more detail the methods for processing natural gas.

Liquid Hydrocarbon Recovery

The following discussion relates only to the broad subject of the use of gas processing plants for recovery of the ethane and heavier hydrocarbons from a gas stream. Such a plant may process the gas associated with oil production, in which case it is known as a casinghead plant. Usually the casinghead gas contains a larger amount of recoverable liquid hydrocarbons than nonassociated gas. Gas from gas reservoirs may also serve as feed to a gas processing plant. Any one of several methods of recovering liquid products from gas may be used to process gas regardless of its source. Products extracted from the gas in liquid recovery processes may include ethane, propane, isobutane and normal butane, and natural gasoline. These products may be fractionated at the plant or leave the plant as a single liquid mixture to be separated elsewhere.

Compression Processing. One type of casinghead gas processing plant is shown schematically in Figure 10-14. The gas may arrive at the plant inlet at very low pressure, 5 psig or even less being common. The gas enters the inlet scrubber where any liquids are removed. The

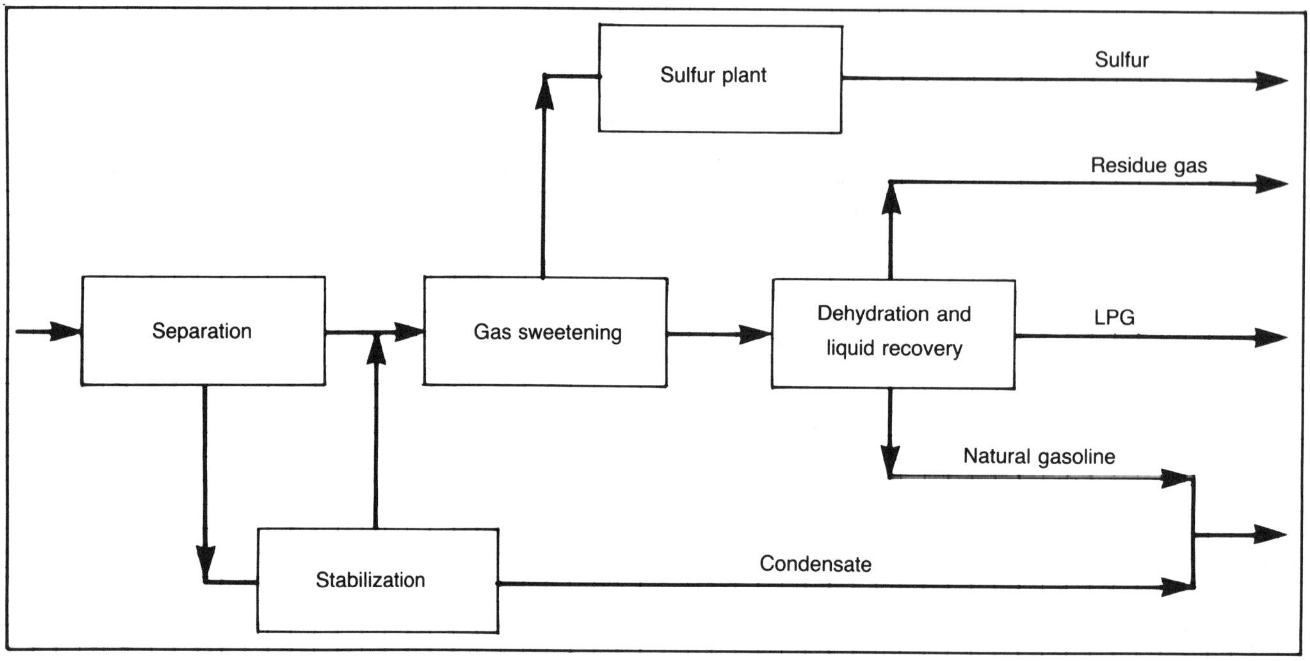

Fig. 10-13. Typical processes combined to form a gas plant.

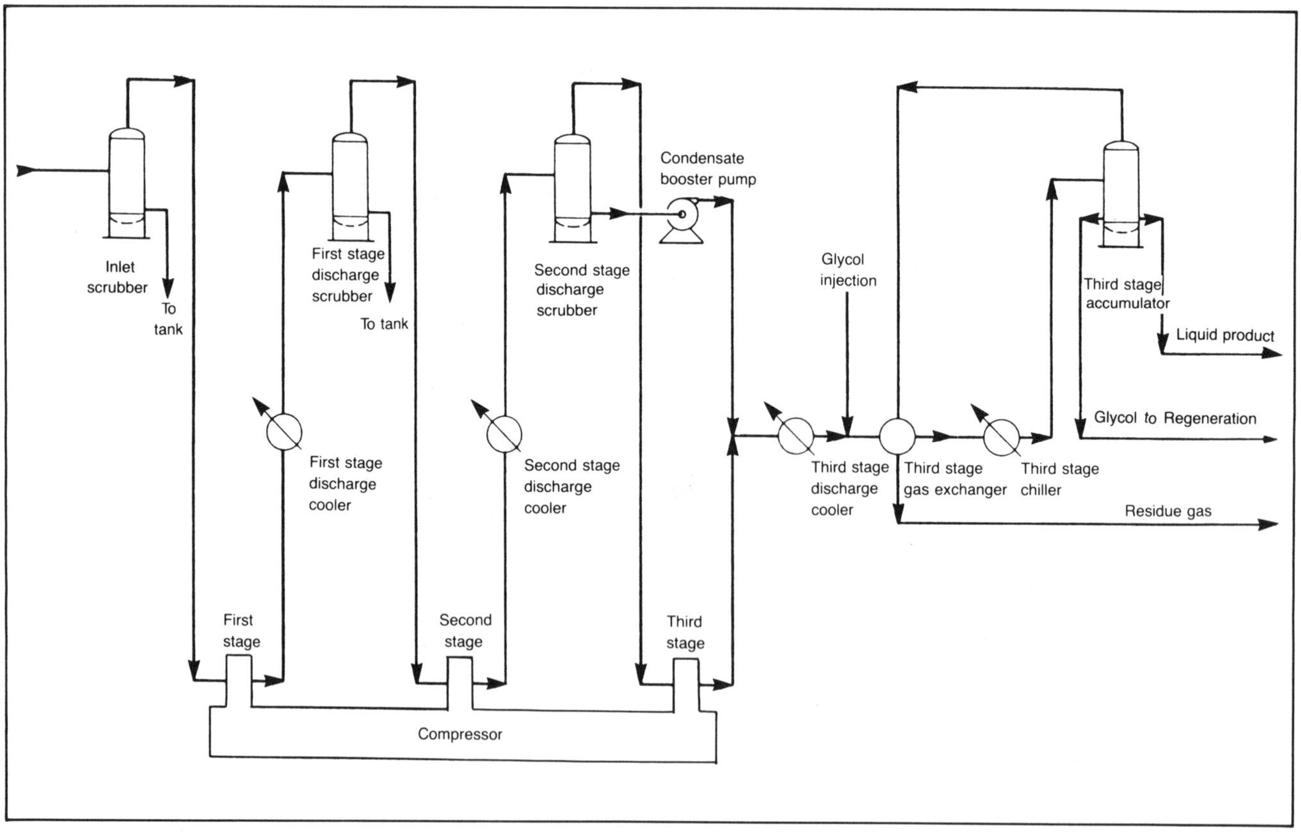

Fig. 10-14. Compresson gasoline plant.

end use of such gas usually requires it to be at a final pressure of 500–900 psig, and accordingly, it will ordinarily be compressed through three stages of compression. Gas compressor equipment is discussed in detail in Chapter 5. The compressed gas is cooled between compressor stages to conserve horsepower and to keep the compressor discharge temperatures within reasonable limits. Either forced air coolers or heat exchangers using water as the cooling medium may be used. Liquids may be condensed between stages depending upon the richness of the gas, its temperature, and pressure.

In a plant such as shown in Figure 10-14, water vapor in the gas condenses as the pressure is increased and the gas cooled. If liquid is formed in any of the coolers, either water or hydrocarbon, then the gas leaving is at its dew point with respect to either water or hydrocarbon, or both. Gas hydrates may cause trouble under these conditions even though dew points are far above freezing. The liquid product from a compression gas processing plant will have a very high vapor pressure and will therefore be difficult and expensive to store without stabilizing the liquid product or taking substantial vapor losses.

Absorption Processing. If economics warrant increasing the recovery of the marketable products, or if gas sales dew point specifications require removal of more of the heavier hydrocarbons, the more complex and expensive absorption process is used in the gas processing plant. Absorption involves contacting the compressed raw gas with a liquid hydrocarbon called lean oil or absorption oil in an absorber where components in the gas dissolve in the lean oil. The heavier components dissolve more easily, and the oil will hold more of them than the lighter components, but some of the lighter components are absorbed also. The bulk of the gas, called residue gas, leaves the top of the absorber while the absorbed components leave with the rich oil from the bottom of the absorber.

The absorption oil process can be carried out at ambient temperatures or process temperatures may be lowered by refrigeration. Although ambient temperature plants are now generally considered obsolete, the design of these older casinghead processing plants usually does not lend itself to economic modification to the modern, more efficient types. However, this operation can be optimized by adjusting process variables.

A flowsheet of a refrigerated light oil absorption gas process is shown in Figure 10-15. The feed gas enters the gas-gas exchanger that serves to recover refrigeration from the plant residue gas stream. The gas then passes through the gas chiller, which is often cooled by vaporizing liquid propane from a mechanical refrigeration system. The cooled gas from the chiller then passes to the absorber. In this type of plant, the lean oil is not pumped directly to the absorber proper, but to a presaturator chiller where the gas from the top of the absorber is mixed with the stripped lean oil. The effluent from the presaturator chiller enters a separator tank from which the separated gas flows to the gas-gas exchanger. The liquid from the presaturator separator is then pumped to the top tray of the absorber. The lean oil is thus saturated with methane and ethane at the chiller temperature before entering the absorber. This minimizes the temperature rise from dissolving gas in the absorber and permits higher absorption of propane.

The rich oil from the bottom of the absorber flows to the rich oil flash tank where some of the undesirable materials picked up in the absorber, particularly methane, are flashed off at a lower pressure. Because of the adiabatic expansion, fluids in the rich oil flash tank will usually be considerably colder than the absorber bottoms. The vapors from the rich oil flash tank are recycled back to the plant inlet after compression and cooling. The cold liquid from the flash enters the lean oil/rich oil exchanger, which conserves refrigeration by further cooling the lean oil from the air-cooled lean oil cooler. Some flashing of the rich oil occurs as the temperature is increased in the exchanger. The mixture is fed to the center section of the rich oil de-ethanizer.

The rich oil de-ethanizer column usually has a larger diameter lower (stripping) section and a smaller diameter top absorption section, the two diameters reflecting the column loadings. Heat enters the column through the reboiler and the side heater. The top (reabsorber) section serves as an absorber where desirable components are dissolved in the lean oil and are washed downward. The reabsorber section is supplied with cold lean oil from the lean oil-rich oil exchanger. The lean oil flow and the bottom temperature determine the amount of propane retained and limit the ethane in the propane as required to meet propane specifications. This same type of column is called a rich oil demethanizer if ethane is being recovered. The overhead product from such columns is gaseous and usually serves as a source of plant fuel gas with any excess going to the recycle recompressor.

The hot rich oil leaves the bottom of the de-ethanizer and moves to the lean oil still where absorbed products are separated from the lean oil. Again, as in the case of the rich oil de-ethanizer, a two-diameter column is used because of the difference in loads between the top and bottom of the column. To supply the necessary stripping vapors in the still, the major part of the still bottoms is pumped to the furnace reboiler. The net lean oil bottom required by the absorption process is withdrawn and returns through the heat exchangers connected to the rich oil de-ethanizer, from which it flows to the lean oil cooler. The products stripped from the lean oil are taken over-

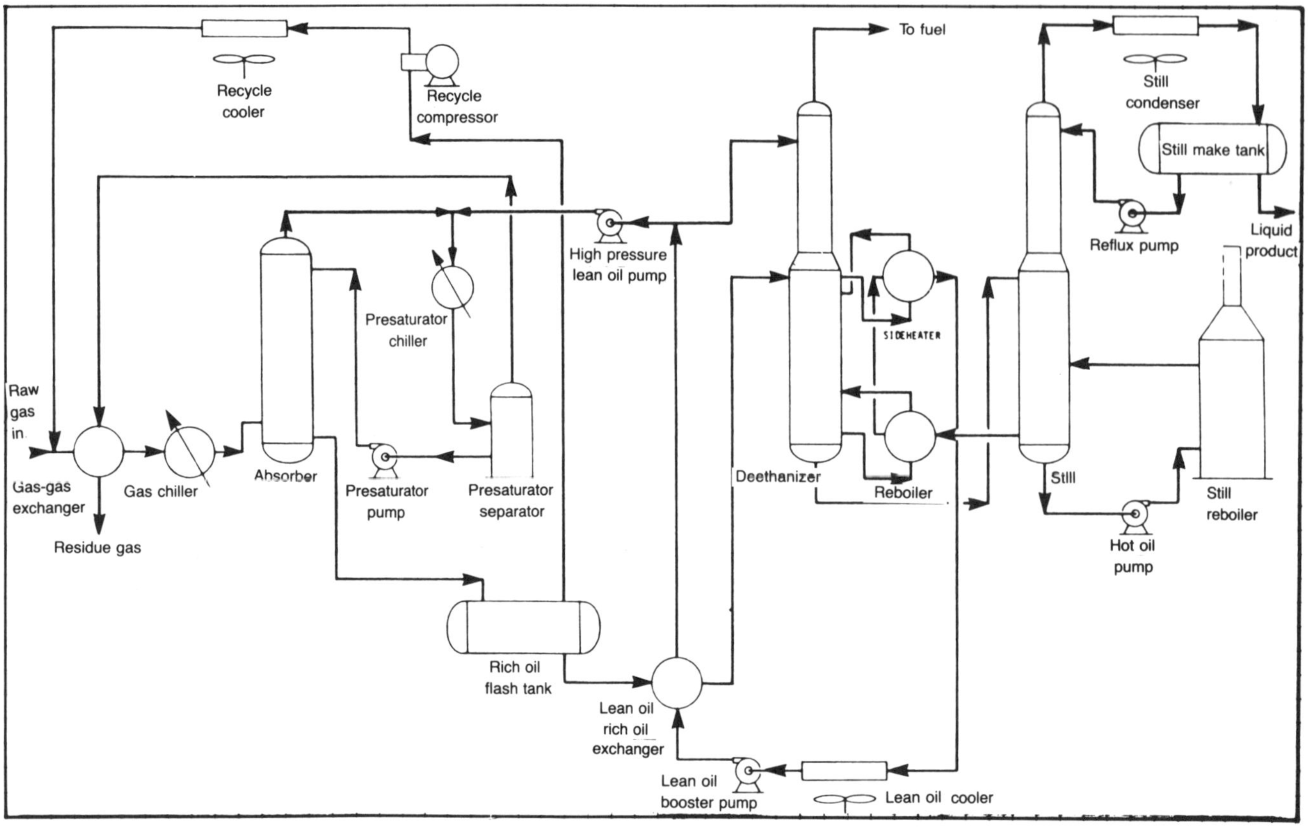

Fig. 10-15. Refrigerated light absorption oil gasoline plant.

head from the still and condensed. Reflux is pumped back to the column. The liquid products from this type plant are quite often sold as a mixed stream or can be fractionated in a separate plant.

Cryogenic Processing. In the previous discussion, it has been shown that relatively high ethane recovery can be achieved by the combination of low temperature and appropriate lean oil circulation in an absorption plant. Under certain circumstances, it has become economic to process gas for high ethane recovery using only extremely low temperatures at moderate pressure. Such plants are in the cryogenic category and have become practical and economic in the natural gas processing industry with the development of the turbine type expander-compressor. This type of process is most appropriate where high pressure gas is available and the end use of the gas is at a low pressure and relatively close to the processing point.

A flowsheet of the process is shown in Figure 10-16. The dehydrated plant feed enters the gas-gas exchanger where, by exchange with the plant residue, the feed temperature is lowered to 80°F. Even with a relatively dry gas, some liquid hydrocarbons will condense at these conditions. The liquid is separated and fed to the demethanizer. The vapor proceeds to the turbine exhaust temperature at about −150°F. The exhaust is a liquid-vapor mixture at these conditions and enters as feed at the top of the unrefluxed demethanizer. The turbine expander, a very high-speed device rotating at speeds up to 25,000 RPM, is directly connected to the centrifugal compressor. Plants such as this normally recover 30–60% of the ethane entering in the plant feed stream.

Adsorption Processing. Hydrocarbon recovery by adsorption is another method that becomes economic under certain conditions. This process depends upon the ability of certain solid materials to adsorb gases or liquids. Materials that have the necessary properties for adsorption due to their large surface area of minute pores are activated carbon, bauxite, activated alumina, silica gel, and synthetic zeolites called molecular sieves. Hydrocarbon recovery units using these materials also remove water vapor from the gas stream.

Solid bed adsorption is a cyclic batch type operation that operates continuously by the use of two (or more) adsorption towers as shown in Figure 10-17. While one tower is on a drying cycle, the other tower is being regenerated by heating the desiccant with hot gas that vaporizes the adsorbed hydrocarbons and water. The hot regeneration gas containing the hydrocarbons and water vapor is cooled, condensing out most of the water and

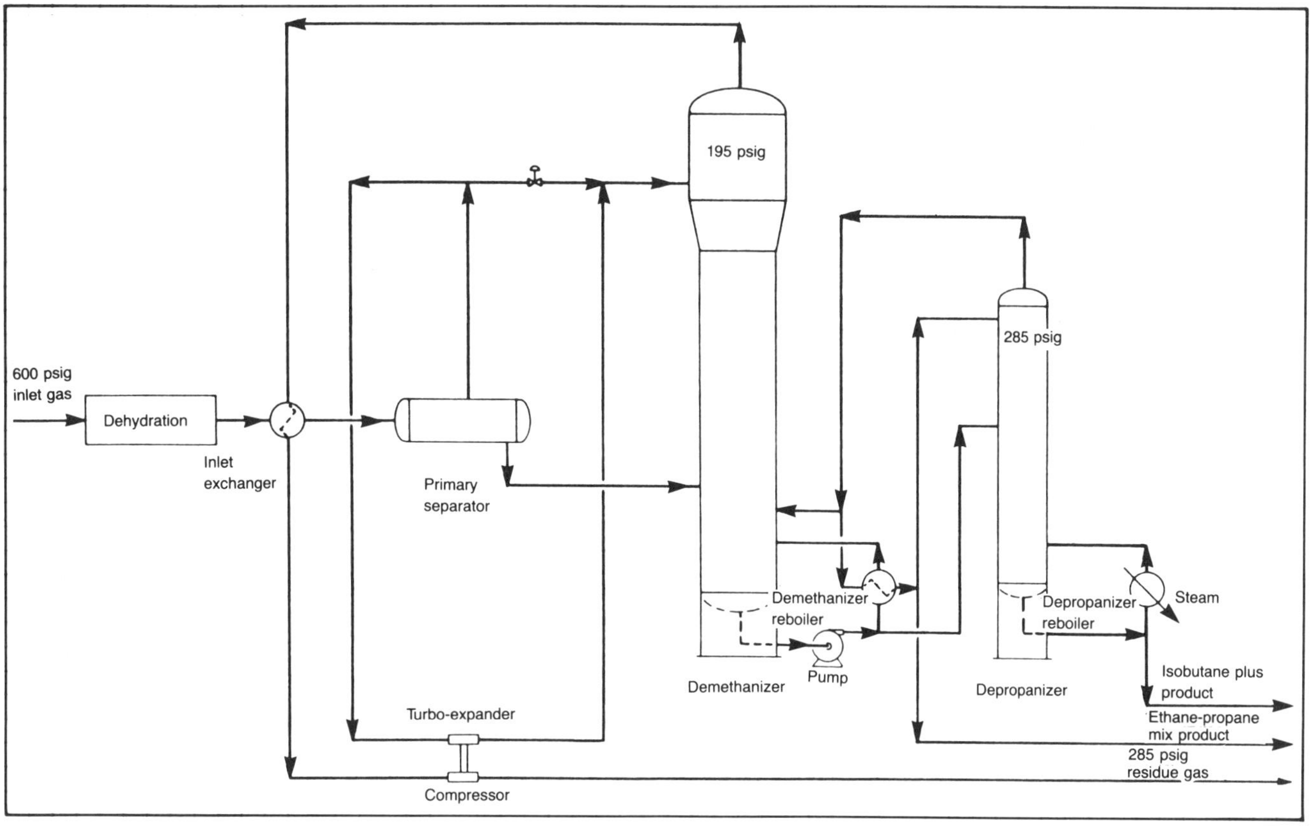

Fig. 10-16. Cryogenic gas plant.

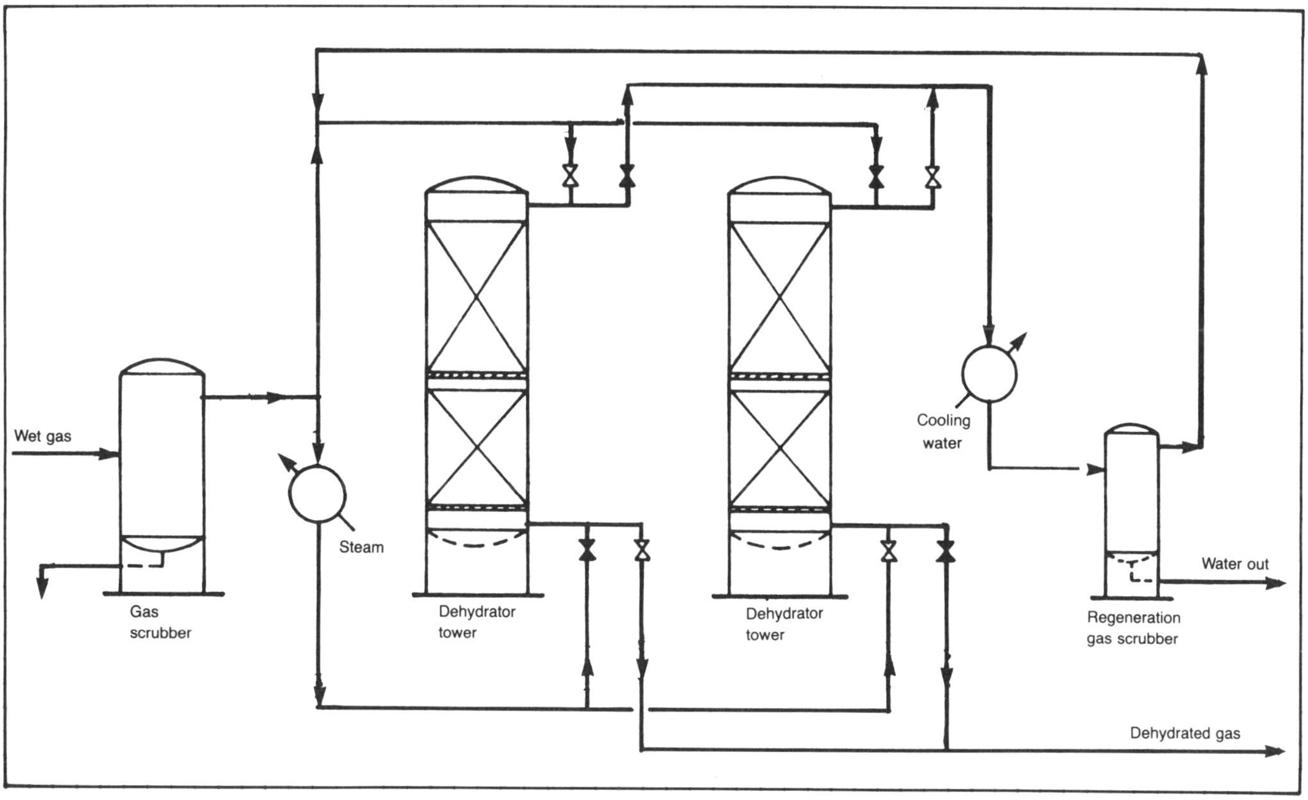

Fig. 10-17. Typical solid desiccant dehydration unit.

hydrocarbons that are drained from the system, and the regeneration gas returns to the wet, main gas stream.

In a large system, the towers may contain several tons of desiccant. Switching of the gas flow through the beds and control of the cycles is usually automatic. The hydrocarbon recovery units are also called fast cycle adsorption units because the cycles are quite short, 20 to 40 minutes. The cycle length is related to the hydrocarbons to be adsorbed and to the desiccant properties. Product recoveries are quite low, usually 70 to 90% of the pentanes and heavier. However, this same principle is used in a process marketed by one company in which an appreciable amount of ethane recovery is achieved. Units using this company's process are reported to be comparable to an adsorption gas processing plant in respect to recovery efficiencies and investment. However, most hydrocarbon recovery units are applicable only to rather dry gases, and are often used where a minimum investment for relatively small gas volumes and short life are involved.

Gas Dehydration

There are three common methods by which water vapor is removed from natural gas. These involve the use of either a solid desiccant, a liquid desiccant, or refrigeration. Solid desiccant dehydration, as in adsorption hydrocarbon recovery, depends upon the adsorption of water on bauxite, activated alumina, silica gel, or molecular sieves. For dehydration, the cycles may be up to 24 hours, and the water being more strongly adsorbed displaces most of the hydrocarbons that were adsorbed. Some hydrocarbon liquid is recovered with the condensed water. Solid desiccant dehydration can produce essentially totally dry gas and is required for the feed gas for the cryogenic type gas processing plant.

Liquid desiccant dehydration of gas uses a continuous contacting process with a very concentrated solution of triethylene glycol usually serving as the desiccant. A typical flow scheme of such a plant is shown in Figure 10-18. While liquid desiccant dehydration does not produce a totally dry gas, it can be used for almost all gas dehydration except that required for cryogenic plants. As shown by the flow diagram, gas enters the absorber, which is a trayed tower, and passes upward countercurrent to the glycol flowing downward. The water is absorbed by the glycol and leaves with it from the bottom of the absorber. The water-rich glycol flows to the regenerator where the water is stripped out by heat.

The water leaves the system as steam from the top of

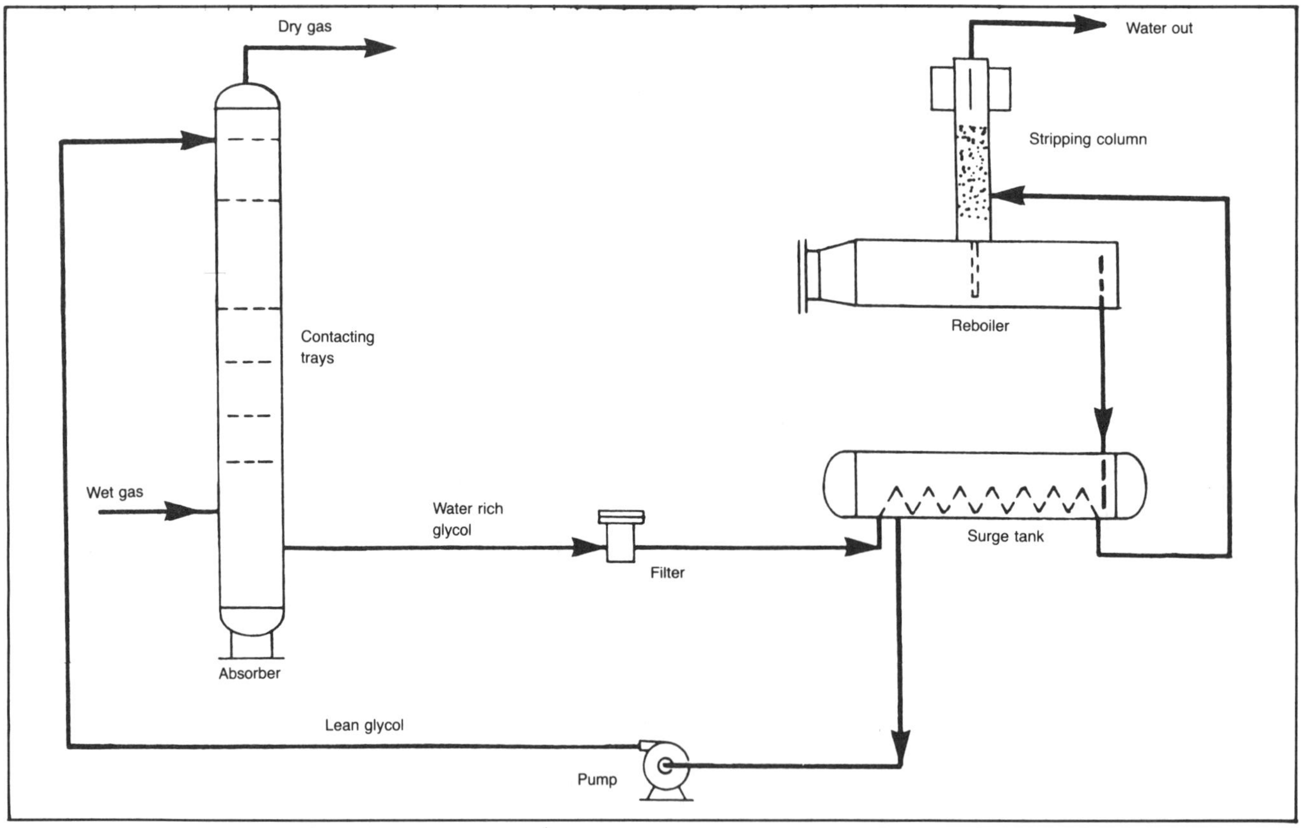

Fig. 10-18. Liquid desiccant dehydration unit.

the column. In some units, stripping gas is introduced into the column to effect a more complete removal of water. The amount of water removed and the dew point depression achieved are governed by the concentration of regenerated glycol, the glycol flow rate, the number of contact trays in the absorber column, and the temperature of the incoming gas. Dew point depressions of 100°F are common and, with careful design, depressions of 150°F can be achieved. Liquid desiccant dehydrators are used extensively with the low temperature absorption type gas processing plants and for field locations since they normally require little attention.

Dehydration by chilling the gas is accomplished in many plants in which hydrocarbon recovery is achieved by either condensation or low temperature absorption. Removal of water vapor by this method requires that the formation of hydrates and ice be prevented during the condensation and removal of the water from the system. This is achieved by injecting a solution of ethylene glycol into the gas stream being chilled. The resultant mixture of cold gas, water-glycol solution, and condensed hydrocarbons, if any, then pass to a tank where the glycol solution that includes the condensed water settles out in the lower layer and is drained from the tank. The water-rich glycol solution is regenerated in equipment similar to that used in a glycol dehydration unit. However, in this unit, regeneration of the glycol is usually limited to about an 85% solution and the flow is adjusted so that the water absorbed dilutes it to about a 75% solution.

Gas Sweetening

Many natural gases available for processing or for sale are termed "sour" because they contain hydrogen sulfide, which is an extremely poisonous gas. Gas sales contracts usually limit the amount of this compound to about 0.25 grains per hundred cubic feet (about four parts per million). If hydrogen sulfide is present in large amounts, it is removed by a sweetening unit of the type shown schematically in Figure 10-19. The chemicals used to remove hydrogen sulfide in this unit also remove carbon dioxide. Both gases are identified by the general term of acid gas.

One of the most commonly used materials for the removal of the acid gases is a water solution of monoethanolamine (MEA). Other common chemicals used to remove acid gases are diethanolamine (DEA) and Sulfinol, which is a mixture of sulfolane, di-isopropanolamine, and water. The first two chemicals remove acid gases by chemical reaction, while Sulfinol works on the

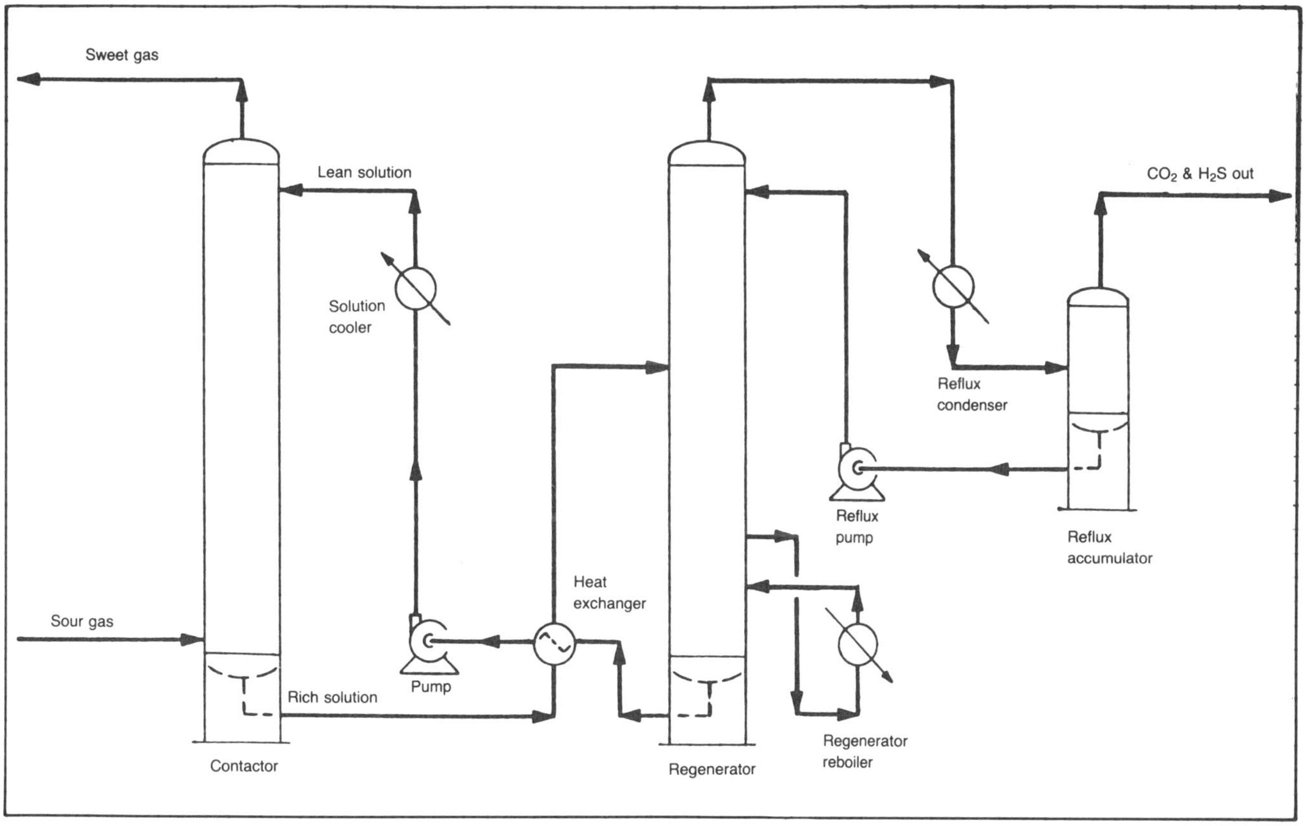

Fig. 10-19. Typical flow diagram of sweetening unit.

basis of chemical reaction plus physical absorption. There are other sweetening processes used commercially, and the flow schemes are all much the same; however, the three mentioned are those most widely used for the removal of acid gas.

Referring to Figure 10-19, the sour natural gas enters the bottom of the contactor and moves upward countercurrent to the flow of MEA solution. Gas meeting the required hydrogen sulfide specification leaves the top of the contactor. The rich MEA solution flows to the regenerator where, by means of heat, the acid gases are stripped from the MEA solution. The lean MEA solution is cooled by heat exchange and pumped back to the contactor. The acid gas and water vapor from the top of the regenerator pass through the acid gas cooler where most of the water vapor is condensed out and is pumped back to the regenerator as reflux. The contactor may operate at pressures from 50 psig to 1000 psig or more, while the regenerator operates at just slightly above atmospheric pressure.

Some gases will contain only trace quantities of hydrogen sulfide, but the concentration may exceed by several times that specified under the sales gas contract. In such cases, either a material called iron sponge or molecular sieves are used to remove hydrogen sulfide. Iron sponge consists of iron oxide deposited on wood chips or shavings. The iron oxide is converted to iron sulfide in sweetening the gas and has a relatively short life. The molecular sieves are regenerated with heat just as they are in other adsorption type processes.

REFERENCES

1. *Field Handling of Natural Gas,* Petroleum Extension Service, University of Texas at Austin, Austin, Texas (1972).
2. *Engineering Data Book,* Natural Gas Processors Association, 9th Edition, Tulsa, Oklahoma (1972).
3. Campbell, J. M.: *Gas Conditioning and Processing,* Campbell Petroleum Series, Norman, Oklahoma (1976).

Appendix A

EQUILIBRIUM CONSTANTS
FOR 5000 psia CONVERGENCE
PRESSURE

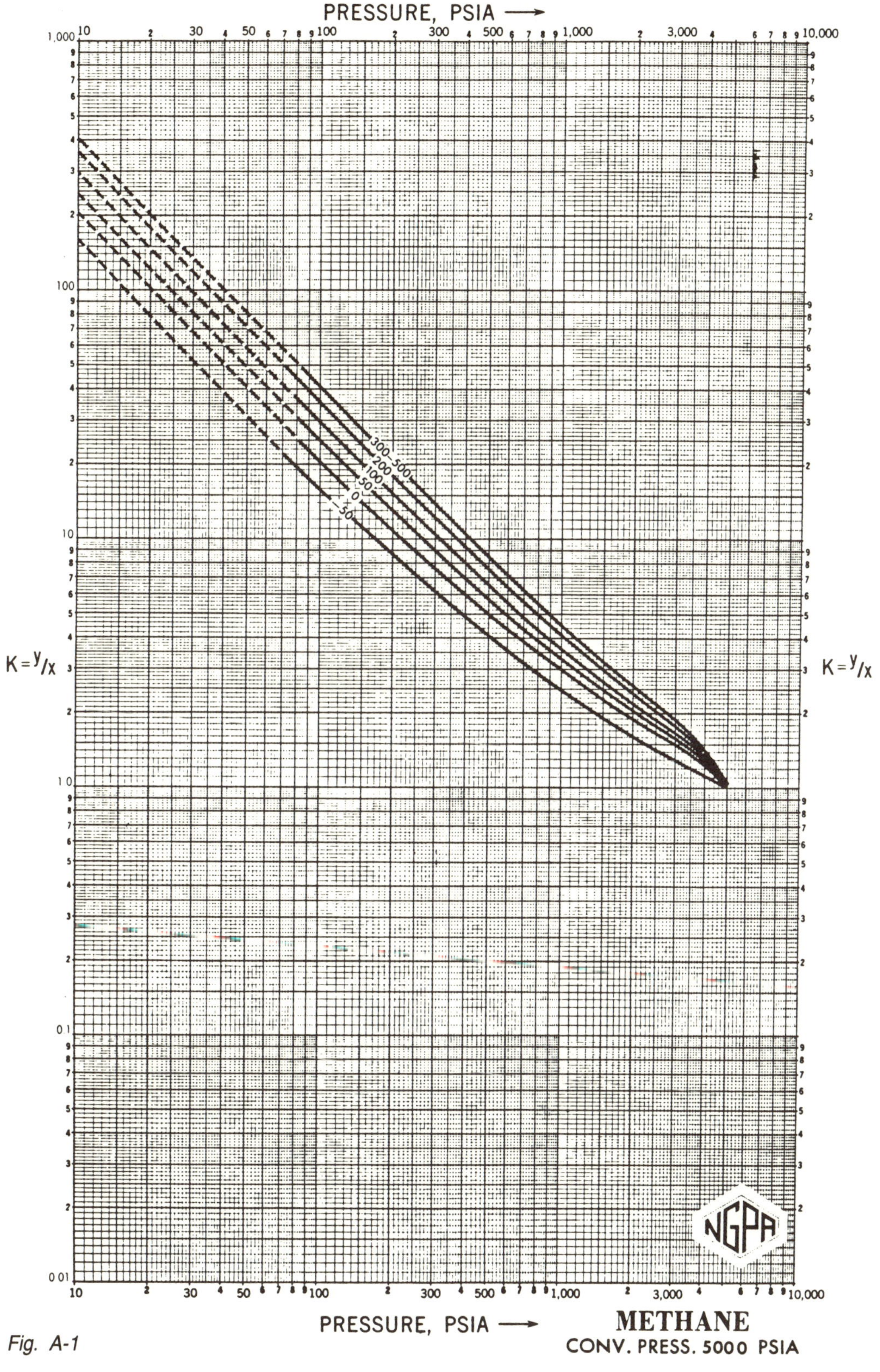
PRESSURE, PSIA →
K = y/x
300–500
200
100
50
0
−50
NGPA
PRESSURE, PSIA →
METHANE
CONV. PRESS. 5000 PSIA

Fig. A-1

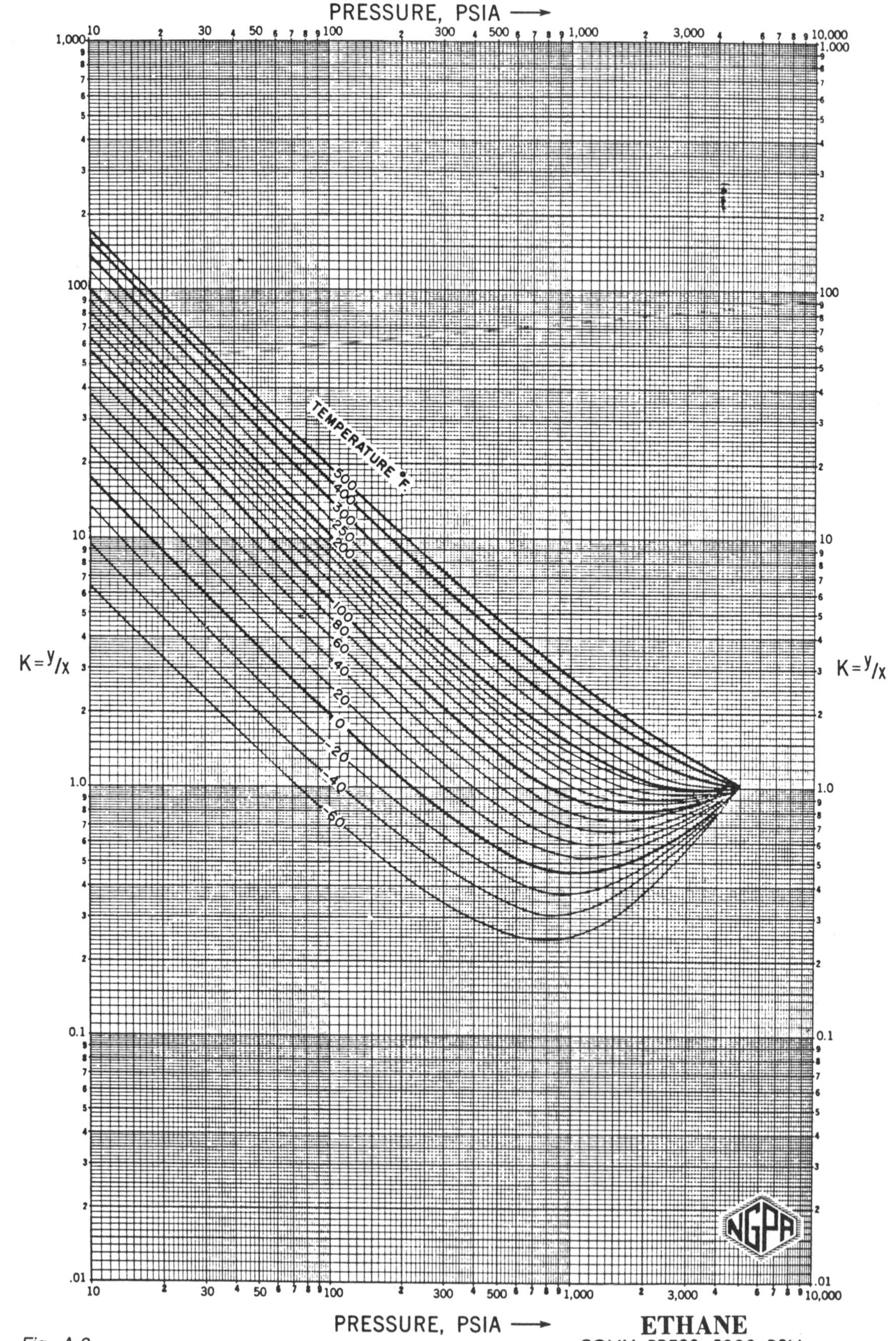
PRESSURE, PSIA
K = y/x
TEMPERATURE °F
500
400
300
250
200
100
80
60
40
20
0
-20
-40
-60
NGPA
ETHANE
CONV. PRESS. 5000 PSIA

Fig. A-2

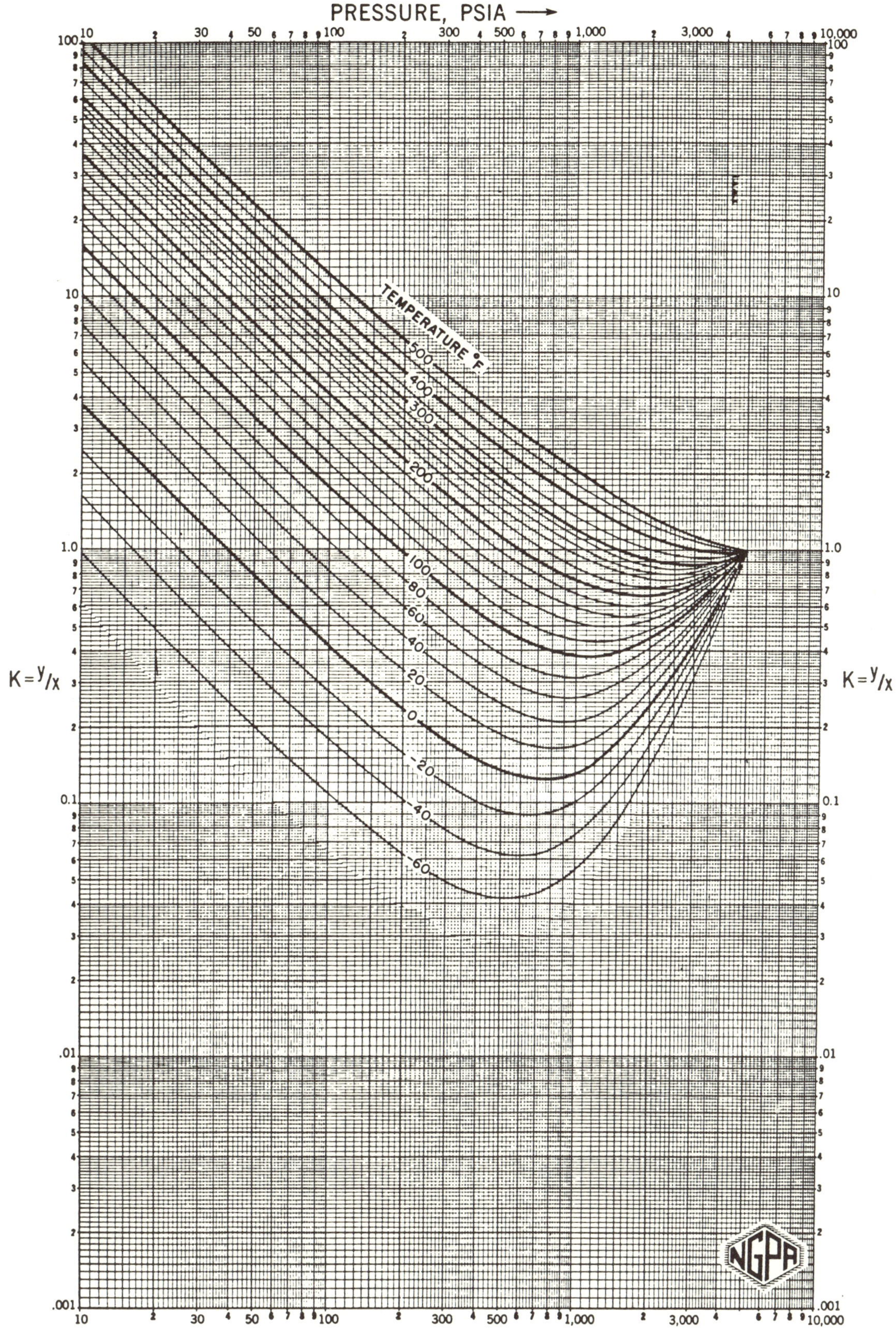

PROPANE
CONV. PRESS. 5000 PSIA

Fig. A-3

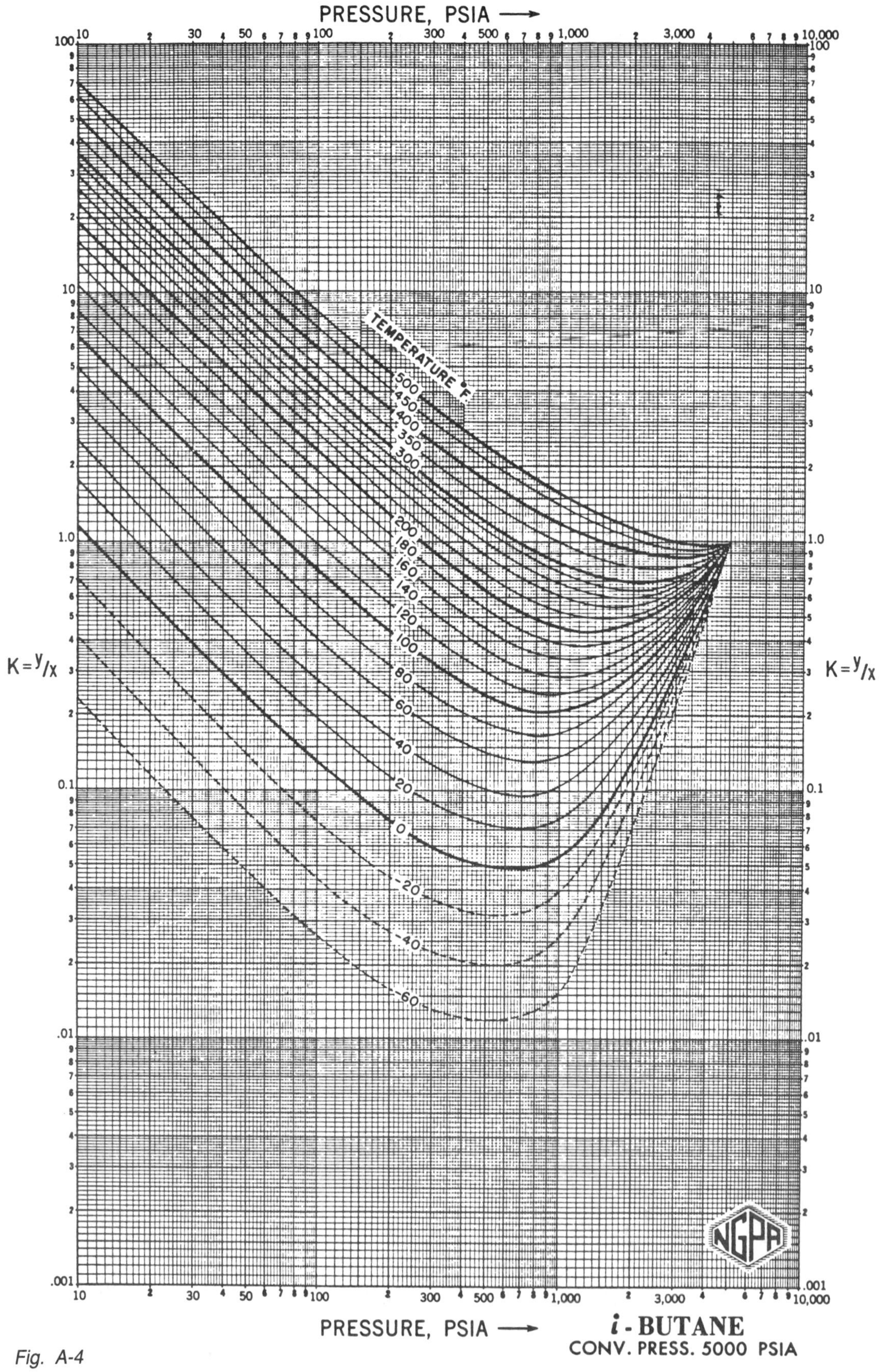
PRESSURE, PSIA
TEMPERATURE °F
500
450
400
350
300
200
180
160
140
120
100
80
60
40
20
0
-20
-40
-60
K = y/x
NGPA
PRESSURE, PSIA
i - BUTANE
CONV. PRESS. 5000 PSIA

Fig. A-4

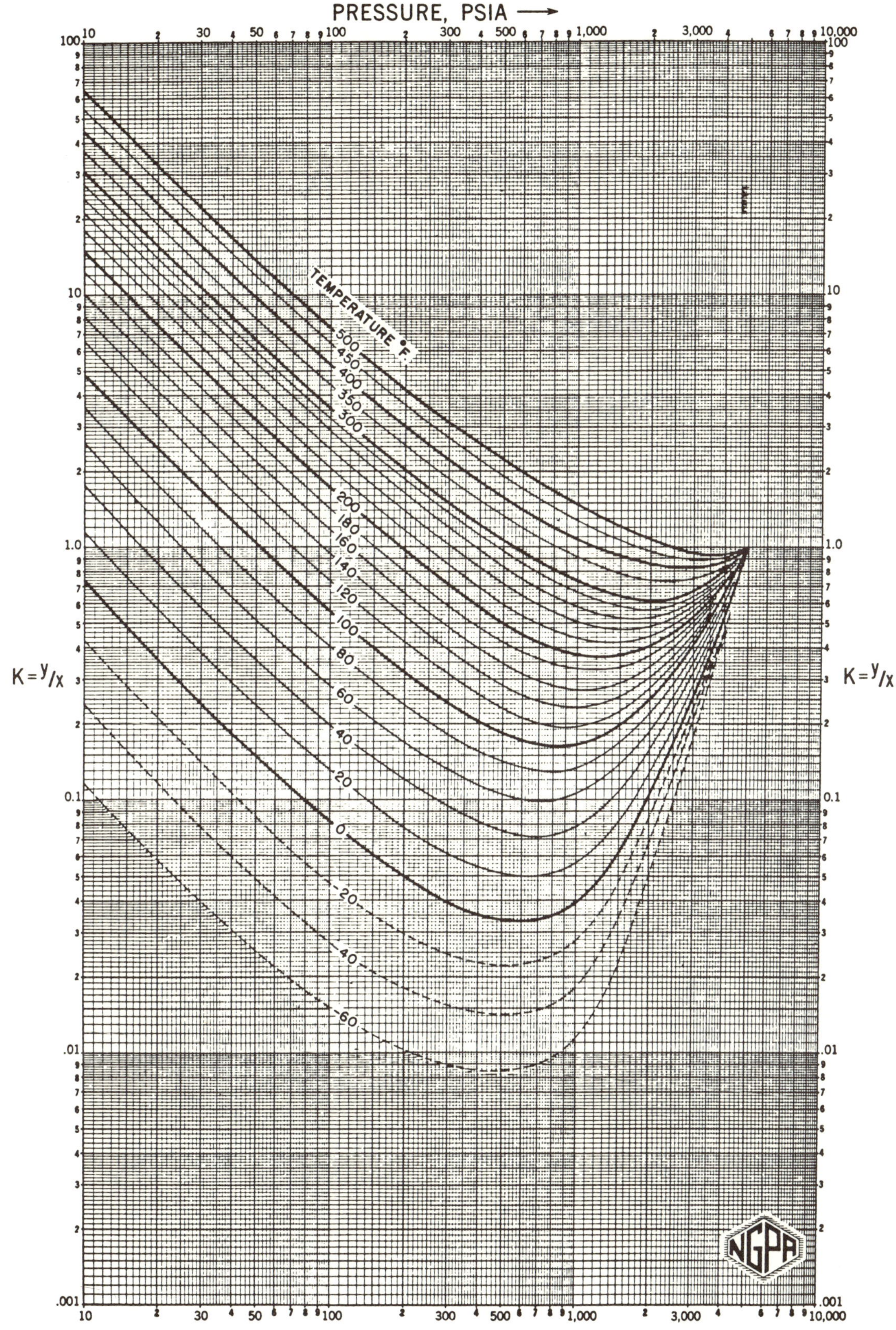

n-BUTANE
CONV. PRESS. 5000 PSIA

Fig. A-5

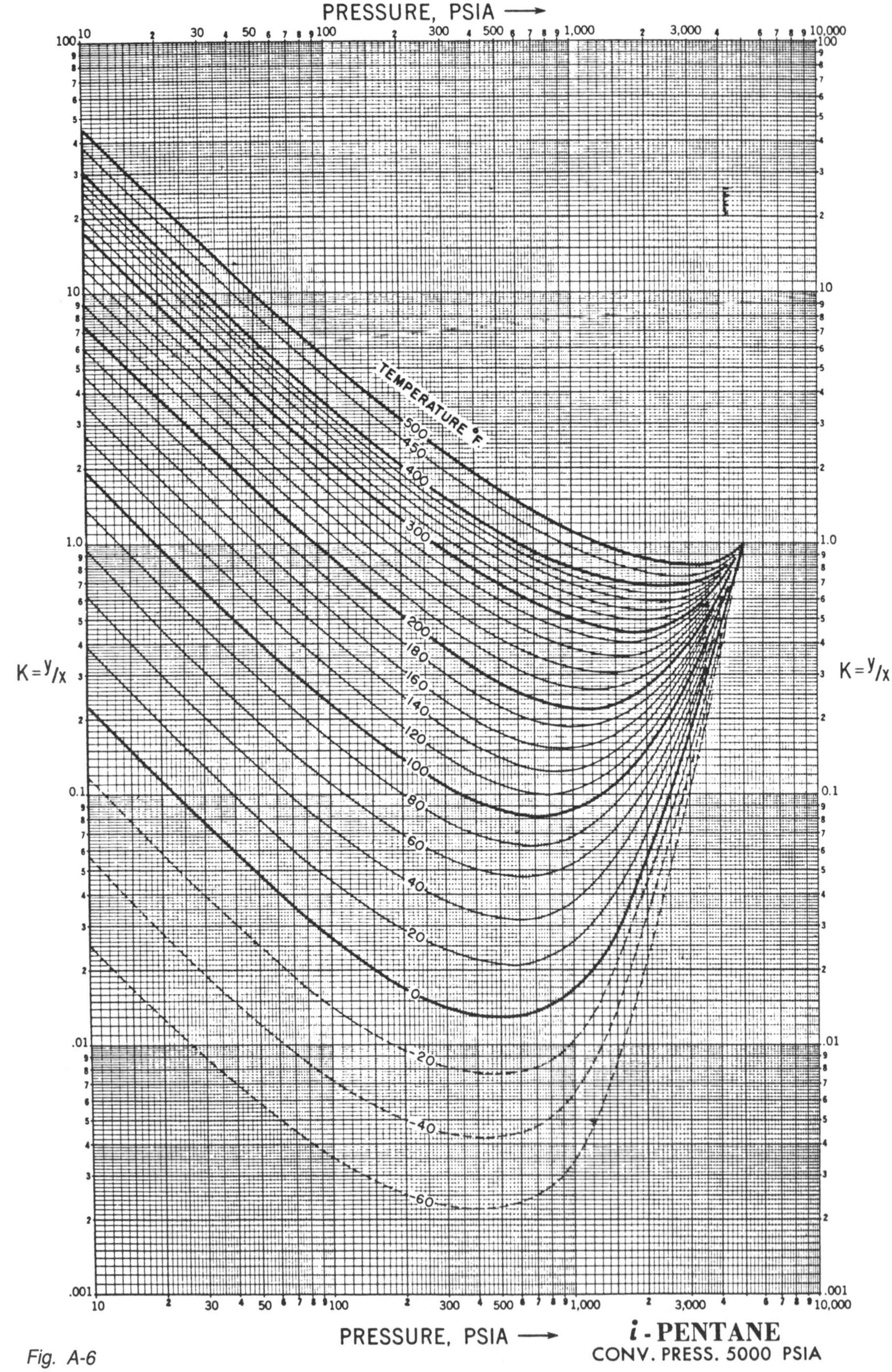
PRESSURE, PSIA ⟶
K = y/x
TEMPERATURE °F
500
450
400
300
200
180
160
140
120
100
80
60
40
20
0
-20
-40
-60
PRESSURE, PSIA ⟶
i-PENTANE
CONV. PRESS. 5000 PSIA

Fig. A-6

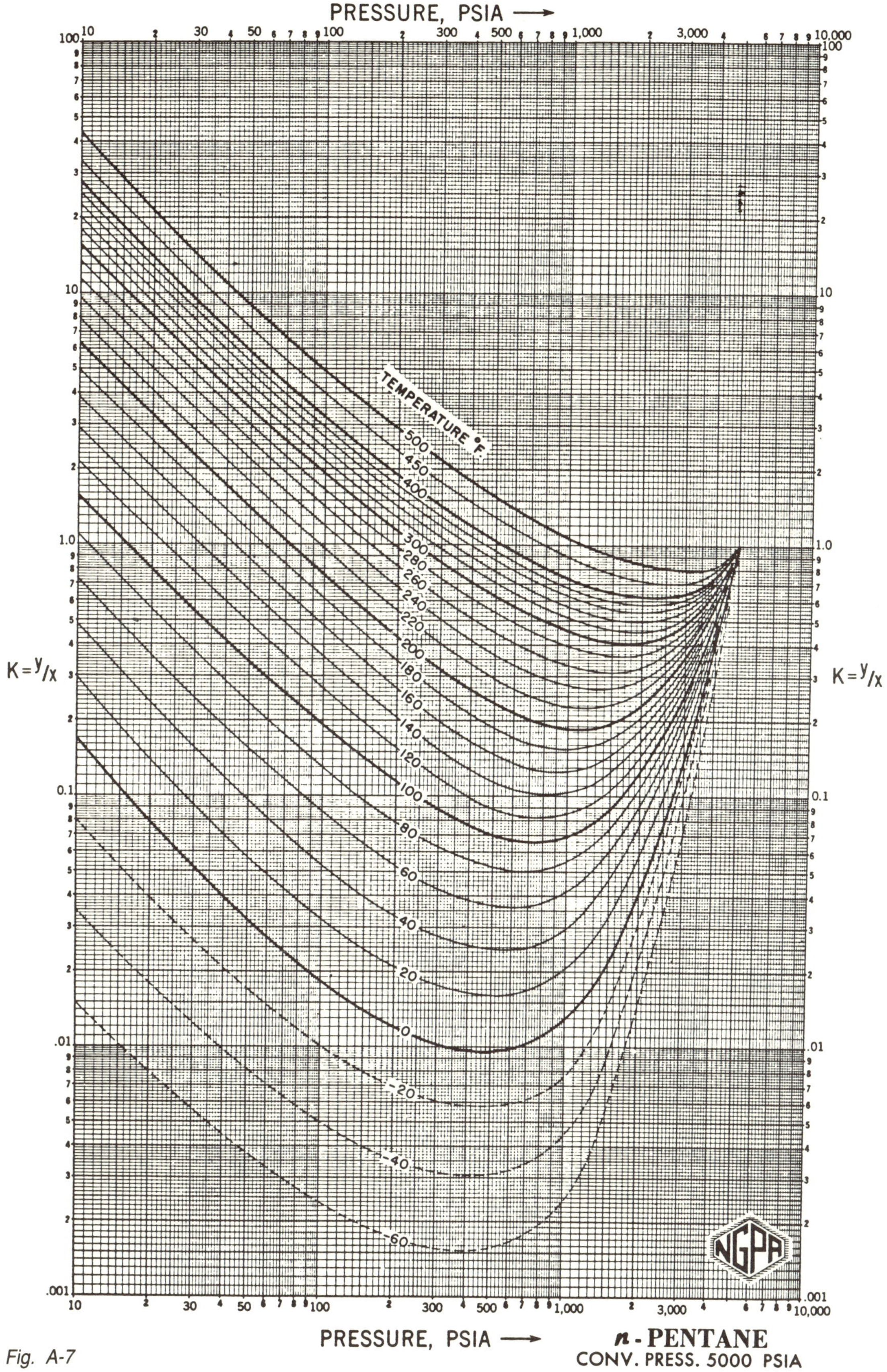
PRESSURE, PSIA
K = y/x
TEMPERATURE °F
500
450
400
300
280
260
240
220
200
180
160
140
120
100
80
60
40
20
0
-20
-40
-60
n-PENTANE
CONV. PRESS. 5000 PSIA

Fig. A-7

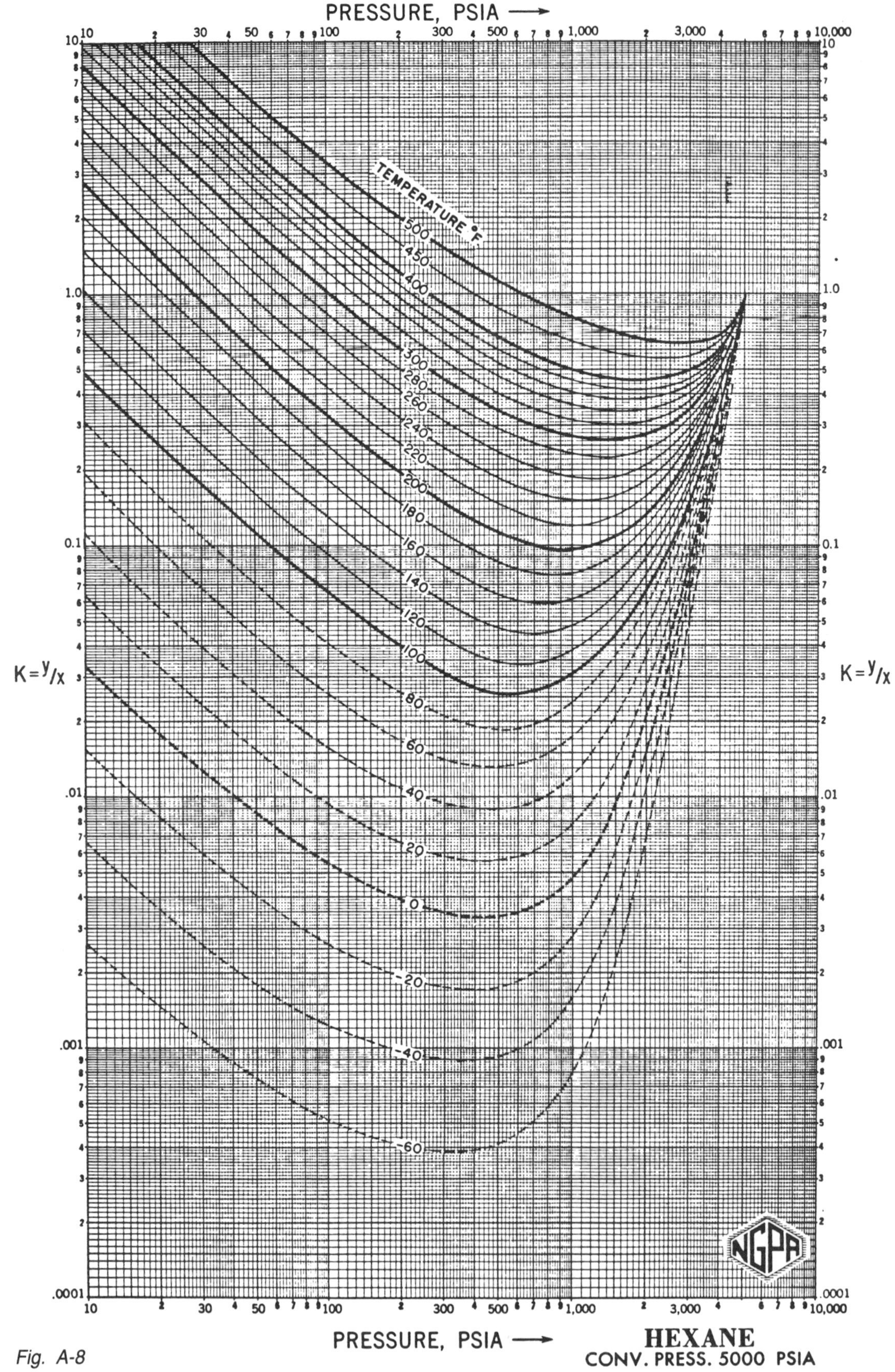
PRESSURE, PSIA ⟶
K = y/x
TEMPERATURE °F
500
450
400
300
280
260
240
220
200
180
160
140
120
100
80
60
40
20
0
-20
-40
-60
NGPA
HEXANE
CONV. PRESS. 5000 PSIA

Fig. A-8

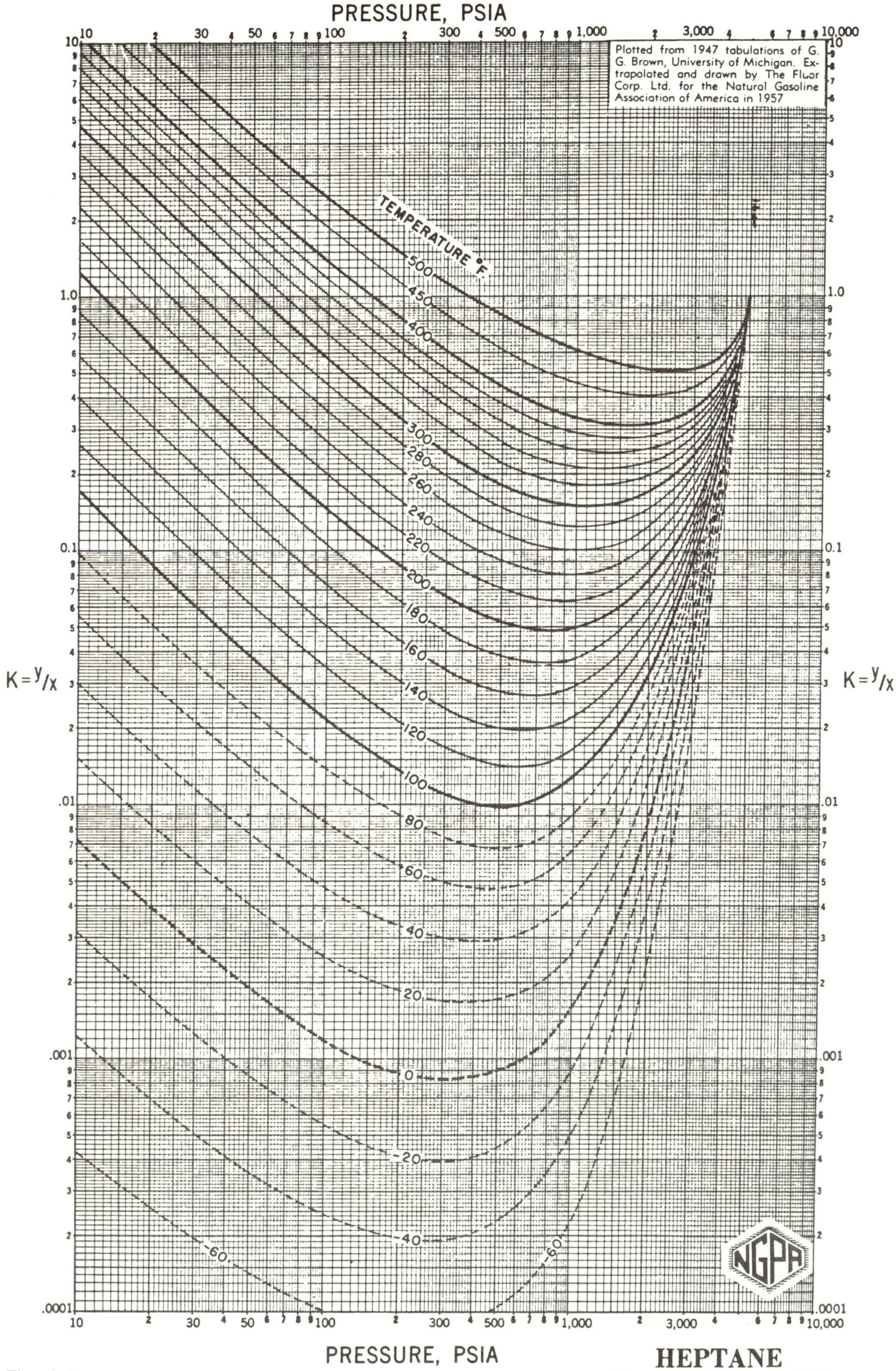

Fig. A-9

HEPTANE
CONV. PRESS. 5000 PSIA

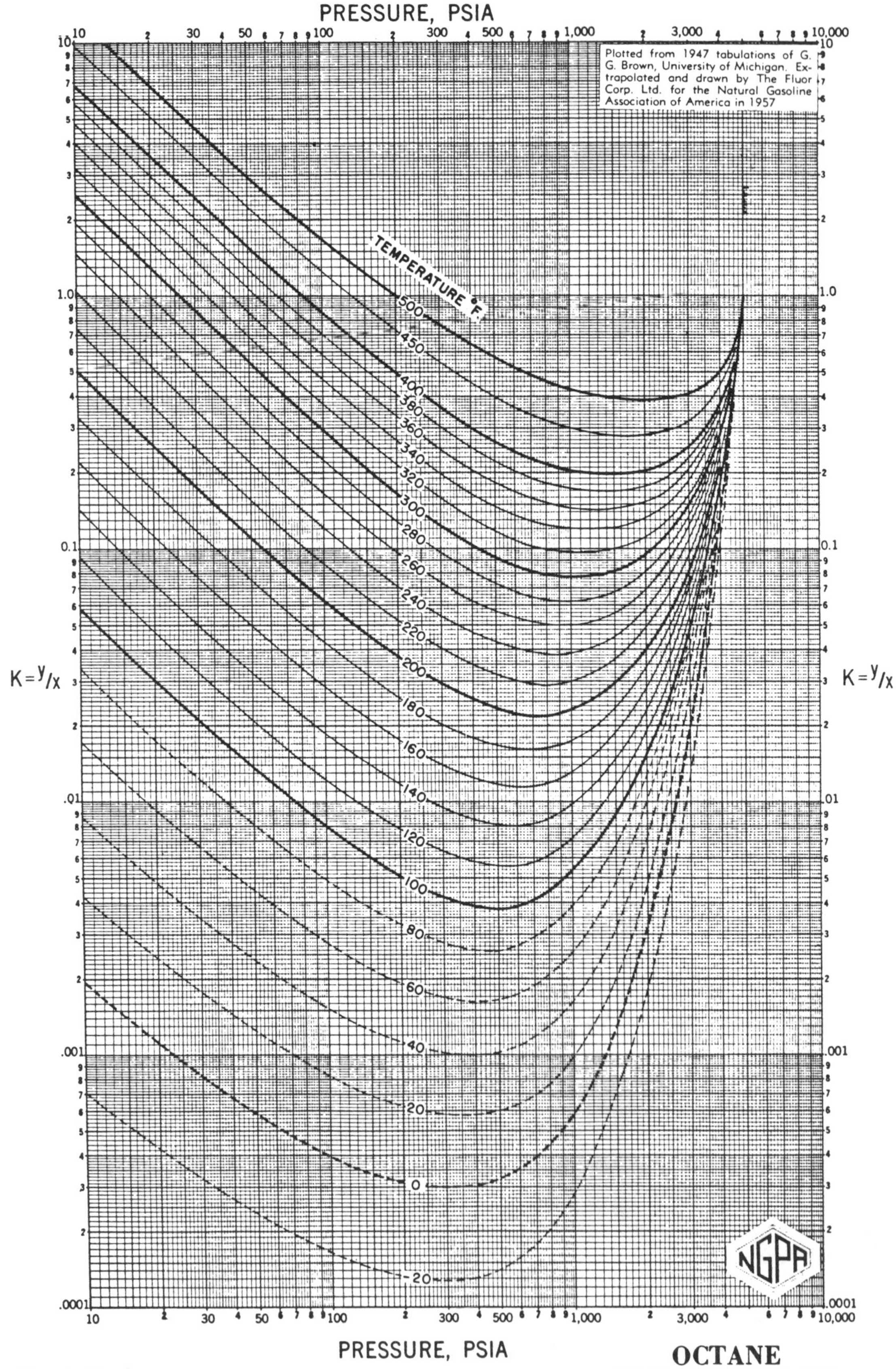

Fig. A-10

OCTANE
CONV. PRESS. 5000 PSIA

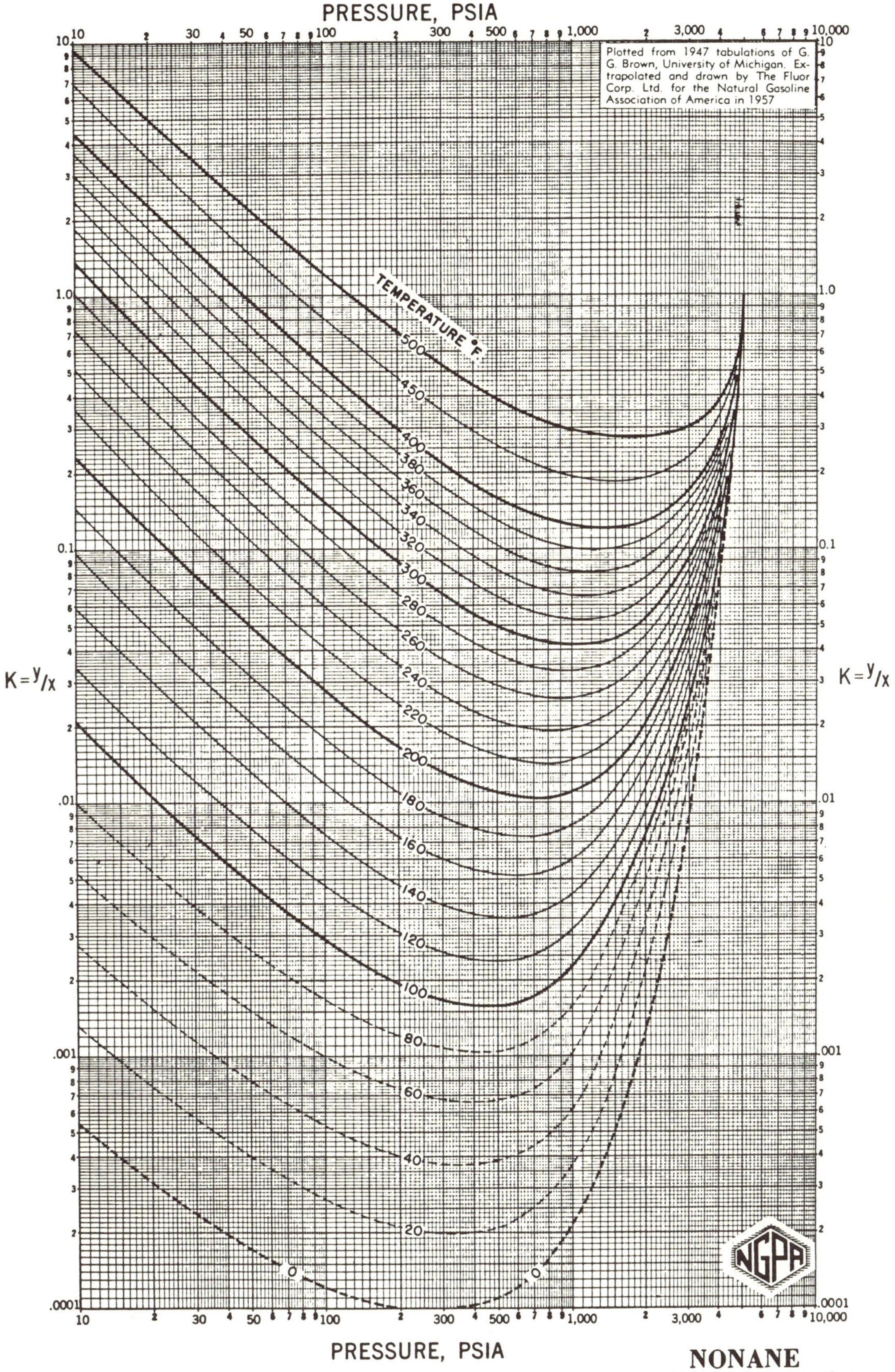

Fig. A-11

NONANE
CONV. PRESS. 5000 PSIA

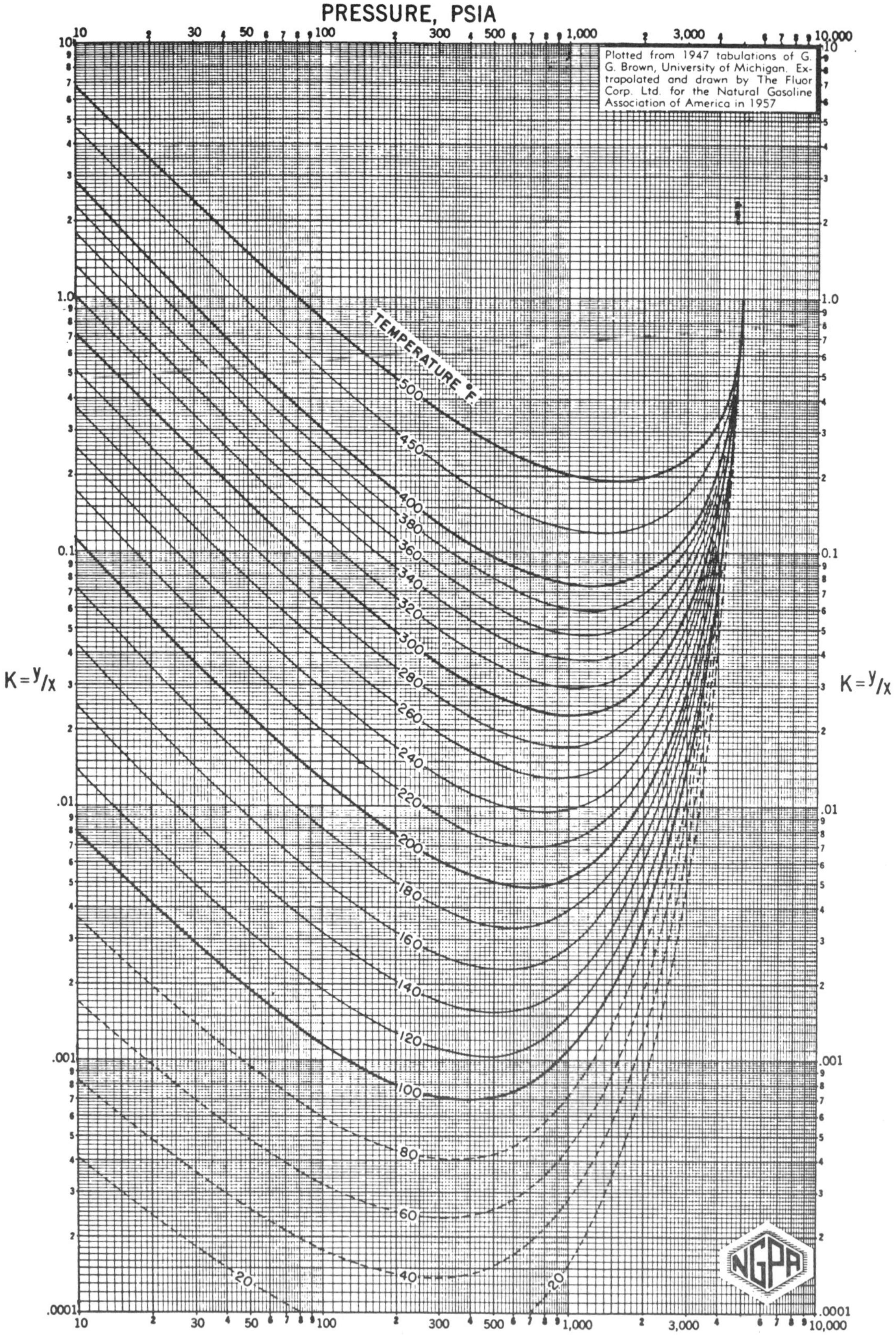

Fig. A-12

DECANE
CONV. PRESS. 5000 PSIA

Appendix B

MATTHEWS–BRONS–HAZEBROOK CURVES FOR VARIOUS RESERVOIR SHAPES

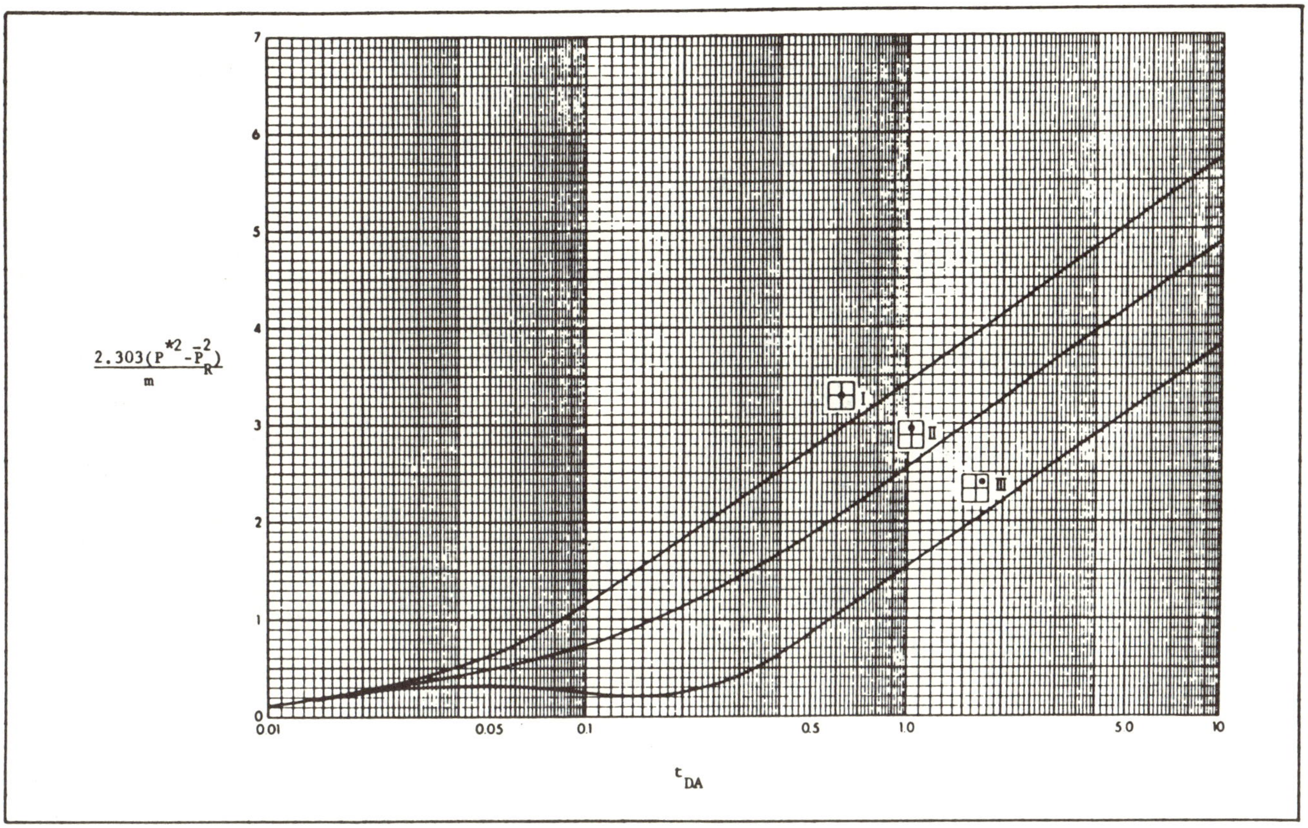

$\frac{2.303(P^{*2}-\bar{P}_R^2)}{m}$
0
1
2
3
4
5
6
7
0.01
0.05
0.1
0.5
1.0
5.0
10
t_{DA}
I
II
III

Fig. B-1

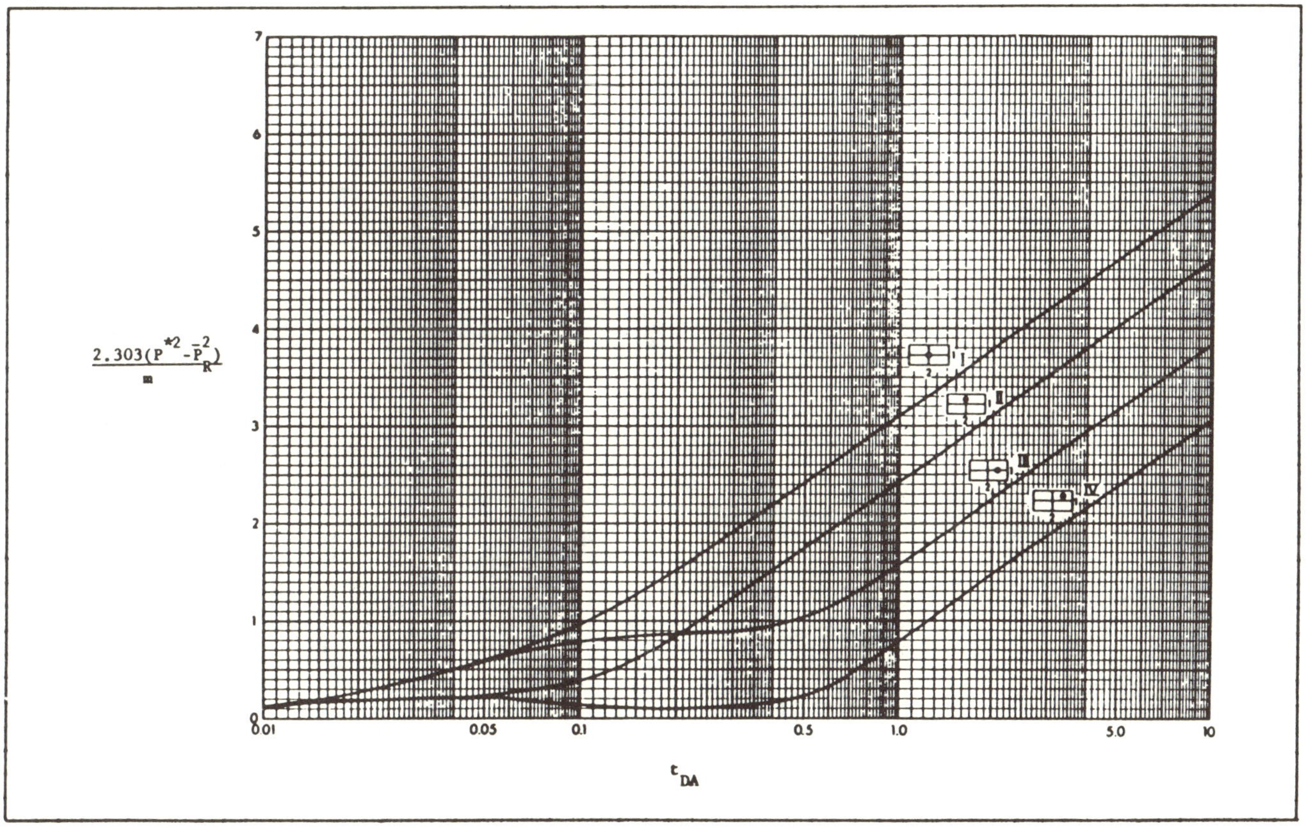

$\frac{2.303(P^{*2}-\bar{P}_R^2)}{m}$
0
1
2
3
4
5
6
7
0.01
0.05
0.1
0.5
1.0
5.0
10
t_{DA}
I
II
III
IV

Fig. B-2

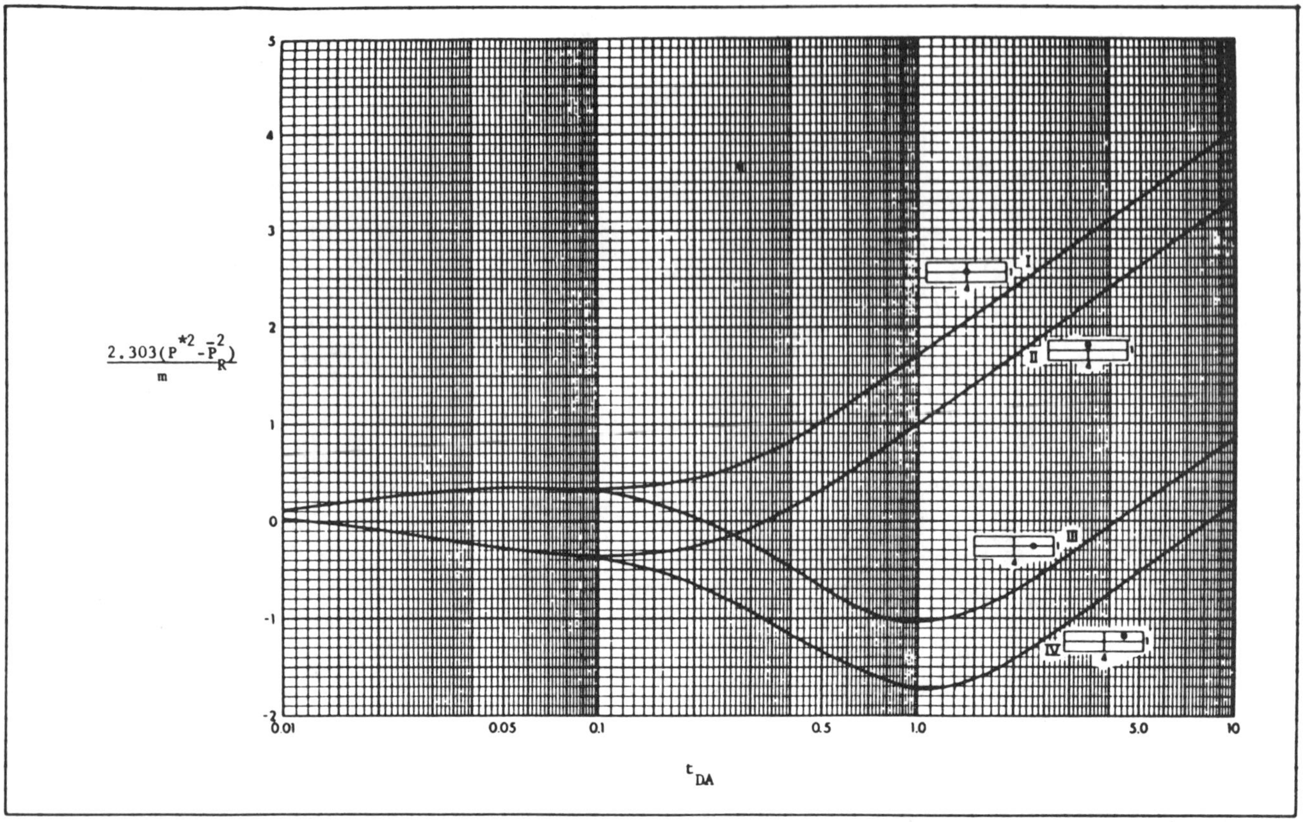
5
4
3
2
1
0
-1
-2
$\frac{2.303(P^{*2}-\bar{P}_R^2)}{m}$
I
II
III
IV
0.01
0.05
0.1
0.5
1.0
5.0
10
t_{DA}

Fig. B-3

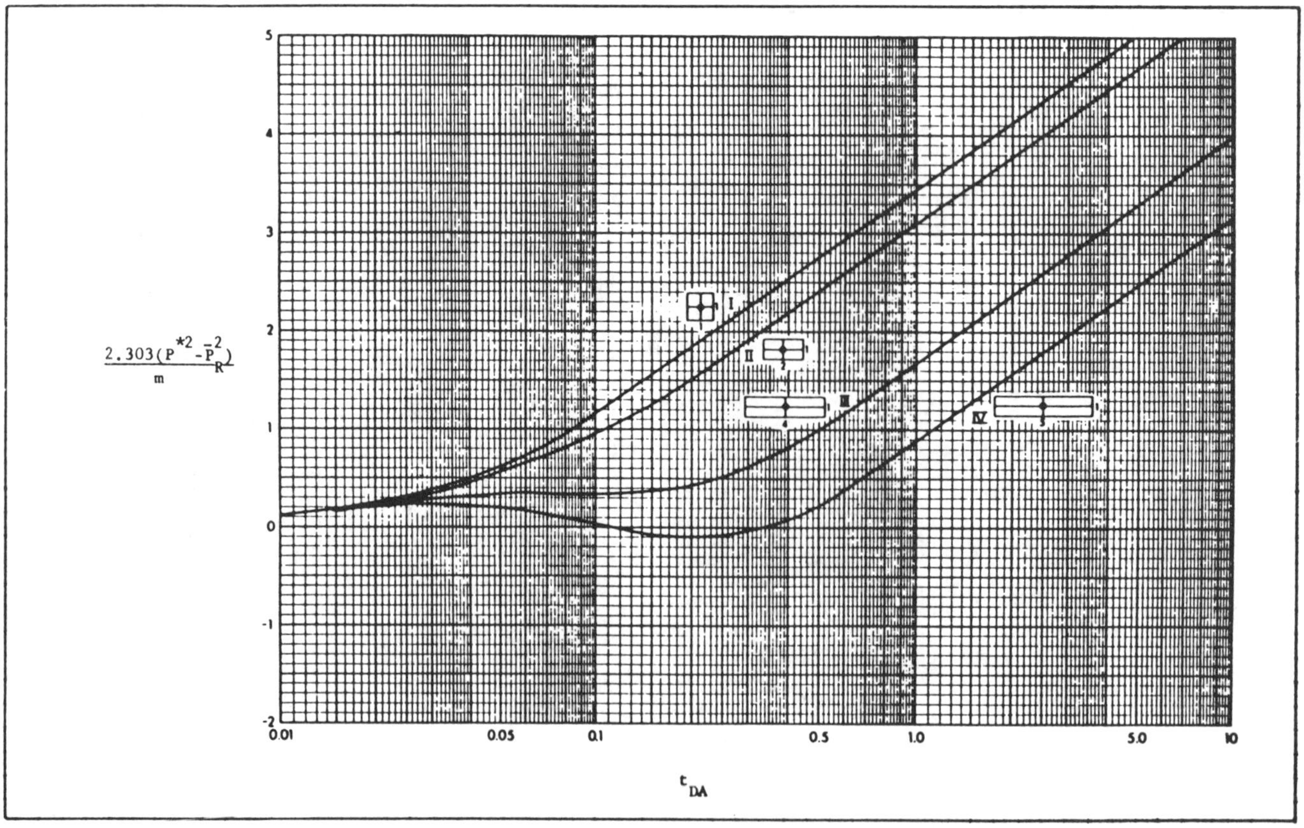
5
4
3
2
1
0
-1
-2
$\frac{2.303(P^{*2}-\bar{P}_R^2)}{m}$
I
II
III
IV
0.01
0.05
0.1
0.5
1.0
5.0
10
t_{DA}

Fig. B-4

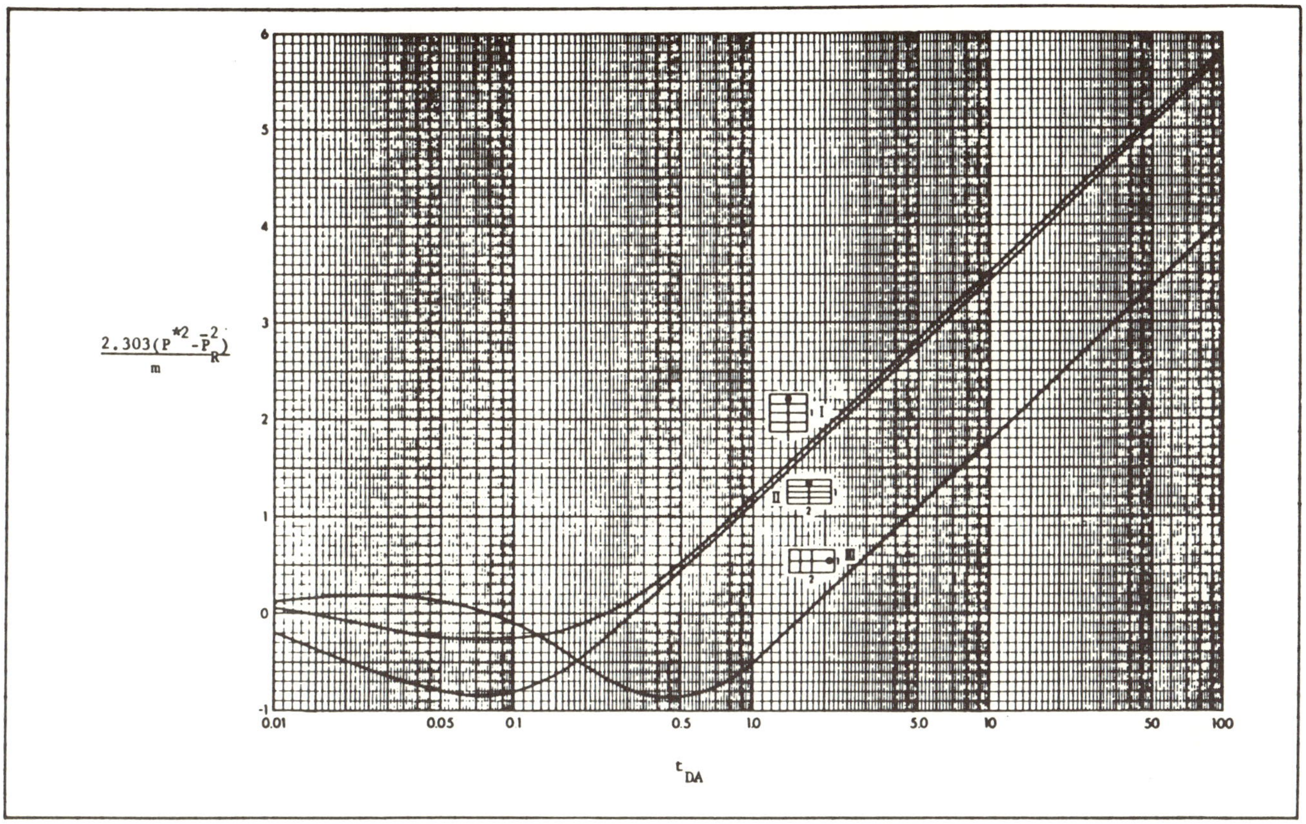
6
5
4
3
2
1
0
-1
2.303(P*2 - P̄R2)/m
0.01
0.05
0.1
0.5
1.0
5.0
10
50
100
tDA
I
II
III

Fig. B-5

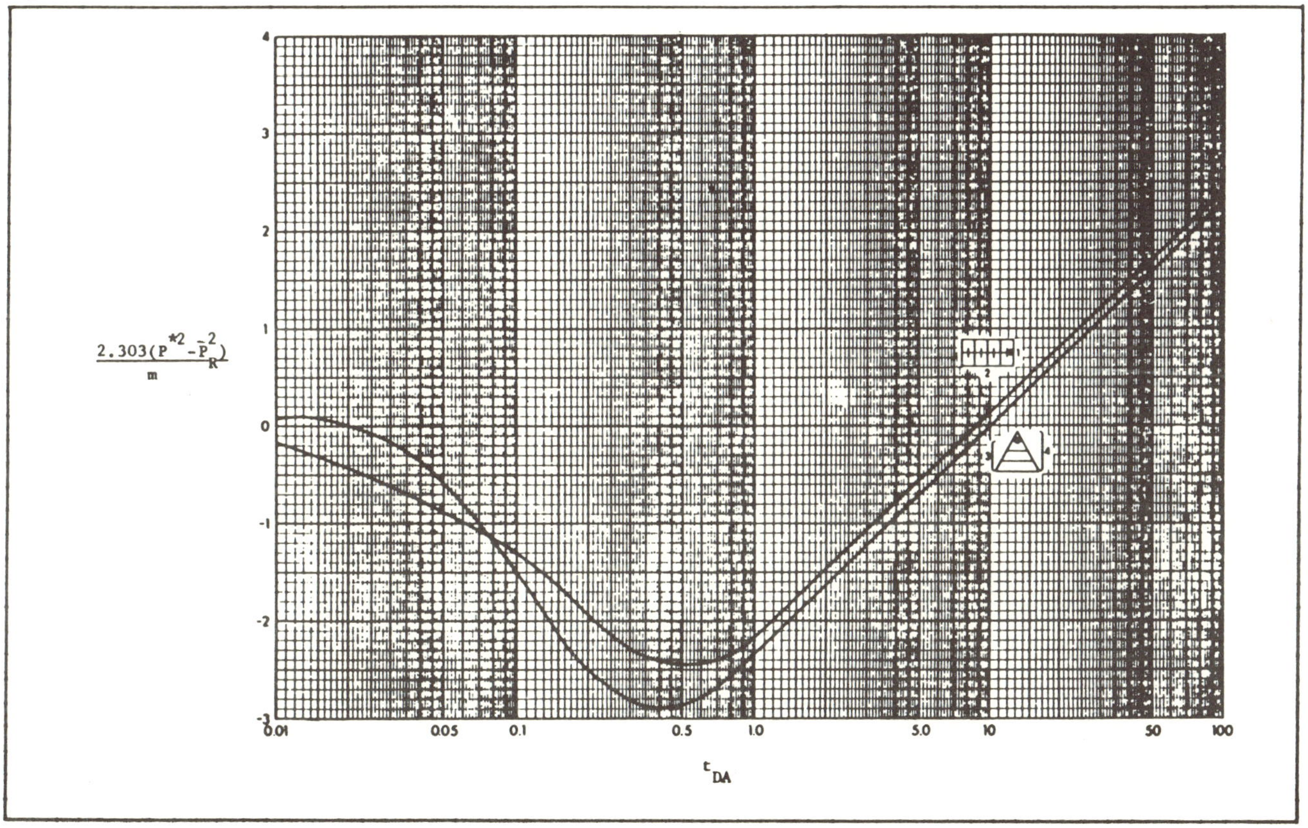
4
3
2
1
0
-1
-2
-3
2.303(P*2 - P̄R2)/m
0.01
0.05
0.1
0.5
1.0
5.0
10
50
100
tDA

Fig. B-6

Appendix C

MOLLIER DIAGRAMS FOR NATURAL GAS

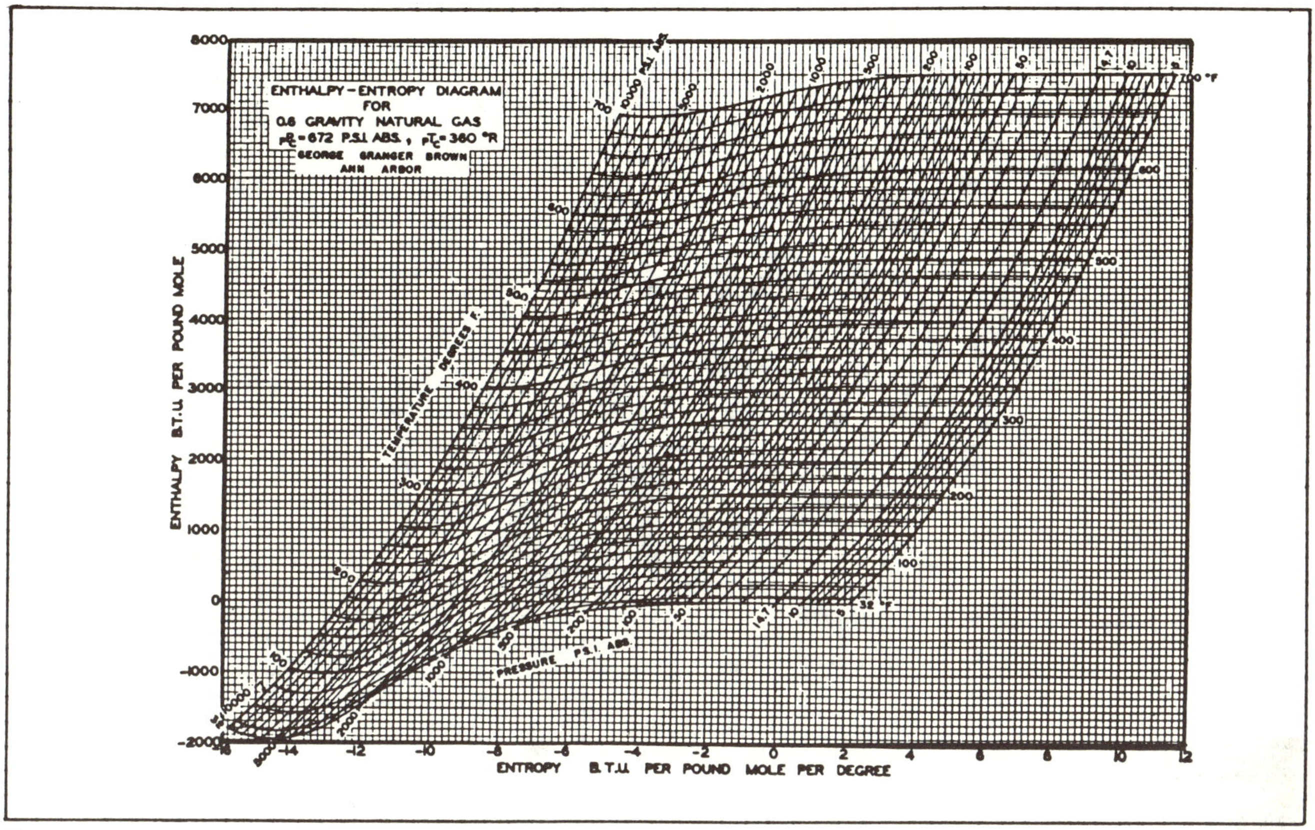
ENTHALPY - ENTROPY DIAGRAM
FOR
0.6 GRAVITY NATURAL GAS
pPc = 672 P.S.I. ABS., pTc = 360 °R
GEORGE GRANGER BROWN
ANN ARBOR
ENTHALPY B.T.U. PER POUND MOLE
ENTROPY B.T.U. PER POUND MOLE PER DEGREE
TEMPERATURE DEGREES F.
PRESSURE P.S.I. ABS

Fig. C-1

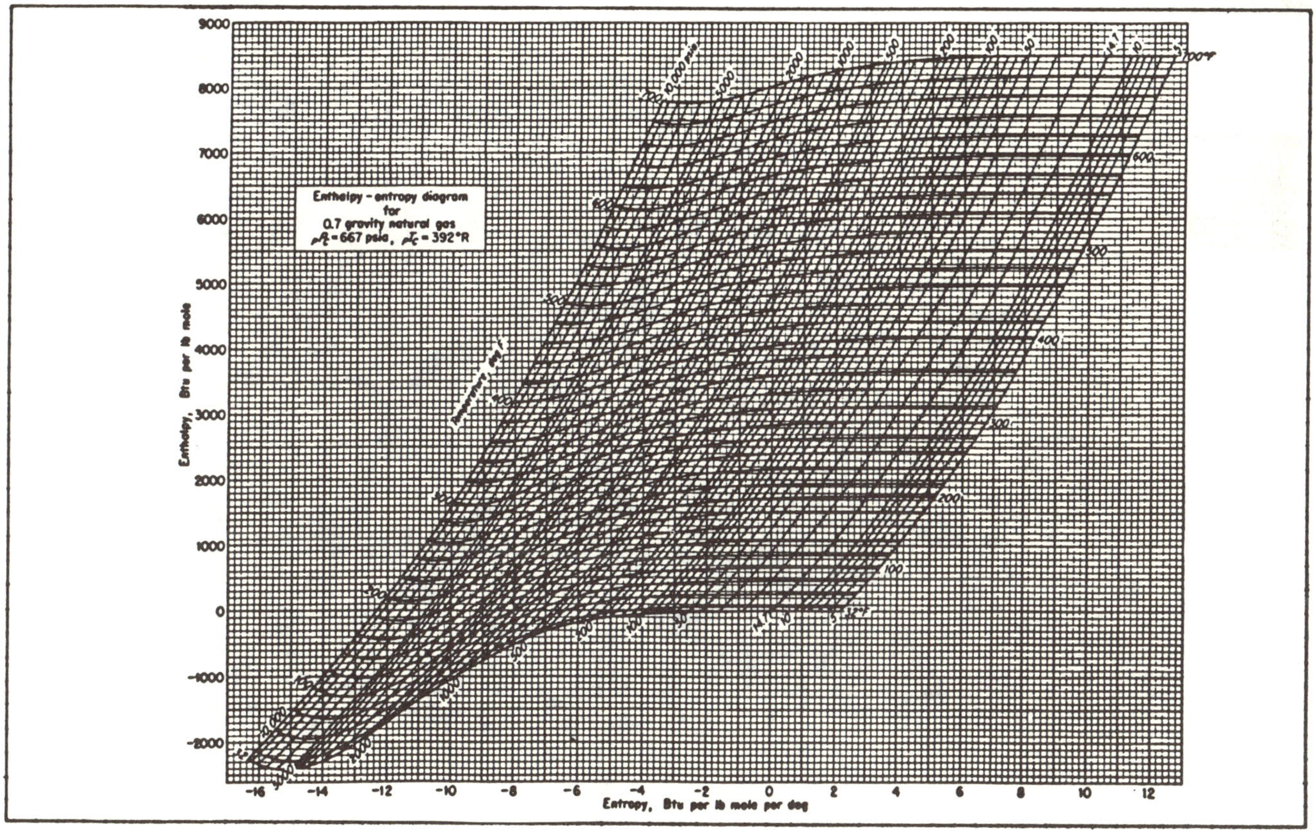
Enthalpy - entropy diagram
for
0.7 gravity natural gas
pPc = 667 psia, pTc = 392°R
Enthalpy, Btu per lb mole
Entropy, Btu per lb mole per deg

Fig. C-2

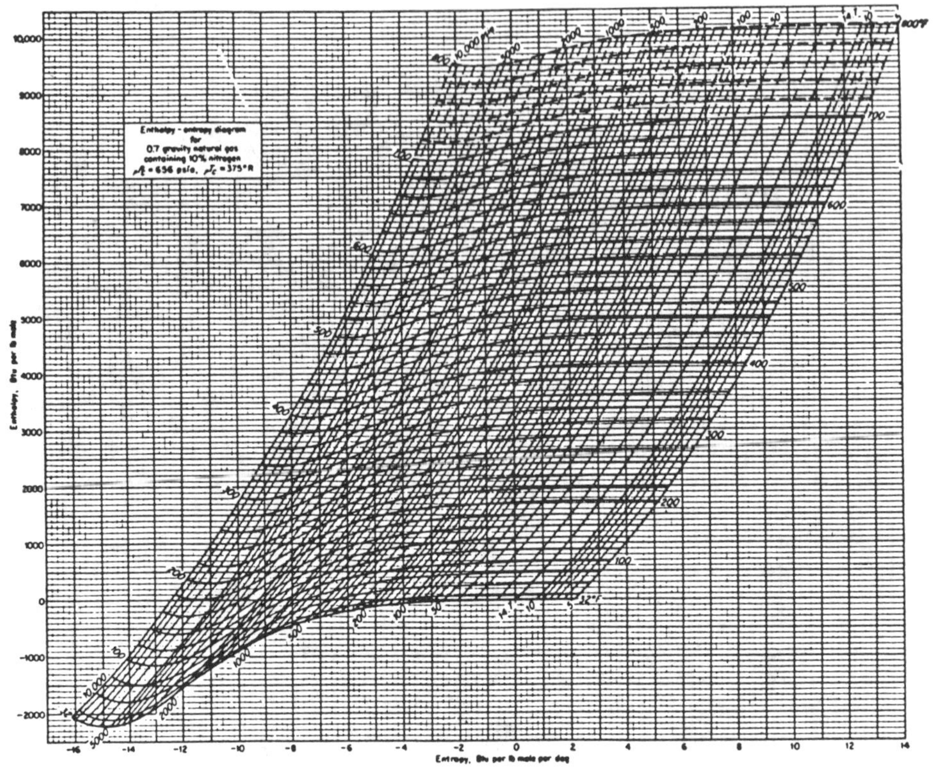
Enthalpy - entropy diagram
for
0.7 gravity natural gas
containing 10% nitrogen
Enthalpy, Btu per lb mole
Entropy, Btu per lb mole per deg

Fig. C-3

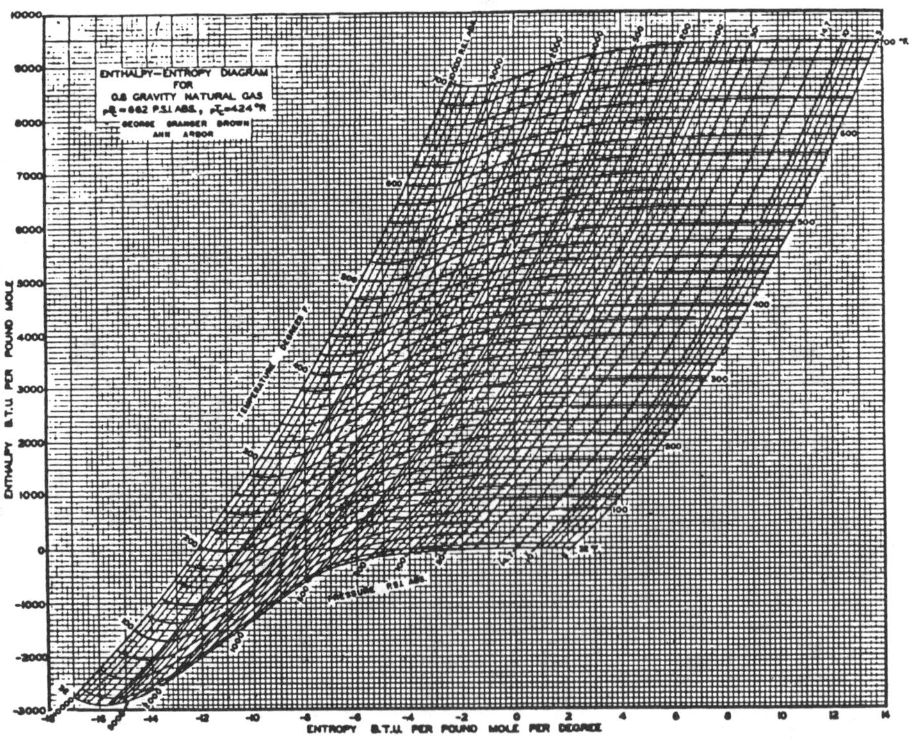
ENTHALPY—ENTROPY DIAGRAM
FOR
0.6 GRAVITY NATURAL GAS
GEORGE GRANGER BROWN
ANN ARBOR
ENTHALPY B.T.U. PER POUND MOLE
ENTROPY B.T.U. PER POUND MOLE PER DEGREE

Fig. C-4

Appendix D

COMPUTER SUBROUTINES

Nomenclature

ACCGR Acceleration pressure gradient, psi/ft
ANG Angle of flow from horizontal, deg
DIA Inside pipe diameter, ft
DENG Gas density, lb_m/ft^3
DENL Liquid density, lb_m/ft^3
DPDL Total pressure gradient, psi/ft (negative), = −(ELGR + PRGR + ACCGR)
ED Relative pipe roughness, −
ELGR Elevation pressure gradient, psi/ft
FRGR Friction pressure gradient, psi/ft
FW In-situ volume fraction water in liquid, −
GLR Producing gas/liquid ratio, scf/STBL
GVIS Gas viscosity, cp
HL Liquid holdup fraction, −
HLNS No-slip liquid holdup fraction, −
IHL Beggs and Brill inclined flow holdup correction factor option: 0 = no correction; ≠ 0 = use Palmer correction factors
IREG Flow regime indicator: 1 = Liquid; 2 = Gas; 3 = Bubble, Distributed, or Two-Phase; 4 = Slug or Intermittent; 5 = Mist or Segregated; 6 = Transition
KHL Hagedorn and Brown liquid holdup selection parameter: 0 = use H & B holdup; ≠ 0 = use no slip holdup if greater than H & B holdup
KREG Hagedorn and Brown flow pattern modification selection parameter: 0 = no modification; ≠ 0 = Griffith and Wallis Correlation for bubble flow calculations
P Pressure, psia
SURL Gas-liquid surface tension, dynes/cm
VISL Liquid viscosity, cp
VM Superficial mixture velocity, ft/sec
XND Dimensionless diameter number, −
XNGV Dimensionless gas velocity number, −
XNL Dimensionless liquid viscosity number, −
XNLV Dimensionless liquid velocity number, −
ED Pipe relative roughness
F Dependent variable array
FF Darcy–Weisbach friction factor
GVIS Gas viscosity, cp
H Horizontal array
HARG Value of horizontal variable
IDEG, IV, IH Degree of interpolation (Use 1 or 2)
NH Number of H entries
NPTS Number of X and Y entries
NV Number of V entries
P Pressure, psia
PR Reduced pressure
REY Reynolds number
SGFG Gas specific gravity (air = 1)
T Temperature, °F
TR Reduced temperature
V Vertical array
X X array
XARG X value
Y Y array
Z Gas compressibility factor

```
      SUBROUTINE BEGBR (ANG,DIA,ED,P,VM,HLNS,DENG,DENL,GVIS,VISL,
     1 XNLV,HL,FRGR,ELGR,ACCGR,DPDL,IREG,IHL,IDH)
C
C     SUBROUTINE TO CALCULATE PRESSURE GRADIENT IN -PSI/FT USING A
C     MODIFIED BEGGS AND BRILL CORRELATION.
C
C     THE HORIZONTAL FLOW REGIMES CORRESPONDING TO IREG ARE:
C     IREG=1  LIQUID
C     IREG=2  GAS
C     IREG=3  DISTRIBUTED
C     IREG=4  INTERMITTENT
C     IREG=5  SEGREGATED
C     IREG=6  TRANSITION
C
C     CONVERT INCLINATION ANGLE TO RADIANS.
      A=ANG*3.1416/180.
C
C     CALCULATE SUPERFICIAL VELOCITIES AND MIXTURE FROUDE NUMBER.
      VSL=VM*HLNS
      VSG=VM- SL
      XNFR=VM**2/(32.2*DIA)
C
C     CHECK FOR SINGLE PHASE FLOW.
      IREG=5
      IF (HLNS.GT..99999) IREG=1
      IF (HLNS.LT..00001) IREG=2
      IF (IREG.GT.2) GO TO 1
      HL=HLNS
      GO TO 12
C
C     DETERMINE FLOW REGIME USING REVISED FLOW PATTERN MAP.
    1 ITRAN=0
      XL1=316*HLNS**.302
      XL2=.0009252/HLNS**2.46842
      XL3=.1/HLNS**1.45155
      XL4=.5/HLNS**6.738
      XDD=XL1
      IF (HLNS.LT..01) GO TO 2
      IF (HLNS.GT..4) XDD=XL4
      IF (XNFR.GE.XL2.AND.XNFR.LT.XL3) ITRAN=1
      IF (XNFR.GE.XL3.AND.XNFR.LT.XDD) IREG=4
      IF (XNFR.GE.XDD) IREG=3
      GO TO 3
    2 IF (XNFR.GE.XL1) IREG=3
C
C     DETERMINE HORIZONTAL FLOW LIQUID HOLDUP AND C-FACTOR COEFFICIENTS
C     FOR UPHILL FLOW.
    3 I=IREG-2
      GO TO (4,5,6),I
C
C     DISTRIBUTED FLOW.
    4 HLO=1.065*HLNS**.5824/XNFR**.0609
      D=1.
      E=0.
      F=0.
      G=0.
      GO TO 7
C
```

```
      D=2.96
      E=.305
      F=-.4473
      G=.0978
      GO TO 7
C
C     SEGREGATED FLOW.
    6 HLO=.98*HLNS**.4846/XNFR**.0868
      D=.011
      E=-3.768
      F=3.539
      G=-1.614
C
C     RESTRICT MINIMUM VALUE OF HLO.
    7 IF (HLO.LT.HLNS) HLO=HLNS
C
C     CHECK FOR HORIZONTAL FLOW.
      IF (A.NE.0.) GO TO 8
      HL=HLO
      GO TO 10
C
C     FLOW IS INCLINED. CALCULATE C-FACTOR.
    8 IF (A.GT.0.) GO TO 9
C     DOWNHILL C-FACTOR COEFFICIENTS.
      D=4.7
      E=-.3692
      F=.1244
      G=-.5056
C
C     CALCULATE THE C-FACTOR.
    9 C=(1.-HLNS)*ALOG(D*HLNS**E*XNLV**F*XNFR**G)
      IF (C.LT.0.) C=0.
C
C     CALCULATE THE ANGLE CORRECTION FACTOR AND THE CORRECTED LIQUID
C     HOLDUP FRACTION.
      XX=SIN(1.8*A)
      FAC=1.+C*(XX-.333*XX**3)
C     CHECK TO BE SURE FAC IS NOT NEGATIVE.
      IF (FAC.LT.0.) FAC=0.
      HL=HLO*FAC
      IF (HL.GT.1) HL=1.
C
C     APPLY PALMER HOLDUP CORRECTION FACTORS IF DESIRED.
      IF (IHL.EQ.0) GO TO 10
      IF (ANG.LT.0.) HL=HL*.541
      IF (ANG.GT.0.) HL=HL*.918
C
C     CHECK FOR TRANSITION FLOW.
   10 IF (ITRAN.LT.1) GO TO 12
      IF (IREG.LT.5) GO TO 11
      HLS=HL
      IREG=4
      GO TO 3
   11 HLI=HL
      AA=(XL3-XNFR)/(XL3-XL2)
      B=1.-AA
      HL=HLS*AA+HLI*B
C
C     CALCULATE MIXTURE FLUID PROPERTIES.
   12 DENNS=DENL*HLNS+DENG*(1.-HLNS)
      DENS=DENL*HL+DENG*(1.-HL)
      VISNS=VISL*HLNS+GVIS*(1.-HLNS)
C
C     CALCULATE MOODY DIAGRAM FRICTION FACTOR.
```

```
C
C     CALCULATE MOODY DIAGRAM FRICTION FACTOR.
      REYN=1488.*DENNS*VM*DIA/VISNS
      FF=1./(1.14-2.*ALOG10(ED+21.25/REYN**.9))**2
      IF (IREG.LE.2) GO TO 13
C
C     CALCULATE TWO PHASE FRICTION FACTOR.
      Y=HLNS/(HL**2)
      X=ALOG(Y)
      S=X/(-.0523+3.182*X-.8725*X**2+.01853*X**4)
     IF (Y.GT.1..AND.Y.LT.1.2) S=ALOG(2.2*Y-1.2)
      FF=FF*EXP(S)
C
C     CALCULATE FRICTION, ELEVATION, ACCELERATION AND TOTAL PRESSURE
C     GRADIENTS.
   13 FRGR=FF*DENNS*VM**2/(2.*32.2*DIA*144.)
      ELGR=DENS*SIN(A)/144.
      IF(IDH.EQ.0)GO TO 16
      IF(A.GT.0.)GO TO 16
      ELGR=DENG*SIN(A)/144.
   16 CONTINUE
      EKK=DENS*VM*VSG/(32.2*P*144.)
      IF (EKK.GT..95) GO TO 14
      DPDL=-(FRGR+ELGR)/(1.-EKK)
      ACCGR=-EKK*DPDL
C
      RETURN
   14 WRITE (108,15)
   15 FORMAT (1X,'APPROACHING CRITICAL FLOW. STOP CALCULATIONS')
      STOP
      END
```

```
      SUBROUTINE HAGBR (ANG,DIA,ED,P,VM,HLNS,DENG,DENL,GVIS,VISL,
     1 XNL,XNLV,XNGV,XND,HL,FRGR,ELGR,ACCGR,DPDL,IREG,KREG,KHL)
C
C     SUBROUTINE TO CALCULATE LIQUID HOLDUP AND PRESSURE GRADIENT
C     USING THE HAGEDORN AND BROWN CORRELATION. THE ACCELERATION
C     PRESSURE GRADIENT IS CALCULATED WITH THE DUNS AND ROS EQUATION.
C
C     THE FLOW REGIMES CORRESPONDING TO IREG ARE:
C     IREG=1  LIQUID
C     IREG=2  GAS
C     IREG=3  BUBBLE
C     IREG=4  SLUG
C
      DIMENSION XHL(12),YHL(12),XCNL(10),YCNL(10),XPSI(12),YPSI(12)
      DIMENSION XHLL(12),XCNLL(10),YCNLL(10)
C
C     ENTER DATA ARRAYS FOR LIQUID HOLDUP CORRELATION.
      DATA XHL/
     1.2,.5,1.,2.,5.,10.,20.,50.,100.,200.,300.,1000./
      DATA YHL/
     1.04,.09,.15,.18,.25,.34,.44,.65,.82,.92,.96,1./
      DATA XCNL/
     1.002,.005,.01,.02,.03,.06,.1,.15,.2,.4/
      DATA YCNL/
     1.0019,.0022,.0024,.0028,.0033,.0047,.0064,.008,.009,.0115/
      DATA XPSI/
     1.01,.02,.025,.03,.035,.04,.045,.05,.06,.07,.08,.09/
      DATA YPSI/
     11.,1.1,1.23,1.4,1.53,1.6,1.65,1.68,1.74,1.78,1.8,1.83/
C
C     CONVERT INCLINATION ANGLE TO RADIANS.
      A=ANG*3.1416/180.
C
C     CALCULATE SUPERFICIAL VELOCITIES.
      VSL=VM*HLNS
      VSG=VM-VSL
C
C     CHECK FOR SINGLE PHASE GAS OR LIQUID FLOW.
      IF (HLNS.LT.1.) GO TO 1
      HL=1.
      DENNS=DENL
      IREG=1
      GO TO 6
    1 IF (HLNS.GT.0.) GO TO 2
      HL=0.
      DENNS=DENG
      IREG=2
      GO TO 6
C
C     CHECK FOR BUBBLE FLOW.
    2 XLB=1.071-.2281*VM**2/DIA
      IF (XLB.LT..13) XLB=.13
      HGNS=1.-HLNS
      IF (HGNS.GT.XLB) GO TO 3
      IREG=3
      IF (KREG.EQ.0) GO TO 3
      VS=.8
      HL=1.-.5*(1.+VM/VS-SQRT((1.+VM/VS)**2.-4.*VSG/VS))
      IF (HL.LT.HLNS) HL=HLNS
      DENS=DENL*HL+DENG*(1.-H@
```

```
      HL=1.-.5*(1.+VM/VS-SQRT((1.+VM/VS)**2.-4.*VSG/VS))
      IF (HL.LT.HLNS) HL=HLNS
      DENS=DENL*HL+DENG*(1.-H@
      REYNB=1488.*DENL*(VSL/HL)*DIA/VISL
      FF=1./(1.14-2.*ALOG10(ED+21.25/REYNB**.9))**2
C
C     CALCULATE ELEVATION AND FRICTION GRADIENTS AND ACCELERATION TERM
C     FOR BUBBLE FLOW.
      ELGR=DENS*SIN(A)/144.
      FRGR=FF*DENL*(VSL/HL)**2/(2.*32.2*DIA*144.)
      EKK=0.
      GO TO 7
C
C     PREPARE HOLDUP CORRELATION ARRAYS FOR INTERPOLATION.
    3 DO 4 K=1,10
      XCNLL(K)=ALOG(XCNL(K))
    4 YCNLL(K)=ALOG(YCNL(K))
      DO 5 K=1,12
    5 XHLL(K)=ALOG(1.E-05*XHL(K))
      IREG=4
C
C     CALCULATE LIQUID HOLDUP.
      XX=ALOG(XNL)
      CNL=EXP(FLAGR(XCNLL,YCNLL,XX,2,10))
      XX=ALOG(XNLV*CNL/(XNGV**.575*XND)*(P/14.7)**.1)
      HL=FLAGR(XHLL,YHL,XX,2,12)
      XX=XNGV*XNL**0.38/XND**2.14
      PSI=FLAGR(XPSI,YPSI,XX,2,12)
      IF (PSI.LT.1.) PSI=1.
      HL=HL*PSI
      IF (HL.LT.0.) HL=0.
      IF (HL.GT.1.) HL=1.
      IF (HL.GT.HLNS) GO TO 6
      IF (KHL.EQ.0) HL=HLNS
C
C     CALCULATE NO-SLIP AND SLIP MIXTURE DENSITIES.
    6 DENNS=DENL*HLNS+DENG*(1.-HLNS)
      DENS=DENL*HL+DENG*(1.-HL)
C
E     CALCULATE FRICTION FACTOR.
      VISS=VISL**HL*GVIS**(1.-HL)
      REYN=1488.*DENNS*VM*DIA/VISS
      FF=1./(1.14-2.*ALOG10(ED+21.25/REYN**.9))**2
C
C     CALCULATE ELEVATION, FRICTION, ACCELERATION, AND TOTAL PRESSURE
C     GRADIENTS.
      ELGR=DENS*SIN(A)/144.
      FRGR=FF*DENNS**2*VM**2/(2.*32.2*DIA*DENS*144.)
      VSG=VM*(1.-HLNS)
      EKK=DENS*VM*VSG/(32.2*P*144.)
      IF (EKK.GT..95) GO TO 8
    7 DPDL=-(ELGR+FRGR)/(1.-EKK)
      ACCGR=-DPDL*EKK
C
      RETURN
    8 WRITE (IO,9)
    9 FORMAT (1X,'APPROACHING CRITICAL FLOW. STOP CALCULATIONS')
      STOP
      END
```

```
 Z
      SUBROUTINE ZFACHY (T,P,SGFG,Z)
C
C     CALCULATE GAS COMPRESSIBILITY FACTOR USING THE HALL AND
C     YARBOROUGH CORRELATION FOR CURVE FITTING THE STANDING-
C     KATZ REDUCED PRESSURE-REDUCED TEMPERATURE Z-FACTOR
C     CHART. (OIL AND GAS JOURNAL, JUNE 18, 1973, PG. 82, AND FEBRUARY
C     18, 1974, PG. 86)
C
C     CALCULATE CRITICAL AND REDUCED TEMPERATURE AND PRESSURE.
      TC=169.0+314.0*SGFG
      PC=708.75-57.5*SGFG
      TR=(T+460.0)/TC
      PR=P/PC
C
C     IF REDUCED TEMPERATURE IS LESS THAN 1.01, CALCULATE A Z-FACTOR
C     FOR A REDUCED TEMPERATURE VALUE OF 1.0.
      IF (TR.GT.1.01) GO TO 1
      RT=1.0
      GO TO 2
    1 RT=1.0/TR
C
C     CALCULATE TEMPERATURE DEPENDENT TERMS.
    2 A=0.06125*RT*EXP(-1.2*(1.-RT)**2)
      B=RT*(14.76-9.76*RT+4.58*RT*RT)
      C=RT*(90.7-242.2*RT+42.4*RT*RT)
      D=2.18+2.82*RT
C
C     CALCULATE REDUCED DENSITY, Y, USING THE NEWTON-RAPHSON METHOD.
      Y=.001
      DO 4 J=1,25
      IF (Y.GT.1.) Y=.6
      F=-A*PR+(Y+Y*Y+Y**3-Y**4)/(1.-Y)**3-B*Y*Y+C*Y**D
      IF (ABS(F).LE.1.E-4) GO TO 5
C
C     IF CONVERGENCE IS NOT OBTAINED IN 25 ITERATIONS, SET Z=1.0 AND
C     RETURN.
      IF (J.LT.25) GO TO 3
      Z=1.0
      RETURN
C
    3 DFDY=(1.+4.*Y+4.*Y*Y-4.*Y**3+Y**4)/(1.-Y)**4
     1-2.*B*Y+D*C*Y**(D-1.)
      Y=Y-F/DFDY
    4 CONTINUE
C
C     CALCULATE Z-FACTOR.
    5 Z=A*PR/Y
C
      RETURN
      END
```

```
      SUBROUTINE GASVIS (T,SGFG,P,GVIS)
C
C     CALCULATE VISCOSITY OF HYDROCARBON GASES USING THE LEE ET AL
C     CORRELATION (TRANSACTIONS AIME, 1966, PG. 997).
C
      TABS=T+460.
      W=SGFG*29.
      AK=(9.4+.02*W)*(TABS**1.5)/(209.+19.*W+TABS)
      X=3.5+(986./TABS)+.01*W
      Y=2.4-.2*X
C
C     CALCULATE GAS DENSITY, GM/CC.
      CALL ZFACHY (T,P,SGFG,Z)
      RHOG=P*W/(10.72*Z*TABS*62.4)
C
C     CALCULATE GAS VISCOSITY,CP.
      GVIS=AK*EXP(X*RHOG**Y)/10000.
C
      RETURN
      END
```

```
 F
      FUNCTION FLAGR2 (V,H,F,NV,NH,IV,IH,VARG,HARG)
C
C     FLAGR2 IS A FUNCTION SUBPROGRAM FOR PERFORMING DOUBLE INTER-
C     POLATION. IT CALLS FLAGR (LISTED ON A FOLLOWING PAGE) FOR EACH
C     INTERPOLATION.
C
C     V IS THE ROW (VERTICAL) ARRAY.
C     H IS THE COLUMN (HORIZONTAL) ARRAY.
C     F IS THE FUNCTION VALUE MATRIX.
C     NV AND NH ARE DIMENSIONS OF THE VERTICAL AND HORIZONTAL ARRAYS.
C     IV AND IH ARE DEGREES OF INTERPOLATION IN THE V AND H ARRAYS.
C     VARG AND HARG ARE ARGUMENTS FOR WHICH INTERPOLATED FUNCTION
C     VALUES ARE DESIRED.
C
      DIMENSION V(2),H(2),F(2),X(50),Y(50)
C
      DO 20 J=1,NH
      DO 10 I=1,NV
      K=I+(J-1)*NV
   10 X(I)=F(K)
   20 Y(J)=FLAGR(V,X,VARG,IV,NV)
C
      FLAGR2=FLAGR(H,Y,HARG,IH,NH)
C
      RETURN
      END
      FUNCTION FLAGR (X,Y,XARG,IDEG,NPTS)
C
C     INTERPOLATION ROUTINE SIMILAR TO FLAGR IN APPLIED NUMERICAL
C     METHODS BY CARNAHAN, LUTHER AND WILKES, JOHN WILEY AND SONS,
C     PG. 31.
C
C     FLAGR USES THE LAGRANGE FORMULA TO EVALUATE THE INTERPOLATING
C     POLYNOMIAL OF DEGREE IDEG FOR ARGUMENT XARG USING THE DATA
C     VALUES X(MIN).....X(MAX) AND Y(MIN).....Y(MAX) WHERE MIN =
C     MAX-IDEG. THE X(I) VALUES ARE NOT NECESSARILY EVENLY SPACED
C     AND CAN BE IN EITHER INCREASING OR DECREASING ORDER.
C
C     X IS THE ARRAY OF INDEPENDENT VARIABLE DATA POINTS.
C     Y IS THE ARRAY OF DEPENDENT VARIABLE DATA POINTS.
C     XARG IS THE ARGUMENT FOR WHICH AN INTERPOLATED VALUE IS DESIRED.
C     IDEG IS THE DEGREE OF INTERPOLATING POLYNOMIAL (1 IS LINEAR,
C     2 IS QUADRATIC, ETC).
C     NPTS IS THE NUMBER OF DATA POINTS IN X AND Y.
C
      DIMENSION X(1),Y(1)
C
      N=IABS(NPTS)
      N1=IDEG+1
      L=1
      IF (X(2).GT.X(1)) GO TO 1
      L=2
C
C     CHECK TO BE SURE THAT XARG IS WITHIN RANGE OF X(I) VALUES
C     FOR INTERPOLATION PURPOSES. IF IT IS NOT, SET FLAGR EQUAL
C     TO THE APPROPRIATE TERMINAL VALUE (Y(1) OR Y(N)) AND RETURN.
C     NOTE THAT THIS PRECLUDES EXTRAPOLATION OF DATA.
```

```
    1 GO TO (2,3),L
    2 IF (XARG.LE.X(1)) GO TO 4
      IF (XARG.GE.X(N)) GO TO 5
      GO TO 6
    3 IF (XARG.GE.X(1)) GO TO 4
      IF (XARG.LE.X(N)) GO TO 5
      GO TO 6
    4 FLAGR=Y(1)
      RETURN
    5 FLAGR=Y(N)
      RETURN
C
C     DETERMINE VALUE OF MAX.
    6 GO TO (10,20),L
C
C     DATA ARE IN ORDER OF INCREASING VALUES OF X.
   10 DO 11 MAX=N1,N
      IF (XARG.LT.X(MAX)) GO TO 12
   11 CONTINUE
C
C     DATA ARE IN ORDER OF DECREASING VALUES OF X.
   20 DO 21 MAX=N1,N
      IF (XARG.GT.X(MAX)) GO TO 12
   21 CONTINUE
C
C     COMPUTE VALUE OF FACTOR.
   12 MIN=MAX-IDEG
      FACTOR=1.
      DO 7 I=MIN,MAX
      IF (XARG.NE.X(I)) GO TO 7
      FLAGR=Y(I)
      RETURN
    7 FACTOR=FACTOR*(XARG-X(I))
C
C     EVALUATE INTERPOLATING POLYNOMIAL.
      YEST=0.
      DO 9 I=MIN,MAX
      TERM=Y(I)*FACTOR/(XARG-X(I))
      DO 8 J=MIN,MAX
      IF (I.NE.J) TERM=TERM/(X(I)-X(J))
    8 CONTINUE
    9 YEST=YEST+TERM
      FLAGR=YEST
C
      RETURN
      END
```

This appendix contains listings of the following programs:

BHOLE —for conversion of wellhead pressures to bottom hole pressures;

P-PSI —for conversion of pressures to pseudo-pressures;

and the following subroutines:

FFCFLO—for use with BHOLE to calculate friction factors;

VISCY —for use with BHOLE and P-PSI to calculate natural gas viscosities;

XLGR4 —for use with VISCY to interpolate viscosity tables;

ZANDC —for use with BHOLE and P-PSI to calculate natural gas compressibility factors. This subroutine also calculates natural gas compressibilities.

The program and subroutines listings are self-explanatory.

Examples D-1 to D-8 given in this appendix illustrate the various options of the program BHOLE.

Description of Input Data for Program BHOLE

Card no.					
KODEA =					
1	2	3	4	Description	Format
1	1	1	1	Title	17A4, A2
2	2	2	2	KODEA, KODEB	2I5
3	—	—	—	Specific gravity of the gas	F10.4
—	3	—	—	Specific gravity of the gas, mole fractions of hydrogen sulphide, carbon-dioxide and nitrogen	4F10.4
—	—	3	3	Complete gas composition	12F6.4
—	—	—	4	Pseudo-critical temperature, pseudo-critical pressure and molecular weight $C7^+$	3F10.2
4	4	4	5	Tubing diameter, well depth, tubing length and absolute roughness of tubing	4F10.4
5	5	5	6	Flow rate, wellhead temperature, bottom hole temperature, wellhead pressure, estimated bottom hole pressure	5F10.4

Description of Input Data for Program P-PSI

Card no.	Description	Format
1	Title	17A4, A2
2	Maximum pressure, pressure increment, gas temperature and specific gravity	4F10.4
3	Pseudo-critical temperature, pseudo-critical pressure, mole per cent of H_2S, CO_2 and N_2	5F10.4

```
C
C     ******************************************************************
C     *                                                                *
C     * PROGRAM NAME    - BHOLE, BOTTOM HOLE PRESSURES IN GAS WELLS     *
C     *                                                                *
C     * PURPOSE         - THIS PROGRAM CONVERTS STATIC AND FLOWING     *
C     *                   WELLHEAD PRESSURES TO CORRESPONDING BOTTOM   *
C     *                   HOLE PRESSURES                               *
C     *                                                                *
C     * METHOD          - BOTTOM HOLE PRESSURES ARE CALCULATED BY THE  *
C     *                   METHOD OF CULLENDER & SMITH (1956), AS       *
C     *                   MODIFIED BY AZIZ (1967A). THE MODIFICATION   *
C     *                   INVOLVES USAGE OF A ONE-STEP, THIRD-ORDER    *
C     *                   NUMERICAL INTEGRATION SCHEME INSTEAD OF      *
C     *                   SIMPSON'S RULE                               *
C     *                                                                *
C     * LANGUAGE        - FORTRAN IV                                   *
C     *                                                                *
C     * SUBROUTINES USED                                               *
C     *                 - FFCFLO                                       *
C     *                 - VISCY                                        *
C     *                 - XLGR4                                        *
C     *                 - ZANDC                                        *
C     *                                                                *
C     * DATA INPUT, OUTPUT AND VARIABLES                               *
C     *                                                                *
C     *  ABSRNS - ABSOLUTE ROUGNESS OF THE PIPE. ASSUMED EQUAL TO      *
C     *           0.00060 IF READ ZERO.                                *
C     *  A(I)   - PARAMETERS IN THE ONE-STEP, THIRD-ORDER NUMERICAL    *
C     *           INTEGRATION SCHEME OF AZIZ(1967A)                    *
C     *  CNCH2S - MOL PERCENT H2S IN NATURAL GAS                       *
C     *  CNCCO2 - MOL PERCENT CO2 IN NATURAL GAS                       *
C     *  CNCN2  - MOL PERCENT N2  IN NATURAL GAS                       *
C     *  DIAT   - TUBING INSIDE DIAMETER, INCHES. FOR PRODUCTION       *
C     *           THROGH CASING REPLACE WITH EFFECTIVE DIAMETER        *
C     *  DPH    - WELL DEPTH, FEET                                     *
C     *  FRCI(I)- MOL FRACTION OF COMPONENT (I). READ IN FOLLOWING     *
C     *           ORDER - H2S, CO2, N2, C1, C2, C3, IC4, NC4, IC5,     *
C     *           NC5, C6, C7+                                         *
C     *  KODEA  = 1 - SWEET GAS, READ SPECIFIC GRAVITY, SPG            *
C     *         = 2 - SOUR GAS, READ SPG AND MOLE PERCENT HYDROGEN     *
C     *               SULPHIDE, CNCH2S, CARBON DIOXIDE, CNCCO2, AND    *
C     *               NITROGEN, CNCN2                                  *
C     *         = 3 - SWEET OR SOUR GAS, READ COMPLETE COMPOSITION,    *
C     *               MOLE FRACTION H2S, CO2, N2, C1, C2, C3, IC4,     *
C     *               NC4, IC5, NC5, C6, C7+. ASSUME PROPERTIES OF     *
C     *               C7+ EQUAL TO C8 PROPERTIES                       *
C     *         = 4 - SAME AS KODEA=3, ALSO READ CRITICAL PRESSURE,    *
C     *               PRSCI(12), CRITICAL TEMPERATURE, TEMCI(12),      *
C     *               AND MOLECULAR WEIGHT, MWTI(12) OF C7+ FRACTION   *
C     *  KODEB  = 1 - STATIC PRESSURE CALCULATION                      *
C     *         = 2 - FLOWING PRESSURE CALCULATION                     *
C     *  LTH    - LENGTH OF FLOW STRING, FEET. FOR VERTICAL WELLS      *
C     *           READ ZERO. NEGATIVE FOR INJECTION WELLS.             *
C     *  MWT    - MOLECULAR WEIGHF OF THE NATURAL GAS                  *
C     *  MWTI   - MOLECULAR WEIGHT OF COMPONENT I.                     *
C     *  NCOMP  - NUMBER OF COMPONENTS IN NATURAL GAS ,USUALLY THE     *
C     *           TWELVE LISTED ABOVE. READ FOR KODEA=3 & KODEA=4      *
C     *  PRSCI  - CRITICAL PRESSURE OF COMPONENT I.                    *
C     *  PRSPC  - PSEUDO-CRITICAL PRESSURE OF THE NATURAL GAS, PSIA    *
C     *  PRST   - STATIC OR FLOWING WELLHEAD (TOP HOLE) PRESSURE,      *
C     *           PSIA. READ MEASURED VALUE.                           *
C     *  PRSW   - STATIC OR FLOWING BOTTOM HOLE PRESSURE,              *
C     *           PSIA. CALCULATED BY THIS PROGRAM.                    *
C     *  PRSWE  - STATIC OR FLOWING BOTTOM HOLE PRESSURE,              *
C     *           PSIA. READ ESTIMATED VALUE. IF CANNOT ESTIMATE,      *
C     *           READ ZERO.                                           *
C     *  RTEG   - GAS FLOW RATE AT STANDARD CONDITIONS (14.65 PSIA,    *
C     *           60 DEGREES F), MMSCFD. READ ZERO FOR STATIC CASE.    *
C     *  STEMCI - WHEN KODEA=4, THE ORIGINAL VALUE OF TEMCI(12) IS     *
C     *           STORED BEFORE THE NEW VALUE IS READ                  *
C     *  SPRSCI - WHEN KODEA=4, THE ORIGINAL VALUE OF PRSCI(12) IS     *
C     *           STORED BEFORE THE NEW VALUE IS READ                  *
C     *  SMWTI  - WHEN KODEA=4, THE ORIGINAL VALUE OF MWTI(12) IS      *
C     *           STORED BEFORE THE NEW VALUE IS READ                  *
C     *  SPG    - SPECIFIC GRAVITY OF THE NATURAL GAS, EITHER READ     *
C     *           OR CALCULATED BY THIS PROGRAM FROM GAS COMPOSITION   *
C     *  TEMCI  - CRITICAL TEMPERATURE OF COMPONENT I.                 *
C     *  TEMPC  - PSEUDO-CRITICAL TEMPERATURE OF THE NATURAL GAS,      *
C     *           DEGREES R                                            *
C     *  TEMT   - STATIC OR FLOWING WELLHEAD (TOP HOLE) TEMPERATURE,   *
C     *           DEGREES F. READ MEASURED VALUE.                      *
C     *  TEMW   - STATIC OR FLOWING BOTTOM HOLE TEMPERATURE,           *
C     *           DEGREES F. READ MEASURED VALUE.                      *
C     *  Y, YY  - INTEGRANDS IN THE CULLENDER AND SMITH METHOD. SEE    *
C     *           ALSO AZIZ(1967A)                                     *
C     *  ZED    - NATURAL GAS COMPRESSIBILITY FACTOR CALCULATED BY     *
C     *           SUBROUTINE ZANDC                                     *
C     *                                                                *
C     ******************************************************************
C
      REAL MWTI, MWT, LTH
C
      DIMENSION FRCI(12), TEMCI(12), PRSCI(12), MWTI(12), A(4),
     A          TITLE(18)
C
      COMMON CNCH2S, CNCCO2, CNCN2, PRSPC, TEMPC, SPG, DIAT, RTEG,
     A       ABSRNS, DPH, LTH
C
      DATA IR/1/, IW/3/, NCOMP/12/
C
      DATA TEMCI /
     A672.37, 547.57, 227.27, 343.04, 549.76, 665.68,
     B734.65, 765.32, 828.77, 845.37, 913.37,1023.89/
C
      DATA PRSCI /
     A1306.0, 1071.0, 493.00, 667.80, 707.80, 616.30,
     B529.10, 550.70, 490.40, 488.60, 436.90, 360.60/
C
      DATA MWTI /
     A34.076, 44.010, 28.013, 16.042, 30.070, 44.097,
     B58.124, 58.124, 72.151, 72.151, 86.178,114.232/
C
      DATA A /0.125, 0.375, 0.375, 0.125/
C
C     ******************************************************************
C     *                                                                *
C     * READ AND WRITE FORMAT STATEMENTS                               *
C     *                                                                *
C     ******************************************************************
C
100   FORMAT (17A4, A2)
101   FORMAT (12F6.4)
102   FORMAT (8I5)
103   FORMAT (8F10.4)
104   FORMAT (3F10.2)
200   FORMAT (1H , 17A4, A2,/)
201   FORMAT (1H0, 'INPUT DATA:'//1H ,5X, 'WELL' ,15X, 'PIPE' ,10X,
     A        'GAS FLOW' ,3X, 'TEMPERATURE' ,4X, 'PRESSURE, PSIA'/1H ,
     B        'DEPTH' ,4X, 'LENGTH' ,3X, 'INSIDE' ,3X, 'ABSOLUTE' ,5X,
     C        'RATE' ,6X, 'DEGREES F' ,12X, 'BOTTOM'/ 1H ,5X, 'FEET' ,
     D        6X, 'DIAMETER' ,2X, 'ROUGHNESS' ,3X, 'MMSCFD' ,3X,
     E        'WELL' ,3X, 'BOTTOM' ,3X, 'WELL' ,4X, 'HOLE'/ 1H ,23X,
     F        'INCHES' ,19X, 'HEAD' ,4X, 'HOLE' ,4X,'HEAD (ASSUMED)'/)
202   FORMAT (1H , F6.0, 3X, F6.0, 4X, F5.3, 5X, F5.4, 5X, F6.3, 3X,
     A        F4.0, 4X, F4.0, 1X, 2(3X, F5.0) //)
203   FORMAT (1H , 'GAS COMPOSITION, MOLE FRACTION'// 1H ,1X, 'H2S' ,
     A        4X, 'CO2' ,5X, 'N2')
204   FORMAT (1H , F5.4, 2(2X, F5.4)/)
205   FORMAT (1H ,2X, 'C1' ,5X, 'C2' ,5X, 'C3' ,4X, 'IC4' ,4X, 'NC4' ,
     A        4X, 'IC5' ,4X, 'NC5' ,5X, 'C6' ,4X, 'C7+')
206   FORMAT (1H , F5.4, 8(2X, F5.4)/)
207   FORMAT (1H , 'PROPERTIES OF C7+')
208   FORMAT (1H0, 'PSEUDO-CRITICAL PRESSURE, PSIA' ,16X, '=' , F7.2)
209   FORMAT (1H0, 'PSEUDO-CRITICAL TEMPERATURE, DEGREES RANKINE  =' ,
     A       F7.2)
210   FORMAT (1H0, 'MOLECULAR WEIGHT' ,30X, '= ', F6.2)
211   FORMAT (1H0, 'GAS GRAVITY' ,35X, '=   ' , F6.4,//)
212   FORMAT (1H0, 'CALCULATED GAS PROPERTIES:')
213   FORMAT (1H0, 79(1H*))
214   FORMAT (1H , 'GAS GRAVITY =' , F7.4)
215   FORMAT (1H0, 'GAS GRAVITY =' , F7.4)
300   FORMAT (1H0, 'NO CONVERGENCE')
301   FORMAT (1H0, 'STATIC BOTTOM HOLE PRESSURE =' , F6.0, 19X,
     A        'NUMBER OF ITERATIONS =', I2)
302   FORMAT (1H0, 'FLOWING BOTTOM HOLE PRESSURE =' , F6.0, 19X,
     A        'NUMBER OF ITERATIONS =', I2)
303   FORMAT (1H1)
C
C     ******************************************************************
C     *                                                                *
C     * READ TITLE, KODEA, KODEB AND THE APPROPRIATE INPUT, AS        *
C     * REQUIRED BY THE VALUE OF KODEA, TO CALCULATE THE PSEUDO-      *
C     * CRITICAL TEMPERATURE, PSEUDO-CRITICAL PRESSURE AND THE        *
C     * MOLECULAR WEIGHT OF THE NATURAL GAS                            *
C     *                                                                *
C     ******************************************************************
C
400   READ   (IR,100,END=415) TITLE
      WRITE  (IW, 303)
      WRITE  (IW, 200) TITLE
      READ   (IR, 102) KODEA, KODEB
      WRITE  (IW, 201)
      IF (KODEA .GE. 3) GO TO 401
      READ   (IR, 103) SPG, (FRCI(I), I = 1, 3)
      CNCH2S = FRCI(1) * 100.
      CNCCO2 = FRCI(2) * 100.
      CNCN2  = FRCI(3) * 100.
      GO TO 403
401   READ   (IR, 101) (FRCI(I), I = 1, NCOMP)
      IF (KODEA .LT. 4) GO TO 402
      STEMCI = TEMCI(12)
      SPRSCI = PRSCI(12)
      SMWTI  = MWTI(12)
      READ   (IR, 104) TEMCI(12), PRSCI(12), MWTI(12)
402   CNCH2S = FRCI(1) * 100.
      CNCCO2 = FRCI(2) * 100.
      CNCN2  = FRCI(3) * 100.
      GO TO 404
C
C     ******************************************************************
C     *                                                                *
C     * CALCULATE THE PSEUDO-CRITICAL TEMPERATURE, PSEUDO-CRITICAL    *
C     * PRESSURE AND THE MOLECULAR WEIGHT OF THE NATURAL GAS          *
C     *                                                                *
C     ******************************************************************
C
403   MWT    = SPG * 28.964
      TEMPC  = 170.491 + 307.344 * SPG
      PRSPC  = 709.604 -  58.718 * SPG
      GO TO 406
404   MWT    = 0.0
      TEMPC  = 0.0
      PRSPC  = 0.0
      DO 405 I = 1, NCOMP
      MWT    = MWT + MWTI(I) * FRCI(I)
      TEMPC  = TEMPC + TEMCI(I) * FRCI(I)
405   PRSPC  = PRSPC + PRSCI(I) * FRCI(I)
      SPG    = MWT / 28.964
C
C     ******************************************************************
C     *                                                                *
C     * READ REMAINING INPUT DATA. WRITE CALCULATED GAS PROPERTIES    *
C     *                                                                *
C     ******************************************************************
C
406   READ   (IR, 103) DIAT, DPH, LTH, ABSRNS
      READ   (IR, 103) RTEG, TEMT, TEMW, PRST, PRSWE
      IF (PRSWE .EQ. 0.) PRSWE = PRST + 200.
      PRSW   = PRSWE
      IF (LTH .EQ. 0.) LTH = DPH
      WRITE  (IW, 202) DPH, LTH, DIAT, ABSRNS, RTEG, TEMT, TEMW,
     A                 PRST, PRSWE
      IF (KODEA .GT. 1) GO TO 407
      WRITE  (IW, 214) SPG
      GO TO 410
407   WRITE  (IW, 203)
      WRITE  (IW, 204) FRCI(1), FRCI(2), FRCI(3)
      IF (KODEA .GT. 2) GO TO 408
      WRITE  (IW, 215) SPG
      GO TO 410
408   WRITE  (IW, 205)
      WRITE  (IW, 206) (FRCI(I), I = 4, NCOMP)
      IF (KODEA .EQ. 4) GO TO 409
      GO TO 410
409   WRITE  (IW, 207)
      WRITE  (IW, 208) PRSCI(12)
      WRITE  (IW, 209) TEMCI(12)
      WRITE  (IW, 210) MWTI(12)
      PRSCI (12)  = SPRSCI
      TEMCI (12)  = STEMCI
      MWTI  (12)  = SMWTI
410   WRITE  (IW, 212)
      WRITE  (IW, 208) PRSPC
      WRITE  (IW, 209) TEMPC
      WRITE  (IW, 210) MWT
      IF (KODEA .GT. 2) WRITE (IW, 211) SPG
      WRITE  (IW, 213)
C
C     ******************************************************************
C     *                                                                *
C     * START OF BOTTOM HOLE PRESSURE CALCULATION                     *
C     *                                                                *
C     ******************************************************************
C
      GL     = 0.01875 * SPG * DPH
      IF (KODEB .EQ. 2) GL = GL * 1000.
      CALL ZANDC (TEMT, TEMPC, PRST, PRSPC, CNCH2S,
     A            CNCCO2, ZED, CMPG, IERR)
      PTZ    = (TEMT + 460.) * ZED / PRST
      ITER   =  0
      IF (KODEB .EQ. 2) CALL FFCFLO (TEMT, PRST, PTZ)
      YY     = A(1) * PTZ
      PRS    = PRST
      TEM    = TEMT
      Y      = YY
411   ITER   = ITER + 1
      PRSINC = (PRSW - PRST) / 3.
      TEMINC = (TEMW - TEMT) / 3.
C
C     ******************************************************************
C     *                                                                *
C     * INTEGRATION LOOP                                               *
C     *                                                                *
C     ******************************************************************
C
      DO 412 I = 2, 4
      PRS    = PRS + PRSINC
      TEM    = TEM + TEMINC
      CALL ZANDC (TEM, TEMPC, PRS, PRSPC, CNCH2S,
     A            CNCCO2, ZED, CMPG, IERR)
      PTZ    = (TEM + 460.) * ZED / PRS
      IF (KODEB .EQ. 2) CALL FFCFLO (TEM, PRS, PTZ)
412   Y      = Y + A(I) * PTZ
C
C     ******************************************************************
C     *                                                                *
C     * NEWTON-RAPHSON ITERATION                                       *
C     *                                                                *
C     ******************************************************************
C
      FUN    = GL - (PRSW - PRST) * Y
      PRSN   = PRSW + FUN / PTZ
      ERROR  = ABS (PRSW - PRSN) / PRSW
      PRSW   = PRSN
      IF (ERROR .LT. 0.002) GO TO 413
      IF (ITER .GT. 20) GO TO 414
      PRS    = PRST
      TEM    = TEMT
      Y      = YY
      GO TO 411
413   IF (KODEB .EQ. 1) WRITE (IW, 301) PRSW, ITER
      IF (KODEB .EQ. 2) WRITE (IW, 302) PRSW, ITER
      GO TO 400
414   WRITE  (IW, 300)
      IF (KODEB .EQ. 1) WRITE (IW, 301) PRSW, ITER
      IF (KODEB .EQ. 2) WRITE (IW, 302) PRSW, ITER
      GO TO 400
415   STOP
      END
```

```
C
C     ******************************************************************
C     *                                                                *
C     * PROGRAM NAME    - P-PSI, PSEUDO PRESSURE CALCULATIONS           *
C     *                                                                *
C     * PURPOSE         - THIS PROGRAM CONVERTS PRESSURE TO THEIR       *
C     *                   CORRESPONDING PSEUDO-PRESSURES               *
C     *                                                                *
C     * METHOD          - PSEUDO-PRESSURES, AS DEFINED BY AL-HUSSAINY  *
C     *                   (1965), ARE CALCULATED BY THE NUMERICAL      *
C     *                   INTEGRATION METHOD ILLUSTRATED BY EXAMPLE    *
C     *                   2-1.                                         *
C     *                                                                *
C     * LANGUAGE        - FORTRAN IV                                   *
C     *                                                                *
C     * SUBROUTINES USED                                               *
C     *                 - VISCY                                        *
C     *                 - XLGR4                                        *
C     *                 - ZANDC                                        *
C     *                                                                *
C     * DATA INPUT, OUTPUT AND VARIABLES                               *
C     *                                                                *
C     *  CNCH2S - MOL PERCENT H2S IN THE NATURAL GAS                   *
C     *  CNCCO2 - MOL PERCENT CO2 IN THE NATURAL GAS                   *
C     *  CNCN2  - MOL PERCENT  N2 IN THE NATURAL GAS                   *
C     *  PRSDEL - PRESSURE INCREMENT FOR NUMERICAL INTEGRATION         *
C     *  PRSMAX - UPPER LIMIT OF PRESSURE, PSIA, UPTO WHICH THE        *
C     *           PSEUDO-PRESSURE CONVERSION IS REQUIRED               *
C     *  PRSPC  - PSEUDO-CRITICAL PRESSURE OF THE NATURAL GAS, PSIA    *
C     *  PRSSI  - DENOTES PSEUDO-PRESSURE, (PSIA)**I/CENTIPOISE        *
C     *  SPG    - SPECIFIC GRAVITY OF THE NATURAL GAS                  *
C     *  TEM    - RESERVOIR TEMPERATURE, DEGREES F                     *
C     *  TEMPC  - PSEUDO-CRITICAL TEMPERATURE OF THE NATURAL GAS,      *
C     *           DEGREES F                                            *
C     *  VISG   - NATURAL GAS VISCOSITY, CENTIPOISE, CALCULATED BY     *
C     *           THE SUBROUTINE VISCY                                 *
C     *  ZED    - NATURAL GAS COMPRESSIBILITY FACTOR, CALCULATED BY    *
C     *           THE SUBROUTINE ZANDC                                 *
C     *                                                                *
C     ******************************************************************
C
      DATA IR/1/, IW/3/
C
      DIMENSION TITLE(18)
C
C     ******************************************************************
C     *                                                                *
C     * READ AND WRITE FORMAT STATEMENTS                               *
C     *                                                                *
C     ******************************************************************
C
100   FORMAT (17A4,A2)
110   FORMAT (8F10.4)
200   FORMAT (1H1)
210   FORMAT (1H , 17A4, A2,/)
220   FORMAT (1H0, 'RESERVOIR TEMPERATURE, DEGREES F       =' ,F10.2,/
     A        1H , 'GAS GRAVITY                             =' ,F10.4,/
     B        1H , 'PSEUDO-CRITICAL TEMP., DEGREES R        =' ,F10.2,/
     C        1H , 'PSEUDO-CRITICAL PRESSURE, PSIA          =' ,F10.2,/
     D        1H , 'MOLE PERCENT - HYDROGEN SULPHIDE        =' ,F10.2,/
     E        1H , 'MOLE PERCENT - CARBON DIOXIDE           =' ,F10.2,/
     F        1H , 'MOLE PERCENT - NITROGEN                 =' ,F10.2,//)
230   FORMAT (1H0, 'PRESSURE' ,16X, 'PSEUDO-PRESSURE'/1H ,1X,
     A        '(PSIA)' ,18X, '(PSIA **2/CP)'//)
240   FORMAT (1H , F7.1 ,18X, E13.7)
C
C     ******************************************************************
C     *                                                                *
C     * READ INPUT DATA                                                *
C     *                                                                *
C     ******************************************************************
C
      READ  (IR, 100, END=400) TITLE
      WRITE (IW, 200)
      WRITE (IW, 210) TITLE
      READ  (IR, 110) PRSMAX, PRSDEL, TEM, SPG
      READ  (IR, 110) TEMPC, PRSPC, CNCH2S, CNCCO2, CNCN2
      WRITE (IW, 220) TEM, SPG, TEMPC, PRSPC, CNCH2S, CNCCO2, CNCN2
      WRITE (IW, 230)
C
C     ******************************************************************
C     *                                                                *
C     * PSEUDO-PRESSURE CALCULATION BY NUMERICAL INTEGRATION           *
C     *                                                                *
C     ******************************************************************
C
      PRS    = 0.0
      PRSSI  = 0.0
      PRSSI2 = 0.0
      GO TO 320
310   PRS    = PRS + PRSDEL
      TEMPRD = (TEM + 460.) / TEMPC
      PRSPRD = PRS / PRSPC
      CALL ZANDC (TEM, TEMPC, PRS, PRSPC, CNCH2S,
     A            CNCCO2, ZED, CMPG, IERR)
      CALL VISCY (TEMPRD, PRSPRD, SPG, TEM, CNCH2S,
     A            CNCCO2, CNCN2, VISGR, VISG, IERR)
      PRSSI2 = 2. * PRS / (VISG * ZED)
      PRSSI  = (PRSSI1 + PRSSI2) / 2. * PRSDEL + PRSSI
320   WRITE (IW, 240) PRS, PRSSI
      PRSSI1 = PRSSI2
      IF (PRS .LT. PRSMAX) GO TO 310
400   STOP
      END
```

```
C
C     ******************************************************************
C     *                                                                *
C     * PROGRAM NAME    - NATURAL GAS VISCOSITY DETERMINATION           *
C     *                                                                *
C     * PURPOSE         - THIS SUBROUTINE WILL DETERMINE THE VISCOSITY *
C     *                   FOR NATURAL GAS INCORPORATING A CORRECTION   *
C     *                   FOR THE PRESENCE OF HYDROGEN SULPHIDE,       *
C     *                   CARBON DIOXIDE AND NITROGEN.                 *
C     *                                                                *
C     * LANGUAGE        - FORTRAN IV                                   *
C     *                                                                *
C     * CALLING SEQUENCE                                               *
C     *                 - CALL VISCY (TEMPRD, PRSPRD, SPG, TEM,        *
C     *                               CNCH2S, CNCCO2, CNCN2, VISGR,    *
C     *                               VISG, IERR)                      *
C     *                                                                *
C     * PARAMETERS      -                                              *
C     *                                                                *
C     *  1. INPUT PARAMETERS                                           *
C     *                                                                *
C     *  TEMPRD - PSEUDO REDUCED TEMPERATURE AT WHICH THE NATURAL      *
C     *           GAS EXISTS                                           *
C     *  PRSPRD - PSEUDO REDUCED PRESSURE AT WHICH THE NATURAL         *
C     *           GAS EXISTS                                           *
C     *  SPG    - SPECIFIC GRAVITY OF THE NATURAL GAS                  *
C     *  TEM    - TEMPERATURE OF THE NATURAL GAS (DEGREES F.)          *
C     *  CNCH2S - MOL PERCENT HYDROGEN SULPHIDE IN NATURAL GAS         *
C     *  CNCCO2 - MOL PERCENT CARBON DIOXIDE IN NATURAL GAS            *
C     *  CNCN2  - MOL PERCENT NITROGEN IN NATURAL GAS                  *
C     *                                                                *
C     *  2. OUTPUT PARAMETERS                                          *
C     *                                                                *
C     *  VISGR  - VISCOSITY RATIO                                      *
C     *  VISG   - VISCOSITY OF THE NATURAL GAS, CENTIPOISE             *
C     *  IERR   - ERROR FLAG INDICATING BOUNDS ERROR                   *
C     *                                                                *
C     * VARIABLES                                                      *
C     *                                                                *
C     *  CORCO2 - CORRECTION FOR CO2                                   *
C     *  CORH2S - CORRECTION FOR H2S                                   *
C     *  CORN2  - CORRECTION FOR N2                                    *
C     *  ICALL  - INDICATES THE POSITION IN THE PROGRAM FROM WHICH     *
C     *           THE CALL TO XLGR4 WAS MADE                           *
C     *  IHITEM - HIGHEST INDEX TO BE CHECKED IN THE TABLE OF          *
C     *           TEMPERATURE INDEXES, TEMTBL                          *
C     *  IHIPRS - HIGHEST INDEX TO BE CHECKED IN THE TABLE OF          *
C     *           PRESSURE INDEXES, PRSTBL                             *
C     *  LOTEM  - LOWEST INDEX TO BE CHECKED IN THE TABLE OF           *
C     *           TEMPERATURE INDEXES, TEMTBL                          *
C     *  LOPRS  - LOWEST INDEX TO BE CHECKED IN THE TABLE OF           *
C     *           PRESSURE INDEXES, PRSTBL                             *
C     *  TEMTBL - TABLE OF INDEXES TO THE APPROPRIATE TEMPERATURE      *
C     *           COLUMN IN THE VISCOSITY TABLE, VISTBL.               *
C     *  PRSTBL - TABLE OF INDEXES TO THE APPROPRIATE PRESSURE ROW     *
C     *           IN THE VISCOSITY TABLE, VISTBL                       *
C     *  VISGA  - VISCOSITY AT 1 ATM PRESSURE (CORRECTED-H2S,CO2,N2)   *
C     *  VISGU  - VISCOSITY AT 1 ATM PRESSURE (UNCORRECTED)            *
C     *  VISI   - INTERPOLATED VISCOSITY RATIO                         *
C     *  VISJ   - INTERPOLATED VISCOSITY RATIO                         *
C     *  VISTBL - TABLE OF VISCOSITY RATIOS                            *
C     *  VISTB1 - EQUIVALENT TO APPROXIMATELY THE FIRST HALF OF        *
C     *           VISTBL.                                              *
C     *         - SET UP BECAUSE VISTBL COULD NOT BE INITIALIZED       *
C     *           WITH 1 DATA STATEMENT                                *
C     *  VISTB2 - EQUIVALENT TO APPROXIMATELY THE LATTER HALF OF       *
C     *           VISTBL.                                              *
C     *                                                                *
C     ******************************************************************
C
      SUBROUTINE VISCY (TEMPRD, PRSPRD, SPG, TEM, CNCH2S, CNCCO2,
     A                  CNCN2, VISGR, VISG, IERR)
C
      DIMENSION TEMTBL(13), PRSTBL(22), VISTBL(22,13), VISTB1(22,7),
     A          VISTB2(22,6)
C
      EQUIVALENCE (VISTB1(1,1), VISTBL(1,1)), (VISTB2(1,1), VISTBL(1,8))
C
      DATA TEMTBL /
     A1.05,1.10,1.15,1.20,1.30,1.40,1.50,1.60,1.75,2.00,2.25,2.50,3.00 /
C
      DATA PRSTBL /
     A 0.1 , 0.2 , 0.3 , 0.4 , 0.5 , 0.6 , 0.7 , 0.8 , 0.9 , 1.0 , 1.2 ,
     B 1.4 , 1.6 , 1.8 , 2.0 , 3.0 , 4.0 , 6.0 , 8.0 ,10.0 ,15.0 ,20.0 /
C
      DATA VISTB1 /
     A1.000,1.012,1.025,1.050,1.075,1.100,1.145,1.195,1.285,1.415,1.760,
     B2.285,2.865,3.290,3.650,4.760,5.500,6.460,7.150,7.680,8.650,9.370,
     C1.000,1.011,1.023,1.043,1.065,1.086,1.120,1.150,1.195,1.255,1.435,
     D1.700,2.070,2.465,2.800,3.850,4.655,5.720,6.500,7.060,8.100,8.880,
     E1.000,1.010,1.021,1.036,1.055,1.073,1.095,1.120,1.145,1.175,1.280,
     F1.420,1.590,1.850,2.160,3.225,3.975,5.030,5.820,6.385,7.410,8.180,
     G1.000,1.009,1.019,1.030,1.045,1.060,1.070,1.085,1.110,1.135,1.195,
     H1.285,1.425,1.570,1.750,2.600,3.350,4.380,5.125,5.740,6.750,7.500,
     I1.000,1.008,1.017,1.027,1.040,1.054,1.063,1.075,1.100,1.120,1.155,
     J1.215,1.285,1.360,1.460,2.020,2.560,3.500,4.185,4.755,5.790,6.500,
     K1.000,1.007,1.015,1.024,1.035,1.048,1.056,1.067,1.089,1.100,1.135,
     L1.185,1.235,1.280,1.335,1.690,2.110,2.790,3.380,3.860,4.790,5.410,
     M1.000,1.006,1.013,1.021,1.030,1.042,1.049,1.059,1.078,1.100,1.120,
     N1.150,1.185,1.220,1.260,1.500,1.785,2.325,2.820,3.230,4.060,4.610/
C
      DATA VISTB2 /
     A1.000,1.005,1.011,1.018,1.025,1.036,1.042,1.051,1.067,1.070,1.095,
     B1.120,1.150,1.180,1.215,1.385,1.595,2.030,2.425,2.770,3.490,4.025,
     C1.000,1.004,1.009,1.015,1.021,1.030,1.035,1.043,1.056,1.065,1.090,
     D1.110,1.125,1.145,1.165,1.280,1.435,1.770,2.095,2.375,2.990,3.500,
     E1.000,1.003,1.007,1.012,1.017,1.024,1.028,1.035,1.045,1.055,1.060,
     F1.070,1.080,1.095,1.110,1.205,1.290,1.500,1.725,1.955,2.480,2.925,
     G1.000,1.002,1.005,1.009,1.013,1.018,1.021,1.027,1.034,1.040,1.045,
     H1.055,1.065,1.075,1.085,1.145,1.210,1.340,1.485,1.665,2.085,2.460,
     I1.000,1.001,1.003,1.006,1.009,1.012,1.015,1.019,1.023,1.025,1.030,
     J1.040,1.050,1.060,1.065,1.105,1.155,1.245,1.360,1.485,1.830,2.150,
     K1.000,1.000,1.001,1.003,1.005,1.007,1.009,1.011,1.013,1.015,1.020,
     L1.025,1.030,1.035,1.040,1.060,1.085,1.140,1.205,1.265,1.495,1.750/
C
C     ******************************************************************
C     *                                                                *
C     * INITIALIZE-UPPER AND LOWER BOUNDS FOR THE TABLES OF            *
C     * PSEUDO REDUCED TEMPERATURE AND PRESSURE INDEXES.               *
C     *           -ERROR FLAG, IERR, TO 0                              *
C     *                                                                *
C     ******************************************************************
C
      LOPRS  =  2
      IHIPRS = 21
      LOTEM  =  3
      IHITEM = 11
      IERR   =  0
C
C     ******************************************************************
C     *                                                                *
C     * CHECK UPPER AND LOWER LIMITS OF THE PARAMETERS SPECIFIED        *
C     * BY THE FOLLOWING RANGES (INCLUSIVE) -                          *
C     *                                                                *
C     *    1. TEMPRD -  1.05   TO   3.00                               *
C     *    2. PRSPRD -  0.01   TO  20.00                               *
C     *    3. SPG    -  0.55   TO   1.50                               *
C     *    4. TEM    - 40.00   TO 400.00                               *
C     *                                                                *
C     * IF ANY VALUE IS OUTSIDE THE RANGE, SET IERR = 1 & VISCG = 0    *
C     *                                  & RETURN TO THE MAINLINE      *
C     *                                                                *
C     ******************************************************************
C
      IF (TEMPRD .LT. 1.05 .OR. TEMPRD .GT.   3.00) GO TO 340
      IF (PRSPRD .LT. 0.01 .OR. PRSPRD .GT.  20.00) GO TO 340
      IF (SPG    .LT. 0.55 .OR. SPG    .GT.   1.50) GO TO 340
      IF (TEM    .LT. 40.0 .OR. TEM    .GT. 400.0) GO TO 340
C
C     ******************************************************************
C     *                                                                *
C     * LOCATE TEMPRD IN THE TEMPERATURE INDEX                         *
C     *                                                                *
C     ******************************************************************
C
      DO 100 J = LOTEM, IHITEM
      IF (TEMTBL(J) .GE. TEMPRD) GO TO 120
100   CONTINUE
      J      = IHITEM + 1
C
C     ******************************************************************
C     *                                                                *
C     * CHECK PRSPRD AGAINST THE PRESSURE INDEX.                       *
C     *     1. IF PRSPRD.LE.THE LOWEST INDEX VALUE, PERFORM THE        *
C     *        LAGRANGE INTERPOLATION USING RATIOS FOR PRSTBL(1).      *
C     *        CALCULATE THE VISCOSITY RATIO, VISGR.                   *
C     *     2. IF PRSPRD.GE.THE HIGHEST INDEX VALUE, PERFORM THE       *
C     *        LAGRANGE INTERPOLATION USING RATIOS FOR                 *
C     *        PRSTBL(IHIPRS + 1).                                     *
C     *        SET THE VISCOSITY RATIO TO THE INTERPOLATED VALUES.     *
C     *     3. OTHERWISE, LOCATE PRSPRD IN THE PRESSURE INDEX.         *
C     *        USING RATIOS DEPENDENT UPON ITS LOCATION IN THE INDEX,  *
C     *        PERFORM THE LAGRANGE INTERPOLATION.                     *
C     *        CALCULATE THE VISCOSITY RATIO USING A STRAIGHT LINE     *
C     *        INTERPOLATION.                                          *
C     *                                                                *
C     ******************************************************************
C
120   IF (PRSTBL(LOPRS - 1) .LT. PRSPRD) GO TO 160
      I      = 1
      ICALL  = 1
      GO TO 320
140   VISGR  = 1.00 + (PRSPRD * (VISI - 1.0))
      GO TO 300
160   IF (PRSTBL(IHIPRS + 1) .GT. PRSPRD) GO TO 200
      I      = IHIPRS + 1
      ICALL  = 2
      GO TO 320
180   VISGR  = VISI
      GO TO 300
200   DO 220 I = LOPRS, IHIPRS
      IF (PRSTBL(I) .GE. PRSPRD) GO TO 240
220   CONTINUE
      I      = IHIPRS + 1
240   ICALL  = 3
      GO TO 320
260   VISJ   = VISI
      I      = I-1
      ICALL  = 4
      GO TO 320
280   I      = I + 1
      VISGR  = VISI + (( (PRSPRD-PRSTBL(I-1))/(PRSTBL(I)-PRSTBL(I-1)) )
     A                 * (VISJ-VISI) )
C
C     ******************************************************************
C     *                                                                *
C     * CALCULATE PARTIAL VISCOSITY, VISCGU.  MODIFY THE               *
C     * VALUE BY APPLYING CORRECTIONS DEPENDENT UPON THE PERCENTAGE    *
C     * CONTENT OF H2S, CO2 & N2 TO OBTAIN THE PARTIAL VISCOSITY       *
C     * AT 1 ATM PRESSURE, VISGA.                                      *
C     * CALCULATE THE OVERALL VISCOSITY, VISG                          *
C     *                                                                *
C     ******************************************************************
C
300   VISGU  = (0.126585E-01) - (0.611823E-02)*SPG + (0.164574E-02)*SPG
     A         *SPG + (0.164574E-04)*TEM - (0.719221E-06)*SPG*TEM
     B         - (0.609046E-06)*SPG*SPG*TEM
      CORH2S = (0.000113*CNCH2S*SPG - 0.000038*CNCH2S + 0.000001)
     A         * ( 1.0/(1.0+SPG) ) + 0.000001
      CORCO2 = (0.000134*CNCCO2*SPG - 0.000004*CNCCO2 + 0.000004*SPG)
     A         * ( 1.0/(1.0+SPG) ) - 0.000003
      CORN2  = (0.000170*CNCN2 *SPG + 0.000021*CNCN2  + 0.000010*SPG)
     A         * ( 1.0/(1.0+SPG) ) - 0.000006
      VISGA  = VISGU + CORH2S + CORCO2 + CORN2
      VISG   = VISGR * VISGA
      GO TO 360
C
C     ******************************************************************
C     *                                                                *
C     * CALL TO LAGRANGE INTERPOLATION ROUTINE                         *
C     *                                                                *
C     ******************************************************************
C
320   CALL XLGR4 (TEMPRD, TEMTBL(J-2), TEMTBL(J-1), TEMTBL(J),
     A            TEMTBL(J+1), VISI, VISTBL(I,J-2), VISTBL(I,J-1),
     B            VISTBL(I,J), VISTBL(I,J+1))
      GO TO (140, 180, 260, 280), ICALL
C
C     ******************************************************************
C     *                                                                *
C     * DATA ERROR PROCEDURE                                           *
C     *                                                                *
C     ******************************************************************
C
340   IERR   = 1
      VISGR  = 0.00
      VISG   = 0.0
C
C     ******************************************************************
C     *                                                                *
C     * RETURN PROCEDURE                                               *
C     *                                                                *
C     ******************************************************************
C
360   RETURN
      END
```

```
C     ********************************************************************
C     *                                                                  *
C     * PROGRAM NAME     - LAGRANGE INTERPOLATION                        *
C     *                                                                  *
C     * PURPOSE          - THIS SUBROUTINE IS USED BY SUBROUTINE VISCY   *
C     *                    TO PROVIDE INTERPOLATED VISCOSITY RATIO       *
C     *                    VALUES CORRESPONDING TO GIVEN VALUES OF       *
C     *                    PSEUDO-CRITICAL TEMPERATURE AND PSEUDO-       *
C     *                    CRITICAL PRESSURE                             *
C     *                                                                  *
C     * METHOD           - THIS SUBROUTINE EVALUATES THE GENERAL         *
C     *                    LAGRANGE EQUATION ( K.L.NEILSON, METHODS IN   *
C     *                    NUMERICAL ANALYSIS, THE MACMILLAN COMPANY,    *
C     *                    (1956) FOR A Y VALUE CORRESPONDING TO A       *
C     *                    GIVEN X VALUE LYING WITHIN THE RANGE OF       *
C     *                    FOUR GIVEN POINTS: (X1,Y1), (X2,Y2),          *
C     *                    (X3,Y3) AND (X4,Y4)                           *
C     *                                                                  *
C     * LANGUAGE         - FORTRAN IV                                    *
C     *                                                                  *
C     * CALLING SEQUENCE                                                 *
C     *                  - CALL XLGR4 (X, X1, X2, X3, X4,                *
C     *                                Y, Y1, Y2, Y3, Y4)                *
C     *                                                                  *
C     ********************************************************************
      SUBROUTINE XLGR4 (X, X1, X2, X3, X4, Y, Y1, Y2, Y3, Y4)
C
      A1     = X1 - X2
      A2     = X1 - X3
      A3     = X1 - X4
      A4     = X2 - X3
      A5     = X2 - X4
      A6     = X3 - X4
      B1     = X  - X1
      B2     = X  - X2
      B3     = X  - X3
      B4     = X  - X4
      Y      = B2 / A1 * B3 / A2 * B4 / A3 * Y1 -
     1         B1 / A1 * B3 / A4 * B4 / A5 * Y2 +
     2         B1 / A2 * B2 / A4 * B4 / A6 * Y3 -
     3         B1 / A3 * B2 / A5 * B3 / A6 * Y4
C
C     ********************************************************************
C     *                                                                  *
C     * RETURN PROCEDURE                                                 *
C     *                                                                  *
C     ********************************************************************
C
      RETURN
      END
```

```
C     ********************************************************************
C     *                                                                  *
C     * PROGRAM NAME     - FRICTION FACTOR DETERMINATION                 *
C     *                                                                  *
C     * PURPOSE          - THIS SUBROUTINE IS USED BY THE BOTTOM HOLE    *
C     *                    PRESSURE CALCULATION PROGRAM FOR COMPUTING    *
C     *                    THE VARIABLE PTZ FOR THE FLOWING PRESSURE     *
C     *                                                                  *
C     * LANGUAGE         - FORTRAN IV                                    *
C     *                                                                  *
C     * CALLING SEQUENCE                                                 *
C     *                  - CALL FFCFLO (TEM, PRS, PTZ)                   *
C     *                                                                  *
C     * PARAMETERS       -                                               *
C     *                                                                  *
C     *   1. INPUT PARAMETERS                                            *
C     *                                                                  *
C     *  TEM    - TEMPERATURE OF THE GAS, DEGREES F                      *
C     *  PRS    - PRESSURE OF THE GAS, PSIA                              *
C     *                                                                  *
C     *   2. OUTPUT PARAMETER                                            *
C     *                                                                  *
C     *  PTZ    - PARAMETER IN THE BOTTOM HOLE PRESSURE                  *
C     *           CALCULATION SCHEME                                     *
C     *                                                                  *
C     * VARIABLES                                                        *
C     *                                                                  *
C     *  FFC    - FRICTION FACTOR CALCULATED BY THE COLEBROOK            *
C     *           EQUATION                                               *
C     *  REYQ   - REYNOLDS NUMBER                                        *
C     *  VISG   - NATURAL GAS VISCOSITY CALCULATED BY                    *
C     *           SUBROUTINE VISCY                                       *
C     *                                                                  *
C     ********************************************************************
C
      SUBROUTINE FFCFLO (TEM, PRS, PTZ)
C
      REAL LTH
C
      COMMON CNCH2S, CNCCO2, CNCN2, PRSPC, TEMPC, SPG, DIAT, RTEG,
     A       ABSRNS, DPH, LTH
C
C     ********************************************************************
C     *                                                                  *
C     * CALCULATE THE REYNOLDS NUMBER FOR THE FLOWING GAS                *
C     *                                                                  *
C     ********************************************************************
C
      PTZ    = 1. / PTZ
      PRSPRD = PRS / PRSPC
      TEMPRD = (TEM + 460.) / TEMPC
      CALL VISCY (TEMPRD, PRSPRD, SPG, TEM, CNCH2S,
     A            CNCCO2, CNCN2, VISGR, VISG, IERR)
      REYQ   = 20011. * SPG * RTEG / (VISG * DIAT)
      IF (ABSRNS .LE. 0.) ABSRNS = 0.00060
      RELRNS = ABSRNS / DIAT
C
C     ********************************************************************
C     *                                                                  *
C     * FRICTION FACTOR FROM COLEBROOK EQUATION AND NEWTON-RAPHSON       *
C     * ITERATION TECHNIQUE                                              *
C     *                                                                  *
C     ********************************************************************
C
      ITER   = 0
      DDEL   = 1. / RELRNS
      C1     = 4.67 * DDEL / REYQ
      C2     = 1.73716 * ALOG(DDEL) + 2.28
      FFC    = 1. / (C2 ** 2)
100   FUNC   = C2 - 1.73716 * ALOG(1. + C1 / SQRT(FFC))
     A         - 1. / SQRT(FFC)
      DER    = 0.8686 * C1 / ( FFC ** 1.5 + C1 * FFC)
     A         + 0.5 / (FFC ** 1.5)
      FFCN   = FFC - FUNC / DER
      IF (ABS((FFCN-FFC)/FFCN)-0.005) 300,300,200
200   FFC    = FFCN
      ITER   = ITER + 1
      GO TO 100
300   FFC    = FFCN
      FSQ    = 2.6665 * FFC * RTEG * RTEG / (DIAT ** 5.0) * LTH / DPH
      PTZ    = PTZ / (FSQ + PTZ * PTZ * 0.001)
C
C     ********************************************************************
C     *                                                                  *
C     * RETURN PROCEDURE                                                 *
C     *                                                                  *
C     ********************************************************************
C
      RETURN
      END
```

```
C     ********************************************************************
C     *                                                                  *
C     * PROGRAM NAME     - NATURAL GAS COMPRESSIBILITY FACTOR            *
C     *                    AND COMPRESSIBILITY DETERMINATION             *
C     *                                                                  *
C     * PURPOSE          - THIS SUBROUTINE DETERMINES COMPRESSIBILITY    *
C     *                    FACTORS AND COMPRESSIBILITIES OF A NATURAL    *
C     *                    GAS. CORRECTION FACTORS FOR THE PRESENCE OF   *
C     *                    HYDROGEN SULPHIDE AND CARBON DIOXIDE ARE      *
C     *                    INCORPORATED.                                 *
C     *                                                                  *
C     * CORRELATIONS     - COMPRESSIBILITY FACTORS ARE CALCULATED FROM   *
C     *                    AN 8-PARAMETER BWR EQUATION OF STATE,         *
C     *                    DEVELOPED BY DRANCHUK,PURVIS & ROBINSON       *
C     *                    (1974), THAT EFFECTIVELY REPRODUCES THE       *
C     *                    STANDING-KATZ Z-FACTOR CHARTS.                *
C     *                    COMPRESSIBILITIES ARE CALCULATED BY THE       *
C     *                    METHOD OF MATTAR,BRAR & AZIZ (1975) WHICH     *
C     *                    USES A DIFFERENTIATED FORM OF THE ABOVE       *
C     *                    EQUATION OF STATE.                            *
C     *                    CORRECTIONS FOR THE PRESENCE OF HYDROGEN      *
C     *                    SULPHIDE AND CARBON DIOXIDE ARE APPLIED BY    *
C     *                    THE METHOD OF WICHERT & AZIZ (1974).          *
C     *                                                                  *
C     * LANGUAGE         - FORTRAN IV                                    *
C     *                                                                  *
C     * CALLING SEQUENCE                                                 *
C     *                  - CALL ZANDC (TEM, TEMPC, PRS, PRSPC, CNCH2S,   *
C     *                                CNCCO2, ZED, CMPG, IERR)          *
C     * PARAMETERS       -                                               *
C     *                                                                  *
C     *  1. INPUT PARAMETERS                                             *
C     *                                                                  *
C     *  TEM    - TEMPERATURE OF THE NATURAL GAS, DEGREES F              *
C     *  TEMPC  - PSEUDO-CRITICAL TEMPERATURE OF THE NATURAL GAS,        *
C     *           DEGREES R                                              *
C     *  PRS    - PRESSURE AT WHICH THE NATURAL GAS EXISTS, PSIA         *
C     *  PRSPC  - PSEUDO-CRITICAL PRESSURE OF THE  NATURAL GAS, PSIA     *
C     *  CNCH2S - MOLE PERCENT HYDROGEN SULPHIDE IN THE NATURAL GAS      *
C     *  CNCCO2 - MOLE PERCENT CARBON DIOXIDE IN THE NATURAL GAS         *
C     *                                                                  *
C     *  2. OUTPUT PARAMETERS                                            *
C     *                                                                  *
C     *  ZED    - COMPRESSIBILITY FACTOR OF THE NATURAL GAS              *
C     *  CMPG   - COMPRESSIBILITY OF THE NATURAL GAS                     *
C     *  IERR   - ERROR FLAG INDICATING BOUNDS ERROR                     *
C     *                                                                  *
C     * VARIABLES                                                        *
C     *                                                                  *
C     *  A      - CONSTANTS IN THE DRANCHUK ETAL. CORRELATION            *
C     *  CMPPRD - PSEUDO-REDUCED COMPRESSIBILITY                         *
C     *  DENRD  - REDUCED DENSITY                                        *
C     *  DZED   - PARTIAL DIFFERENTIAL OF ZED WITH RESPECT TO DENRD      *
C     *                                                                  *
C     *  EPS    - WICHERT AND AZIZ ADJUSTMENT FOR H2S AND CO2            *
C     *  FRCA   - MOLE FRACTION H2S AND CO2                              *
C     *  FRCB   - MOLE FRACTION H2S                                      *
C     *  PRSPCA - ADJUSTED PSEUDO-CRITICAL PRESSURE,                     *
C     *  PRSPRD - PSEUDO-REDUCED PRESSURE                                *
C     *  TEMPCA - ADJUSTED PSEUDO-CRITICAL TEMPERATURE                   *
C     *  TEMPRD - PSEUDO-REDUCED TEMPERATURE                             *
C     *                                                                  *
C     ********************************************************************
C
      SUBROUTINE ZANDC (TEM, TEMPC, PRS, PRSPC, CNCH2S,
     A                  CNCCO2, ZED, CMPG, IERR)
C
      DIMENSION A(8)
C
      DATA A /
     A  0.31506237 , -1.04670990 , -0.57832729 ,  0.53530771 ,
     B -0.61232032 , -0.10488813 ,  0.68157001 ,  0.68446549 /
C
C     ********************************************************************
C     *                                                                  *
C     * CALCULATE ADJUSTMENT FOR PRESENCE OF H2S & CO2                   *
C     *                                                                  *
C     ********************************************************************
C
      FRCA   = (CNCH2S + CNCCO2) / 100.
      FRCB   = CNCH2S / 100.
      EPS    = 120. * (FRCA ** 0.9 - FRCA ** 1.6) + 15. * (FRCB ** 0.5
     A         - FRCB ** 4.0)
C
C     ********************************************************************
C     *                                                                  *
C     * ADJUST PSEUDO CRITICAL TEMPERATE AND PRESSURE. CALCULATE         *
C     * PSEUDO REDUCED TEMPERATURE AND PRESSURE                          *
C     *                                                                  *
C     ********************************************************************
C
      TEMPCA = TEMPC - EPS
      PRSPCA = PRSPC * TEMPCA / (TEMPC + FRCB * (1. - FRCB) * EPS)
      TEMPRD = (TEM + 460.) / TEMPCA
      PRSPRD = PRS / PRSPCA
C
C     ********************************************************************
C     *                                                                  *
C     * SET ERROR FLAG, IERR, TO 0                                       *
C     * CHECK UPPER AND LOWER LIMITS OF THE PARAMETERS SPECIFIED         *
C     * BY THE FOLLOWING RANGES (INCLUSIVE) -                            *
C     *                                                                  *
C     *  1. TEMPRD -  1.05  TO  3.00                                     *
C     *  2. PRSPRD -  0.00  TO 15.00                                     *
C     *  3. FRCA   -  0.00  TO  0.85                                     *
C     *                                                                  *
C     * IF ANY VALUE IS OUTSIDE THE RANGE, SET IERR = 1.0, ZED = 0.0,    *
C     *                     AND CMPG = 0.0 AND RETURN TO THE MAINLINE    *
C     *                                                                  *
C     ********************************************************************
C
      IERR   = 0.0
      IF (TEMPRD .LT. 1.05 .OR. TEMPRD .GT.  3.00) GO TO 140
      IF (PRSPRD .LT. 0.00 .OR. PRSPRD .GT. 15.00) GO TO 140
      IF (FRCA   .LT. 0.00 .OR. FRCA   .GT.  0.85) GO TO 140
C
C     ********************************************************************
C     *                                                                  *
C     * INITIALIZE REDUCED DENSITY AND CALCULATE COMPRESSIBILITY         *
C     * FACTOR AND COMPRESSIBILITY OF THE NATURAL GAS                    *
C     *                                                                  *
C     ********************************************************************
C
      ITER   = 0.0
      T1     = A(1) * TEMPRD + A(2) + A(3) / (TEMPRD * TEMPRD)
      T2     = A(4) * TEMPRD + A(5)
      T3     = A(5) * A(6)
      T4     = A(7) / (TEMPRD * TEMPRD)
      T5     = A(8)
      DENRD  = 1.0
      DO 120 ITER=1,10
      DENRD2 = DENRD  * DENRD
      DENRD3 = DENRD2 * DENRD
      DENRD4 = DENRD2 * DENRD2
      DENRD5 = DENRD3 * DENRD2
      P      = (TEMPRD + T1 * DENRD + T2 * DENRD2 + T3 * DENRD5)
     A         * DENRD + T4 * DENRD3 * (1.0 + T5 * DENRD2) * EXP
     B         (-T5 * DENRD2)
      DP     = TEMPRD + 2.0 * T1 * DENRD + 3.0 * T2 * DENRD2
     A         + 6.0 * T3 * DENRD5 + T4 * DENRD2 * EXP
     B         (-T5 * DENRD2) * (3.0 + 3.0 * T5 * DENRD2
     C         - 2.0 * T5 * T5 * DENRD4)
      DENRD1 = DENRD - (P - 0.270 * PRSPRD) / DP
      IF (DENRD1 .GT. 0.0) GO TO 100
      DENRD1 = 0.5 * DENRD
100   IF (DENRD1 .LT. 2.2) GO TO 110
      DENRD1 = DENRD + 0.9 * (2.2 - DENRD)
110   IF (ABS(DENRD - DENRD1) .LT. 0.00001) GO TO 130
120   DENRD  = DENRD1
130   ZED    = 0.270 * PRSPRD / (DENRD1 * TEMPRD)
      DENRD  = DENRD1
      DENRD2 = DENRD  * DENRD
      DENRD4 = DENRD2 * DENRD2
      DZED   = T1 / TEMPRD + 2.0 * T2 / TEMPRD * DENRD
     A         + 5.0 * T3 * DENRD4 / TEMPRD + (1.0 + T5 * DENRD2
     B         - T5 * T5 * DENRD4) * 2.0 * T4 / TEMPRD * DENRD
     C         * EXP(-T5 * DENRD2)
      CMPPRD = 1.0 / PRSPRD - 0.270 * DZED / (ZED * ZED * TEMPRD
     A         * (1.0 + DENRD / ZED * DZED))
      CMPG   = CMPPRD / PRSPCA
      GO TO 150
C
C     ********************************************************************
C     *                                                                  *
C     * DATA ERROR PROCEDURE                                             *
C     *                                                                  *
C     ********************************************************************
C
140   IERR   = 1
      ZED    = 0.0
      CMPG   = 0.0
C
C     ********************************************************************
C     *                                                                  *
C     * RETURN PROCEDURE                                                 *
C     *                                                                  *
C     ********************************************************************
C
150   RETURN
      END
```

EXAMPLE D-1. (KODEA=1,KCDEB=1)

INPUT DATA:

WELL DEPTH FEET	LENGTH	PIPE INSIDE DIAMETER INCHES	PIPE ABSOLUTE ROUGHNESS	GAS FLOW RATE MMSCFD	TEMPERATURE DEGREES F WELL HEAD	TEMPERATURE DEGREES F BOTTOM HOLE	PRESSURE, PSIA WELL HEAD	PRESSURE, PSIA BOTTOM HOLE (ASSUMED)
10471.	10471.	2.441	.0018	0.0	146.	242.	2685.	2885.

GAS GRAVITY = 0.9800

CALCULATED GAS PROPERTIES:

PSEUDO-CRITICAL PRESSURE, PSIA = 652.06

PSEUDO-CRITICAL TEMPERATURE, DEGREES RANKINE = 471.69

MOLECULAR WEIGHT = 28.38

****

STATIC BOTTOM HOLE PRESSURE = 3961. NUMBER OF ITERATIONS = 4

EXAMPLE D-2. (KODEA=1,KCDEB=2)

INPUT DATA:

WELL DEPTH FEET	LENGTH	PIPE INSIDE DIAMETER INCHES	PIPE ABSOLUTE ROUGHNESS	GAS FLOW RATE MMSCFD	TEMPERATURE DEGREES F WELL HEAD	TEMPERATURE DEGREES F BOTTOM HOLE	PRESSURE, PSIA WELL HEAD	PRESSURE, PSIA BOTTOM HOLE (ASSUMED)
10471.	10471.	2.441	.0018	11.716	146.	242.	2685.	2885.

GAS GRAVITY = 0.9800

CALCULATED GAS PROPERTIES:

PSEUDO-CRITICAL PRESSURE, PSIA = 652.06

PSEUDO-CRITICAL TEMPERATURE, DEGREES RANKINE = 471.69

MOLECULAR WEIGHT = 28.38

****

FLOWING BOTTOM HOLE PRESSURE = 4557. NUMBER OF ITERATIONS = 3

```
EXAMPLE D-3.        (KODEA=2,KCDEB=1)

INPUT DATA:
     WELL                    PIPE            GAS FLOW    TEMPERATURE        PRESSURE, PSIA
DEPTH     LENGTH       INSIDE   ABSOLUTE       RATE       DEGREES F                 BOTTOM
     FEET            DIAMETER  ROLGHNESS     MMSCFD     WELL    BOTTOM     WELL      HOLE
                           INCHES                       HEAD     HOLE      HEAD  (ASSUMED)

10471.    10471.     2.441      .0018        0.0       146.     242.      2685.    2885.

GAS COMPOSITION, MOLE FRACTION

 H2S    CO2     N2
.1706  .0320  .0192

GAS GRAVITY = 0.9800

CALCULATED GAS PROPERTIES:

PSEUDO-CRITICAL PRESSURE, PSIA                      = 652.06

PSEUDO-CRITICAL TEMPERATURE, CEGREES RANKINE        = 471.69

MOLECULAR WEIGHT                                    =  28.38
****************************************************************************************
STATIC BOTTOM HOLE PRESSURE = 3870.                         NUMBER OF ITERATIONS = 4
```

```
EXAMPLE D-4.        (KODEA=2,KCDEB=2)

INPUT DATA:
     WELL                    PIPE            GAS FLOW    TEMPERATURE        PRESSURE, PSIA
DEPTH     LENGTH       INSIDE   ABSOLUTE       RATE       DEGREES F                 BOTTOM
     FEET            DIAMETER  ROLGHNESS     MMSCFD     WELL    BOTTOM     WELL      HOLE
                           INCHES                       HEAD     HOLE      HEAD  (ASSUMED)

10471.    10471.     2.441      .0018       11.716     146.     242.      2685.    2885.

GAS COMPOSITION, MOLE FRACTION

 H2S    CO2     N2
.1706  .0320  .0192

GAS GRAVITY = 0.9800

CALCULATED GAS PROPERTIES:

PSEUDO-CRITICAL PRESSURE, PSIA                      = 652.06

PSEUDO-CRITICAL TEMPERATURE, CEGREES RANKINE        = 471.69

MOLECULAR WEIGHT                                    =  28.38
****************************************************************************************
FLOWING BOTTOM HOLE PRESSURE = 4510.                        NUMBER OF ITERATICNS = 3
```

```
EXAMPLE D-5.       (KODEA=3,KCDEB=1)

INPUT DATA:
      WELL                     PIPE              GAS FLOW    TEMPERATURE       PRESSURE, PSIA
DEPTH     LENGTH        INSIDE    ABSOLUTE         RATE       DEGREES F                BOTTOM
     FEET              DIAMETER  ROLGHNESS       MMSCFD     WELL     BOTTOM    WELL      HOLE
                            INCHES                          HEAD      HOLE     HEAD (ASSUMED)

10471.    10471.     2.441      .0018          0.0       146.      242.      2685.    2885.

GAS COMPOSITION, MOLE FRACTION

 H2S    CO2     N2
.1706  .0320  .0192

  C1      C2      C3      IC4     NC4     IC5     NC5      C6      C7+
.5903   .0727   .0297   .0079   .C144   .0059   .0062   .0120   .0391

CALCULATED GAS PROPERTIES:

PSEUDO-CRITICAL PRESSURE, PSIA                        = 767.88

PSEUDO-CRITICAL TEMPERATURE, CEGREES RANKINE          = 476.78

MOLECULAR WEIGHT                                      =  28.39

GAS GRAVITY                                           =   0.9803

*****************************************************************************************
STATIC BOTTOM HOLE PRESSURE = 3953.                        NUMBER OF ITERATIONS = 4
```

```
EXAMPLE D-6.       (KODEA=3,KCDEB=2)

INPUT DATA:
      WELL                     PIPE              GAS FLOW    TEMPERATURE       PRESSURE, PSIA
DEPTH     LENGTH        INSIDE    ABSOLUTE         RATE       DEGREES F                BOTTOM
     FEET              DIAMETER  ROLGHNESS       MMSCFD     WELL     BOTTOM    WELL      HOLE
                            INCHES                          HEAD      HOLE     HEAD (ASSUMED)

10471.    10471.     2.441      .0018        11.716      146.      242.      2685.    2885.

GAS COMPOSITION, MOLE FRACTION

 H2S    CO2     N2
.1706  .0320  .0192

  C1      C2      C3      IC4     NC4     IC5     NC5      C6      C7+
.5903   .0727   .0297   .0079   .C144   .0059   .0062   .0120   .0391

CALCULATED GAS PROPERTIES:

PSEUDO-CRITICAL PRESSURE, PSIA                        = 767.88

PSEUDO-CRITICAL TEMPERATURE, CEGREES RANKINE          = 476.78

MOLECULAR WEIGHT                                      =  28.39

GAS GRAVITY                                           =   0.9803

*****************************************************************************************
FLOWING BOTTOM HOLE PRESSURE = 4563.                       NUMBER OF ITERATIONS = 3
```

```
EXAMPLE D-7.       (KODEA=4,KCDEB=1)

INPUT DATA:
     WELL                     PIPE             GAS FLOW    TEMPERATURE        PRESSURE, PSIA
DEPTH    LENGTH       INSIDE    ABSOLUTE        RATE        DEGREES F                 BOTTOM
   FEET              DIAMETER  ROUGHNESS       MMSCFD     WELL    BOTTOM     WELL      HOLE
                          INCHES                           HEAD    HOLE       HEAD (ASSUMED)

10471.   10471.     2.441      .0018          0.0       146.     242.       2685.    2885.

GAS COMPOSITION, MOLE FRACTION

 H2S    CO2     N2
.1706  .0320  .0192

  C1     C2     C3     IC4    NC4    IC5    NC5     C6     C7+
.5903  .0727  .0297  .0079  .0144  .0059  .0062  .0120  .0391
PROPERTIES OF C7+

PSEUDO-CRITICAL PRESSURE, PSIA                     = 360.22

PSEUDO-CRITICAL TEMPERATURE, DEGREES RANKINE      =1045.22

MOLECULAR WEIGHT                                   = 124.52

CALCULATED GAS PROPERTIES:

PSEUDO-CRITICAL PRESSURE, PSIA                     = 767.86

PSEUDO-CRITICAL TEMPERATURE, DEGREES RANKINE      = 477.61

MOLECULAR WEIGHT                                   =  28.80

GAS GRAVITY                                        =   0.9942

******************************************************************************
STATIC BOTTOM HOLE PRESSURE = 3976.                      NUMBER OF ITERATIONS = 4


EXAMPLE D-8.       (KODEA=4,KODEB=2)

INPUT DATA:
     WELL                     PIPE             GAS FLOW    TEMPERATURE        PRESSURE, PSIA
DEPTH    LENGTH       INSIDE    ABSOLUTE        RATE        DEGREES F                 BOTTOM
   FEET              DIAMETER  ROUGHNESS       MMSCFD     WELL    BOTTOM     WELL      HOLE
                          INCHES                           HEAD    HOLE       HEAD (ASSUMED)

10471.   10471.     2.441      .0018        11.716      146.     242.       2685.    2885.

GAS COMPOSITION, MOLE FRACTION

 H2S    CO2     N2
.1706  .0320  .0192

  C1     C2     C3     IC4    NC4    IC5    NC5     C6     C7+
.5903  .0727  .0297  .0079  .0144  .0059  .0062  .0120  .0391
PROPERTIES OF C7+

PSEUDO-CRITICAL PRESSURE, PSIA                     = 360.22

PSEUDO-CRITICAL TEMPERATURE, DEGREES RANKINE      =1045.22

MOLECULAR WEIGHT                                   = 124.52

CALCULATED GAS PROPERTIES:

PSEUDO-CRITICAL PRESSURE, PSIA                     = 767.86

PSEUDO-CRITICAL TEMPERATURE, DEGREES RANKINE      = 477.61

MOLECULAR WEIGHT                                   =  28.80

GAS GRAVITY                                        =   0.9942

******************************************************************************
FLOWING BOTTOM HOLE PRESSURE = 4594.                     NUMBER OF ITERATIONS = 4
```

```
EXAMPLE D-9.

INPUT DATA:
     WELL                   PIPE               GAS FLOW    TEMPERATURE       PRESSURE, PSIA
DEPTH     LENGTH     INSIDE    ABSOLUTE      RATE        DEGREES F                 BOTTOM
    FEET            DIAMETER  ROUGHNESS     MMSCFD     WELL    BOTTOM      WELL      HOLE
                         INCHES                        HEAD     HOLE       HEAD (ASSUMED)

10000.    10000.     2.441       .0006        0.0        35.     245.      2500.    2700.

GAS GRAVITY = 0.7500

CALCULATED GAS PROPERTIES:

PSEUDO-CRITICAL PRESSURE, PSIA                      = 665.57

PSEUDO-CRITICAL TEMPERATURE, DEGREES RANKINE        = 401.00

MOLECULAR WEIGHT                                    =  21.72

**************************************************************************************

STATIC BOTTOM HOLE PRESSURE = 3388.                       NUMBER OF ITERATIONS = 5


EXAMPLE D-10.

INPUT DATA:
     WELL                   PIPE               GAS FLOW    TEMPERATURE       PRESSURE, PSIA
DEPTH     LENGTH     INSIDE    ABSOLUTE      RATE        DEGREES F                 BOTTOM
    FEET            DIAMETER  ROUGHNESS     MMSCFD     WELL    BOTTOM      WELL      HOLE
                         INCHES                        HEAD     HOLE       HEAD (ASSUMED)

10000.    10000.     2.441       .0006       4.915     110.     245.      2000.    2200.

GAS GRAVITY = 0.7500

CALCULATED GAS PROPERTIES:

PSEUDO-CRITICAL PRESSURE, PSIA                      = 665.57

PSEUDO-CRITICAL TEMPERATURE, DEGREES RANKINE        = 401.00

MOLECULAR WEIGHT                                    =  21.72

**************************************************************************************

FLOWING BOTTOM HOLE PRESSURE = 2725.                      NUMBER OF ITERATIONS = 4


EXAMPLE D-11.

RESERVOIR TEMPERATURE, DEGREES F        =     120.00
GAS GRAVITY                             =     0.6100
PSEUDO-CRITICAL TEMP., DEGREES R        =     357.00
PSEUDO-CRITICAL PRESSURE, PSIA          =     664.00
MOLE PERCENT - HYDROGEN SULPHIDE        =       0.0
MOLE PERCENT - CARBON DIOXIDE           =       0.0
MOLE PERCENT - NITROGEN                 =       0.0

PRESSURE                   PSEUDO-PRESSURE
 (PSIA)                     (PSIA **2/CP)

   0.0                      0.0
 400.0                      0.1416346E 08
 800.0                      0.5636334E 08
1200.0                      0.1251145E 09
1600.0                      0.2175370E 09
2000.0                      0.3304471E 09
```

Appendix E

PRESSURE TRAVERSE
CURVES

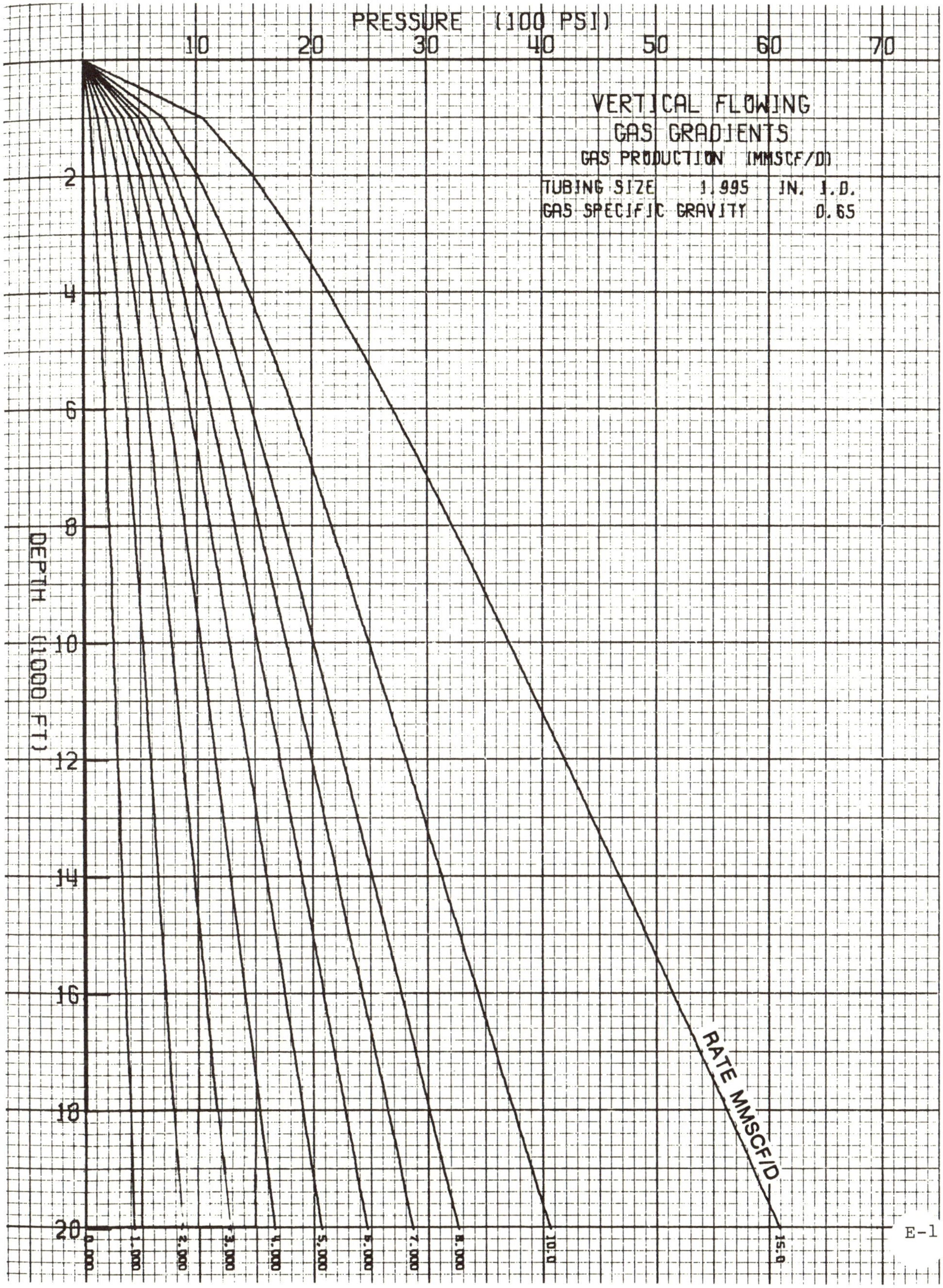
PRESSURE (100 PSI)
10 20 30 40 50 60 70
VERTICAL FLOWING
GAS GRADIENTS
GAS PRODUCTION (MMSCF/D)
TUBING SIZE 1.995 IN. I.D.
GAS SPECIFIC GRAVITY 0.65
DEPTH (1000 FT)
2 4 6 8 10 12 14 16 18 20
RATE MMSCF/D
0.000 1.000 2.000 3.000 4.000 5.000 6.000 7.000 8.000 10.0 15.0
E-1

Fig. E-1

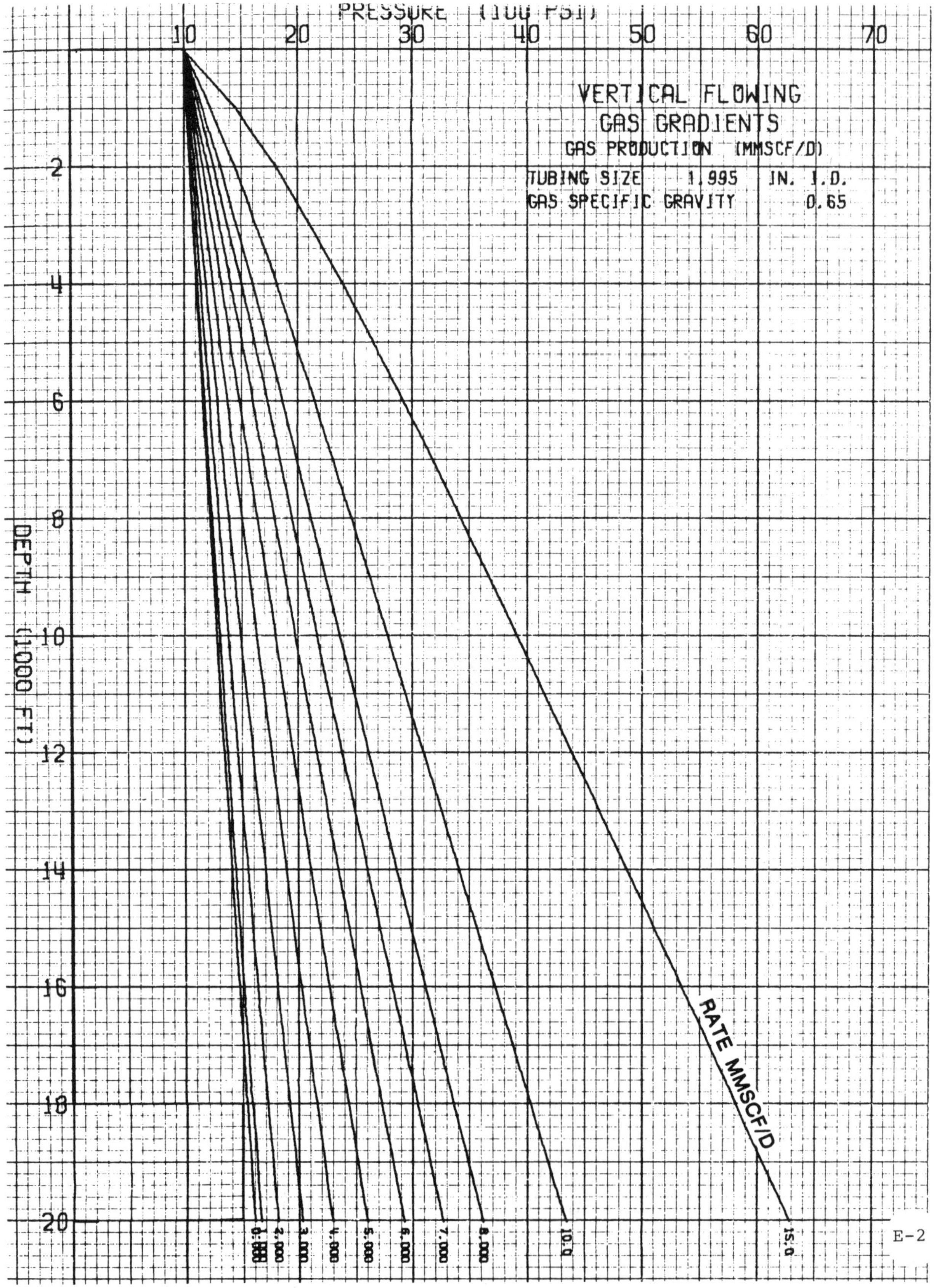
PRESSURE (100 PSI)
10 20 30 40 50 60 70
VERTICAL FLOWING
GAS GRADIENTS
GAS PRODUCTION (MMSCF/D)
TUBING SIZE 1.995 IN. I.D.
GAS SPECIFIC GRAVITY 0.65
DEPTH (1000 FT)
2 4 6 8 10 12 14 16 18 20
RATE MMSCF/D
2.000 3.000 4.000 5.000 6.000 7.000 8.000 10.0 15.0
E-2

Fig. E-2

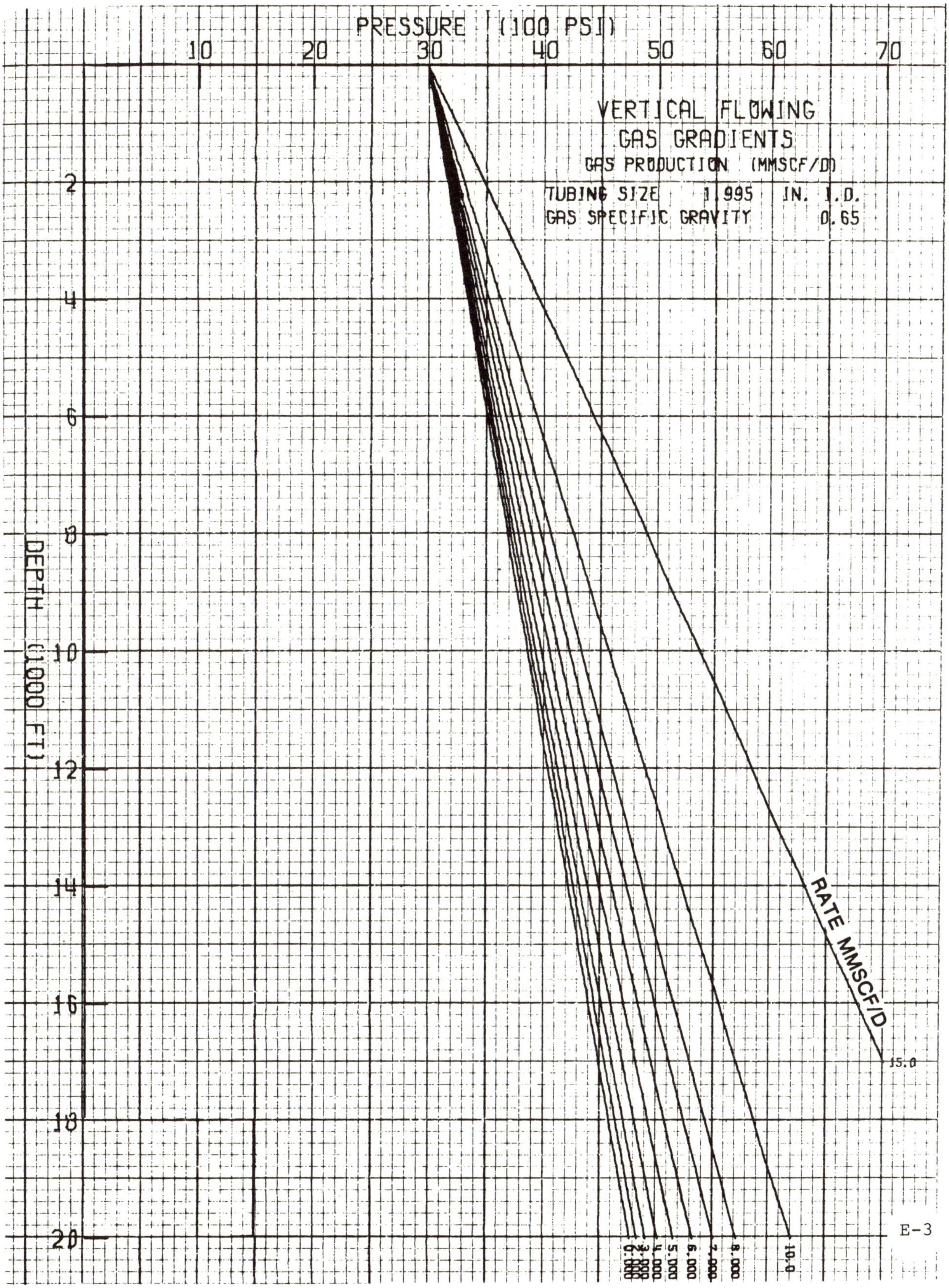
PRESSURE (100 PSI)
10
20
30
40
50
60
70
VERTICAL FLOWING
GAS GRADIENTS
GAS PRODUCTION (MMSCF/D)
TUBING SIZE 1.995 IN. I.D.
GAS SPECIFIC GRAVITY 0.65
DEPTH (1000 FT)
2
4
6
8
10
12
14
16
18
20
RATE MMSCF/D
15.0
3.000
4.000
5.000
6.000
7.000
8.000
10.0
E-3

Fig. E-3

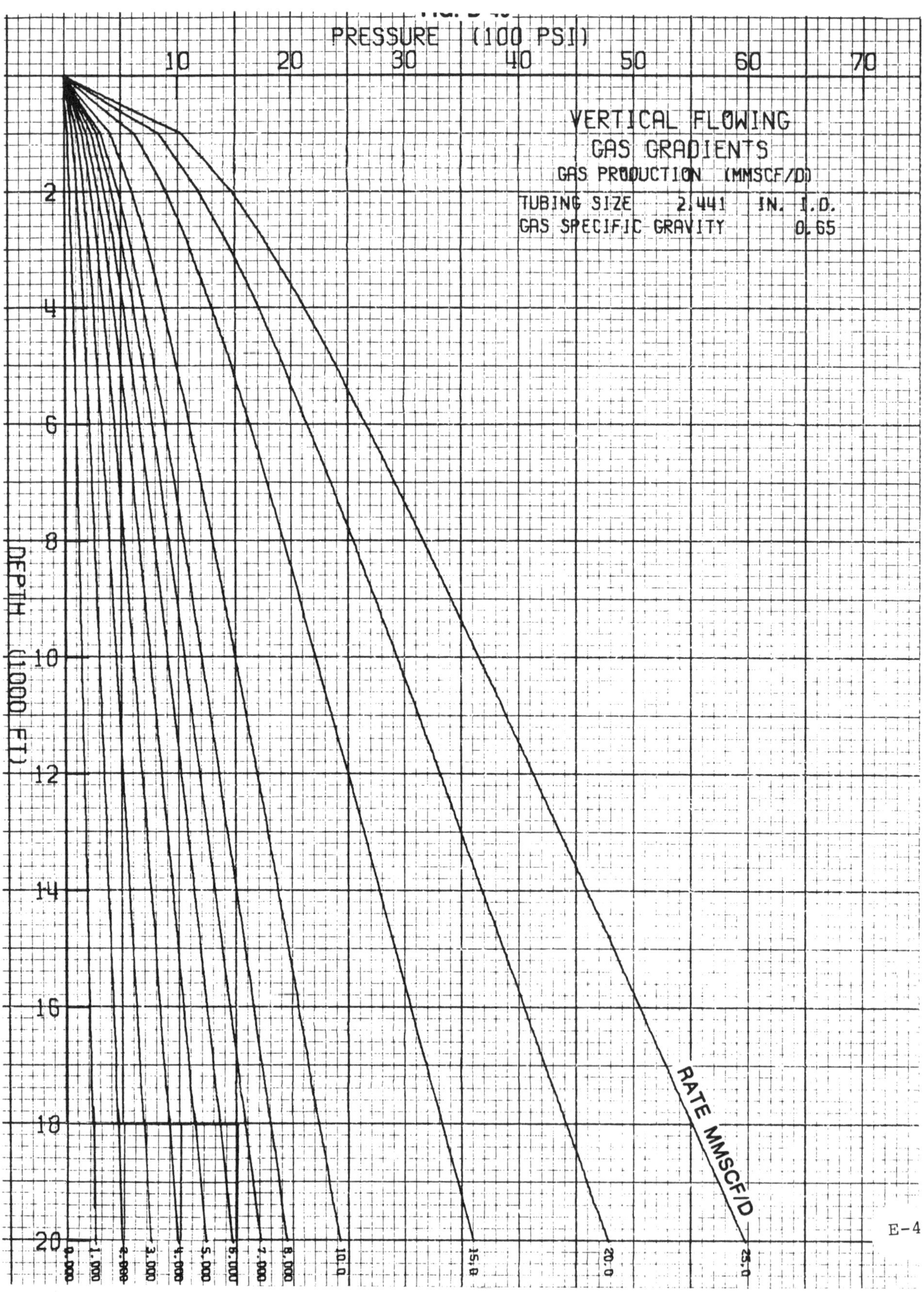
PRESSURE (100 PSI)
10 20 30 40 50 60 70
VERTICAL FLOWING
GAS GRADIENTS
GAS PRODUCTION (MMSCF/D)
TUBING SIZE 2.441 IN. I.D.
GAS SPECIFIC GRAVITY 0.65
DEPTH (1000 FT)
2 4 6 8 10 12 14 16 18 20
RATE MMSCF/D
0.000 1.000 2.000 3.000 4.000 5.000 6.000 7.000 8.000 10.0 15.0 20.0 25.0
E-4

Fig. E-4

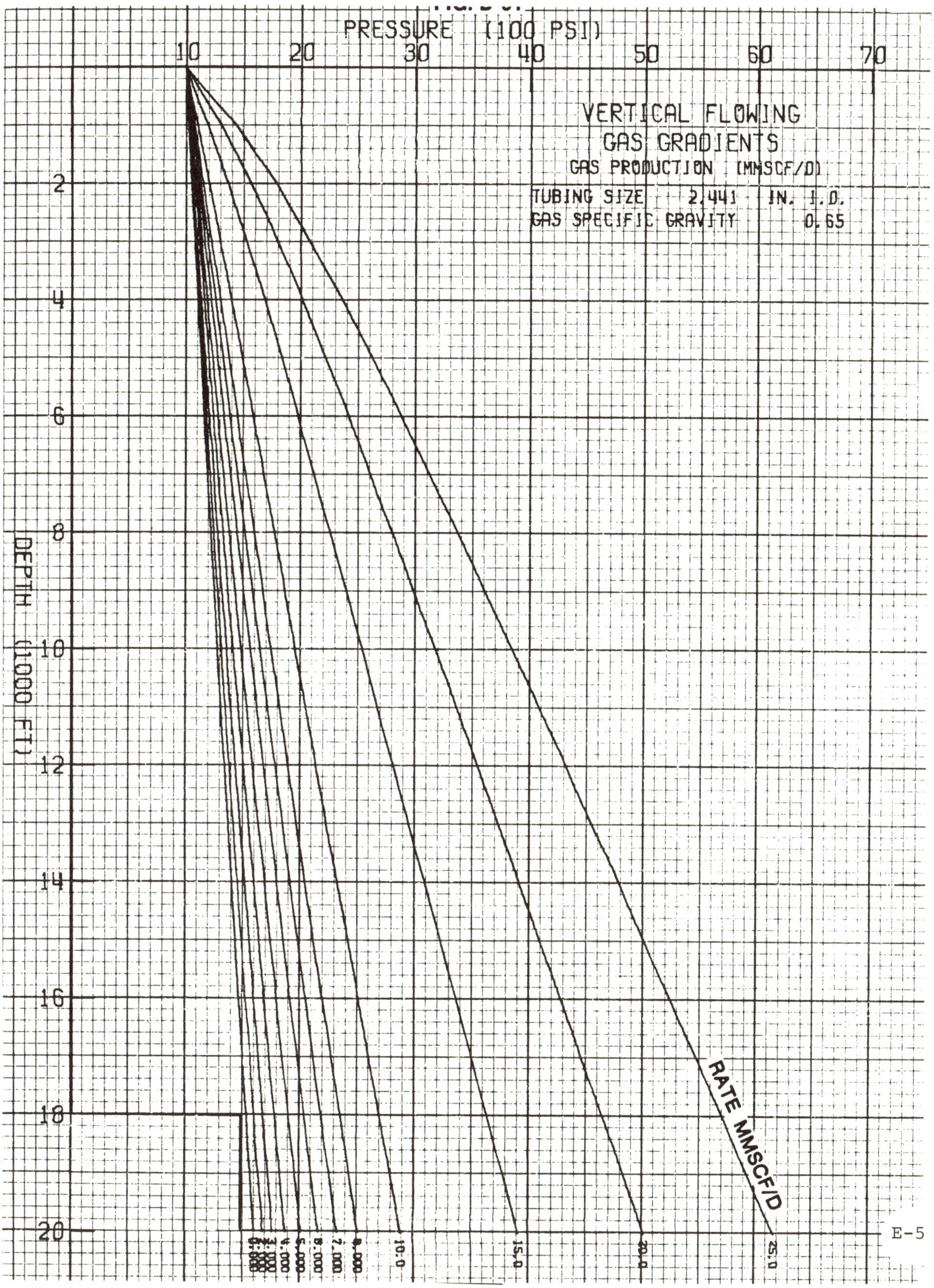
PRESSURE (100 PSI)
10
20
30
40
50
60
70
VERTICAL FLOWING
GAS GRADIENTS
GAS PRODUCTION (MMSCF/D)
TUBING SIZE 2.441 IN. I.D.
GAS SPECIFIC GRAVITY 0.65
DEPTH (1000 FT)
2
4
6
8
10
12
14
16
18
20
RATE MMSCF/D
4.000
5.000
6.000
7.000
10.0
15.0
20.0
25.0

E-5

Fig. E-5

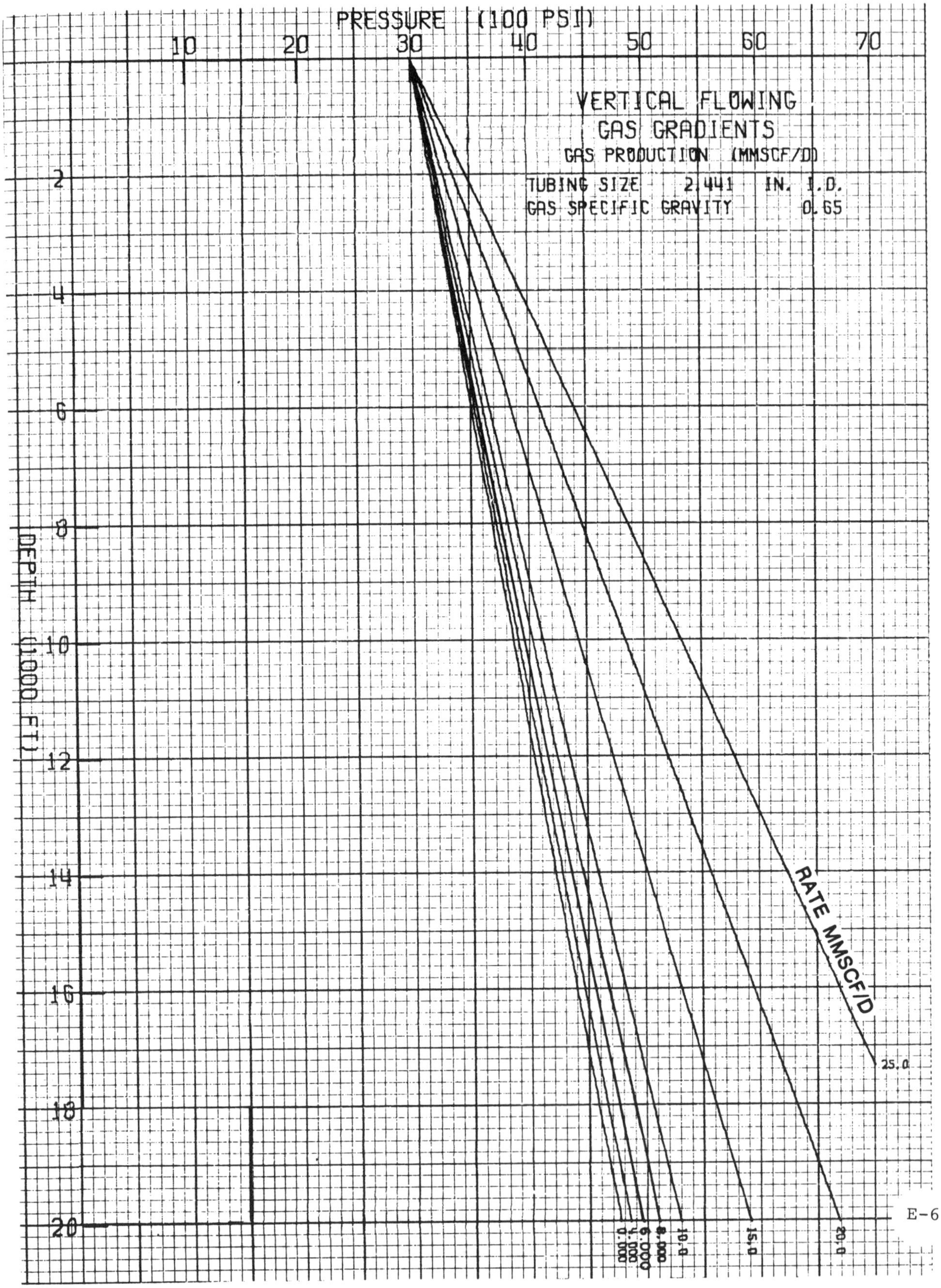
PRESSURE (100 PSI)
10 20 30 40 50 60 70
VERTICAL FLOWING
GAS GRADIENTS
GAS PRODUCTION (MMSCF/D)
TUBING SIZE 2.441 IN. I.D.
GAS SPECIFIC GRAVITY 0.65
DEPTH (1000 FT)
2 4 6 8 10 12 14 16 18 20
RATE MMSCF/D
25.0
20.0
15.0
10.0
8.000
6.000
4.000
0.000
E-6

Fig. E-6

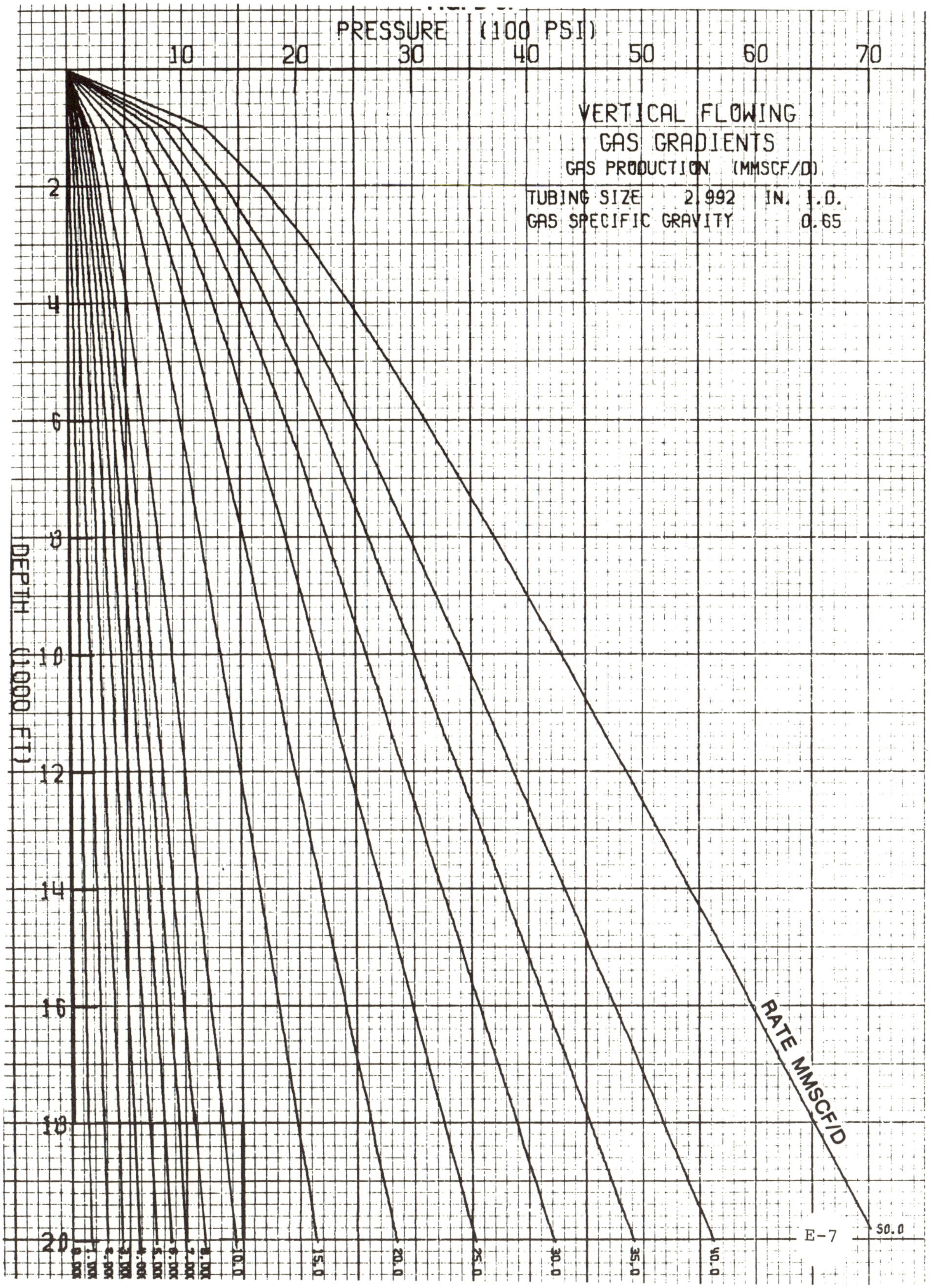
PRESSURE (100 PSI)
10
20
30
40
50
60
70
VERTICAL FLOWING
GAS GRADIENTS
GAS PRODUCTION (MMSCF/D)
TUBING SIZE 2.992 IN. I.D.
GAS SPECIFIC GRAVITY 0.65
DEPTH (1000 FT)
2
4
6
8
10
12
14
16
18
20
RATE MMSCF/D
10.0
15.0
20.0
25.0
30.0
35.0
40.0
50.0
E-7

Fig. E-7

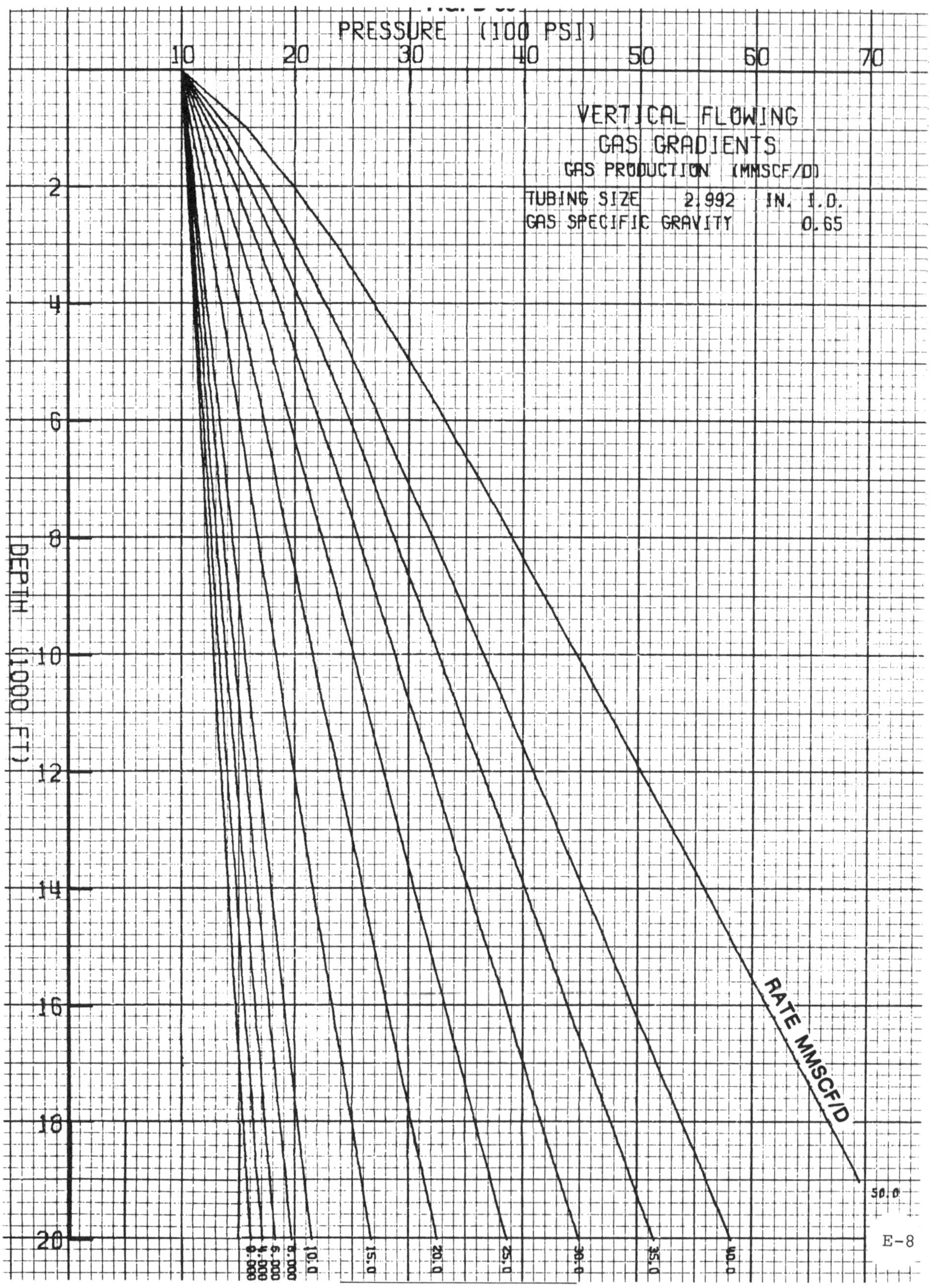
PRESSURE (100 PSI)
10
20
30
40
50
60
70
VERTICAL FLOWING
GAS GRADIENTS
GAS PRODUCTION (MMSCF/D)
TUBING SIZE 2.992 IN. I.D.
GAS SPECIFIC GRAVITY 0.65
DEPTH (1000 FT)
2
4
6
8
10
12
14
16
18
20
RATE MMSCF/D
50.0
10.0
15.0
20.0
25.0
30.0
35.0
40.0
E-8

Fig. E-8

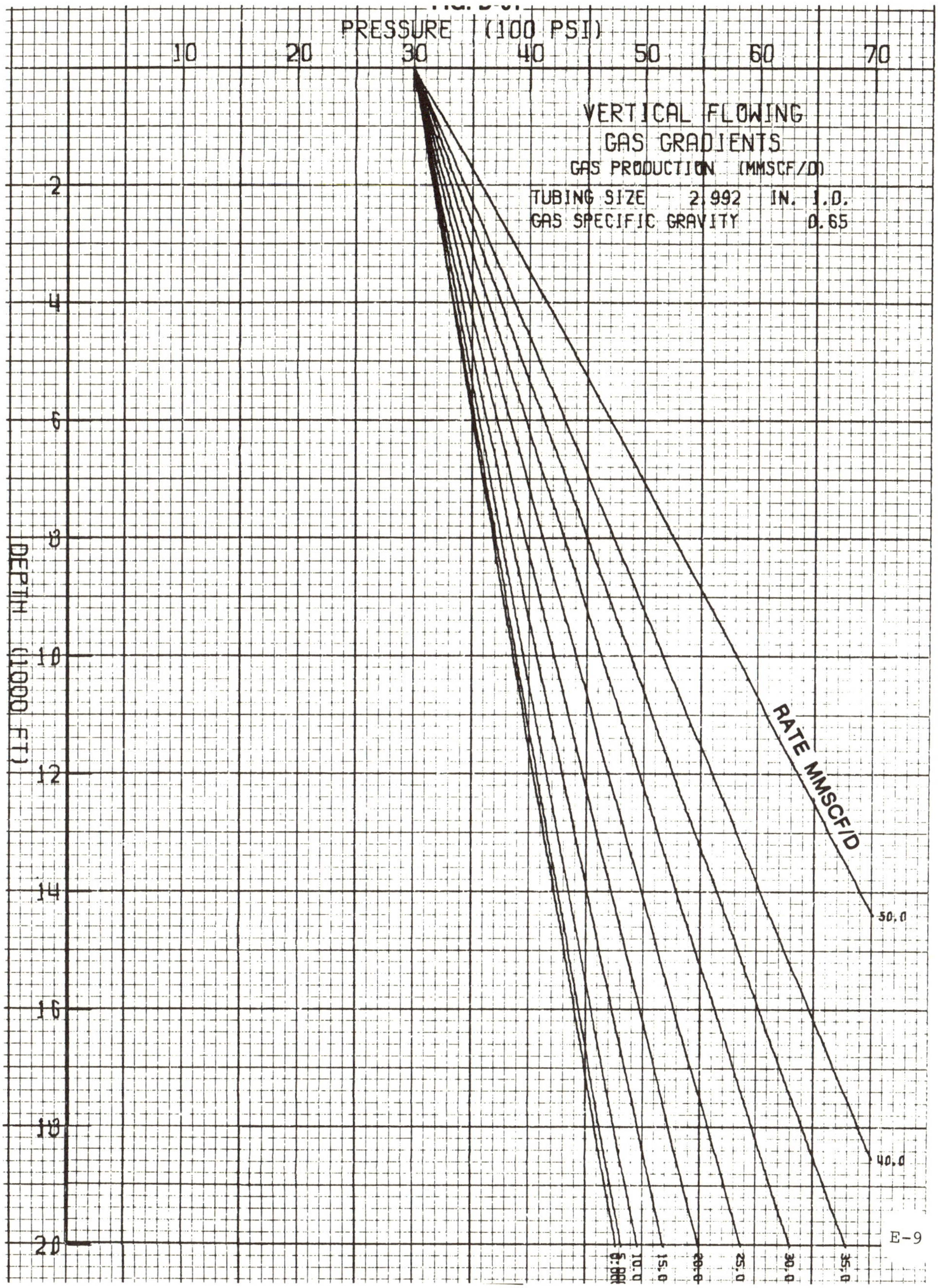
PRESSURE (100 PSI)
10 20 30 40 50 60 70
VERTICAL FLOWING
GAS GRADIENTS
GAS PRODUCTION (MMSCF/D)
TUBING SIZE 2.992 IN. I.D.
GAS SPECIFIC GRAVITY 0.65
DEPTH (1000 FT)
2 4 6 8 10 12 14 16 18 20
RATE MMSCF/D
50.0
40.0
10.0 15.0 20.0 25.0 30.0 35.0
E-9

Fig. E-9

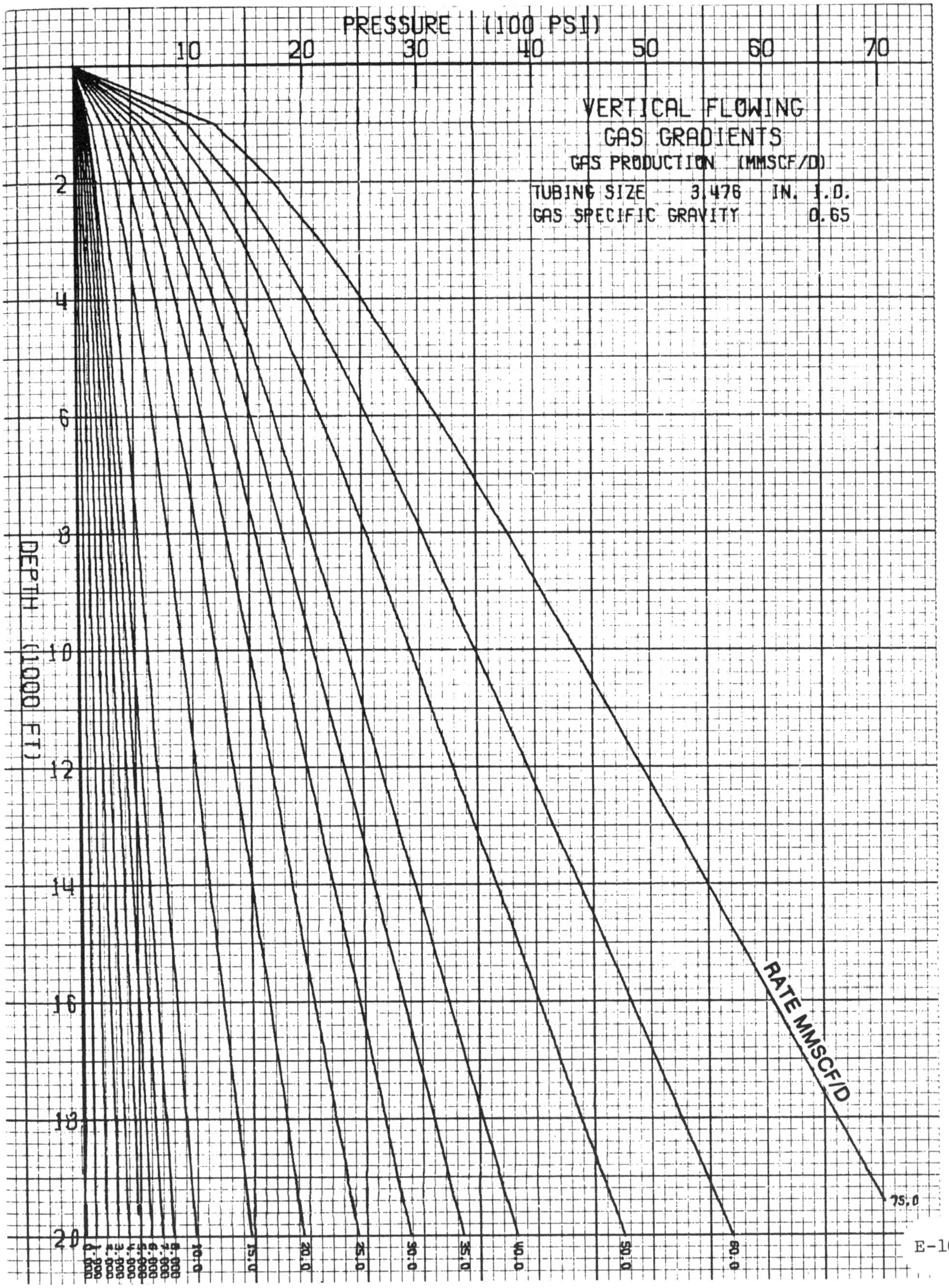
PRESSURE (100 PSI)
10
20
30
40
50
60
70
VERTICAL FLOWING
GAS GRADIENTS
GAS PRODUCTION (MMSCF/D)
TUBING SIZE 3.476 IN. I.D.
GAS SPECIFIC GRAVITY 0.65
DEPTH (1000 FT)
2
4
6
8
10
12
14
16
18
20
RATE MMSCF/D
10.0
15.0
20.0
25.0
30.0
35.0
40.0
50.0
60.0
75.0
E-10

Fig. E-10

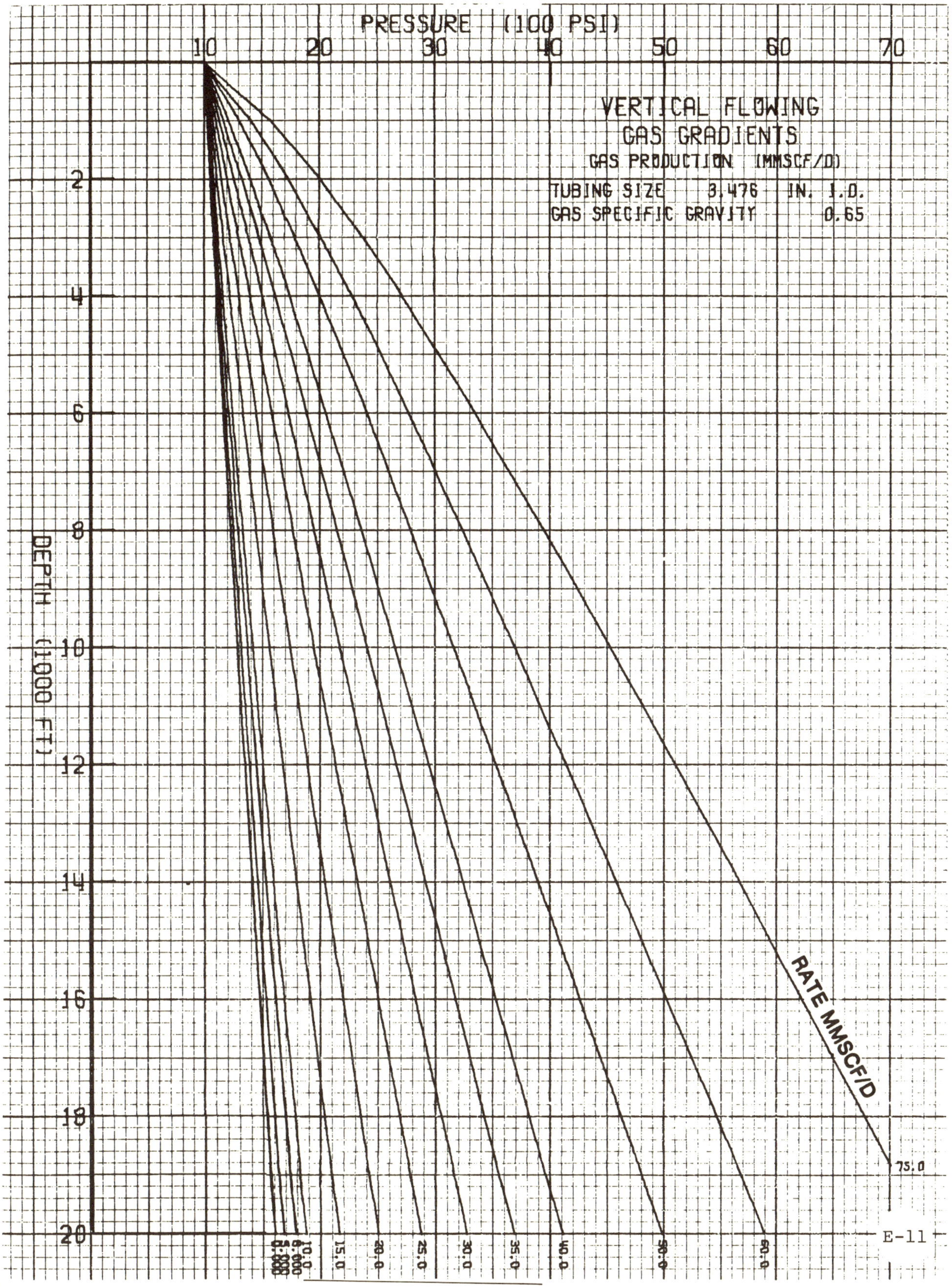
PRESSURE (100 PSI)
10
20
30
40
50
60
70
VERTICAL FLOWING
GAS GRADIENTS
GAS PRODUCTION (MMSCF/D)
TUBING SIZE 3.476 IN. I.D.
GAS SPECIFIC GRAVITY 0.65
DEPTH (1000 FT)
2
4
6
8
10
12
14
16
18
20
RATE MMSCF/D
10.0
15.0
20.0
25.0
30.0
35.0
40.0
50.0
60.0
75.0
E-11

Fig. E-11

Field to Metric Conversions

1 acre = 4047 m^2
1 bbl = 0.159 m^3
1 btu = 1055 J
1 cp = 0.001 Pa-sec
1 darcy = 0.987 μ m^2
1 °F = °C/1.8
1 °R = K/1.8
1 ft = 0.3048 m
1 ft^3 = 0.0283 m^3
1 gal = 0.003785 m^3
1 HP = 745.7 W

1 in. of water = 0.2488 kPa
1 mile = 1609.3 m
1 lbm = 0.4536 kg
1 lbm/ft^3 = 16.02 kg/m^3
1 lbf = 4.448 N
1 psi = 6.895 kPa

To convert temperatures:

°C = (°F–32)/1.8

Liquid specific gravity:

$\gamma_L = 141.5/(131.5 + API)$

Summary of Equations

Reservoir

Stabilized (Pseudo Steady-State) Flow

$$q_{sc} = \frac{C_1\, kh\, (\bar{p}_R^2 - p_{wf}^2)}{T\, \bar{\mu}\, \bar{Z}\, [\ln(.472 r_e/r_w) + S + Dq_{sc}]}$$

or

$$\bar{p}_R^2 - p_{wf}^2 = A_R q_{sc} + R_R\, q_{sc}^2$$

Where
$$A_R = \frac{C_2\, T \bar{\mu}\, \bar{Z}}{kh}\, [\ln(.472 r_e/r_w) + S]$$

$$B_R = \frac{C_3\, \gamma g\, \bar{Z}\, T\, \beta}{r_w\, h^2} = \frac{C_2\, T\, \bar{\mu}\, \bar{Z}}{kh} D$$

$$\beta = \frac{C_4}{k^{1.2}}$$

$$D = \frac{C_5\, \gamma g}{\bar{\mu}\, h\, r_w\, k^{0.2}}$$

An alternate form of the equation is:

$$q_{sc} = C(\bar{p}_R^2 - p_{wf}^2)^n$$

Transient (Unsteady-State) Flow

Drawdown Testing

$$p_i^2 - p_{wf}^2 = m\left[\log t + \log\left(\frac{k}{\phi\,\bar{\mu}\,\bar{c}\,r_w^2}\right) - C_6 + .869S'\right]$$

$$S' = 1.151\left[\frac{p_i^2 - p_{1\,hr}^2}{m} - \log\left(\frac{k}{\phi\,\bar{\mu}\,\bar{c}\,r_w^2}\right) + C_6\right]$$

Buildup Testing

$$p_{ws}^2 = p_i^2 - m\log\left(\frac{t + \Delta t}{\Delta t}\right)$$

$$m = \frac{C_7\,q_{sc}\,T\,\bar{\mu}\,\bar{Z}}{kh}$$

$$S' = 1.151\left[\frac{p_{1\,hr}^2 - p_{wf}^2}{m} - \log\left(\frac{k}{\phi\,\bar{\mu}\,\bar{c}\,r_w^2}\right) + C_6\right]$$

$$t_{DA} = \frac{C_8\,kt}{\phi\,\bar{\mu}\,\bar{c}\,A}$$

Perforations

Perforated Completions

$$\bar{p}_R^2 - p_{wf}^2 = (A_R + A_p)\,q_{sc} + (B_R + B_p)\,q_{sc}^2$$

Where $A_p = \dfrac{C_2\,T\,\bar{\mu}\,\bar{Z}}{kh}\,(S_p + S_{dp}).$

$$B_p = \frac{C_3\,\gamma_g\,\bar{Z}\,T\,\beta_{dp}}{r_p\,L_p^2\,N^2}$$

$$\beta_{dp} = \frac{C_4}{k_{dp}^{\,1.2}}$$

Gravel-Packed Completions

$$p_{wfs}^2 - p_{wf}^2 = A_G\, q_{sc} + B_G\, q_{sc}^2$$

Where $$A_G = \frac{C_9\, T\, \bar{\mu}\, \bar{Z}\, L}{k_G\, N\, r_p^2}$$

$$B_G = \frac{C_{10}\, \gamma_g\, \bar{Z}\, T\, L\, \beta_G}{N^2\, r_p^2}$$

$$\beta_G = \frac{C_{11}}{k_G^{0.55}}$$

Units

Variable	Field	SI
q_{sc}	Mscfd	m^3/day
k	md	md
h	ft	m
p	psia	kPa
T	°R	K
μ	cp	Pa.s
r_e	ft	m
r^w	ft	m
D	$Mscfd^{-1}$	day/m^3
β	ft^{-1}	m^{-1}
c	$psia^{-1}$	kPa^{-1}
t	hrs	hrs
A	ft^2	m^2
m	$psia^2$	kPa^2
r_p	ft	m
L_p	ft	m
L	ft	m
C_1	703×10^{-6}	763.2×10^{-9}
C_2	1422	1.3103×10^6
C_3	3.161×10^{-12}	2.915×10^{-15}
C_4	2.33×10^{10}	7.64×10^{10}
C_5	5.18×10^{-5}	1.70×10^{-10}
C_6	3.23	8.098
C_7	1637	1.509×10^6
C_8	2.64×10^{-4}	3.557×10^{-9}
C_9	2844	5.5135×10^4
C_{10}	1.263×10^{-11}	7.463×10^{-14}
C_{11}	1.47×10^7	4.823×10^7

Piping Systems

$$N_{Re} = \frac{C_1 \gamma_g q_{sc}}{\mu d}$$

Flow in Wells

Average Pressure and Temperature Method:

$$P_{ws} = P_{ts} \text{ EXP } (C_2 \gamma_g H/\bar{T}\bar{Z})$$

$$P_{wf}^2 = P_{tf}^2 \text{ EXP } (S) + \frac{C_3 \gamma_g \bar{T}\bar{Z} f \text{ (MD) (EXP (S) } -1) q_{sc}^2}{Sd^5}$$

Where $S = C_4 \gamma_g \text{ (TVD)}/\bar{T}\bar{Z}$

Cullender and Smith Method:

$$P_{ms} = P_{ts} + C_2 \gamma_g/(I_{ms} + I_{ts})$$

$$P_{ws} = P_{ms} + C_2 \gamma_g/(I_{ms} + I_{ws})$$

Where $I = \frac{TZ}{P}$

$$P_{mf} = P_{tf} + C_2 \gamma_g \text{ (MD)}/(I_{mf} + I_{tf})$$

$$P_{wf} = P_{mf} + C_2 \gamma_g \text{ (MD)}/(I_{wf} + I_{mf})$$

Where $I = \dfrac{\dfrac{P}{TZ}}{\left(\dfrac{P}{TZ}\right)^2 \left(\dfrac{TVD}{MD}\right) + F^2}$

$$F^2 = C_5 f q_{sc}^2/d^5$$

Flow in Pipelines

$$q_{sc} = \frac{C_6\,T_b}{P_b}\left[\frac{P_1^2 - P_2^2}{\gamma_g f\,\bar{T}\,\bar{Z}\,L}\right]^{0.5} d^{2.5}$$

Flow Through Restrictions

Critical Flow

$$q_{sc} = \frac{C_7\,A\,P_1\,C_d}{(T\,\gamma_g)^{0.5}}$$

Sub-Critical Flow

$$P_1 - P_2 = \frac{C_8\,\gamma_g\,Z_1\,T_1\,(1 - \beta^4)}{P_1\,d^4\,C_d^2\,Y^2}\,q_{sc}^2$$

UNITS

Variable	Field	SI
q_{sc}	Mscfd	m^3/day
μ	cp	Pa.s
d	in.	m
P	psia	kPa
H	ft	m
T	°R	K
MD	ft	m
TVD	ft	m
I	°R/psia	K/kPa
L	ft	m
A	in^2	m^2
C_1	20	1.796×10^{-5}
C_2	0.01875	0.0342
C_3	2.5×10^{-5}	9.256×10^{-2}
C_4	0.0375	0.0683
C_5	6.67×10^{-7}	1.356×10^{-15}
C_6	5.634	1.149×10^{6}
C_7	605.4	2.87×10^{6}
C_8	1.048×10^{-6}	4.661×10^{-14}

Molal heat capacity MC_p° (ideal-gas state), BTU/lb mol/°R

Data source: Selected Values of Properties of Hydrocarbons, API Research Project 44

Gas	Chemical formula	Mol wt	MC_p 0°F	50°F	60°F	100°F	150°F	200°F	250°F	300°F
Methane	CH_4	16.042	8.233	8.414	8.456	8.651	8.947	9.277	9.638	10.01
Ethyne (Acetylene)	C_2H_2	26.036	9.683	10.23	10.33	10.71	11.13	11.54	11.88	12.22
Ethene (Ethylene)	C_2H_4	28.052	9.324	10.02	10.16	10.72	11.40	12.08	12.75	13.14
Ethane	C_2H_6	30.068	11.44	12.17	12.32	12.95	17.33	14.63	15.49	16.34
Propene (Propylene)	C_3H_6	42.078	13.63	14.69	14.90	15.57	16.80	17.85	18.87	19.89
Propane	C_3H_8	44.094	15.64	16.88	17.13	18.17	19.52	20.89	22.25	23.56
1-Butene (Butylene)	C_4H_8	56.104	17.96	19.59	19.91	21.17	22.71	24.25	25.70	27.15
cis-2-Butene	C_4H_8	56.104	16.54	18.04	18.34	19.54	21.04	22.53	24.00	25.47
trans-2-Butene	C_4H_8	56.104	18.84	20.02	20.50	21.61	22.99	24.37	25.72	27.06
iso-Butane	C_4H_{10}	58.120	20.40	22.15	22.50	23.95	25.77	27.59	29.39	31.11
n-Butane	C_4H_{10}	58.120	20.80	22.38	22.71	24.07	25.81	27.54	29.23	30.90
iso-Pentane	C_5H_{12}	72.146	24.93	27.16	27.61	29.42	31.66	33.87	36.03	38.14
n-Pentane	C_5H_{12}	72.146	25.64	27.61	28.01	29.70	31.86	33.99	36.07	38.12
Benzene	C_6H_6	78.108	16.41	18.38	18.75	20.46	22.46	24.46	27.08	29.71
n-Hexane	C_6H_{14}	86.172	30.17	32.78	33.30	35.36	37.91	40.45	42.91	45.36
n-Heptane	C_7H_{16}	100.198	34.96	37.00	38.61	41.01	43.47	46.93	49.77	52.60
Ammonia	NH_3	17.032	8.516	8.518	8.519	8.521	8.523	8.525	8.527	8.530
Air		28.966	6.944	6.951	6.952	6.960	6.973	6.990	7.009	7.033
Water	H_2O	18.016	7.983	8.006	8.010	8.033	8.075	8.116	8.171	8.226
Oxygen	O_2	32.000	6.970	6.997	7.002	7.030	7.075	7.120	7.176	7.232
Nitrogen	N_2	28.016	6.951	6.954	6.954	6.956	6.963	6.970	6.984	6.998
Hydrogen	H_2	2.016	6.782	6.856	6.871	6.905	6.929	6.953	6.953	6.977
Hydrogen sulfide	H_2S	34.076	8.00	8.091	8.109	8.18	8.27	8.36	8.455	8.55
Carbon monoxide	CO	28.010	6.952	6.957	6.958	6.963	6.975	6.986	7.007	7.028
Carbon dioxide	CO_2	44.010	8.380	8.698	8.762	9.004	9.282	9.559	10.31	10.05

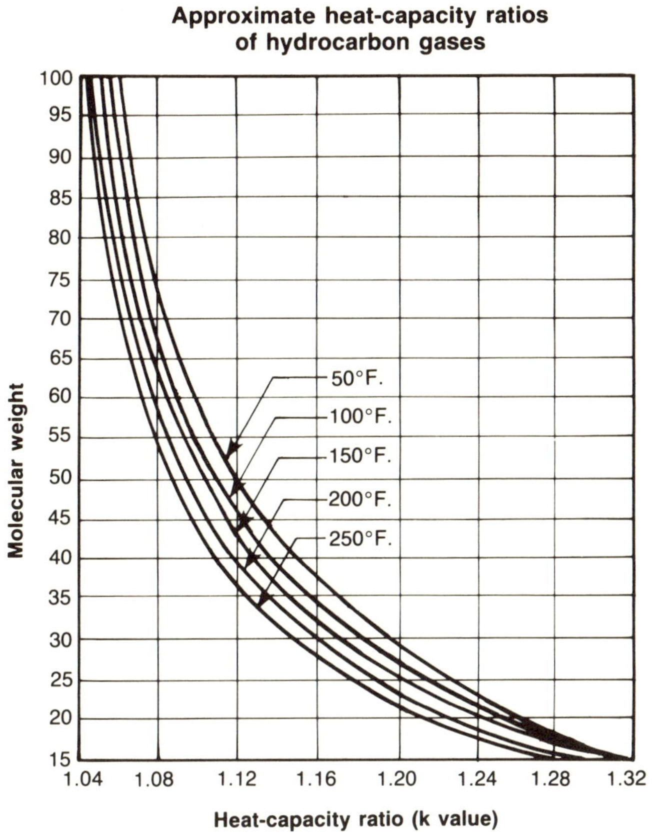
Approximate heat-capacity ratios
of hydrocarbon gases
Molecular weight
100
95
90
85
80
75
70
65
60
55
50
45
40
35
30
25
20
15
50°F.
100°F.
150°F.
200°F.
250°F.
1.04
1.08
1.12
1.16
1.20
1.24
1.28
1.32
Heat-capacity ratio (k value)

Index

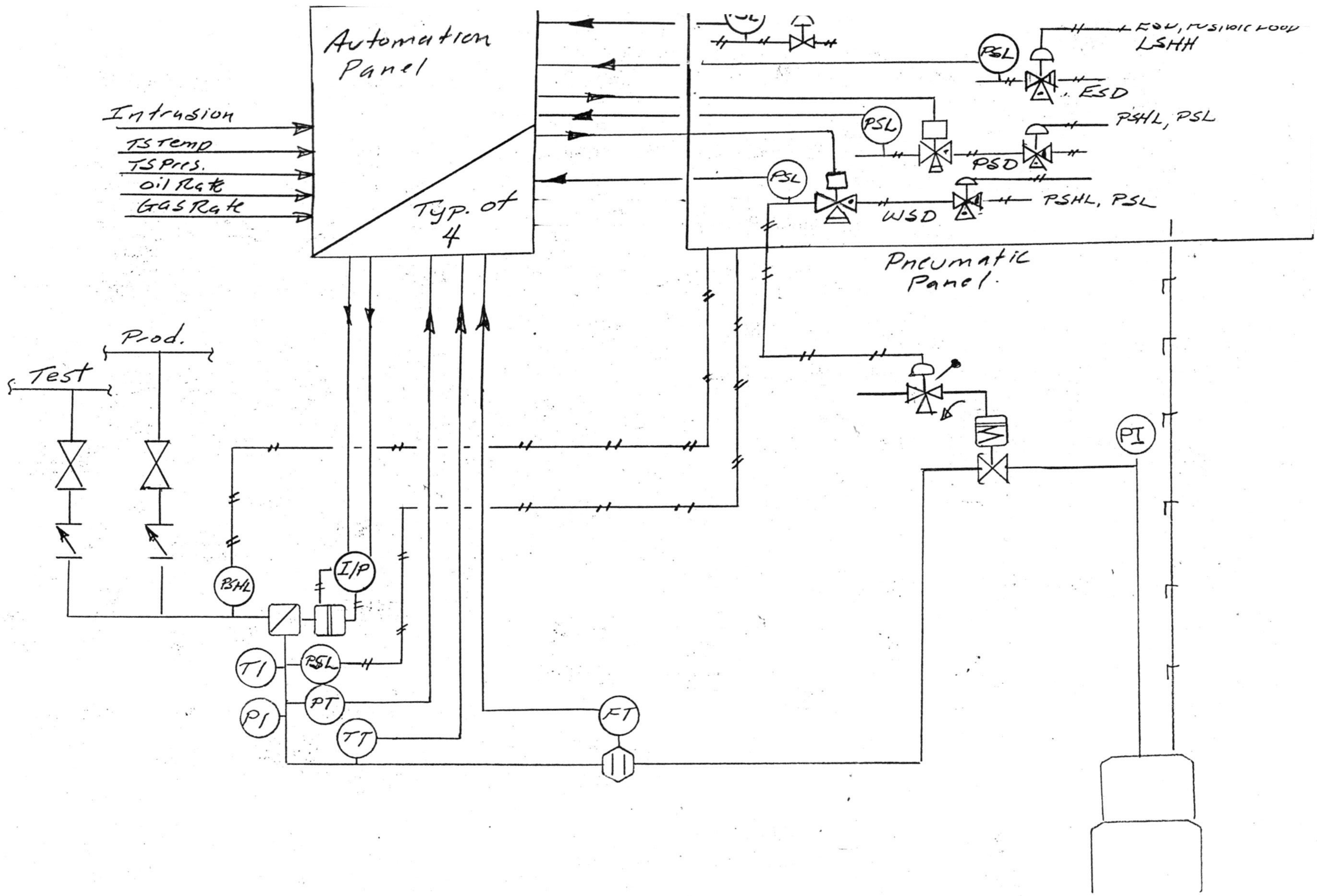
Automation Panel
Intrusion
TS Temp
TS Pres.
Oil Rate
Gas Rate
Typ. of 4
PSL
ESD
PSHL, PSL
PSD
WSD
LSHH
Pneumatic Panel
Prod.
Test
PSHL
I/P
TI
PSL
PT
PI
TT
FT
PI